D0916976

Mechanics,
Molecular Physics,
Heat, and Sound

THE "Melancholia" is one of the three most famous copper engravings of the great Nuremberg painter, draftsman, and engraver Albrecht Dürer (1471–1528). Like other masterpieces, it suggests much more than it clearly expresses, and endless meanings, rightly or wrongly, have been read into it. The authors of this book hope, therefore, that they may be permitted to take the brooding figure, sitting amidst a litter of mechanical tools, scientific instruments, and mathematical symbols, as the embodiment of the spirit of physical science.

Physics is an *experimental science*, as suggested by the tools — hammer, plane, saw — at the feet of the winged figure. Thus, by means of a few simple *experiments* with string, balls, and wax, GALILEO did more to discover the actual *facts* of motion than had centuries of mere *observation*. Physics is a *science of exact measurement*, as suggested by the balance, the dividers, and the hourglass. Indeed, these three instruments are designed to measure the quantities *mass, length,* and *time,* the units for which enter in general into the units for all the quantities dealt with in physics. Physics *makes use of* the tools, methods, and results of *mathematics*, as suggested by the magic square, the sphere, and the cube, so close at hand. But physics is *more than experiment, more than exact measurement, more than mathematics* — these are but its tools; it is *careful, profound, and exact thinking*, as suggested by the attitudes and delineation of the brooding figures, — the younger learning the methods of the elder through imitation and association, — both winged and able to soar to otherwise inaccessible heights, but secure in the "plainness and soundness of observations on material and obvious things," and perchance using by preference the ladder that leans against the house and has its base upon the solid earth.

As Hermann Schubert points out in his *Mathematical Essays and Recreations*, the term *melancholy* meant in Dürer's time "thought or thoughtfulness." Thus "Dürer's melancholy does not represent the gloominess of thought, but the power of invention. Soberness and even a certain sadness are considered only as an element of this melancholy, but on the whole the genius of thought appears bright, self-possessed and strong. . . . At a distance a bat-like creature, being the gloom of melancholy, hovers in the air like a dark cloud, but the sun rises above the horizon, and at the happy middle between these two extremes stands the rainbow of serene hope and cheerful confidence."

●

Mechanics, Molecular Physics, Heat, and Sound

by

Robert Andrews Millikan

Duane Roller

Earnest Charles Watson

THE M.I.T. PRESS

Massachusetts Institute of Technology
Cambridge, Massachusetts

First published by Ginn and Company, 1937

First M.I.T. Press Paperback Edition, August, 1965

Printed in the United States of America

PREFACE

THIS is a textbook for the serious student who seeks a thorough training in science or engineering, who has already mastered trigonometry, and who has had the equivalent of a good secondary-school course in physics. The material provided is designed to arouse the interest and test the mettle of even the ablest beginners; yet we know from several years' trial in the first half of the two-year general physics course at the California Institute of Technology that it can be mastered by any reasonably able and industrious student.

We have gone far beyond our original plan, which was simply to revise the senior author's *Mechanics, Molecular Physics and Heat* so as to bring it up to date. A thorough treatment of fundamental principles rather than the presentation of a large mass of facts has been stressed, as before; but the treatment is more comprehensive and general than in the earlier book. Comprehensiveness, however, has not been gained simply by making the book bulky but rather by a correlation of a large amount of related material with a relatively few general principles. There is considerable detailed analysis of important physical situations, both in the text proper and in the solved examples, so as to develop the student's analytical ability and physical intuition, and to lead him to seek insight into phenomena and to demand fundamental explanations.

The present revision makes possible a closer relationship between class and laboratory work than even the earlier book afforded. The carefully selected experiments are grouped in such a manner that the laboratory and class work can be kept in step without expensive duplication of apparatus. Because of its comprehensive nature, the book will serve also as a class text for courses in which laboratory and class work are separated. Some of the experiments provide useful and important material for class discussion even if the experiments are not actually performed. Our aim has been to develop a realization on the part of the student that the essential strength of physics lies in its continual resort to experiment and that it is only through the constant interplay of experiment and theory that the science has obtained its enormous successes.

A radical departure from the plan of the earlier book consists in the inclusion of some of the historical and humanistic background of the subject. Most of the attempts to humanize the teaching of physics have reached non-science majors rather than those who later will teach and work in the sciences. We have tried to keep the historical material and references on the same plane as the analytical treatment and have felt free to follow the logical rather than the historical order where it offers real advantages. Much of the historical material appears in the form of chapter introductions and plates, and so does not affect the continuity of the analytical treatment.

Great emphasis is placed on the source literature of the field and on the value of collateral reading as well as on the desirability of depending on original sources rather than on the secondary material found in ordinary textbooks. Wherever possible, the references, quotations, and historical material have been taken directly from the original sources. Although references available in English may have been overemphasized, we have cited the most important sources found in other languages. It is not expected that the student will have either the time or the degree of maturity needed to make use of more than a small part of the many references; but they are here for his future use and for the use of the instructors in this and similar courses.

We believe that there is very little in the text that will have to be "unlearned" by the student when he goes on in physics. Where it has been necessary to simplify or specialize the treatment of a topic in order to bring it within the scope of the course, care has been taken to indicate clearly the assumptions or restrictions involved. The terminology and notation are consistent throughout the book and have been chosen in the light of the best modern practice. We have not hesitated to introduce the calculus in places where its value and importance can be made obvious, for this encourages the student from the start to attack problems by methods which he ultimately must use, and vitalizes the mathematics instruction for those who study the calculus concurrently with the course in physics. Experience has shown, however, that able students who have not had the calculus can master the text without undue difficulty.

Great emphasis is placed on problems, for it is only by practice in applying physical principles to many different situations that any real understanding of these principles can be obtained. In some cases important material has been introduced through the medium of solved examples. Sometimes detailed solutions are given, sometimes only hints for solution. Significant figures have been taken into account in stating the problems and their answers; hence questions of the accuracy of data and of calculated results can, and should, be stressed continually.

We are grateful to many of our colleagues and graduate students who have made suggestions of great value. We also wish to thank the authors, publishers, and libraries who have kindly given us permission to reproduce material.

ROBERT ANDREWS MILLIKAN
DUANE ROLLER
EARNEST CHARLES WATSON

CONTENTS

○

List of Experiments

ix

○

List of Optional Laboratory Problems

○

Appendixes

SOME IMPORTANT REFERENCES

IT IS strongly recommended that every student make the acquaintance of many of the original memoirs of the great physicists. NIELS ABEL (1802–1829), who died before he was twenty-seven years of age, but still "left mathematicians enough to keep them busy for five hundred years," when asked how he had done all this, replied, "By studying the masters, not the pupils"; and J. CLERK MAXWELL observes in the preface to his *A Treatise on Electricity and Magnetism*, "It is of great advantage to the student of any subject to read the original memoirs on that subject, for science is always most completely assimilated when it is in the nascent state." Carefully selected excerpts from the original papers of more than ninety physicists will be found in W. F. MAGIE'S *A Source Book in Physics*. Every serious student should also have eventually some first-hand knowledge of the following original works. They will not be found easy reading, — no genuine scientific work can be, — but their value to the student will increase as he matures. A more comprehensive list of references appears in the Bibliography on pages 435–456. Many references to periodical literature, treatises, and other books of importance as source material or collateral reading will be found in footnotes throughout the book. Asterisks in the footnotes mark references that are particularly suitable for beginners.

GALILEO GALILEI, *Two New Sciences*, translation by H. CREW and A. DE SALVIO.

> This, "the first work on modern physics," contains practically all that the man who founded the science of kinetics had to say on the subject of physics.

ISAAC NEWTON, *Mathematical Principles of Natural Philosophy*, translation by ANDREW MOTTE, revised by F. CAJORI.

> This is without exception the most important work on physical science extant and is the starting point of most of the modern treatises on dynamics. The "Principia," as it is usually called, discusses a host of questions that are still alive.

ARCHIMEDES, "On the Equilibrium of Planes," Books I and II; "On Floating Bodies," Books I and II.

> In the first two books the general law of the lever is developed and propositions are proved for finding the centers of gravity of numerous plane figures. In the last two the foundations of hydrostatics are laid. Translations will be found in *The Works of Archimedes*, translated and edited by T. L. HEATH.

CHRISTIAAN HUYGENS, "On the Motion of Bodies through Impact" (*De Motu Corporum ex Percussione*);
"On Centrifugal Force" (*De Vi Centrifuga*);
"The Pendulum Clock" (*Horologium Oscillatorium*).

In the last of these papers the first application of dynamics to bodies of finite size and not merely to particles is made and the important and difficult problem of the physical pendulum is solved. It is a work of importance second only to Newton's *Principia*. These three papers will be found in the original Latin and in French translation in *Œuvres complètes de Christiaan Huygens*, Vols. XVI and XVIII. German translations are available in OSTWALD'S *Klassiker der Exakten Wissenschaften*.

o

The student will find that it also will add greatly to his interest in, and understanding of, physics if he will read biographies of the pioneer investigators and good histories of the periods under consideration. The Bibliography on pages 435–456 lists a number of such works, but the following are especially recommended:

H. CREW, *The Rise of Modern Physics*.

Modern, scholarly, and readable. The best book of its kind in English.

W. C. D. DAMPIER WHETHAM, *A History of Science*.

The best single volume treating the general history of science available today. Clear, interesting, and not too difficult.

E. MACH, *The Science of Mechanics*, translation by T. J. McCORMACK.

A fascinating critical presentation of the historical development of mechanics which will repay careful study.

J. J. FAHIE, *Galileo, His Life and Work*.

The best life in English of the man who started physics in the direction in which it is now traveling.

L. T. MORE, *Isaac Newton, A Biography*.

A vivid but critical narrative of the life and character of the greatest man of science England has produced. This is easily the best and the only really adequate biography of NEWTON so far written.

L. CAMPBELL and W. GARNETT, *The Life of James Clerk Maxwell*.

An excellent biography by two men who knew him intimately, one a schoolfellow and lifelong friend, the other his assistant at Cambridge. The poetic feeling and overflowing humor of this great physicist are shown in his letters and verses, many of which are quoted.

B. JONES, *The Life and Letters of Faraday*.

The standard biography of "the greatest experimental philosopher the world has ever seen."

Mechanics, Molecular Physics, Heat, and Sound

CHAPTER ONE

LINEAR MOTION

THERE IS, in nature, perhaps nothing older than motion, concerning which the books writ-ten by philosophers are neither few nor small; nevertheless I have discovered by experiment some properties of it which are worth knowing and which have not hitherto been either observed or demonstrated. Some superficial observations have been made, as, for instance, that the free motion of a heavy falling body is continuously accelerated; but to just what extent this accelera-tion occurs has not yet been announced; for so far as I know, no one has yet pointed out that the distances traversed, during equal intervals of time, by a body falling from rest, stand to one an-other in the same ratio as the odd numbers beginning with unity.

It has been observed that missiles and projectiles describe a curved path of some sort; how-ever no one has pointed out the fact that this path is a parabola. But this and other facts, not few in number or less worth knowing, I have succeeded in proving; and what I consider more important, there have been opened up to this vast and most excellent science, of which my work is merely the beginning, ways and means by which other minds more acute than mine will ex-plore its remote corners.

GALILEO GALILEI, Two New Sciences, " Third Day " [1]

○

The physical phenomena that first attract attention are those pre-sented by the motion of objects about us. The science of *mechanics* is concerned ultimately with the conditions which govern motion and changes in motion. Before, however, we can hope to *explain* the motion of a body, we must find out *how* the body moves. This necessity of investigating the "how" of phenomena rather than attempting to answer the question "why" was emphasized by GALILEO GALILEI (1564–1642), who has on this account been well called the founder of modern physics.[2] The introduction of this point of view really marks the beginning of modern science, and it is to it that the remarkable scientific developments since the sixteenth century have been largely due.

The description of how bodies move is usually called *kinematics*, whereas the more fundamental investigation of the causes govern-ing motion constitutes what now is called *dynamics*. Kinematics

---

[1] From *Galileo, *Two New Sciences*, tr. by H. Crew and A. de Salvio (1914), pp. 153–154. By permission of The Macmillan Company, publishers.

[2] See *E. Mach, *The Science of Mechanics* (Open Court, 1893), pp. 128–155; also *J. Cox, *Mechanics* (Cambridge University Press, 1919), pp. 69–78. The whole of MACH's classical treatment of the concepts and laws of mechanics should be read eventually by every serious student.

differs from ordinary geometry only in that it introduces the idea of motion; it takes into consideration the element of time as well as the three dimensions of space.

Of the various kinematical concepts, acceleration is the most important; yet the concept of acceleration was not clearly defined until the late sixteenth century, when GALILEO attacked the problem of describing the way in which bodies fall toward the earth. This is not surprising, for even given the "scientific view" of nature — that is, the conviction that all experience can be described objectively — the problem of expressing these observations in terms of exact measurements still remains. This is no easy matter; the simplest physical phenomenon is so modified and conditioned by other related phenomena that great intelligence is required to determine its *essential features* and great skill to measure them and only them.

The first problem of the physicist when he uncovers a new phenomenon is to determine *what* he is to measure; the next to devise means to measure it. Take for example the relatively simple and now thoroughly familiar problem of a freely falling body. It is a matter of ordinary observation that a stone, a feather, a snow-flake, and a droplet of fog fall through the air at different rates, and it is not obvious that these different rates result from differences in the effects of the air on the various objects. Moreover, objects fall so swiftly that experiments with them are difficult to make. It required, therefore, a man of GALILEO's insight into the essential points to be investigated and of his ingenuity in overcoming experimental difficulties to clarify this fundamental problem and to work out the laws of falling bodies. It is easy enough, looking back, to see that others besides GALILEO were acquainted to some extent with these laws, but they were inarticulate; they possessed no good methods of description and notation for conveying the meanings of the laws to other people.

o

The Problem of Falling Bodies

1. Galileo's Contributions. GALILEO's pioneer experiments on motion were published in 1638 in his treatise *Discorsi e dimostrazioni matematiche, intorno a due nuove scienze.*[1] He had no means of pro-

---

[1] Translated into English by H. Crew and A. de Salvio under the title *Two New Sciences* (Macmillan, 1914). See especially the earlier parts of the dialogue of the "Third Day" and the latter half of that of the "First Day," beginning on page 61.

ducing a vacuum, but by considering various cases of objects falling in a series of fluids of different densities and finding that, as the density of the fluid diminished, the objects fell more nearly at the same rate, he concluded that "in a medium totally devoid of resistance all bodies would fall with the same speed." By making the hypothesis that the speed acquired by a falling object is proportional to the time of fall, he correctly deduced that the vertical distance traversed would be proportional to the square of the time of fall. Before he could verify this conclusion experimentally he had to find a way to avoid the high speeds with which bodies fall, for the most accurate time-measuring instrument at his disposal was the water clock, invented by the Babylonians or Egyptians at least three thousand years earlier. A series of experiments showed him that a ball rolling down an inclined plane would follow the same kind of rule as a ball falling vertically, and hence that he could transfer the results obtained at diminished speed on the inclined plane over to a freely falling body. In this way, GALILEO arrived at a genuine experimental law for the fall of a body.

2. A Modern Experiment. Although many experiments on falling bodies have been made since, it was not until the present century that GALILEO's discovery was subjected to an accurate *direct* test with heavy bodies falling from a great height. In this modern experiment illuminated bombs were dropped at night from an airplane and their paths were photographed upon the plates of two cameras placed on the ground at the ends of a long base line. The cameras took the place of surveying instruments and may be regarded as equivalent to two transits. Times were registered on the plates by simultaneously closing the shutters of the cameras at intervals known to an accuracy of about 0.003 sec. The position of the bomb in space at any instant could be determined to within two feet approximately, and the speed at any point of the trajectory could be found with a similar degree of accuracy.[1] Some of the resulting data are reproduced in Table I (on the next page), and the student should satisfy himself that they lead to the following assertions:

a. The data for the first six seconds show that the vertical distance fallen was proportional to the square of the time, or that $y_t = 16\,t^2$, thus confirming GALILEO's law.

b. The distances fallen in successive seconds increased by increments of 32 feet during each of the first six seconds, but as the speed of the bomb became greater, these increments diminished. Since,

---

[1] *Problems in Physics*, War Department Committee on Education and Special Training, edited by H. L. Dodge.

AMONG all the portraits of GALILEO, the most precious, whether for the excellence of the artist, or for the exquisiteness of his work, or for its resemblance (which all contemporaries have declared to be perfect), is that which we owe to Giusto Sustermans.'' For information regarding the interesting history and vicissitudes of this celebrated picture, see J. J. Fahie's *Memorials of Galileo Galilei* (Courier Press, 1929), p. 32, from which the foregoing quotation was taken. In 1635, when this portrait was painted, GALILEO was just completing the writing of his *Two New Sciences*, the foundations for which, however, were laid during the eighteen best years of his life, those spent in Padua.

Along with the portrait it is interesting and useful to have the following word picture, taken from the excellent description given by one of GALILEO's most famous pupils, VINCENZIO VIVIANI.

GALILEO was square of frame, well proportioned, and above medium height. In countenance he was cheerful and pleasant, with eyes that were blue and sparkling, and hair and beard that were abundant and of a reddish hue. His constitution was sound until the age of thirty, when he experienced a serious illness, and from then on he was subject to various complaints which increased in gravity and frequency with age. Doubtless this suffering and the troubles, public and private, from which he was seldom free contributed to his shortness of temper. Yet, though he was easily ruffled, he also was easily pacified.

His industry was extraordinary. It was said that no one ever saw him idle, and one of his favorite sayings was that occupation is the best medicine for mind and body. Gardening was about his only relaxation from his studies. He was a connoisseur of wines and attended to his own vineyard. He was perhaps too fond of wine for his health and temper; even in old age his taste for it apparently was as keen as ever.

Although conservative in most expenditures, he spared no cost for the success of his many experiments, gave freely to charity, and helped those in whom he saw promise. He liked friends and received them cordially though simply. Except with intimate friends, he seldom conversed on scientific and philosophic subjects; when others broached such subjects he skillfully turned the conversation into other channels, usually in such a way as to satisfy the curiosity of the inquirer.

His general demeanor was modest and unassuming. He did not envy the talents of others, but gave to each his just dues. As a teacher he was no less loved than as a friend. He gave freely of his knowledge and, no matter how clear a subject might be in his own mind, was not satisfied till he made it clear to his pupils.

○

GALILEO GALILEI, 1564–1642

From the portrait by Sustermans, in the Uffizi Gallery, Florence

for the first six seconds, the downward speed v_t increased at a constant rate of 32 feet per second during each second, it follows that $v_t = 32\,t$, again confirming one of GALILEO's assertions. Now this rate of change of speed, v_t/t, was called by GALILEO the *acceleration* in the downward direction. If we denote the magnitude of this acceleration by g, we have $v_t = gt$, where, within the limits of accuracy of the measurements, g had the constant value 32 feet per second per second during the first few seconds that the bomb fell.

c. After about six seconds the acceleration began to decrease in magnitude; this can be attributed to air resistance, which increased rapidly with the increasing speed of the bomb.

TABLE I · *Data for Bomb Dropped from an Airplane*

Time t in seconds	Horizontal displacement x in feet at time t	Vertical displacement y in feet at time t
0.000	0	0
1.000	98	16
2.000	196	64
3.000	294	144
4.000	392	256
5.000	490	400
6.000	587	576
7.000	684	782
19.000	1783	5499
19.075	1790	5539

3. Acceleration Due to Gravity. It can therefore be concluded that, so long as the air resistance is negligible, a body near the earth falls with a constant acceleration of approximately[1] 32 ft · sec^{-2}, or 980 cm · sec^{-2}. This *acceleration due to gravity,* as it is called, has since the time of JEAN BERNOULLI (1667–1748) been denoted by the symbol g. Eventually we shall see that the value of g can be regarded as constant only for distances of fall that are small compared to the radius of the earth, and that it depends on both the gravitational attraction and the rotation of the earth and hence on

---

[1] The expressions 32 ft · sec^{-2} and 980 cm · sec^{-2} are to be read "32 feet per second per second" and "980 centimeters per second per second" respectively. Obviously they might also be written as 32 ft/sec^2 and 980 cm/sec^2. Other examples of the use of exponents in the abbreviations for units are: cm^2, for "square centimeter"; ft · sec^{-1}, for "feet per second"; and lb · ft^{-3}, for "pounds per cubic foot."

the latitude and the height above sea level. There are also local variations in g of comparatively small amount.[1]

The results of careful experiments show that the value at sea level, in latitude[2] θ, is represented by the formula

$$g_0 = 978.04(1 + 0.00529 \sin^2 \theta), \qquad [1]$$

where g_0 is expressed in centimeters per second per second; thus the total variation from equator to pole is about 0.5 percent. The variation with altitude is given by the formula

$$g_H = g_0 - 0.0003086\, H, \qquad [2]$$

where g_H is the acceleration due to gravity (cm · sec^{-2}) at a height H meters above sea level. This variation is unimportant for most purposes.

In scientific literature, unless it is otherwise specified, the value $g_s = 980.665$ cm · sec^{-2} is used in computations and tabulated data which involve the acceleration due to gravity. This value, g_s, called *standard acceleration due to gravity*, has been adopted as a conventional constant by the International Committee on Weights and Measures.

o

Some Preliminary Ideas about Motion

It has been seen that the first problem which GALILEO set himself to solve was one of *describing* a particular kind of motion, and this independently of any attempt to find out the cause of the motion. This process of abstraction, of dealing with one difficulty at a time, is an essential part of the scientific method. Thus in the present chapter some relatively simple types of motion will be discussed in detail; but the nature of the bodies which move and the explanations of their motions are left for future inquiry.

4. Frame of Reference. When we say that an automobile has moved a mile, we usually mean that it has moved a mile relative to the earth; relative to the sun, the automobile may be thousands of miles from where it started; and relative to us, if we happen to be the passengers, the automobile has not moved at all. Evidently by the *position* or by the *motion* of a body, we can mean only its

---

[1] See the *International Critical Tables* (1926), Vol. I, pp. 395–402, for values of g at various stations and g_0 at various latitudes.

[2] See Appendix 11 for names of Greek letters used throughout this text.

position or its change of position *relative* to another body or framework of lines which must be considered as fixed. The way in which we describe the position or motion of a given body will depend on our choice of this fixed *frame of reference*; by choosing wisely, the problem of description is often greatly simplified. In all that follows it will be understood, unless some other choice is indicated, that our frame of reference is fixed in the earth.

5. Translation and Rotation. In the up-and-down motion of an elevator car, or in the motion of a railway car on a straight track, all points in the car move along parallel lines. The same kind of motion is also to be observed in a piece of straight wire that is held, say, in a vertical position but otherwise moved in any irregular way whatever. A body that moves in this way without turning is said to have a motion of *translation* or a *linear motion*. If, on the other hand, a certain line in a body remains stationary as the body moves, the motion is called a *rotation* or *angular motion*. Such is the motion of the flywheel of a stationary engine, or of a door as it opens and closes. All points in the body describe concentric circles about an axis and all the radii of the body turn through equal angles.

However complex the motion of a rigid body may be, it can always be resolved into pure translations and pure rotations, and then these can be studied one at a time. The motion of a nut on a bolt, of the wheels of a passing automobile, or of an airplane in a tail spin, may be regarded as a combination of these two kinds of motion; these objects move as a whole with respect to the earth, and they also turn.

Bodies in motion also undergo changes of size and shape. These changes are often extremely complicated, but in practice do not always produce important effects. In order to simplify our study of motion, we will disregard such changes for the present by confining our attention to bodies that do not change in size or shape. Such perfectly *rigid bodies*, of course, exist only in the imagination, although objects like a billiard ball or a piece of steel may be regarded as perfectly rigid when the forces acting upon them are small.

o

The Kinematics of a Particle

We have said that when a body undergoes pure translation, every line of points in the body retains its direction relative to the frame of reference. This means that the motion of the body will be completely described just as soon as the motion of any one of its points is given. In other words, in describing the motion of a body that has translation without rotation, the body may be treated as if it were a single particle. The term *particle* is here used in its technical sense to refer to a moving point considered as defining the position from time to time of a very small part of a body.

6. Position of a Particle. One familiar way to locate a particle on the earth's surface is to give its latitude and longitude and its altitude from sea level. Since the particle may change its position with time, the description is made more complete if the time is also given. Four data, then, are needed for the complete location of a particle in ordinary space and time: three of these data locate the particle in space of three dimensions relative to some fixed frame of reference; the fourth gives the time of location with respect to some "zero" time.

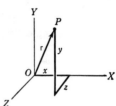

FIG. 1. Method of locating a particle in space by means of a right-handed rectangular coordinate system

In the familiar rectangular system of coordinates, Fig. 1, the frame of reference consists of three mutually perpendicular *coordinate axes* OX, OY, OZ drawn from a common point O called the origin. The position of a particle P is given in this system by its distances x, y, z, positive or negative, from the three coordinate planes YOZ, ZOX, XOY. These distances x, y, z are called the *rectangular coordinates* of the particle. It is important to note that the coordinate axes in Fig. 1 are designated in such a manner that upon looking from a point on the positive Z-axis we see that a counterclockwise rotation is needed to carry the positive X-axis by the shortest way around to the direction of the positive Y-axis. This type of rectangular-coordinate system is the one now generally used in physics. It is called a *right-handed* system, and if we indicate the X-axis with the thumb, the Y-axis with the forefinger, and the Z-axis with the middle finger, we can represent it with the right hand. It is evident that if the left hand is used, another system of rectangular coordinates is represented which is the mirror-image of the right-handed system and thus can never be made to coincide with the latter.

Another way of describing the position of a particle that often proves simpler than rectangular coordinates is to locate it as the terminal point of a *directed line* drawn from the origin. In Fig. 1, for example, the position of P is completely and uniquely determined by giving the *direction* and *length*, measured from the origin, of the straight line $\overline{OP}$, or r. Three data in addition to the time are, of course, needed for this purpose.

7. Linear Displacement of a Particle. Suppose that a particle moves from the point A_0 to the point A_1, Fig. 2, by any path. Such a change of position, made without reference to the time involved, is called a *linear displacement*, and the easiest way to describe it is by the directed straight line $\overline{A_0A_1}$. Evidently change of position, like position itself, is completely described by specifying (*a*) its magnitude, or the number of units of length represented by the straight line $\overline{A_0A_1}$, and (*b*) its direction, or the direction of $\overline{A_0A_1}$. To say that a particle has moved 10 cm from a certain point merely locates it as somewhere on the surface of a sphere, but to specify that this motion was, say, eastward is to have described completely its change of position.

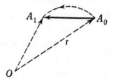

Fig. 2. The vector $\overline{A_0A_1}$ represents a change of position from A_0 to A_1

○

Vector Quantities

8. Vector and Scalar Quantities. A physical quantity, such as displacement, that is completely determined only when a *direction* as well as a *magnitude* is given, is called a *vector quantity*. Evidently such a quantity can be represented graphically by a straight arrow drawn in the proper direction, the length of the arrow being made proportional, upon some convenient scale, to the magnitude of the quantity represented. Symbolically a vector may be denoted by two letters with a bar over them, the first one indicating the initial point or tail of the arrow, and the second one the end point or head. More usually, it is represented by a single letter in special type (in this book Gothic type will be used). Thus in Fig. 2, either $\overline{OA_0}$ or r means the going of the distance OA_0 in the direction O to A_0.

Two vectors a and b are said to be *equal* if they have the same magnitude and the same direction; this is symbolically expressed by the equation a = b. It follows from this definition of equality that the location of a given vector in space may be changed without changing the value of the vector, provided, of course, that the magnitude and direction of the vector are left unaltered.

Some physical quantities, such as time, volume, distance, mass, work, and temperature, are not vector quantities; 25 cm^3 of water and 10 sec of time require no specification of direction in space to render them complete. Quantities of this kind are completely described by an algebraic magnitude alone and can be represented simply by points on a scale. Hence they are called *scalar quantities*, a term which originated with WILLIAM ROWAN HAMILTON [1] (1805–1865).

The addition and multiplication of scalar quantities are of course accomplished by the methods of ordinary algebra. Vectors, on the other hand, are not ordinary algebraic numbers, and consequently they cannot be added and multiplied by ordinary methods. Vector addition and vector multiplication are accomplished by methods that have been developed in *vector analysis*, the branch of mathematics devoted to the study of vectors.[2] We shall find, however, that the various vector operations have been so defined that in certain special cases they may be identified with the similarly named scalar processes.

9. Addition of Vectors. It is immaterial whether we "cut across" a vacant lot or walk around it, so far as the *resulting* displacement is concerned. Also it does not matter which way we walk around the lot. If, in Fig. 2, a particle goes the distance OA_0 in the direction O to A_0 and, at the same time or later, the distance A_0A_1 in the direction A_0 to A_1, it has then undergone two linear displacements, $\overline{OA_0}$ and $\overline{A_0A_1}$. But the final position of the particle would have been the same if it had gone the single distance OA_1 in the direction O to A_1; that is, the displacement $\overline{OA_1}$ is exactly equivalent to the two displacements $\overline{OA_0}$ and $\overline{A_0A_1}$ taken together. Thus $\overline{OA_1}$ is to be regarded as the *vector sum*, or *resultant*, of $\overline{OA_0}$ and $\overline{A_0A_1}$; briefly,

$$\overline{OA_0} + \overline{A_0A_1} = \overline{OA_1}. \qquad [3]$$

The vectors $\overline{OA_0}$ and $\overline{A_0A_1}$ are called *components* of the vector sum $\overline{OA_1}$. It is to be emphasized that Eq. [3] is a *vector equation*; in ordinary addition, which deals only with scalar quantities,

$$OA_0 + A_0A_1 > OA_1.$$

---

[1] *Lectures on Quaternions* (Dublin, 1853); also *Elements of Quaternions* (Longmans, Green, 1866). There are still earlier references in his correspondence; see R. P. Graves, *Life of Sir William Rowan Hamilton* (Longmans, Green, 1882–1889), Vol. III, p. 354. The term *vector* also is due to HAMILTON.

[2] Vector analysis, in the form in which it is used today in mathematical physics, is largely due to the great American theoretical physicist JOSIAH WILLARD GIBBS (1839–1903), and the treatment of the subject in Gibbs's *Collected Works* (Longmans, Green, 1928), Vol. II, Part 2, pp. 17–90, is one of the classics in this field. An extensive treatise on the subject, based on the lectures of GIBBS, was published in 1901 by E. B. Wilson. Treatments that are suitable for the ambitious beginner will be found in *J. G. Coffin, *Vector Analysis* (Wiley, 1911), and *H. B. Phillips, *Vector Analysis* (Wiley, 1933).

It is apparent that the operation expressed by Eq. [3] may be performed by a graphical method. In fact, any number of vector quantities may be added in this way. The quantities to be added are first represented by arrows drawn to scale (Fig. 3), and these arrows are then placed end to end in a chain, in any order but without altering their lengths and directions; finally, a new arrow is drawn to join the initial point of the first arrow to the terminus of the last. This new arrow, which is the closing side of the polygon thus formed, is the vector that represents the sum of all the other vectors.

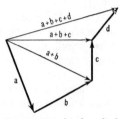

FIG. 3. Graphical method of adding several vectors

A little consideration of Fig. 3 will convince one that it does not matter in what order the vectors are added or how they are grouped;

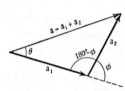

FIG. 4. Sum of two vectors which make an angle ϕ with each other. Note carefully that when the second vector is drawn from the extremity of the first vector, the angle ϕ between the two vectors is taken as the angle between the second vector and the first vector produced as shown

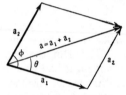

FIG. 5. Alternative method of representing the sum of two vectors which make an angle ϕ with each other. Note that the sum is the diagonal of the parallelogram formed with the two vectors as sides. This statement of the rule for the addition of two vectors is called the *parallelogram law*; it tells us no facts that are not included in the method of Fig. 4 and is not so readily extended to include more than two vectors

thus $a + b + c + d = d + b + a + c = (a + b) + c + d$, etc. We express these facts in mathematical language by saying that vector addition is both *commutative* and *associative*.

EXAMPLE. Let a_1 and a_2 be the respective magnitudes of two vectors that make an angle ϕ with each other (Fig. 4 or 5). Show that if these two vectors are added, the magnitude a of their sum is given by

$$a^2 = a_1^2 + a_2^2 + 2\,a_1 a_2 \cos\,\phi.$$

Also show that the angle θ which the sum a makes with the vector $\mathbf{a_1}$ is given by $\theta = \sin^{-1}\left(\dfrac{a_2}{a}\sin\,\phi\right)$.

10. Vector Subtraction. The *subtraction* of vectors offers no difficulties, for it is merely the inverse of addition; that is, by the difference of two vectors is meant the third vector which must be added to one of them to obtain the other. Thus, in Fig. 2, $\overline{A_0A_1}$ is the vector that must be added to $\overline{OA_0}$ to obtain $\overline{OA_1}$; hence, $\overline{A_0A_1}$ is the difference between the latter two vectors, or

$$\overline{A_0A_1} = \overline{OA_1} - \overline{OA_0}. \tag{4}$$

It will be recalled that the subtraction of an ordinary algebraic quantity is accomplished by changing the sign of the quantity and adding. In an analogous manner, a vector is subtracted by reversing its direction and adding. For example, $\overline{OA_0}$ may be subtracted from $\overline{OA_1}$ by adding $\overline{A_0O}$ to $\overline{OA_1}$; this means that Eq. [4] may be rewritten in the form $\overline{A_0A_1} = \overline{OA_1} + \overline{A_0O}.$

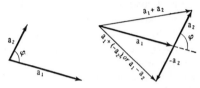

Fig. 6. Illustrating the distinction between vector addition and vector subtraction

Obviously $\overline{A_0O} = -\overline{OA_0}$. In other words, when a vector is multiplied by the scalar quantity -1, its direction is changed $180°$ but its magnitude is unaffected.

> EXAMPLE. Given that a_1 and a_2 are two vectors that make an angle ϕ with each other (Fig. 6), show that the magnitude of $a_1 - a_2$ is $\sqrt{a_1{}^2 + a_2{}^2 - 2\,a_1a_2\cos\phi}.$

○

Linear Speed

By the *speed* of a particle is meant the time-rate at which the particle moves along its path. Speed is evidently a scalar quantity, for the direction of motion does not enter into its definition.

11. Average Speed. Suppose that the distances of a particle from some point O on its path are s_0 at a time t_0, and s_1 at some later time t_1. The *average*, or *effective*, speed v_{av} of the particle is then defined by the equation

$$v_{av} = \frac{s_1 - s_0}{t_1 - t_0}, \tag{5}$$

the distance $s_1 - s_0$ being measured along the path. More concisely, if s is the number of units of distance traversed in t units of time,

$$v_{av} = \frac{s}{t}. \tag{5a}$$

12. Instantaneous Speed. If we are told that an automobile has traveled 220 km in 3.2 hr, we are able to compute its average speed as 69 km · hr^{-1}, but we cannot say how its speed varied during this time. Evidently if we wish to know the character of the motion more precisely, we must divide the whole distance traversed by the automobile into parts and ascertain the average speed in each part. The smaller these parts, the more nearly does the average speed in any one of them represent the speed at some instant t or point s in that part.

Let s_0, measured from some reference point O, represent the distance traversed by a particle at the end of some time t_0, and let s_1, measured from the same reference point, be the distance traversed at the end of a later time t_1. The average speed in the interval $t_1 - t_0$ is $(s_1 - s_0)/(t_1 - t_0)$, and the limit which this approaches as $t_1 - t_0$ is made to approach zero is defined as the *instantaneous speed* of the particle at a time t in the interval $t_1 - t_0$. This may be written

$$v_t = \lim_{(t_1 - t_0) \to 0} \frac{s_1 - s_0}{t_1 - t_0}, \qquad [6]$$

and is read "the instantaneous speed at the time t is the limit of $(s_1 - s_0)/(t_1 - t_0)$ as $t_1 - t_0$ approaches zero." Clearly v_t may also be spoken of as the speed at a point s in the space interval $s_1 - s_0$. It is this instantaneous speed v_t that is measured by a correctly calibrated automobile speedometer.

To state the foregoing definition more concisely, let Δs represent the distance moved by the particle in an interval of time Δt, where Δt includes the time t; then

$$v_t = \lim_{\Delta t \to 0} \frac{\Delta s}{\Delta t}. \qquad [7]$$

In the differential calculus this equation is written in the still more abbreviated form $v_t = ds/dt$ and is read "v_t equals the derivative of s with respect to t." Rules for "differentiating," or finding the value of this derivative, are developed in the differential calculus.

In the case of a particle that is moving with *constant speed* the average speed v_{av} and the instantaneous speed v_t are the same. If this constant speed be denoted by v, one evidently may write the useful relation

$$s = vt. \qquad [8]$$

EXAMPLE. A certain particle starts from a point O on its path and moves in such a way that its distance s from O is proportional to the square of the time that has elapsed since it started. How does the speed of this particle vary with the time?

Solution. By hypothesis, the distance traversed by the particle in the time t is $s = kt^2$, where k is a constant, and the distance traversed in the time $t + \Delta t$ is $s + \Delta s = k(t + \Delta t)^2$. By subtracting the first expression from the second, one obtains

$$\Delta s = 2 \, kt \, \Delta t + k(\Delta t)^2,$$

and therefore

$$\frac{\Delta s}{\Delta t} = 2kt + k \, \Delta t.$$

Then, from Eq. [7], $\quad v_t = \lim_{\Delta t \to 0} (2 \, kt + k \, \Delta t) = 2 \, kt.$

The speed of this particle is thus seen to be proportional to the time that has elapsed since it started. Note that the limit has a definite finite value at any time t, even though $\Delta t = 0$, and that it is this limiting value which we call the instantaneous speed.

The differential calculus gives a set of rules for finding this limiting value in any given case; in the briefer notation of the calculus, the foregoing solution would be written $v_t = \dfrac{ds}{dt} = \dfrac{d}{dt} (kt^2) = 2 \, kt.$

EXAMPLE. By making use of the result of the preceding example and the data in Table 1, show that the downward speed of an airplane bomb 3.0 sec after it is released is 96 ft · sec⁻¹.

EXAMPLE. Find the expression for the speed of a particle that starts from 0 and moves in such a way that $s = kt^3$.

Solution. From Eq. [6], $v_t = \lim\limits_{\Delta t \to 0} \dfrac{k(t + \Delta t)^3 - kt^3}{\Delta t} = 3 \, kt^2.$

13. Scalar Diagrams. A graph in which the distance traversed by a body is plotted as a function of the time is often of importance in the study of motion. To see how such a *distance-time curve* is constructed and interpreted, let us examine Fig. 7, where some of the experimental data from the first and last columns of Table I have been plotted. The plotted points have been connected by a smooth curve, and the position of the bomb at *any* given instant is represented by the ordinate of the curve corresponding to that instant. At the time t_0 the distance fallen is represented by the ordinate $t_0 p_0$, and at a later time t_1 this distance has increased

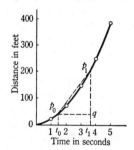

FIG. 7. Distance-time curve for falling airplane bomb

by an amount represented by $q p_1$. The *average* speed during the interval $t_1 - t_0$ is then represented by $q p_1 / p_0 q$ or $\tan q p_0 p_1$; that is, by the slope of the chord $p_0 p_1$. If we imagine the interval of time $t_1 - t_0$ to be reduced indefinitely, the point p_1 moves along the curve toward the point p_0 and finally coalesces with it; the chord p_0p_1 then becomes the tangent line to the curve at p_0 and the average speed becomes the speed at the instant t_0. Hence *the speed at any instant is represented by the slope of the distance-time curve at that instant.* What is the meaning of a positive slope? of a negative slope? What if the distance-time curve is a straight line? The slope gives the speed in arbitrary units, the nature of which depends on the scales adopted for plotting the graph. If distances of equal length are used to represent the foot and the second, the slope expresses the speed in feet per second.

Another kind of curve that is useful in the study of motion is the *speed-time curve*, obtained by observing the speeds of a body at various instants and plotting these speeds as a function of the time. It is not always easy to measure speeds directly, but they can be obtained indirectly by plotting the space-time curve for the motion and measuring the slope at different points of this curve. The student should experience no difficulty in proving that the area bounded by the speed-time curve, the axis of time, and any two given ordinates represents the distance traversed by the moving object during the interval of time defined by the two ordinates.

○

Linear Velocity

No matter how much information is available concerning the speed of a moving object, its position or destination remains unpredictable until the direction of motion is known. There is need, then, for a concept that involves direction as well as speed, and this we call velocity.

14. Definition of Velocity. The *velocity* of a particle is defined as its time-rate of displacement or time-rate of change of position. Like displacement, velocity is a vector quantity, because it requires for its complete description the specification of both a direction and a magnitude. To the *magnitude* of a velocity we have already given the special name *speed*, just as the special name *distance* is often given to the magnitude of a displacement. Two automobiles going 90 km · hr^{-1} in opposite directions on a straight roadway have the same speed, but their velocities are different.

In Fig. 8 let us suppose that a particle is moving along the curved path $A_0A_1A_2$ and that it is at the point A_0 at a time t_0 and at the point A_1 at a slightly later time t_1. Then in the interval $t_1 - t_0$ the particle undergoes a displacement $\overline{A_0A_1}$, and hence its time-rate of displacement is given by

$$v_t = \lim_{(t_1-t_0)\to 0} \frac{\overline{A_0A_1}}{t_1-t_0}. \qquad [9]$$

This is the defining equation for the velocity v_t at an instant t in the interval $t_1 - t_0$. The velocity v_t has the direction of the tangent to the curve described by the particle; this follows from the fact that the vector $\overline{A_0A_1}$ in Eq. [9] has the direction of the secant A_0A_1, and that this secant approaches the tangent in direction as A_1 approaches A_0.

If r_0 and r_1 be the two vectors which give the positions of the points A_0 and A_1 relative to the origin O, Fig. 8, then $\overline{A_0A_1} = r_1 - r_0$, and Eq. [9] may be written in the form

$$v_t = \lim_{(t_1-t_0)\to 0} \frac{r_1-r_0}{t_1-t_0}. \qquad [10]$$

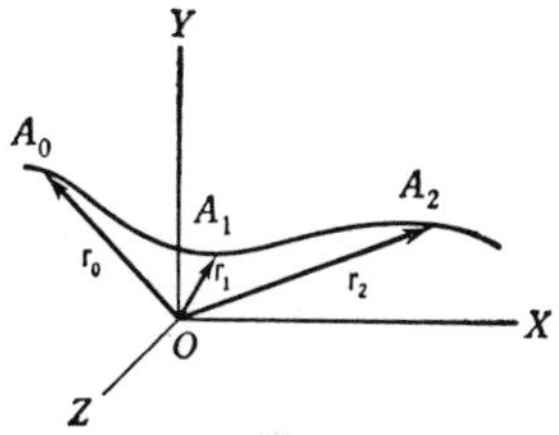

Fig. 8. Velocity is a vector quantity

15. Motion in a Straight Line. If a particle is moving in a straight line, one direction along the line can be taken as positive and the other as negative, and it is then no longer necessary to distinguish between velocity and speed. In the still more restricted case of *constant* velocity, in which both the speed and the direction are constant, Eq. [8] may be used to describe the motion.

16. Digression on the Multiplication and Division of a Vector by a Scalar. We have seen that the vector quantity *displacement* divided by the positive scalar quantity *time* is another vector quantity called *velocity*. Both the displacement and the velocity obtained from it have the same direction, but they have different magnitudes. The magnitude of the displacement is simply the distance, whereas the magnitude of the velocity is the distance divided by the time. This illustrates the rule in vector analysis that if a vector be multiplied or divided by a *positive* scalar, the resulting product or quotient is itself a vector; this new vector has the same direction as the original one, but its magnitude is the original magnitude multiplied or divided by the scalar. Thus an automobile undergoing a displacement of 216 km northward in 3.0 hr has an average velocity of 72 km · hr^{-1} northward.

The result of multiplying or dividing a vector by a *negative* scalar is apparent when it is recalled (Sec. **10**) that the multiplication of a vector by -1 simply reverses the direction of the vector without changing its magnitude.

17. Addition of Simultaneous Velocities. The way in which we describe the motion of a given body depends, of course, on our choice of the frame of reference. A motorboat going upstream against a current will have one velocity with respect to the water and another with respect to the banks of the stream. Thus a body may have a certain velocity relative to a frame of reference, while the frame of reference itself has another velocity relative to a second frame. The vector sum of these simultaneous velocities gives the velocity of the body in the second frame.

EXAMPLE. It is observed from the ground that an airplane is flying north-east with an average speed of 3.0 km · min^{-1} relative to the ground and that the wind is blowing from the south with a speed of 52 km · hr^{-1}. Find the average velocity of the air-plane relative to the air.

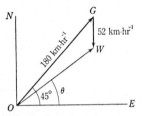

Solution. The airplane has a velocity of 3.0 km · min^{-1} northeast, or 180 km · hr^{-1} northeast, relative to the ground. The ground itself has a velocity of 52 km · hr^{-1} south relative to the second frame, the air. The velocity of the airplane relative to the air is the vector sum of these two simultaneous

FIG. 9. Addition of velocities

velocities. In Fig. 9, this sum is $\overline{OG} + \overline{GW} = \overline{OW}$. The magnitude of $\overline{OW}$ is, by the theorem of cosines,

$$OW = \sqrt{(180)^2 + (52)^2 - 2 \cdot 180 \cdot 52 \cdot \cos 45°} = 1.5 \times 10^2 \text{ km · hr}^{-1}.$$

The direction of $\overline{OW}$ is obtained with the help of the theorem of sines; thus

$$\frac{\sin (45° - \theta)}{52} = \frac{\sin 45°}{1.5 \times 10^2}, \text{ or } \theta = 31°.$$

The velocity of the airplane relative to the air is therefore[1] 1.5×10^2 km · hr^{-1} at 31° north of east. This result should be checked by drawing Fig. 9 to scale and measuring OW and θ with scale and protractor.

○

Linear Acceleration

Mention has already been made of the fact that the concept of acceleration was introduced into mechanics by GALILEO; this was a most important step in the development of the science, for, as we shall see later, it made possible the modern conception of force and thus marks the beginning of the science of dynamics. In ordinary language the expression *accelerated motion* commonly refers only to

---

[1] See Appendix 1, *Significant Figures and Notations by Powers of Ten.*

increases of speed, but in physical science it is employed in a much wider sense to denote *any* change of *velocity*. Thus a physicist would say that an automobile undergoes acceleration when it slows down or turns a corner, as well as when it speeds up. The foot-throttle of an automobile is an "accelerator," but so is the steering wheel or the brake.

18. Definition of Acceleration. *Linear acceleration* is defined as the time-rate of change of linear velocity. It is the quotient of a vector and a scalar, and therefore, like displacement and velocity, is a vector quantity. Hence the statement that a particle has an acceleration may refer to a change in speed, to a change in direction of motion, or to a change in both speed and direction of motion. Let us consider these three possible cases separately:

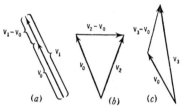

Fig. 10. The change of velocity is the vector difference obtained graphically by reversing the second velocity and adding it to the first

Case 1. If, as in Fig. 10 (*a*), the velocity of a particle at a time t_0 is represented by the vector v_0, and if its velocity at a later time t_1 is represented by v_1, which has the same direction as v_0, then there has been no change in the direction of motion, but only a change in speed, and the average acceleration has been $(v_1 - v_0)/(t_1 - t_0)$. The expression for the acceleration of the particle at an instant t in the interval $t_1 - t_0$ is then the limit of this quantity as the interval approaches zero, or[1]

$$a_t = \lim_{(t_1 - t_0) \to 0} \frac{v_1 - v_0}{t_1 - t_0} = \lim_{\Delta t \to 0} \frac{\Delta v}{\Delta t} = \frac{dv}{dt}. \qquad [11]$$

It will be noted that in this case the direction of both the average and the instantaneous acceleration is parallel to that of the velocity itself, being positive with respect to the velocity if $v_1 > v_0$, negative if $v_1 < v_0$.

Case 2. The second case is illustrated by Fig. 10 (*b*). The velocity after the interval of time $t_1 - t_0$ is represented by v_2, a vector having the same length as v_0 but a different direction; there has been no change in speed, but only a change in the direction of the velocity. The average acceleration of the particle is here $(v_2 - v_0)/(t_1 - t_0)$ and,

---

[1] Also, since $v = \dfrac{ds}{dt}$, we have $a_t = \dfrac{d}{dt}\left(\dfrac{ds}{dt}\right) = \dfrac{d^2 s}{dt^2}$, which is called the second derivative of s with respect to t and means that the differentiation is performed twice.

ON THE second floor of the Museum of Physics and Natural History in Florence, Italy, there has been erected an inspiring monument to GALILEO,—the Tribuna di Galileo. It is profusely decorated with frescoes, medallions, busts, and drawings illustrating GALILEO's discoveries and inventions, and in its niches may be seen, through glass frames, two of his telescopes, his geometrical and military compass, and a loadstone, or natural magnet, fashioned by him. At the top of *Plate 2* a drawing of a fresco from one of the lunettes above the walls of the central hall is reproduced. It shows GALILEO in Pisa, demonstrating the laws of falling bodies by experiments on an inclined plane. In the center, just behind the inclined plane, down which a ball is rolling, stands GALILEO, explaining the effect to two of the spectators. Seated on the right are Don Giovanni dei Medici (a jealous opponent of GALILEO) and the Head of the University of Pisa, both looking very much displeased. On the left are two Peripatetics, bent over a large open book on a table, seeking anxiously for some passage in ARISTOTLE in disproof of GALILEO's argument. In the background are the Baptistery, the Cathedral, and the Leaning Tower.

VINCENZIO VIVIANI, GALILEO's last and well-beloved disciple, formed the front of his house in Florence into a mural monument to his master. One of the sculptured plaques, a drawing of which is reproduced in the lower part of *Plate 2*, shows a man observing the sunspots through GALILEO's telescope, while opposite is another figure observing the curve described by a ball fired from a cannon ; between the two men is a heavy beam supported at one end, and breaking under its own weight. These refer to GALILEO's use of the telescope and to his studies on projectile motion and the strength of beams.

The invention of the thermometer is also due to GALILEO (about 1602). The date is uncertain, but CASTELLI, writing to FERDINANDO CESARINI, on September 20, 1638, says: " I remember an experiment which our Signor Galileo had shown me more than thirty-five years ago. He took a small glass bottle about the size of a hen's egg, the neck of which was two palms long (about 22 inches), and as narrow as a straw. Having well heated the bulb in his hand, he inserted its mouth in a vessel containing a little water, and, withdrawing the heat of his hand from the bulb, instantly the water rose in the neck more than a palm above its level in the vessel. It is thus that he constructed an instrument for measuring the degrees of heat and cold.''

These summarize in pictorial form GALILEO's more important contributions: his epoch-making work on falling bodies and the laws of motion, his astronomical discoveries, his pioneer work on the strength of beams, and his invention of the thermometer. The reproductions, together with the foregoing description of them, were taken, with the permission of the author's executor, from J. J. Fahie's *Memorials of Galileo Galilei*.

o

GALILEO's Inclined-Plane Experiment

Fresco in the Tribuna di Galileo, after an old engraving

Drawing
of the Sculptured Plaque
on the Front of VIVIANI's
House in Florence

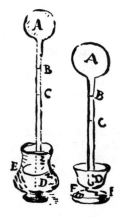

CASTELLI's Sketch
of
GALILEO's
Thermometer

as before, the instantaneous acceleration a_t is the limit of this quantity as the interval $t_1 - t_0$ approaches zero. Since, as $t_1 - t_0$ approaches zero, $v_2 - v_0$ becomes more and more nearly perpendicular to v_0, it is evident that in the limit the acceleration a_t is at right angles to the velocity v_t. An example of this is the acceleration of a particle that is moving in a circle with a constant speed. Although the treatment of this case offers no particular difficulties, further discussion of it will be postponed until Chapter 14.

General Case. Fig. 10 (c) represents the case in which both the speed and the direction of motion of the particle are changing. The average acceleration is $(v_3 - v_0)/(t_1 - t_0)$, and it has the direction of the vector $v_3 - v_0$. The instantaneous linear acceleration is then

$$a_t = \lim_{(t_1 - t_0) \to 0} \frac{v_3 - v_0}{t_1 - t_0} = \lim_{\Delta t \to 0} \frac{\Delta v}{\Delta t} = \frac{dv}{dt}. \qquad [12]$$

This constitutes the most general definition of linear acceleration possible, the others being obviously special cases of it. For the remainder of the present chapter we will apply these definitions in the treatment of the restricted, though very important, case of motion with *constant* linear acceleration.

○

Constant Linear Acceleration

The acceleration of a particle is said to be constant when the velocity changes by equal amounts in equal intervals of time; that is to say, when the *changes* in velocity in equal intervals of time all have the same magnitude and direction. It is possible to have a constant linear acceleration either with or without a change in the direction of motion.

19. Constant Linear Acceleration Without Change in the Direction of Motion. The motion in this case is not only in a straight line (Sec. 18, *Case 1*), but there is the additional restriction that the speed changes by equal amounts in equal intervals of time. If v_0 denotes the speed of the particle at any given moment and v_t denotes its speed at a time t units later, the magnitude a of a constant acceleration is given by the equation

$$a = \frac{v_t - v_0}{t}. \qquad [13]$$

This is often written in the form

$$v_t = v_0 + at. \qquad [14]$$

The best illustration in nature of a constant acceleration is the motion of a falling body, provided it is a heavy, dense body, so that the air resistance is not important (Sec. 2).

In Sec. 13 it was shown that the speed at any instant is represented by the slope of the distance-time curve at that instant. Similarly one can show that if the *speed* is plotted as a function of time, the slope of the resulting speed-time curve represents the acceleration. In a case where the acceleration is constant, the slope will of course be constant and the speed-time curve will be a straight line (Fig. 11). What is the nature of the *acceleration-time curve* in such a case?

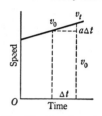

Fig. 11. Speed graph of a constantly accelerated body

EXAMPLE. A northbound train has its speed reduced from 45 to 35 mi·hr^{-1} in 5.5 sec. Show that the direction of the acceleration is south and that its magnitude is 1.8 mi · hr^{-1} · sec^{-1}, or 2.6 ft · sec^{-2}.

20. The Equations of Constantly Accelerated Rectilinear Motion. Eq. [14] provides a means for calculating the speed v_t at any time t of a particle moving in a straight line with a constant linear acceleration. To calculate also the distance s traversed by the particle in the time t, we note that for a constant acceleration (Fig. 11) the average speed is $\frac{1}{2}(v_0 + v_t)$; therefore, by Eq. [5a],

$$s = \frac{v_0 + v_t}{2} t. \tag{15}$$

Eqs. [14] and [15] are independent and together enable us to find any two of the five quantities a, v_0, v_t, s, and t if the other three are known. Although these two equations are all that are needed to describe the motion of a particle in a straight line when the acceleration is constant, other useful equations can be derived by eliminating one of the quantities v_0, v_t, t, which occur in both of them.

Thus if v_t is eliminated, we obtain

$$s = \frac{v_0 + (v_0 + at)}{2} t,$$

or $$s = v_0 t + \tfrac{1}{2} at^2. \tag{16}$$

In this equation $v_0 t$ is the distance that the particle would have traveled with a constant speed v_0, and $\frac{1}{2} at^2$ is the additional distance due to the acceleration. Eq. [16] can also be obtained directly from Fig. 11, if it be recalled that the area under the speed-time curve represents the distance covered.

If, instead of v_t, we eliminate t from Eqs. [14] and [15], there results

$$v_t^2 = v_0^2 + 2\ as. \qquad [17]$$

This equation is useful when the interval of time t is not given or required.

Although the defining equations for velocity and acceleration, namely Eqs. [9] and [12], are equally applicable whether the particle considered moves in a straight or a curved path or with constant or varying acceleration, the three equations derived from them in this and the preceding section, namely Eqs. [14], [16], and [17], apply only to motion for which the acceleration is constant. The student must therefore never attempt to apply these three equations to an actual motion until he has made certain that the acceleration is constant. Motion with varying acceleration will be treated in Chapter 14. When a particle is moving without acceleration $(a = 0)$, Eqs. [14], [16], and [17] obviously reduce to the equations for constant linear velocity.

21. Constant Linear Acceleration With Change in the Direction of Motion. Path of a Projectile. If the direction of the acceleration is not the same as that of the velocity, a change in the direction of motion will result even though the acceleration is constant. The most familiar illustration of this is the case of a projectile shot into space in any direction except vertically. We will treat for the present only the problem of a projectile shot horizontally; this is the case of the airplane-bomb experiment discussed in Sec. 2. In that experiment, it will be recalled, the frame of reference was fixed to the earth and the origin of coordinates was taken to be at the point where the bomb was released, with the positive Y-axis extending vertically downward (Fig. 12). For the *vertical* part of the motion of the bomb, $v_0 = 0$, $s = y$, and, for the first few seconds, a has the constant value g; therefore, Eq. [16] is applicable and takes the form $y = \frac{1}{2} g t^2$. As for the *horizontal* part of the motion, the second column of Table I reveals that the horizontal speed remained constant for about five seconds, after which it decreased slowly. Hence for this direction, $s = x$ and $v_t = v_0$, where v_0 denotes the horizontal speed of the airplane at the moment when the bomb was released. Since the ac-

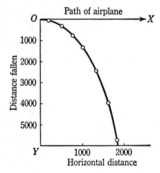

Fig. 12. Trajectory of a bomb in air

celeration had the constant value zero during the first few seconds, Eq. [16] is applicable and takes the form $x = v_0 t$. Between these equations for x and y one may eliminate t and thus obtain the path of the bomb for the first few seconds of its motion; namely,

$$y = \frac{g}{2\,v_0{}^2}\,x^2,$$

an equation which will be recognized as that of a parabola.

○

EXPERIMENT I. MOTION WITH CONSTANT ACCELERATION

The purpose is to study one example of motion with constant acceleration, namely, that of a freely falling body, and to make an approximate measurement of the acceleration due to gravity in a given locality.

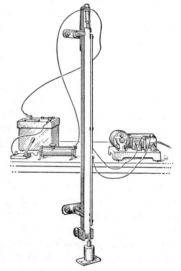

The experimental method consists in obtaining on a long strip of paper a permanent record of the successive positions of a freely falling object at instants separated by equal short intervals of time (Fig. 13). The only resistance which the falling body encounters is that of the air, which is very small in this case. Why? The record is obtained by means of a series of equally timed sparks which jump from a projecting edge of the falling weight to a vertical flat metal plate over which a strip of recording paper has been placed. The sparks are produced by means of a commutator mounted on a constant-speed shaft, so arranged that there is one spark for each

Fig. 13. Free-fall apparatus

rotation of the shaft. The commutator charges a condenser of large capacitance and immediately discharges it through the primary of a small high-potential transformer. The apparatus is so made that no spark can pass until the weight begins to fall, and thus the first mark is made on the paper strip after the weight has attained a small speed.

Adjustment and Operation of Apparatus. Level the apparatus. If the strip of paper is in place, remove it by rewinding on the lower reel.

Hold the weight against the electromagnet and close the switch. With the hand placed under the weight, decrease the current in the electromagnet by means of a rheostat until the weight falls into the hand; then increase the current slightly and again place the weight in contact with the magnet. Now start the timing apparatus and the spark, and then release the weight by opening the magnet switch. If the weight appears to fall properly, you are ready for the final adjustment. Stop the spark. Thread the paper strip through the rolls at the lower end, draw it up over the recording surface, and thread it through the upper rolls. The strip must be stretched smooth.

To obtain a record of the fall of the weight, replace it on the magnet and, as soon as it hangs perfectly still, start the spark and open the magnet switch. After the weight has fallen, open the spark switch, and then make a small circle with a pencil around each spot on the paper while it is still in place. Remove the strip by pulling it back through the upper rolls and cutting it off at the bottom.

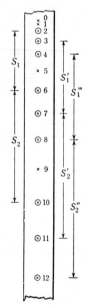

Finally, measure the speed of the timing-apparatus motor by means of a watch and speed-counter.

All of the measurements, derived data, and final results should be recorded systematically in tabular form just as soon as they are obtained. These data and results, together with the answers to the questions which follow, are to be included in the final report.

PART I. **Graphical Study of the Motion.** *a.* Make a distance-time graph (Sec. 13) of the motion of the weight. To insure accuracy in measuring the distances, place the paper strip against a window or other highly illuminated glass, and at each of the very minute perforations produced by the spark make a cross with a sharp metal pointer, the intersection marking exactly the location of the perforated point. Number these points 0, 1, 2, etc., beginning at the top (Fig. 14), and then measure the distances 0–1, 0–2, 0–3, etc., to a tenth of a milli-

Fig. 14. Record strip for falling weight

meter. Eliminate parallax by placing the meter stick on edge. In plotting the distance-time graph, measure time as well as distance from the perforation 0 and use as the unit of time the easily calculated time τ required for the weight to move from one perforation to the next.

1. Why does the distance-time curve have a slope different from zero at the origin?

b. Determine the speeds at various points, including the point 0, by constructing tangents to the distance-time curve and measuring their slopes. Use these data to construct the speed-time curve for the falling weight.

2. What does the y-intercept of your speed-time curve represent? What does the slope represent? Is there any evidence that the motion of the weight during the initial moments was affected by its closeness to the magnet? Write the equation for the straight portion of your speed-time curve.

c. Extend the speed-time curve backward until it meets the time axis.

3. What does the x-intercept represent? In making the foregoing extrapolation, what do you assume? How can you make use of the extrapolated part to obtain the approximate distance through which the weight fell before it reached the perforation 0 (Fig. 14)?

d. Returning to your distance-time curve, construct the portion of the curve lying between the origin and the point of zero slope.

e. Find the acceleration of the falling weight from the speed-time curve and then construct the acceleration-time curve.

4. What is the equation of the acceleration-time curve? Express the magnitude of the acceleration g both in the units used in drawing the graph and in centimeters per second per second. In view of the data used and the accuracy with which you constructed the curves, how many significant figures[1] are you justified in retaining in this value of g?

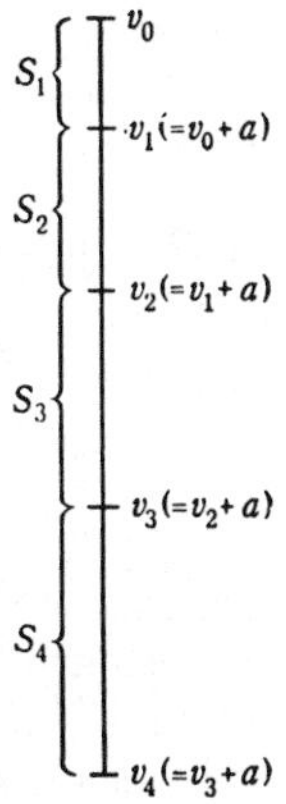

PART II. Calculation of g. This calculation is based on the assertion that if a body moves with constant acceleration, the difference between the distances traveled in any two successive equal intervals of time gives the acceleration. To prove this, let the straight line (Fig. 15) be the path of a body moving with uniform acceleration, let S_1, S_2, S_3, etc. be the distances traversed by the body during the 1st, 2d, 3d, etc. units of time respectively, and let v_1, v_2, v_3, etc. be the speeds at the ends of units of time 1, 2, 3, etc. The space passed over must always be the *mean* speed multiplied by the number of

v_0

$S_1 \Big\{$

$v_1 (= v_0 + a)$

$S_2 \Big\{$

$v_2 (= v_1 + a)$

$S_3 \Big\{$

$v_3 (= v_2 + a)$

$S_4 \Big\{$

$v_4 (= v_3 + a)$

Fig. 15.
$S_2 - S_1 = a$

---

[1] See Appendix 1.

units of time in the interval. In case the speed increases uniformly this mean speed is the half-sum of the speeds at the beginning and at the end of the interval. Hence, for example (Fig. 15),

$$S_4 = \frac{v_3 + v_4}{2} \times 1 = \frac{v_3 + (v_3 + a)}{2} = v_3 + \frac{a}{2}.$$

Similarly, $$S_3 = v_2 + \frac{a}{2}.$$

Therefore, $$S_4 - S_3 = v_3 - v_2.$$

But $v_3 - v_2$ is, by definition, a. Hence,

$$S_4 - S_3 = a.$$

Similarly for $S_3 - S_2$, $S_2 - S_1$, etc. The acceleration can therefore be most directly determined by measuring *distances* traversed in successive units of time. The numerical value of a thus obtained will depend, of course, upon the time interval chosen.

a. To obtain the data for calculating g, proceed as follows: First discard the points on the paper strip marked 0 and 1 (Fig. 14). Why? If the perforation at the bottom of the strip is an odd number, it will be found convenient to discard it also.

For identification, draw a circle around each of the first three and the last three points, as in Fig. 14. Locate still another group of three circles by observing the last number used (in Fig. 14, it happens to be 12), dividing this number by two, and letting the resulting quotient determine the first of the middle group of three circles.

You now have marked with circles three groups of three points each. Carefully measure the distance between the first point of the first group and the first point of the second group; call this S_1. Next measure the distance between the first point of the second group and the first point of the third group, calling this S_2. Then $S_2 - S_1 = a$.

Repeat for the distance between the second point of the first group and that of the second group, and between that of the second group and that of the third group, calling these distances S'_1 and S'_2; then proceed similarly for the third points, calling the distances this time S''_1 and S''_2.

By subtracting S'_1 from S'_2, and S''_1 from S''_2, compute three values of a and average them. Also, from your knowledge of the time τ required for the weight to move from one perforation to the next, calculate the length of the time unit involved in your value of a.

b. In order to see what connection exists between the numerical value of the acceleration and the length of the unit of time used, repeat the measurements and calculation of Part II, *a*, using, however, a time unit just one half as large. Thus, if in Part II, *a*, S_1 was the distance between point 2 and point 6 (Fig. 14), so that the time unit used was $4\,\tau$, find a again for a time unit of $2\,\tau$ by letting S_1 be the distance between points 2 and 4, S_2 the distance between points 4 and 6, S_3 the distance between points 6 and 8, S'_1 the distance between points 3 and 5, S'_2 the distance between points 5 and 7, etc. As before, $a = S_2 - S_1 = S_3 - S_2 = S'_2 - S'_1 = S'_3 - S'_2$, etc. Compute in this way a number of values of a, and average.

5. Compare the mean values of the accelerations for the two different units of time used and state the rule that connects the numerical value of the acceleration with the length of the unit of time.

c. Express in centimeters per second per second the value which you have obtained experimentally for the acceleration due to gravity. Ascertain the accepted value of g for your locality and compute the percentage of error.

6. How many significant figures [1] are you entitled to retain in your calculated value of g? How does this compare with the number retained in the value of g obtained graphically?

7. Show that the definition of acceleration also yields the rule which you have discovered experimentally; namely, that the numerical value of the acceleration varies directly as the square of the interval of time taken as the unit.

8. The justification for the rather elaborate method used to obtain $S_2 - S_1$ etc. is not only that the distances measured are large but that every measurement made on the strip is rendered significant in obtaining the final average value of a. Suppose that you had not used this method but, instead, had denoted the distances 2–3, 3–4, 4–5, etc. (Fig. 14) by S_1, S_2, and S_3, so that $a = S_2 - S_1 = S_3 - S_2$, etc. Show that the average of this series of values of a depends only on the initial and final values of S and hence that the intermediate observations are wasted.

9. State as clearly and concisely as possible the conclusions which you draw from this experiment.

○

QUESTION SUMMARY

The student should formulate definite answers in his own words to the questions that appear at the end of each chapter. By so doing he will make for himself a valuable summary of the important topics covered and at the same time will materially increase his mastery of the subject. The answers thus formulated, obviously, will also be useful for purposes of review.

---

[1] See Appendix 1.

1. What is meant by a *vector quantity*? by a *scalar quantity*? Give illustrations of each.

2. How is the sum of a number of vectors found graphically? the difference of two vectors?

3. How is the sum of two vectors computed? the difference?

4. What is the result of multiplying or dividing a vector by a scalar?

5. What data are needed to specify completely the position of a particle?

6. Define *linear displacement*, and explain how it is measured.

7. Define *average speed, instantaneous speed,* and *linear velocity,* and explain how each is measured.

8. Define, and tell how to measure, *linear acceleration.*

9. State and derive the equations required to find the instantaneous speed v_t acquired in any time t or in any distance s; the average speed v_{av} during any interval of time; the distance s traversed in any time t; and the acceleration (a) for a case in which the velocity is constant, and (b) for a case in which the acceleration is constant.

○

PROBLEMS

If the definitions and principles of physics are to be fully understood, they must be used in the solution of problems. The ability to solve problems is not only a dependable test of one's mastery of the science but is an index of the growth of one's powers to meet and solve original situations and to use the subject as a tool in future thinking.

The student is advised to form the habit in solving problems of working them through first in an *implicit* or indicated form before actually carrying out the numerical work. Common factors should be canceled and the results reduced to the simplest terms before the numerical values are substituted and the actual computations made. In this way much tedious numerical calculation will often be avoided and comparisons of the results of alternative methods of solution can be made accurately.

1. An object undergoes four successive displacements in the XY-plane. The magnitudes of the displacements are 4.0 km, 9.0 km, 3.0 km, and 2.0 km, and their directions relative to the X-axis are 25°, 320°, 180°, and 60° respectively. (*a*) Find the vector sum of these displacements by plotting them on a convenient scale in the order given. (*b*) Plot them in some other order and verify the truth of both the commutative and associative laws for vector addition.   *Ans.* 8.8 km at 345°.

2. Two velocities, each of magnitude 30 cm · sec^{-1}, include an angle of 60° between them. Find the magnitude of their vector sum by two different methods.   *Ans.* 52 cm · sec^{-1}.

3. A steamship has a velocity of 28 km · hr^{-1} northward, and the smoke from its funnels lies 35° south of west. If the wind is blowing from due east, what is its speed?   *Ans.* 40 km · hr^{-1}.

4. A body moving in a straight line with a speed of 50 ft · sec^{-1} begins to lose speed at a constant rate. At the end of 3.0 sec its speed is 20 ft · sec^{-1}. (*a*) How long a time will elapse before the body will come to rest? (*b*) How far will the body move before coming to rest?
          Ans. (*a*) 5.0 sec; (*b*) 1.3 × 10^2 ft.

5. If an engineer has orders not to stop his train with an acceleration of more than 8.0 ft · sec^{-2}, how much time must he allow as a minimum for stopping (*a*) from 30 mi · hr^{-1}? (*b*) from 60 mi · hr^{-1}? (*c*) What distance does the train travel during each of these times?
          Ans. (*a*) 5.5 sec; (*b*) 11 sec; (*c*) 1.2 × 10^2 ft; 4.8 × 10^2 ft.

6. If the length of a railroad rail is 33 ft, show that the speed of a train in miles per hour is equal numerically to three eighths of the number of rails passed over in 1 min.

7. The unit of time usually employed in physics is the *mean solar second*, defined as 1/86,400 of the mean solar day. If the second were redefined as 1/28,800 of the mean solar day, how would this affect the numerical value of g?   *Ans.* 9 times as large.

8. At a certain point the speedometer of an automobile reads 12 mi · hr^{-1}, and 650 ft farther along the road it reads 45 mi · hr^{-1}. If the acceleration is constant, find its value.   *Ans.* 3.1 ft · sec^{-2}.

9. What initial speed in the vertical direction must a stone have in order to fall back to its starting point in 7.0 sec? What is the total vertical distance traversed during the fourth second?   *Ans.* 34 m · sec^{-1}; 2.5 m.

10. Suppose that the event described in Prob. 9 occurred at sea level, in latitude 30°. (*a*) Would you be justified in using the value 980 cm · sec^{-2} for the acceleration due to gravity? (*b*) What if the stone fell back to its starting point in 7.00 sec instead of 7.0 sec?

11. A stone thrown from the top of a cliff with a horizontal speed of 100 m · sec^{-1} strikes the ground at a horizontal distance of 300 m. Compute the height of the cliff.   *Ans.* 44.1 m.

12. An arrow is shot with a speed of 40.0 m · sec^{-1} from the top of the Eiffel Tower, which is 335 m high. What time will elapse before the arrow reaches the ground if it is shot (*a*) vertically upward? (*b*) vertically downward? (*c*) horizontally? (*d*) In each case, what will be the velocity on hitting the ground? Neglect any effects due to the air.
          Ans. (*a*) 13.3 sec; (*b*) 5.15 sec; (*c*) 8.27 sec; (*d*) 90.4 m · sec^{-1}
          vertically; same; 90.4 m · sec^{-1} at 26° 20′ with vertical.

13. A train has a maximum speed of 50 mi · hr⁻¹, which it gains or loses by constant acceleration in 5 min. Compared with a nonstop run of 70 mi, how much traveling time is lost by making a one-minute stop midway in the journey?   *Ans.* 6 min.

14. An automobile passes an intersection with a speed of 30 mi · hr⁻¹ and continues on at this rate. Ten seconds later a traffic policeman starts from rest at the intersection with a constant acceleration of 4.0 ft · sec⁻² in the same direction. When and where will he overtake the car?
Ans. 29 sec after he starts, 1.7 × 10³ ft from the intersection.

15. A ditch 2 ft wide has been cut across a level highway. Will it make any difference to a car traveling at 30 mi · hr⁻¹ whether the ditch has been filled up to within 0.5 in. of the surface of the road or not? How will the flattening of the tires because of the weight upon them affect your answer?

16. Two automobiles have the same initial speed of 8.0 mi · hr⁻¹ and start from the same point. One moves due west with an acceleration of 3.0 ft · sec⁻² and the other moves due south with an acceleration of 5.0 ft · sec⁻². (*a*) Find the velocity of each after 9.0 sec. (*b*) Find the velocity of the first automobile relative to the second at that time. (*c*) What is their relative acceleration?
Ans. (*a*) 39 ft · sec⁻¹ west, 57 ft · sec⁻¹ south; (*b*) 69 ft · sec⁻¹ at 34° 20′ west of north; (*c*) 5.8 ft · sec⁻² at 31° west of north.

17. One ship is sailing south with a speed of 15√2 mi · hr⁻¹ and another southeast at the rate of 15 mi · hr⁻¹. Find the apparent speed and direction of motion of the second vessel to an observer on the first vessel.
Ans. 15 mi · hr⁻¹ northeast.

18. A body stops with an acceleration of 5.0 ft · sec⁻². How long does it take to move a distance of 48 ft if the speed at the mid-point is 17 ft · sec⁻¹?
Ans. 3.2 sec.

19. Two motorboats having the same speeds v relative to the water start simultaneously from an island in a river. One makes the round trip to a buoy located at a distance d directly upstream and the other to a buoy located at the same distance d cross-stream. Which boat wins the race? How do the times compare with the time the boats would make in traveling the same distance on a lake?

20. It can be shown that the distances traversed during successive equal intervals of time by a body falling from rest stand to each other in the same ratio as the odd numbers beginning with unity. Prove this for yourself; then consult *Two New Sciences* to see how GALILEO did it.

CHAPTER TWO

THE FOUNDATIONS OF DYNAMICS

I. Every body continues in its state of rest or of uniform motion in a straight line, except in so far as it may be compelled by force to change that state.

II. Change of motion is proportional to force applied, and takes place in the direction of the straight line in which the force acts.

III. To every action there is always an equal and contrary reaction: or, the mutual actions of any two bodies are always equal and oppositely directed.

> NEWTON'S three laws of motion as translated from the *Principia*
> by WILLIAM THOMSON (LORD KELVIN) and PETER GUTHRIE TAIT
> in their *Treatise on Natural Philosophy* [1]

○

The ideas as to the causes of motion which prevailed in early times were so confused and conflicting that we may disregard them and consider the science of dynamics as beginning with GALILEO in the sixteenth century. Previous to GALILEO's time it was believed that wherever motion existed, force had to exist to explain it. For the origin of the notion of force we must go back to the feeling of exertion that accompanies the use of our muscles in changing the positions of actual bodies. This notion was naturally extended to explain the similar effects arising from the interaction of any two bodies in contact with one another. Thus we say, in keeping with this primitive anthropomorphic conception of force, that a guy wire "exerts" a force on a telephone pole or that a railway train "experiences" a force which is "exerted" by the locomotive.

The motion of a falling body had long been attributed to its *weight,* that is, to the force urging it toward the earth. Just as soon, then, as GALILEO discovered that bodies do not fall with a constant speed but with a constant increase in speed (Sec. 1), the way was opened for the conception that the steady pull of the earth must be producing this steady increase in velocity. This was an entirely new idea, for it pointed to *change of motion* rather than to motion itself as the criterion of the existence of force. For the ancient world the absence of force meant the absence of motion; the new conception was that unless there is force there will be no change of motion, or acceleration.

---

[1] This treatise contains in modern terms a remarkably complete statement of NEWTON's dynamical contributions. Chapter 2 of the same authors' *Elements of Natural Philosophy* (Cambridge University Press, 1912) will repay careful study at this time. See the Bibliography, p. 449.

THE original of this painting has been characterized by Professor Antonio Favaro of Padua University, the devoted and scholarly editor-in-chief of the monumental national edition of GALILEO's works (*Le Opere di Galileo Galilei*, 20 vols., Florence, 1890–1909) and the author of nearly five hundred separate studies on matters relative to the life, times, and activities of GALILEO, as follows:

" Of all the studies of Galileo, the most admired and, I will say, the most deservedly popular is *Galileo in Arcetri*, by Niccolò Barabino, which was first exhibited at the Turin Exposition of 1880, and is now preserved in the Palazzo Orsini in Genoa. Galileo is shown in the last days of his life, ill in bed, and demonstrating some geometrical problem to Torricelli (in front), Viviani (writing), and his [Galileo's] son Vincenzio (listening). The attitudes of the three young men, intent on the words of the Master, contrasted with his own hieratic calmness is very notable, and, however much art critics may find fault with it, the composition is, undoubtedly, the best work on Galileian subjects by which Italian Art has honored itself."

The following description of the scene and characters depicted in the painting is given in J. J. Fahie's *Memorials of Galileo Galilei*, pp. 100–101. It is reproduced by permission of the author's executor.

" Torricelli was with Galileo for the last few months, and, on the latter's death in 1642, succeeded him as First Mathematician to the Grand Duke. . . . Viviani, aged eighteen, joined Galileo in 1639, and remained to the end — his last well-beloved disciple. . . . Vincenzio was a rolling stone, a grief and disappointment to his father. . . .

" Viviani tells us that in those last days, and amid much suffering, the Master's mind was ever occupied with mechanical and mathematical problems. Thus, one day Viviani drew his attention to a flaw in his statement on the falling of a body down an inclined plane. That night he lay sleepless in bed, but found the necessary correction. He had the idea of preparing two other dialogues, to be added to the four already published in his great work ' Intorno A Due Nuove Scienze,' Leyden, 1638. In the first, he would give new reflections and demonstrations on passages in the first four dialogues, and the solution of many problems in Aristotle's ' Physics.' In the second, he proposed to open up an entirely new science — the wonderful force of percussion — which he claimed to have discovered, and which, he said, exceeded by a long way his previous speculations. These were left unfinished, but were put together afterwards by Viviani. . . ."

o

GALILEO and His Scholars

From the painting by Niccolò Barabino in the Palazzo Orsini, Genoa

Mass and Momentum

22. Galileo's Principle of Inertia. But if force is not needed to explain motion but only to explain acceleration, why is it that some kind of engine or other active agent is always needed to keep a body in steady motion? GALILEO'S conjecture was that this is because a body left to itself is slowed down, accelerated, by friction and other retarding forces. Such hindering influences are indeed always present, although they can be diminished, and it is a matter of common experience that with each diminution less effort is needed to keep a body in steady motion. Thus it is harder to drag a stone over a rough floor than over smooth ice, and if the stone is given a push along the floor it is soon stopped by friction, but on the ice it keeps moving for a considerable time. One begins to speculate on what would happen, or rather not happen, if the ice were perfectly smooth and of indefinite extent and there were no resistance due to the air.

By reasoning along similar lines with bodies rolling up and down inclined planes,[1] GALILEO became convinced that a *body will persist in its state of rest or of constant speed in a straight line unless it is compelled by some force to change that state.* The importance of this conclusion in the general theory of motion was recognized by CHRISTIAAN HUYGENS (1629–1695) and later by ISAAC NEWTON (1642–1727). Since actual bodies are never entirely free from external influences, the principle is necessarily an idealization from observed facts. Its proof is that the innumerable deductions which have since been made from it are in harmony with experience.

23. The Concept of Mass. That it requires effort to accelerate a body is described by saying that all matter has *inertia*. The amount of inertia evidently is not the same for all bodies. It is easy to tell, for example, whether a passing motor truck is loaded or empty. The loaded truck does not bounce around so much and it is less easily turned aside from a straight course; it has more inertia than the empty truck. Again, in trying to set in motion the massive though nicely balanced door of a bank vault one is sensible of a powerful resistance, whereas the same effort applied to an ordinary room door will quickly set it in motion. This difference in one's experiences with the two doors must be due to a difference in their inertias rather than in their weights, for neither door, if properly balanced, will assume any particular position by virtue of its weight.

---

[1] See *E. Mach, *The Science of Mechanics* (Open Court, 1893), pp. 140–141, or *J. Cox, *Mechanics* (Cambridge University Press, 1919), p. 78. The latter book is the more elementary; it incorporates many of MACH's ideas.

Both CHRISTOPHER WREN (1632–1723) and NEWTON made experiments with balls of unequal weights swung on cords and correctly
determined their relative changes in velocity when they collided.[1]
It remained for NEWTON, however, to be the first to see clearly that
it was not the difference in the weights of the balls that determined
their relative accelerations upon collision but the differences in their
inertias. In order to compare these inertias quantitatively, NEWTON
introduced the concept of *mass*.[2] A series of experiments with hollow
pendulums filled with various materials showed him that in any
given locality the masses, or amounts of inertia, of bodies are proportional to their weights; moreover, he found the ratio of mass to
weight to be independent of the chemical composition of the bodies.[3]
This furnished him with a method for measuring the mass of one
body in terms of the mass of another body taken as standard: *the
masses of the two bodies are compared by putting them on an ordinary
beam balance and comparing their weights.* Here we have not only
an approved method for comparing masses but in practice also the
best method, for the beam balance surpasses all other ordinary instruments of the laboratory in accuracy, and it is convenient to
operate.

24. Units of Mass. Scientists have adopted as the standard of
mass a piece of platinum-iridium deposited at the Bureau International des Poids et Mesures, Sèvres, France. It is called the *standard
kilogram* and was designed to be, and is very nearly, equal to the mass
of 1000 cm³ of water at a temperature of 4° C. Physicists usually
employ as the unit of mass the *gram*, which is one thousandth part of
the standard kilogram. The *pound* is defined as the mass of a certain
block of platinum which is preserved in London; it is equivalent to
453.5924 g.

25. Linear Momentum. Certain properties of moving bodies depend not alone on the mass but jointly on the mass and velocity.
For example, the effort required to bring a motor truck to rest within
a given time depends both on how much it is loaded and on how fast
it is traveling. Thus GALILEO and NEWTON were led to give a special
name to the product obtained by multiplying the mass of a body by
its velocity. They called this product *momentum* or "quantity of
motion," to distinguish it from rate of motion.

---

[1] See Chapter 5.

[2] It is probable that HUYGENS, as a result of his work on centripetal force and his
pendulum experiments, had by 1673 arrived independently at a clear conception of
mass as distinct from *weight*. See *H. Crew, *The Rise of Modern Physics* (Williams and
Wilkins, 1928), pp. 121–126; ed. 2 (1935), pp. 124–129.

[3] *Principia*, Bk. III, Prop. VI.

The *linear momentum* of a body is defined as the product of the mass m and the linear velocity $\mathbf{v}$; that is,

$$\mathbf{p} = m\mathbf{v}. \tag{18}$$

Since mass is a scalar quantity and velocity is a vector quantity, momentum must be a vector quantity (Sec. **16**); its direction is that of the velocity $\mathbf{v}$ and its magnitude is mv, the product of the mass m and the speed v. From the definition it is obvious that the momentum of a body changes whenever either the velocity or the mass of the body changes. Experience shows, however, that the mass of a given body may be regarded as an invariable quantity except when dealing with extraordinary cases, such as the motions of electrons or of certain astronomical bodies moving with enormous speeds. Hence, in all cases of concern to us for some time to come, any change in the momentum of a body is to be attributed to a change in its velocity alone.

Let us suppose that a particle undergoes a change of momentum of amount $\Delta \mathbf{p}$ in the time Δt. The average time-rate of change of its momentum is then $\Delta \mathbf{p}/\Delta t$. By choosing Δt indefinitely small and passing to the limit, we obtain for the instantaneous time-rate of change of momentum

$$\lim_{\Delta t \to 0} \frac{\Delta \mathbf{p}}{\Delta t} = \frac{d\mathbf{p}}{dt}. \tag{19}$$

Since the mass of a particle ordinarily may be regarded as constant, it is obvious that the foregoing expression may be written in any of the equivalent forms

$$\lim_{\Delta t \to 0} \frac{\Delta \mathbf{p}}{\Delta t} = \lim_{\Delta t \to 0} m \frac{\Delta \mathbf{v}}{\Delta t} = m\mathbf{a}, \tag{20}$$

where $\mathbf{a}$ is the instantaneous acceleration of the particle. Excepting, then, the rare cases where the mass of a body varies appreciably, *the time-rate of change of linear momentum equals the product of the mass and the linear acceleration.*

o

Newton's Laws of Motion

We owe to NEWTON the consolidation of the views on physics which were current in his time into one coherent system. NEWTON is by universal acclaim the greatest figure in all the history of science. Born within a year after GALILEO's death, his creative genius began to appear early in life, and while still an undergraduate at Cambridge he formulated the binomial theorem in algebra, developed the methods of infinite series, and began the invention of his differential calculus. NEWTON published the first full account of his dynamical results in 1687, in his monumental *Philosophiae Naturalis Principia Mathematica* ("Mathematical Principles of Natural Philosophy"), usually called simply the *Principia*, a work which the great LAPLACE (1749–1827) characterized as the supreme exhibition of individual intellectual effort in the history of the human race.[1]

26. Newton's Three Laws of Motion. In the Introduction to the *Principia*, NEWTON formulates the three fundamental laws of motion that are still called by his name. These laws and the accompanying corollaries and scholium are so comprehensive that the whole of the GALILEO-NEWTON dynamics is built upon them. Indeed, for over two hundred years after their publication progress in dynamics consisted largely in deducing from these laws other principles which are more useful for particular classes of motion.

Stated in modern terms and in a form applicable only to motions of translation, NEWTON's three laws are as follows:

1. *If a body in translation is not acted upon by any external force, its linear momentum remains constant.*
2. *The time-rate of change of the linear momentum of a body is proportional to the force acting on the body and is in the direction of the force.*
3. *To every force acting upon a body there always exists a corresponding force exerted by the body, and these two forces are equal in magnitude but opposite in direction.*

As THOMSON and TAIT remark and as NEWTON himself asserts, "these laws must be considered as resting on convictions drawn from observation and experiment, not on intuitive perception."

27. Newton's First Law. The first law is GALILEO's principle of inertia (Sec. **22**). It asserts that force is that which changes the momentum of a body, and implies that this is true regardless of the nature of the body or of the body's motion. It is therefore essentially a *qualitative* definition of force. The law obviously presupposes the existence of a fixed framework, with respect to which momentum is constant or changing. This presents a difficulty, for in the uni-

---

[1] The *Principia* was written in Latin, the international language of learning until a few generations ago, but translations are available, the best being that by Andrew Motte, first published in 1729 and recently revised by *F. Cajori (University of California Press, 1934). The section on "Definitions" and that on "Axioms, or Laws of Motion" (pp. 1–28 of Cajori's revision) should be read by every student in connection with this and subsequent chapters on dynamics.

A N EXCELLENT description of NEWTON's appearance, habits, and character
will be found in L. T. More's *Isaac Newton, a Biography* (Scribner, 1934),
pp. 126–137. The number of original portraits of him made from the life is not large,
and unfortunately most of them were painted when he was an old man and are so
evidently idealized as to give only a vague and unsatisfactory impression of the real
man. Concerning the portrait reproduced here, Professor D. E. Smith of Columbia
University, who has probably the largest collection of Newtoniana in existence,
says : " For real humanity, however, the painting by Gandy, executed in 1706, is
one of the most noteworthy. It represents the savant in the prime of life, without a
wig, and with an expression of determination tempered with kindness that is not
seen in the more conventional portraits of Kneller and Vanderbank." A critical
listing of the other important portraits of NEWTON and the engravings made from
them is given by Professor Smith in *Isaac Newton, 1642–1727*, edited by W. J.
Greenstreet (Bell, 1927), pp. 171–178.

The foundation upon which the whole of the science of mechanics rests was laid
by GALILEO and NEWTON, together with ARCHIMEDES (c. 287–212 B.C.) and CHRIS-
TIAAN HUYGENS (1629–1695). It is instructive to note that each of these pioneers
made his chief contribution in connection with the *quantitative* solution of a *particular*
problem. Thus ARCHIMEDES treated the problem of the lever, GALILEO that of the
freely falling body, HUYGENS the period of the physical pendulum, and NEWTON the
motion of the heavenly bodies. In the course of the solution of these *specific* and *limited* problems the *general* principles required for the treatment of any mechanical
situation became clear, and were definitely formulated by NEWTON in his laws of
motion. All scientific investigations which have led to real progress have begun
in just this way, by the treatment of simple and specific problems with quantitative
exactness, not by making deductions from general philosophic schemes or a priori
principles.

O

When Newton saw an apple fall, he found
  In that slight startle from his contemplation —
'Tis said (for I'll not answer above ground
  For any sage's creed or calculation) —
A mode of proving that the earth turn'd round
  In a most natural whirl, called "gravitation";
And this is the sole mortal who could grapple,
Since Adam, with a fall or with an apple.

Man fell with apples, and with apples rose,
  If this be true; for we must deem the mode
In which Sir Isaac Newton could disclose
  Through the then unpaved stars the turnpike road,
A thing to counterbalance human woes :
  For ever since immortal man hath glow'd
With all kinds of mechanics, and full soon
Steam-engines will conduct him to the moon.

LORD BYRON, *Don Juan*, Canto the Tenth, I and II

ISAAC NEWTON, 1642–1727

From the portrait painted by William Gandy in 1706. Reproduction made from a lithograph copy, executed by G. Black in 1848 and reproduced by Day & Son, in the collection of Professor David Eugene Smith of Columbia University, through the courtesy of Professor Smith

verse there is no fixed framework.[1] In practice, however, when the velocities involved are not too large, it is generally possible to choose a reference frame which, without being fixed in an absolute sense, can be regarded as such for the purposes of any given problem. This reference system consists of a set of axes fixed relative to the average position of the so-called fixed stars, and it is designated as the *primary inertial system*. The errors introduced by applying NEWTON's laws of motion to a set of axes fixed in the earth's surface are, however, either small or very easily corrected, and we shall proceed at present by neglecting them altogether.

28. Newton's Second Law. The second law asserts that force is proportional to, and therefore may be measured by, the time-rate of change of momentum. Its mathematical statement is, accordingly,

$$f \propto \lim_{\Delta t \to 0} \frac{\Delta p}{\Delta t},$$

or $$f = K \lim_{\Delta t \to 0} \frac{\Delta p}{\Delta t} = K \frac{dp}{dt},$$ [21]

where K is a constant of proportionality the value of which depends on the units used in measuring force, momentum, and time. In the ordinary cases where the mass of the body under consideration does not change during the time interval Δt, we have, in view of Eq. [20],

$$f = Kma.$$ [22]

This equation tells us that *force is proportional to the product of mass and linear acceleration*; in other words, the direction of the force is that of the acceleration which it produces, and the magnitude of the force is proportional to ma, where m is the mass of the body and a the magnitude of the acceleration.

The second law thus gives us a *quantitative* definition of force, but we shall find that it is a definition whose usefulness presupposes experimental knowledge.

29. Units of Force. The constant K in the equation $f = Kma$ obviously becomes unity if we arbitrarily choose as the unit of force *that force which is required to impart to a unit mass one unit of linear acceleration*, or, what is the same thing, that force which is required to change the linear momentum of a body at the rate of one unit per second. When this unit of force is employed, the defining equation for force becomes simply $f = ma$. If the unit of acceleration

---

[1] For a critical discussion of the principle of inertia, see E. Mach, *The Science of Mechanics* (Open Court, 1893), pp. 140–143, 232.

be taken as 1 cm · sec^{-2} and the unit of mass 1 g, this unit of force is called the *dyne*; the dyne is the unit of force in the *centimeter-gram-second*, or *cgs*, system of units. If, on the other hand, the unit of acceleration be taken as 1 ft · sec^{-2} and the unit of mass as 1 lb, this unit of force is called the *poundal*; the poundal is the unit of force in the *foot-pound-second*, or *fps*, system.

Another unit of force in common use is the so-called *gravitational* unit, which is defined as the weight of a unit mass at a place where the acceleration due to gravity g has the standard value g_s (Sec. 3); that is, it is the force required to impart to a body of unit mass an acceleration of amount g_s. When this unit is used, the equation f = Kma becomes $1 = K \cdot 1 \cdot g_s$, thus giving $K = 1/g_s$ numerically. If the unit of mass is taken as 1 g, the gravitational unit of force is called a *gram weight*; if the unit of mass is 1 lb, it is called a *pound weight*. To avoid confusion these units should never be referred to as the "gram" and the "pound"; the latter are the names of units of mass, whereas "gram weight" (abbreviation, "gwt") and "pound weight" (abbreviation, "lbwt") designate units of force.

To summarize:

$$f \text{ (dynes)} = m \text{ (gram)} \cdot a \text{ (cm} \cdot \text{sec}^{-2}),$$

$$f \text{ (poundals)} = m \text{ (lb)} \cdot a \text{ (ft} \cdot \text{sec}^{-2}),$$

$$f \text{ (gwt)} = \frac{1}{g_s} m \text{ (gram)} \cdot a \text{ (cm} \cdot \text{sec}^{-2}), \quad g_s = 980.665 \quad [23]$$

$$f \text{ (lbwt)} = \frac{1}{g_s} m \text{ (lb)} \cdot a \text{ (ft} \cdot \text{sec}^{-2}), \qquad g_s = 32.174$$

It is to be noted that g_s is used in the last two equations to denote pure numbers; ordinarily it denotes standard acceleration due to gravity, but here it represents only the numerical values of this quantity.

Units of force that are defined by making K a pure number in the equation f = Kma are called *absolute* units. Thus the dyne and the poundal are the *absolute cgs* and *fps* units of force, and the gram weight and pound weight are *absolute gravitational* units of force.

30. What the Second Law Implies. The second law contains certain implications that must not pass unnoticed.

a. It implies that the effect of a force on a body is independent of the state of motion of the body at the time when the force begins to act. For example, gravitational attraction at any given place always imparts to a projectile the same downward acceleration, regardless of whether the projectile is shot from a gun or simply dropped from a balloon.

b. When several forces act on a body at the same time, the law implies that each force produces its own effect independently of the actions of the other forces. It follows from this that the vector sum of the accelerations due to the several forces gives the acceleration of the body. A special case of this arises when, as a result of the action of two forces, a body remains at rest or else continues to move with constant velocity. In this case the accelerations due to the two forces must be equal in magnitude but opposite in direction; hence the forces producing them are equal in magnitude but opposite in direction. For example, when an object is suspended by a cord it is acted upon by two forces: one is its weight and the other is the force exerted by the cord. If the cord is cut, the second force disappears and the object falls with an acceleration.

31. How the Second Law Is Applied. The application of the second law of motion to the solution of particular problems can best be shown by means of several examples. No difficulties will be encountered by the student in making this application if he recognizes clearly what body it is whose motion he wishes to determine, then lists completely *all* the forces which act *on* that body, and finally applies the laws of motion to that body alone.

> EXAMPLE. A sled of mass 400 lb is pulled along a level road by a team of horses. If the horizontal force which the tugs exert on the sled is 25.0 lbwt and the force of friction is 20.0 lbwt, will the velocity be constant? If not, what is the acceleration? How long will it take the sled starting from rest to acquire a speed of 4.0 mi · hr^{-1}?

Solution. The body whose motion is to be determined is the sled. Therefore we are not concerned with the weight of the horses or with the forces which they exert on the ground. The forces acting *on* the sled are the forward pull of the tugs, amounting to 25.0 lbwt; the backward pull of friction, amounting to 20.0 lbwt; and the downward pull of gravity and the upward push of the ground, each amounting to 400 lbwt. Since the pull of gravity and the upward push of the ground have no effect in producing motion (Why?), the total unbalanced force acting on the sled is 5.0 lbwt or (5.0 × 32) poundals, and this will cause an acceleration. Since the mass of the sled is 400 lb, the second law gives 5.0 × 32 = 400 a, or $a = 0.40$ ft · sec^{-2}, in a horizontal direction. Also, since this acceleration will be constant, we have, by Eq. [14], Chap. 1,

$$t = \frac{v}{a} = \frac{4.0 \times 5280}{3600 \times 0.40} = 15 \text{ sec.}$$

> EXAMPLE. To the ends of a cord passing over a light, delicately balanced pulley are attached two bodies of unequal masses, one of 203 g, the other something more than 203 g (Fig. 16). The acceleration imparted to the masses is observed to be 245 cm · sec^{-2}. Find the mass of the second body.

Solution. We will assume that the mass of the pulley is negligibly small. We will also assume that the stretching force in the cord is the same throughout; this amounts to saying that there is no friction between the pulley and the cord, and that the mass of the cord is negligible in comparison with the masses of the two bodies. Let m denote the unknown mass and let F be the force in the cord. The forces acting on m are mg vertically downwards and F vertically upwards; hence the total force acting on m is $(mg - F)$ dynes, and the equation of motion of m is by Eq. [22], $mg - F = m \cdot 245$. The forces acting upon the 203-gram mass are $203 \cdot g$ vertically downwards and F vertically upwards; its equation of motion is therefore $203 g - F = -203 \cdot 245$. By eliminating F between these two equations, we find $m(g - 245) = 203(g + 245)$. If, finally, g at the place under consideration is 980 cm $\cdot$ sec^{-2}, then $m = 338$ grams.

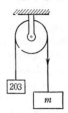

Fig. 16. A problem involving the second law of motion

In the foregoing solution the equation of motion of each body was obtained separately and then the unknown quantity F was eliminated by solving these two equations simultaneously. A second method of solution consists in finding a single equation of motion for the system as a whole. Thus the total force acting on the system is $(m - 203)g$ dynes and the total mass accelerated is $(203 + m)$ grams; hence, from Eq. [22], $(m - 203)g = (m + 203)245$, or $m = 338$ grams. Generally speaking, the first method of solution will be found to be more useful.

EXAMPLE. In the preceding example show that the force F exerted by the cord is 2.49×10^5 dynes, or 254 gwt.

Solution. Either one of the equations of motion given in the first solution of the preceding example may be used to obtain F, or one can deduce its value from first principles, as the following considerations will show. Since all *freely falling* bodies have the same acceleration g, the *weight* of any mass of m grams is mg dynes. If, because of some retarding force, such as a force exerted by a cord, the *downward* acceleration is not g but some smaller quantity a, it is at once clear from the statement of the second law that the value of the upward, or retarding, force is $m(g - a)$ dynes. If, on the other hand, the body has an *upward* acceleration of amount a, the upward force exerted by the cord is $m(g + a)$ dynes. If the body is at rest or is moving upward or downward with *constant speed*, the force exerted by the cord is mg dynes. In the present example, the force F exerted by the cord is evidently $338(g - a)$ dynes, or $\frac{1}{g_s} 338(g - a)$ grams weight; it is also $203(g + a)$ dynes, or $\frac{1}{g_s} 203(g + a)$ grams weight.

32. Newton's Third Law. A force acting upon a particle of matter is always due to the action of another particle upon the one in question. The third law asserts that in such an interaction of two particles the first particle exerts a force on the second and the second exerts a force on the first; these two forces have the same magnitude but opposite directions, and their line of action is the line joining the two particles. It is to be noted that the two forces concerned in this action and reaction do not act on the same body; either body *experiences* one of the two forces and *exerts* the other. There are many familiar illustrations of the law: a man sitting in a chair cannot lift himself by pulling on the chair; an athlete throwing a hammer must steady himself or the hammer will throw him; when a gun is fired, it recoils; it is more tiring to walk in loose sand or melting snow than on a brick pavement. The third law also holds when the interacting bodies are not in contact. When one body attracts another from a distance, this other attracts it with a force that is equal in magnitude but opposite in direction. NEWTON showed this to be true for gravitational attractions, and he also verified it experimentally for the case of magnetic attractions.

Indeed it is only by means of experiments that one can justify either the third law or the implications contained in the second law. These experiments can be made in any laboratory. But the most convincing evidence of their correctness is furnished by astronomical observations; predictions of eclipses, made years in advance, are wholly based upon these laws of motion taken in connection with the law of gravitation.

o

The Planetary Motions and Gravitation [1]

NEWTON'S formulation of the laws of motion was an incident in the course of his investigation of the motions of the heavenly bodies. A hundred years before the birth of NEWTON, NICOLAUS COPERNICUS had published the immortal *De Revolutionibus Orbium Coelestium*,[2] in

---

[1] * Annotated extracts from the original papers of NEWTON, CAVENDISH, and others will be found in *The Laws of Gravitation*, ed. by A. S. Mackenzie (American Book Co., 1900). Pages 88–98 of J. Cox, *Mechanics* (Cambridge University Press, 1919) also will be found interesting.

[2] Published at Nuremberg in 1543. This is the first of the three greatest masterpieces of modern astronomical literature, the other two being Newton's *Principia* (1687) and Galileo's *Dialogo dei due massimi sistemi del mondo, Tolemaico e Copernicano* (Florence, 1632), of which there is a translation under the title *Dialogues concerning the two Chief Systems of the World, the Ptolemaic and the Copernican,* by T. Salusbury in his *Mathematical Collections and Translations,* 2 vols. (London, 1661 and 1665). Translations of portions of all three of these works will be found in *H. Shapley and H. E. Howarth, *A Source Book in Astronomy* (McGraw-Hill, 1929).

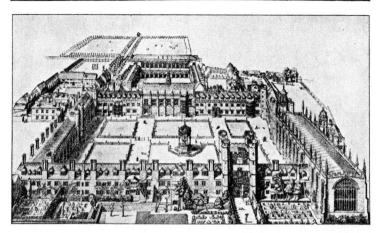

Trinity College, Cambridge,
at the Time of NEWTON

From David Loggan's print of 1690, one of the beautiful and accurate prints of the colleges and halls of Cambridge University which appeared in his *Cantabrigia Illustrata* (Cambridge. 1690)

THE ROOMS in which NEWTON wrote the *Principia* were probably on the second floor of the entry to the right of the Great Gate as one enters the Great Court. The small garden laid out between these rooms and the street seems to have been reserved for his private use, and here in all probability he paced up and down for the only exercise he allowed himself.

After an Illustration from HENRY PEMBERTON'S
A View of Sir Isaac Newton's Philosophy
(London, 1728)

HENRY PEMBERTON was a friend of NEWTON and the editor of the third edition of the *Principia*. The illustration reproduced here is based on one executed by J. Pine after J. Grison. It is an artist's attempt to describe a part of NEWTON's experimental work in mechanics, particularly his very important experiments on impact (*Principia*, scholium to "Axioms, or Laws of Motion"), from which he undoubtedly obtained his clearest and most definite conceptions of mass and momentum; his experiments on fluid resistance (*Principia*, Book II, scholiums to sections VI and VII); and his application of the laws of motion to the pulley, wheel and axle, and other "mechanical powers."

The setting for these experiments shown in the illustration is hardly one that would be selected as typical today, but it must be remembered that physical laboratories are a modern development and that GALILEO, NEWTON, BOYLE, YOUNG, HOOKE, and all the great experimenters of the past carried on their work in their own private quarters and wherever else they could. Thus some of NEWTON's experiments on air resistance were carried out in St. Paul's Cathedral in London, as were important experiments of WREN and HOOKE. It is improbable that the setting shown above was intended to represent St. Paul's, but it might well have been.

which he proclaimed the heliocentric theory of the solar system as against the then accepted geocentric, or Ptolemaic, view. The essential idea of the heliocentric theory, from the modern point of view, is that the complicated motions of the planets reduce to comparative order and simplicity when the sun, rather than the earth, is taken as the reference body. The Copernican doctrine had a profound influence on man's views of the universe and his place in it, but it was by no means a perfect system, as COPERNICUS himself realized. Among its hostile critics was TYCHO BRAHE (1546–1601), who proposed a theory of his own,[1] and in order to test it spent twenty-five years making astronomical observations of extraordinary accuracy. The data obtained by this remarkable experimentalist evidently were not vitiated by prejudice, for in the hands of the mathematician JOHANN KEPLER (1571–1630) they became the means of placing the Copernican doctrine on a firm foundation.

33. Kepler's Laws. COPERNICUS, in the absence of data to the contrary, had retained the ancient notion that the planets moved in circles with constant speeds, but KEPLER found that such an assumption led to a difference of as much as eight minutes of arc in the computed and observed positions of Mars. "Out of these eight minutes," said KEPLER, "we will construct a new theory that will explain the motions of all the planets." It took a lifetime of the most patient study of TYCHO's data, but he finally arrived at the three empirical laws which bear his name and which paved the way for modern astronomy. KEPLER found that the motions of the planets could be predicted by making the following assumptions[2]:

1. *The orbit of each planet is an ellipse with the sun at one focus.*
2. *The speed in the orbit varies in such a manner that the radius vector joining the planet with the sun sweeps over equal areas in equal times.*
3. *The cubes of the semi-major axes of the elliptical orbits are proportional to the squares of the periods of revolution or lengths of planetary years.*

34. Newton's Law of Gravitation. When KEPLER'S laws were published, in the early part of the seventeenth century, attempts were

---

[1] *De Mundi Aetherei recentioribus Phaenomenis* (1588).

[2] The first two laws, together with an account of the laborious process by which they had been found, were published by KEPLER in Chapter 59 of his great work, *Astronomia Nova* ΑΙΤΙΟΛΟΓΗΤΟΣ, *seu Physica Coelestis* (1609). The third law was published in Kepler's *Harmonices Mundi Libri V* (1619), Bk. V, Part 3.

already being made to relate the planetary motions to the more familiar motions of objects on the earth. KEPLER himself, and also DESCARTES and GALILEO, speculated on the notion of gravitation, but it remained for NEWTON to give the first clear quantitative explanation of the relations contained in KEPLER'S laws. By considering these laws in the light of the mechanical principles which he had developed, NEWTON was able to reduce them to the single law that the acceleration, a, of every planet is always directed toward the sun, depends only on the distance of the planet from the sun, and is of magnitude

$$a = \frac{C_1}{r^2},$$ [24]

where r is the distance of the planet from the sun and C_1 is a constant which is the same for all the planets.

NEWTON found, moreover, that this same law described the motions of the satellites around their respective planets and that the acceleration of the earth's moon could be explained in terms of the gravitational pull of the earth on the moon. He thus arrived at his great generalization that the forces exerted by the heavenly bodies on one another and by the earth on the bodies at its surface are merely special cases of a gravitational attraction existing between all particles of matter. For any two particles of masses m_1 and m_2, at a distance r apart, this gravitational force of attraction was found to be

$$f = G \frac{m_1 m_2}{r^2}.$$ [25]

NEWTON assumed that G is a constant for all bodies, regardless of their nature or of their location in space.

Eq. [25] is the mathematical statement of *Newton's Law of Gravitation*, which states that *any* two *particles* attract each other with a force that is directly proportional to the product of their masses and inversely proportional to the square of the distance between them. The word *particle*, rather than body, must be used in the statement of the law because of the uncertainty as to what is meant by the distance r between two extended bodies. NEWTON was able to show that a spherical body which may be regarded as homogeneous, or as made up of shells each of which is homogeneous, will attract an outside body as if all the particles of the sphere were concentrated at its center.[1]

---

[1] *Principia*, Bk. I, Prop. LXXI. See also *J. Cox, *Mechanics* (Cambridge University Press, 1919), p. 253.

35. The Constant of Gravitation. The constant G in Eq. [25] is called the *Newtonian constant of gravitation*.[1] It is one of the so-called *universal* constants of physics, for its value depends only on the units used and not on time, locality, or any particular properties of matter. To find the magnitude of G it is necessary to measure the gravitational attraction in some case where m_1, m_2, and r are known. This difficult measurement was first accurately made by HENRY CAVENDISH[2] in 1798 by means of a type of apparatus known as a torsion balance. A very reliable determination has been made recently by HEYL[3] at the United States Bureau of Standards. His result is $G = 6.670 \times 10^{-8}$ cm³ · g⁻ⁱ · sec⁻². This is therefore the force in dynes with which a particle of mass 1 g attracts a similar particle at a distance of 1 cm.

○

The Proportionality of Gravitational Force and Mass

According to the law of gravitation, the gravitational attraction between two bodies is proportional to the masses of the bodies and is entirely independent of their physical and chemical nature. This means, for example, that the earth exerts the same pull on all bodies of 1-kg mass placed at a given distance from its center, regardless of whether the bodies consist of lead, water, ice, steam, wood, or what not; and for 2 kg of any of these materials the force is just twice as much. We have already learned how NEWTON and others demonstrated the correctness of this remarkable conclusion and that it is the proportionality of mass to weight in a given locality which makes possible the comparison of masses by comparing their weights on a beam balance (Sec. 23).

The mass of a body as determined by weighing on a beam balance is sometimes referred to as the *gravitational mass* in order to distinguish it from the *inertial mass* of the body. The former is a measure of the ability of the body to attract other bodies by gravitation, whereas the latter is a measure of the inertia of the body. The assertion that the gravitational mass and the inertial mass are equal is a consequence of the *principle of equivalence*. This important principle, which is supported by many experiments, is one of the three fundamental postulates upon which EINSTEIN bases his general theory of relativity.

---

[1] For an excellent account of the methods which have been used to determine this important constant see *J. H. Poynting and J. J. Thomson, *Properties of Matter* (Griffin, 1913), Chap. 3.

[2] See Plate 7.

[3] P. R. Heyl, *Journal of Research of the National Bureau of Standards* 5, 1243 (1930).

TWO lead balls x, x, each 2 in. in diameter, were hung at the ends of a light wooden rod hh, 6 ft long, which was suspended horizontally from a long thin wire lg attached to its center. The whole was enclosed in a case $ABCDDCBAEFFE$ resting on posts fixed firmly into the ground. Two large attracting spheres of lead W, W, each 12 in. in diameter, hung outside the case from an arm which could turn around an axis Pp in the line of gl. To insure constancy of temperature the whole apparatus was placed in a closed room $GGGG$, and all adjustments and readings were made from outside.

The method of taking observations was to turn the two large spheres W, W by means of the pulley MM into positions as close as possible to the small spheres x, x, one on each side, so that their attractions both tended to twist the rod the same way. They were then moved round so as to produce an opposite twist, and the total twist was measured by telescopes T, T, which were sighted at a scale on the torsion rod hh.

CAVENDISH's own account of this justly famous experiment is worth studying. It " is frequently given for detailed study to young physicists in order to train them in the art of reading for themselves periodical scientific literature. Certainly no better piece of work could be used for the purpose, whether one considers the intrinsic importance of the subject-matter, the keenness of argument and the logical presentation in detail, or the use and design of apparatus and the treatment of sources of error " (from the preface of A. S. Mackenzie's *The Laws of Gravitation*). The account was published in the *Philosophical Transactions of the Royal Society* **88**, 469 (1798), and has been reproduced in *The Scientific Papers of the Honorable Henry Cavendish*, F.R.S. (Cambridge University Press, 1921), Vol. II, pp. 249–286, and in *The Laws of Gravitation*, ed. by A. S. Mackenzie (American Book Co., 1900), pp.59–107. Excerpts from it are also given in W. F. Magie's *A Source Book in Physics* (1935), pp. 106–111.

◎

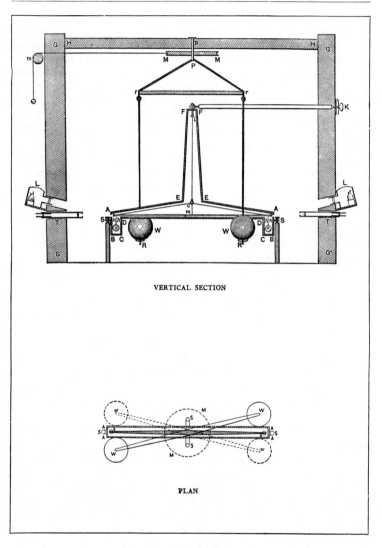

VERTICAL SECTION

PLAN

Diagram of the Apparatus
Used by CAVENDISH in his Determination
of the Constant of Gravitation

Mass and Force as Dynamical Concepts

In the preceding pages the essential principles of the dynamics of translatory motion have been presented with rather close regard for the order and form in which they were developed historically. It rarely happens, however, that a really new theory achieves its simplest and most logical form in the first development. This usually comes only after a philosophical study of the nature of the postulates and definitions involved. The essential principles of classical dynamics, as they were laid down by NEWTON, have been but little affected by such studies, but several important modifications have been introduced to make the theory a logical conceptual scheme.[1] The remainder of this chapter is devoted to such a systematic presentation, in a form that is suitable for elementary instruction.

36. A Dynamical Definition of Mass. The beam balance affords the best practical method for comparing masses. Yet it is evident that if the mass of a body is to be the measure of its inertia (Sec. **23**), then a more direct and fundamental method of comparing two masses is to compare their accelerations when both are acted upon by the same external influence.

Suppose, for example, that the two cars in Fig. 17 run with little friction. The spring between the cars is compressed, and the two cars are tied together by means of a piece of cord. One car contains a lead block and the other a wooden block of the same size and color. Which is which becomes apparent just as soon as the cord is cut, for the two cars acquire velocities v_1 and v_2 that differ in magnitude to an

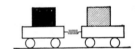

FIG. 17. A dynamical method for measuring mass

extent far exceeding any difference in frictional effects. The question now arises as to what kind of relation exists between these velocities acquired by the cars due to the action of the spring or, for that matter, between the changes in velocities of any two bodies when they interact. NEWTON concluded, and this is borne out by experiments, that the ratio of their changes in velocities, $\Delta v_1/\Delta v_2$, is a constant, regardless of whether the interaction is due to a spring, gravitational attraction, impact, or other cause. We may therefore assign a number m_1 to one body, taken as the standard body, and

---

[1] The following references are important in this connection: E. Mach, *The Science of Mechanics* (Open Court, 1893), pp. 187–255; J. H. Poincaré, *Science and Hypothesis* (Scott, 1905), Chap. 6; K. Pearson, *Grammar of Science* (Black, 1900), Part I.

define a corresponding number m_2 for the other by means of the equation

$$\frac{\Delta v_1}{\Delta v_2} = -\frac{m_2}{m_1}. \tag{26}$$

It is this number m which NEWTON called the measure of the *mass* of the body. To compare the masses of two bodies we have, therefore, only to take the ratio of the changes of velocity Δv when these changes are due to the same external influence. Since we have agreed to call the product mv the linear momentum (Sec. 25), Eq. [26] is equivalent to the statement that two given bodies interacting in any way undergo changes of linear momentum that are equal in magnitude but opposite in sign.

37. The Definition of Force. GALILEO avoided the great speeds with which bodies fall by using an inclined plane. GEORGE ATWOOD [1] in 1784 accomplished the same end by suspending two weights by a thread passing over a light, delicately balanced pulley. We can make use of this simple device (Fig. 18) in our further investigation of the relation between mass and change of velocity.

FIG. 18. Atwood's machine

Two bodies of the same weight and of masses which we will designate by m_1 and m_2 are hung from the ends of the cord. Since they have the same weights, they balance each other. If a small body of mass m_3 is now added to one of the two balanced bodies, say to m_2, the whole system of three bodies will be given an acceleration. The cause of this acceleration is the weight of the rider. The total mass m experiencing the acceleration is $m = m_1 + m_2 + m_3$, if the effect of the rotating pulley is neglected.

At R is a horizontal ring through which m_2 can pass but which is too small for the rider m_3, and so the latter is caught by the ring while m_2 passes on alone. With the removal of the rider, it is found that the system now moves with a constant speed v which can be measured by timing one of the bodies over known distances. The time of action of the accelerating force can be changed by changing the position of the ring R, and if this is done it will be found that the final constant speed v is proportional to the time t during which the force acts, or

$$v \propto t, \tag{27}$$

as long as the force and mass are the same.

---

[1] G. Atwood, *A Treatise on the Rectilinear Motion and Rotation of Bodies; with a Description of Original Experiments Relative to the Subject* (Cambridge, 1784).

If the total mass of the system is now increased by adding two more bodies of equal weight to m_1 and m_2, but not changing m_3, it will be found that when this same force acts for a given time, the speed v produced is inversely proportional to the mass m of matter set in motion, or

$$v \propto \frac{1}{m}. \tag{28}$$

By combining the two experimental relations (Eqs. [27] and [28]) into a single expression, we obtain

$$v \propto \frac{t}{m}, \quad \text{or} \quad \frac{mv}{t} = \text{constant}, \tag{29}$$

for a given accelerating force. Thus we arrive experimentally at the result that the *time-rate of change of momentum* is the quantity which remains constant when the *same force* acts on various bodies in succession. It is this quantity, then, that is to be regarded as the proper description of force. We therefore define the force acting on a body as a vector quantity that is proportional to the time-rate at which the linear momentum of the body is changing (Eq. [21]).

○

EXPERIMENT II. ACCELERATION, MASS, AND FORCE

The purpose is to become familiar with the relations between mass, acceleration, and force, and to measure the masses of several bodies by the dynamical method.

PART I. Determination of the Acceleration Imparted to a Given Mass by a Given Force. The apparatus consists essentially of a car and a horizontal track. The track consists of a pair of parallel rails which are mounted on bases provided with leveling screws. A bracket at one end of the track supports an insulated spark point and a light transverse roller (Fig. 19). The car is provided with a clip for attaching the end of a strip of sensitized paper which passes over the roller. To the other end of the strip is attached a similar clip with a weight-holder.

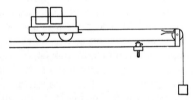

Fig. 19. An apparatus for studying the relations between mass, acceleration, and force

The force is transmitted to the car through the paper strip.

The record of the motion is made by the passage of electric sparks from the spark point to the roller, a dot being made on the surface of the paper by each spark passage. The source of the spark is an induction coil supplied with current from a 6-volt storage battery through a vibrating spark-timer. The time intervals between sparks are measured with a watch and impulse-counter. The sweep hand of the timer makes one circuit of the dial in 60 impulses, and the small hand indicates the total number of rotations of the sweep hand. The counter is started and stopped by means of a push button. A difference of potential of more than 6 volts should not be applied to the terminals of the timer.

a. Place the car [1] on the track and attach to it a strip of the paper. Pass the strip over the roller and attach the clip for holding the weights to the free end of the paper at such a point that an attached weight hangs clear of the floor when the car is at the roller end of the track.

Adjust the level of the track sidewise so that the paper runs smoothly over the roller for its whole length. Move the spark point to a position near one edge of the paper; connect this point to one secondary terminal of the induction coil and ground the other terminal to the track. Tilt one end of the track slightly until the car, when given a push, runs down the track with constant speed. To test the constancy of speed, start the spark-timer bar vibrating and let the car run the length of the track. If the spots on the paper are not uniformly spaced, adjust and test again, the spark point being moved over about 5 mm for this next record.

b. Secure four time-records on the paper strip as follows: (1) place a mass m' on the weight-hanger, start the spark-timer bar vibrating, and let the car run down the incline; (2) move the spark over about 5 mm and repeat; (3) place a different mass m' on the weight-hanger, move the spark over again and repeat; and (4) move the spark over once more and repeat. Beginning with the first distinct puncture on the paper, number the successive points and then proceed to find for each spark record the observed acceleration in centimeters per second per second as in Exp. I.

c. Compare the observed value of the acceleration with the calculated value. To obtain the latter, measure with a beam balance the total mass M of the car and the mass m of the clip to which m' is attached. Then, from Eq. [22],

$$m'g = (M + m + m')a, \qquad [30]$$

where g is known from Exp. I. Compute the percentage of difference between the observed and the calculated values of the acceleration a.

---

[1] In order to minimize the effect of the roller it may be advisable to load the car to its maximum capacity.

Part II. Comparison of Masses. In this part of the experiment two cars of smaller size are employed (Fig. 20). The spark recording mechanism will not be needed. The two cars are set in motion by means of a spring mechanism which is placed between them and compressed, the cars initially being held together by a cord. The cars are of approximately the same mass, but one of them is provided with a rack upon which slotted weights may be placed. Each end of the track is equipped with friction brakes for stopping and holding the cars.

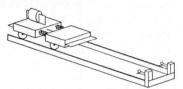

Fig. 20. An apparatus for measuring mass by the dynamical method

a. Carefully level the track both longitudinally and transversely with the aid of a spirit level.

Select a 500-g slotted weight as your *standard mass* and place it on the car provided with the weight rack. Call this car No. 1 and the other car No. 2. Place the two cars near the middle of the track, compress the springs between them, and fasten the cars together.

b. When the cars are released by burning or cutting the string, they are acted upon by the same external influence, the spring, for the same short time, and in this time the cars acquire certain speeds v_1 and v_2. If the friction is negligible, these speeds will remain constant until the cars are stopped. Release the cars and note the comparative distances traveled before the less massive car reaches the friction brake. Repeat this procedure, the location of the cars on the track being changed until they strike the friction brakes simultaneously. Then measure the distances d_1 and d_2 through which the cars have traveled with constant speeds and compute their ratio d_1/d_2.

> **1.** What are the important sources of error in determining d_1/d_2? Can you find methods for eliminating any of them?

Repeat the experiment and again determine d_1/d_2. If this value is not in good agreement with the previous one, make several trials and take the average.

> **2.** Does the amount which the spring is compressed affect the values of d_1 and d_2? of d_1/d_2? If in doubt, resort to experiment to obtain an answer.

Since the distances d_1 and d_2 are traversed in the same time with the constant speeds v_1 and v_2, it follows that

$$\frac{d_1}{d_2} = \frac{v_1}{v_2}. \qquad [31]$$

But if m_1 and m_2 denote the respective masses of the two empty cars, the definition of mass, Eq. [26], gives

$$\frac{v_1}{v_2} = \frac{m_2}{m_1 + 500};$$

hence Eq. [31] may be written

$$\frac{d_1}{d_2} = \frac{m_2}{m_1 + 500}. \qquad [32]$$

This equation contains two unknown quantities, namely m_1 and m_2.

c. Remove the standard mass from car No. 1 and find the ratio d'_1/d'_2 for the empty cars. Write the equation for this case that is analogous to Eq. [32].

d. By means of the two equations which you have obtained, calculate the unknown masses m_1 and m_2.

e. If time permits, measure the masses of various other objects by this dynamical method.

3. If the two cars could be set in motion by some other kind of interaction, say by means of an electrical repulsion, how would the resulting values of v_1/v_2 compare with those that you obtained with the spring? What if a different spring were used?

4. Why can masses also be compared by comparing their weights on a beam balance? How do the measurements of mass that you have made by the dynamical method compare in accuracy with those that you could make on the ordinary balance in your laboratory?

○

OPTIONAL LABORATORY PROBLEM

Density of a Solid by Weighing and Measuring, and a Study of Errors. Determine the average density of steel, glass, or some other substance of which you have a sample in the form of a long, thin cylinder. The average density is found by dividing the mass of the cylinder by its volume. A beam balance, a meter stick, a micrometer caliper, and a vernier caliper [1] will be available. You are to make the most accurate determination that you can with such of these instruments as are needed and are to record the total time required for the preliminary estimations, measurements, and calculations. Remember that some preliminary estimations, in which the number of significant figures [2] involved in the various measurements is taken into account, may save you considerable time in the end.

In the event that your determination of the density involved making a series of measurements of the diameter of the cylinder, calculate the probable error of a single measurement and the probable error of the mean.[3] State in words what each implies.

---

[1] See Appendixes 4 and 5.   [2] Appendix 1.   [3] Appendix 2.

QUESTION SUMMARY

1. What is meant by the *inertia* of a body?

2. What is meant by the *mass* of a body? How can masses be measured?

3. Define *gram of mass*; *pound of mass*.

4. Define *momentum*.

5. Define *force*. How can force be measured?

6. Define *dyne*; *poundal*; *gram weight*; *pound weight*. Distinguish between a *gram of mass* and a *gram weight*.

7. State KEPLER's three laws of planetary motion.

8. State NEWTON's law of gravitation. What is meant by the statement that the Newtonian constant of gravitation is a *universal constant*?

9. What is the principle of equivalence?

○

PROBLEMS

1. A man pushes steadily upon a car of mass 1100 kg which is initialiy at rest. After 5.0 sec the car is moving with a speed of 45 cm · sec^{-1}. Find the force exerted by the man. *Ans.* 10 kgwt.

2. A particle of mass 12 g is moving in a straight line with a momentum of 480 g · m · sec^{-1}. A force of 2000 dynes is applied to the particle so as to oppose its motion. (*a*) How soon will it be brought to rest? (*b*) How soon would an 18-g particle moving with the same speed be brought to rest?
Ans. (*a*) 24 sec; (*b*) 36 sec.

3. A force of 1 megadyne (= 10^6 dynes) acts upon a mass of 1 metric ton for 1 min. Find (*a*) the speed acquired; (*b*) the distance passed over in this time; (*c*) the momentum acquired; (*d*) the rate of change of momentum.
Ans. (*a*) 6 × 10 cm · sec^{-1}; (*b*) 2 × 10^3 cm;
(*c*) 6 × 10^7 g· cm · sec^{-1}; (*d*) 10^6 g · cm · sec^{-2}, or dynes.

4. A man of mass 75 kg jumps out of a balloon with a parachute. After the man has fallen freely for 100 m the parachute opens, and in the next 3.0 sec he is slowed down to a velocity of 5.0 m · sec^{-1} downward. What average force, in addition to the weight of the man, does the parachute exert during the 3.0 sec? *Ans.* 10^2 kgwt.

5. Careful measurements show that the mass of a certain cube of metal is 96.242 g. (*a*) Find its weight in grams weight at a place where $g = 980.572$ cm · sec^{-2}; (*b*) at a place where $g = g_s$. (*c*) If the cube is 2.200 cm on a side, what is its average density? (*d*) The ratio g/g_s usually differs from unity by less than 0.25 percent. When may the mass of a body and its weight in the corresponding gravitational units be considered as numerically equal?
Ans. (*a*) 96.233 gwt; (*b*) 96.242 gwt; (*c*) 9.039 g · cm^{-3}.

6. Over a light, delicately balanced pulley (Fig. 16) are suspended masses of 200 g each. A mass of 100 g is added to one side. (*a*) What is the resulting acceleration? (*b*) How far does each mass move from rest in 1.50 sec? (*c*) What is the force in the cord while the masses are moving freely?   *Ans.* (*a*) $\frac{1}{5} g$; (*b*) 220 cm; (*c*) 240 gwt.

7. Find the force exerted upon the floor of an elevator by a man weighing 150 lbwt, if the elevator is (*a*) ascending with an acceleration of 3.22 ft · sec⁻²; (*b*) descending with an acceleration of 3.22 ft · sec⁻²; (*c*) moving with constant speed; (*d*) descending with the acceleration due to gravity.
          Ans. (*a*) 165 lbwt; (*b*) 135 lbwt; (*c*) 150 lbwt; (*d*) 0.

8. What is the least acceleration with which a man of mass 75 kg can slide down a fire-escape rope which can only sustain a weight of 50 kgwt? And what will be the man's speed after sliding 20 m?
          Ans. 3.3 m · sec⁻²; 11 m · sec⁻¹.

9. Three automobiles, each weighing 3500 lbwt, are tied together (Fig. 21) and the engine of the first car is used to speed them up from rest to 20 mi · hr⁻¹. If this is accomplished while traveling 200 ft, what is the average force causing this acceleration? Neglecting friction, find the forces exerted by the towlines connecting the cars.   *Ans.* 7.0 × 10² lbwt; 4.7 × 10² lbwt; 2.3 × 10² lbwt.

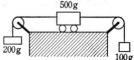

Fɪɢ. 21. Problem 9

10. Over a light, delicately balanced pulley are suspended masses of 5 kg and 2 kg. A force of 2 × 10⁵ dynes pulls down on the 2-kg mass. What are the direction and magnitude of the acceleration, and what is the force in the cord?   *Ans.* The 2-kg mass upwards; 4 × 10² cm · sec⁻²; 3 kgwt.

11. Given the conditions represented in Fig. 22, find the acceleration and find the force in each section of the cord. Neglect friction. *Ans.* $\frac{1}{8} g$; 175 gwt; 113 gwt.

12. Two masses *M* and *m* are connected by a string passing over a light frictionless pulley. What vertical acceleration must be given to the pulley to keep the mass *M* at rest?   *Ans.* $(M - m)g/2\,m$, upwards.

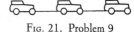

Fɪɢ. 22. Problem 11

13. A train moving in a horizontal direction with a constant speed of 50 mi · hr⁻¹ acquires additional mass at the rate of 100 lb · sec⁻¹ from a vertically falling rain. Find the force, in addition to that necessary to overcome friction, required to maintain the speed of the train at 50 mi · hr⁻¹.
          Ans. 2.3 × 10² lbwt.

14. The distance of the planet Neptune from the sun is 30.0 times that of the earth from the sun. What is its period of revolution?   *Ans.* 164 yr.

15. The mass of the earth is 80 times that of the moon and its diameter is $3\frac{2}{3}$ times that of the moon. Assuming that his initial speed is always the same, find how high a man could jump on the moon if he can jump to a height of 1 yd on the earth.   *Ans.* 6 yd.

16. The experimental determination of the numerical value of the constant of gravitation G (Eq. [25]) enables the mass of the earth to be calculated. Assuming that the earth is a sphere of diameter 8000 mi, calculate its mass in metric tons. Calculate also its average density.

Ans. 6×10^{21} metric tons; $5.3 \text{ g} \cdot \text{cm}^{-3}$.

17. In mechanics only three units ordinarily are chosen arbitrarily, namely, those of length $[L]$, mass $[M]$, and time $[T]$. The degrees to which these three so-called *fundamental units* enter into any *derived unit* are called the dimensions of that unit.[1] For example, the dimensional formula for volume is $[V] = [L]^3$, and that for linear momentum is $[p] = [M][L][T]^{-1}$; the squared brackets indicate that it is not the numerical measures of the quantities but merely the dimensions of their units that are involved in these equations. Write the dimensional formulas for (*a*) area; (*b*) linear speed; (*c*) linear acceleration; (*d*) force; (*e*) pressure; (*f*) density; (*g*) angle; (*h*) cosine of an angle.

18. Use the second law of motion and GALILEO'S discovery that g is the same for all bodies in a given locality to show that the weights of bodies are proportional to their masses at a given place. Does this furnish support for the principle of the equivalence of inertial and gravitational mass?

19. Prove that the total gravitational attraction exerted by a thin-walled hollow sphere of uniform density upon a body anywhere inside of it is zero. Hence prove that the gravitational attraction exerted by any hollow spherical body of uniform density upon a body anywhere inside of it is zero. After you have done this, read NEWTON'S solution in the *Principia*, Bk. I, Prop. LXX.

20. Prove that the gravitational attraction of the earth upon a body below its surface is directly proportional to the distance of the body from the center of the earth. (*Principia*, Bk. I, Prop. LXXIII.)

○

THE BEST and safest method of philosophising certainly seems to be, first to inquire diligently into the properties of things and to establish these properties by experiments; and then to proceed more slowly to hypotheses for the explanation of them. For *hypotheses* ought to be used only in explaining the properties of things, and ought not to be assumed for determining them, except where they are able to furnish experiments. For if from the possibility of *hypothesis* alone anyone makes a conjecture concerning the true nature of things, I do not see by what means it is possible to determine certainty in any science; since it is always possible to devise any number of hypotheses, which will seem to overcome new difficulties.

ISAAC NEWTON, *Philosophical Transactions* **7**, 5014 (1672)

---

[1] For an extensive and masterly discussion of the theory of dimensions see P. W. Bridgman, *Dimensional Analysis* (Yale University Press, 1922). A brief summary of some of the important points brought out in Bridgman's discussion is given by J. C. Oxtoby, "What Are Physical Dimensions?" *The American Physics Teacher* **2**, 85 (1934).

CHAPTER THREE

EFFECT OF SEVERAL FORCES
ON A PARTICLE

A BODY *by two forces conjoined will describe the diagonal of a parallelogram, in the same time that it would describe the sides, by those forces apart.*

If a body in a given time, by the force M impress'd apart in the place A, should with an uniform motion be carried from A to B; and by the force N impress'd apart in the same place, should be carried from A to C: compleat the parallelogram ABCD, and by both forces acting together, it will in the same time be carried in the diagonal from A to D. For since the force N acts in the direction of the line AC, parallel to BD, this force (by the second law) will not at all alter the velocity generated by the other force M, by which the body is carried toward the line BD. The body therefore will arrive at the line BD in the same time, whether the force N be impress'd or not;

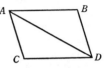

and therefore at the end of that time it will be found somewhere in the line BD. By the same argument, at the end of the same time it will be found somewhere in the line CD. Therefore it will be found in the point D, where both lines meet. But it will move in a right line from A to D by Law I.

Corollary I to NEWTON's laws of motion as translated from the *Principia* in 1729 by ANDREW MOTTE

○

In his study of the equilibrium of bodies on the inclined plane, SIMON STEVIN (1548–1620) arrived at the correct method for determining the effect of several forces acting at a point.[1] He did not, however, expressly formulate the method for adding forces. This was done by NEWTON, who also demonstrated the general dynamical validity of the method, in the corollary to the laws of motion which is quoted at the beginning of this chapter. This principle of the parallelogram of forces, as it is called, was developed independently about 1687 by PIERRE VARIGNON,[2] who applied it to all sorts of statical problems.

38. Addition of Forces. The second law of motion implies that each force acting on a particle produces its own effect independently of the action of any other force and regardless of whether the particle is at rest or in motion (Sec. **30**). In the case of actual bodies this principle of the independence of forces is not self-evident; it

---

[1] See Plate 8; also the Bibliography, p. 452.

[2] *Projet d'une Nouvelle Mécanique* (Paris, 1687); *Nouvelle Mécanique ou Statique* (Jombert, 1725).

B ECAUSE of the effects of friction and the difficulties of all exact measurements, direct experiments made by the student on an inclined plane will not give him the same conviction of the truth of the law of the parallelogram of forces as does the proof of STEVIN. Does this mean that scientific truths rest after all upon *a priori reasoning* rather than upon *observation* and *experience*? The answer to this question given by E. MACH is worth quoting [E. Mach, *The Science of Mechanics* (Open Court, 1893), p. 26, by permission of the publishers]:

" Unquestionably in the assumption from which Stevinus [Latin form of Stevin] starts, that the endless chain does not move, there is contained primarily only a *purely instinctive* cognition. He feels at once, and we with him, that we have never observed anything like a motion of the kind referred to, that a thing of such a character does not exist. This conviction has so much logical cogency that we accept the conclusion drawn from it respecting the law of equilibrium on the inclined plane without the thought of an objection, although the law if presented as the simple result of experiment, or otherwise put, would appear dubious. We cannot be surprised at this when we reflect that all results of experiment are obscured by adventitious circumstances (as friction, etc.), and that every conjecture as to the conditions which are determinative in a given case is liable to error. That Stevinus ascribes to instinctive knowledge of this sort a higher authority than to simple, manifest, direct observation might excite in us astonishment if we did not ourselves possess the same inclination. The question accordingly forces itself. upon us: Whence does this higher authority come? If we remember that scientific demonstration, and scientific criticism generally can only have sprung from the consciousness of the individual fallibility of investigators, the explanation is not far to seek. We feel clearly that we ourselves have contributed *nothing* to the creation of instinctive knowledge, that we have added to it nothing arbitrarily, but that it exists in absolute independence of our participation. Our mistrust of our own subjective interpretation of the facts observed, is thus dissipated."

Vignette from the Title Page

of SIMON STEVIN'S DE BEGHINSELEN DER WEEGHCONST,

or *Principles of Statics* (Leiden, 1586)

"*Wonder en is gheen wonder*" ("A miracle, and yet it is no miracle"), said STEVIN. The endless chain is in equilibrium in any position, and the hanging portion is by itself in equilibrium; therefore, he reasoned, the two inclined sections must balance each other, and either would be balanced by a section of vertical chain of length equal to the altitude of the triangle. Since the weights of the various portions of the chain are proportional to their lengths and since the force in the chain at the top of the incline is evidently equal to the weight of a vertical section, it follows that the force required to support a body resting on an inclined plane is to the weight of the body as the height of the plane is to its length. The "wonder" is the simplicity of the rule; that it really is "no wonder" follows from the fact, admitted at once as self-evident, that the endless chain would never of itself start to slide on the plane, however frictionless, by virtue of its own weight.

The remarks of E. MACH in his *Science of Mechanics* (Open Court, 1893), pp. 24–33, upon the nature of STEVIN's reasoning will repay careful study. A translation of a portion of STEVIN's original paper will be found in W. F. Magie's *A Source Book in Physics* (1935), pp. 23–37. See also H. Crew's *The Rise of Modern Physics* (Williams & Wilkins, 1935), pp. 86–92, and J. Cox's *Mechanics* (Cambridge University Press, 1919), pp. 41–47.

can be established only by experiment. If the principle is accepted as an experimental fact, it follows that several forces acting simultaneously on a particle impart to the particle one definite acceleration which is the vector sum of the several accelerations produced by the separate forces. Since each force is proportional to the acceleration that it produces and is in the direction of this acceleration, it then follows that the vector sum of the separate forces is proportional to, and has the direction of, the vector sum of the separate accelerations. Thus the *vector sum*, or *resultant*, of any number of forces is *that single force which would produce the same acceleration as is produced by the joint action of the several forces.*

The forces acting on a particle are therefore added in the same manner as other vector quantities; namely, by placing end to end the directed lines which represent them and joining the extremities of the broken line so formed (Sec. 9). The magnitude and direction of the sum can be found by the graphical method or, more

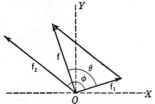

Fɪɢ. 23. Addition of two forces

conveniently and accurately, by the use of trigonometry. Thus the sum f of two forces f_1 and f_2 which include the angle ϕ, Fig. 23, has a magnitude f and a direction θ, measured from f_1, given by

$$f^2 = f_1{}^2 + f_2{}^2 + 2 f_1 f_2 \cos \phi,$$
$$\theta = \sin^{-1}\left(\frac{f_2}{f} \sin \phi\right).$$

[33]

The vector sum of three or more forces may be found by adding the sum of the first and second to the third, this sum to the fourth, and so on. A better method, however, is given in Sec. 40.

☙

The Resolution of Vectors into Components

We have seen that vector addition enables us to replace any number of vectors by a single vector called the vector sum. This process may be reversed and a single vector replaced by any number of vectors which, added together, give the original one. This is called *resolving a vector into components.*

39. Rectangular Components. In practice one usually resolves a vector into components that have the directions of the axes of a coordinate system. If rectangular coordinates are employed, these components will be at right angles to one another, and they are then referred to as the *x*-, *y*-, and *z-components* or *rectangular components* of the vector relative to the chosen axes.

For the sake of definiteness, let us suppose that a certain force $\overline{OF}$, Fig. 24, acts for an *infinitesimal* interval of time on a particle at O. If OA represents the distance which the particle would move in one second by virtue of the velocity acquired through the action of the given force, then Ox evidently represents the distance moved in the direction OX during this second; that is, if $\overline{OA}$ represents the actual acceleration, $\overline{Ox}$ represents the acceleration in the direction OX. Now the component of the force $\overline{OF}$ in the direction OX is $\overline{Ox'}$, and it is ob-

vious from Fig. 24 that this component is equivalent, so far as motion along OX is concerned, to the given force $\overline{OF}$. In other words, if the particle were free to move only in the direction OX, its motion under the action of the force $\overline{OF}$ would be precisely the same as though a force $\overline{Ox'}$ were acting instead of $\overline{OF}$. To help visualize this latter statement, one may think of a car on overhead rails which is accelerated by means of

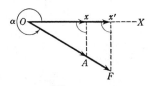

Fig. 24. $\overline{Ox}$ is the component of $\overline{OA}$ in the direction OX

a rope held by a man on the ground. It is apparent from the foregoing considerations that the component of a *force* in any specified direction could be defined as that force which, acting in the direction specified, would produce the same effect, so far as the motion in this direction is concerned, as is produced by the action of the given force.

The process of finding the component of a vector quantity in any direction evidently consists in finding the *orthogonal projection* in the required direction of the directed line which represents the given quantity. Thus, in Fig. 24 the component of $\overline{OA}$ in the direction OX has the magnitude $\overline{Ox} = \overline{OA} \cos \alpha$. In general, then, *the rectangular component in the direction of any chosen coordinate axis is the product of the magnitude of the given vector and the cosine of the positive angle which the vector makes with the positive direction of the chosen axis.* Hence if α, β, γ represent the three direction angles which a vector F makes with a certain set of rectangular axes in space, the magnitudes F_x, F_y, F_z of the rectangular components are

$$F_x = F \cos \alpha, \quad F_y = F \cos \beta, \quad F_z = F \cos \gamma. \qquad [34]$$

Conversely, if we know the rectangular components F_x, F_y, F_z of a vector F, the magnitude and direction of the vector can be obtained by means of the relations

$$F^2 = F_x{}^2 + F_y{}^2 + F_z{}^2,$$

$$\cos \alpha = \frac{F_x}{F}, \quad \cos \beta = \frac{F_y}{F}, \quad \cos \gamma = \frac{F_z}{F}. \qquad [35]$$

EXAMPLE. A bullet is fired with a muzzle velocity of 1.70×10^3 ft · sec^{-1} in an eastward direction and at an angle of $31°\,0'$ with the horizontal. Find the horizontal and vertical components of the muzzle velocity.

Solution. Take the muzzle of the gun as the origin of rectangular axes and the eastward direction as OX (Fig. 25). Denote the horizontal and vertical components of the velocity by V_x and V_y respectively. Then

$V_x = 1.70 \times 10^3 \cos 31° = 1.46 \times 10^3$ ft · sec^{-1},
$V_y = 1.70 \times 10^3 \cos (90° - 31°) = 1.70 \times 10^3 \sin 31° = 876$ ft · sec^{-1}.

As a check obtain V_x and V_y by a graphical method. What would V_z represent in this solution, and what is its value?

EXAMPLE. Suppose that the gun in the foregoing example is mounted on a truck which has a velocity of 88 ft · sec^{-1} east. Find the horizontal and vertical components of the muzzle velocity relative to the earth.

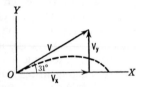

FIG. 25. A velocity resolved into x- and y-components

Solution. One method of solution is to add the velocity of the bullet relative to the truck to that of the truck relative to the ground, thus obtaining the velocity of the bullet with reference to the ground; the horizontal and vertical components of the latter are the required components. The student should complete this solution and then compare it with the following one, which is briefer.

Denote the required components by V_x and V_y; then

$$V_x = 1.70 \times 10^3 \cos 31° + 88 = 1.55 \times 10^3 \text{ ft} \cdot \text{sec}^{-1},$$
$$V_y = 1.70 \times 10^3 \sin 31° = 876 \text{ ft} \cdot \text{sec}^{-1}.$$

40. Analytic Method of Adding Vectors. A very simple method of finding the vector sum of *any* number of vectors is to choose rectangular axes and resolve each vector into rectangular components along these axes. The components parallel to any one axis are then added algebraically to obtain the component in that direction of the required sum. To illustrate the method, let us suppose that we wish to add just three vectors, f_1, f_2, f_3, and that they all lie in the same plane. Choose any convenient pair of rectangular axes in the plane of these vectors, and let the angles which the vectors make with the X-axis be denoted by α_1, α_2, α_3 (Fig. 26). If F represents the sum of the given vectors and F_x and F_y are the rectangular components of this sum referred to the chosen axes, then [1]

$$F_x = f_1 \cos \alpha_1 + f_2 \cos \alpha_2 + f_3 \cos \alpha_3 = \sum_{i=1}^{i=3} f_i \cos \alpha_i,$$

$$F_y = f_1 \sin \alpha_1 + f_2 \sin \alpha_2 + f_3 \sin \alpha_3 = \sum_{i=1}^{i=3} f_i \sin \alpha_i.$$

[36]

The magnitude F of the required vector sum and the angle α which this sum makes with the X-axis can then be obtained by means of the relations

$$F^2 = F_x{}^2 + F_y{}^2,$$

$$\tan \alpha = \frac{F_y}{F_x}.$$

[37]

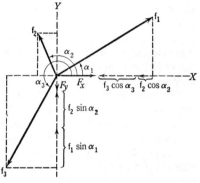

When the vectors to be added do not all lie in one plane, it is necessary to use three rectangular axes and to resolve each vector into its components in these three directions. Suppose that the vectors f_1, f_2, f_3 do not lie in the same plane and that they make angles α_1, α_2, α_3 with the X-axis, angles β_1, β_2, β_3

Fig. 26. Illustrating the analytic method of adding vectors

with the Y-axis, and angles γ_1, γ_2, γ_3 with the Z-axis. If F is the sum of the given vectors and F_x, F_y, F_z are its components referred to the chosen axes,

$$F_x = f_1 \cos \alpha_1 + f_2 \cos \alpha_2 + f_3 \cos \alpha_3 = \sum_{i=1}^{i=3} f_i \cos \alpha_i,$$

$$F_y = f_1 \cos \beta_1 + f_2 \cos \beta_2 + f_3 \cos \beta_3 = \sum_{i=1}^{i=3} f_i \cos \beta_i,$$

[38]

$$F_z = f_1 \cos \gamma_1 + f_2 \cos \gamma_2 + f_3 \cos \gamma_3 = \sum_{i=1}^{i=3} f_i \cos \gamma_i.$$

The magnitude F of the required sum and its direction angles α, β, γ are then obtained by means of Eqs. [35].

---

[1] The symbol $\sum_{i=1}^{i=3} f_i \cos \alpha_i$ is merely an abbreviated way of writing the sum $f_1 \cos \alpha_1$ $+ f_2 \cos \alpha_2 + f_3 \cos \alpha_3$, and is read "the sum of the terms $f_i \cos \alpha_i$, where i has all the integral values from 1 to 3."

The Statics of a Particle

41. The Meaning of Equilibrium. *Statics* is the subdivision of dynamics that deals with the conditions under which bodies are in equilibrium. A *particle* is said to be in *equilibrium* when its momentum is constant. Since the effect of a force is to change the momentum, it follows that a particle can be in equilibrium only when the vector sum of the forces acting on it is zero. It is important to see that equilibrium does not necessarily mean rest; rest is merely one case of constant momentum. A weight hung from a string and swinging like a pendulum is in equilibrium only when it is at the middle of its path, and this is just the moment when it is moving most rapidly; at the ends of its path, where it is motionless, and at all other points, it is acted upon by unbalanced forces and hence is accelerated. Thus the criterion of equilibrium is not zero velocity but zero acceleration and hence zero force.

42. Conditions for the Equilibrium of a Particle. When a particle is acted on by two forces that have the same magnitude but opposite directions, the particle must be in equilibrium, for the sum of the forces is zero; each force produces its own effect independently of the other, and these two effects just neutralize each other. Conversely, if a particle is in equilibrium under the action of two forces, the forces necessarily must be equal in magnitude and opposite in direction, for otherwise their vector sum could not be zero.

STEVIN was the first to prove that three forces acting at a point are in equilibrium when the directed lines representing them can be arranged to form a triangle. In general, when any number of forces acting on a particle are of such a nature that the directed lines representing them form a closed polygon, the particle is in equilibrium, for the vector sum of the forces then is zero. If the directions of *three* forces producing equilibrium are known, the relative magnitudes of the forces can be found by constructing a triangle with its sides in the directions of the forces.

If the particle under consideration is made the órigin of a rectangular system of coordinates, the conditions for its equilibrium under the action of *n* forces are, by Eqs. [38],

$$F_x = \sum_{i=1}^{i=n} f_i \cos \alpha_i = 0,$$

$$F_y = \sum_{i=1}^{i=n} f_i \cos \beta_i = 0, \qquad [39]$$

$$F_z = \sum_{i=1}^{i=n} f_i \cos \gamma_i = 0;$$

that is, the algebraic sum of the force components along each axis must be zero. This is evident from Eqs. [35]; for when $F_x = F_y = F_z = 0$, then $F = 0$.

EXAMPLE. Show by means of Eqs. [35] that the converse of the foregoing proposition also is true; that is, if a particle is in equilibrium, then $F_x = F_y = F_z = 0$.

EXAMPLE. Show that if all the forces acting on a particle lie in one plane, then $F_z = 0$, and Eqs. [38] reduce to Eqs. [36]; also show that the latter equations can then be used to express the conditions for the equilibrium of a particle.

Although a problem involving the equilibrium of a particle usually can be solved by finding the conditions for which the vector sum of all the forces will be zero, or those for which the force polygon will be closed, the most general method, and the most useful one, is to set the algebraic sum of the x-, y-, and z-components of all the forces equal to zero, as in Eqs. [39], and solve the resulting equations. In order to become thoroughly familiar with this method, the student should use it in solving most of his problems on equilibrium.

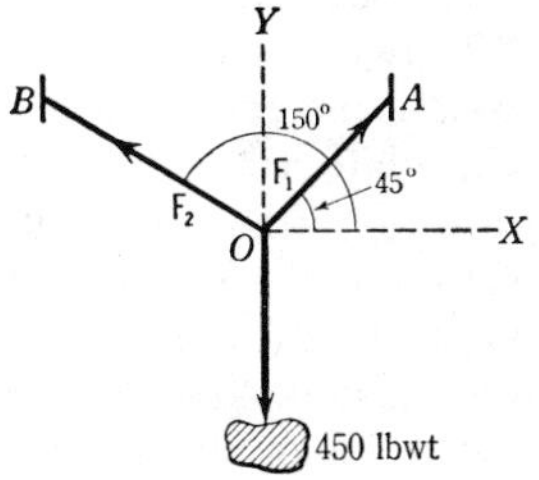

FIG. 27. The particle of the wire at O is in equilibrium under the action of three forces

EXAMPLE. A rock weighing 450 lbwt is fastened to a wire that is stretched between two posts, as shown in Fig. 27. Find the forces in the parts OA and OB of the wire.

Solution. Take the point O as the origin of coordinates, and let F_1 and F_2 denote the forces in the wire. The forces involved in the problem are then F_1, F_2, and the weight of 450 lbwt. All three of these act on the same particle O of the wire, and this particle is, by hypothesis, in equilibrium. By writing down the conditions for equilibrium, one obtains

$$F_x = F_1 \cos 45° + F_2 \cos 150° = 0,$$
$$F_y = F_1 \sin 45° + F_2 \sin 150° + 450 \sin 270° = 0.$$

These equations give $F_1 = 4.0 \times 10^2$ lbwt and $F_2 = 3.3 \times 10^2$ lbwt. If these results do not look reasonable, they should be checked by solving the problem graphically.

EXAMPLE. Assume that the wire AOB, Fig. 27, could be stretched perfectly straight, and find F_1 and F_2 for this case. In view of your results, is it possible to stretch a wire perfectly straight between two points on the same level?

EXPERIMENT III. ADDITION AND RESOLUTION
OF FORCES

The object is to show that forces follow the rules for the addition and resolution of vector quantities.

PART I. The Force Table. Fig. 28 is a device for applying at a common point several known forces that make known angles with one another. The forces are applied by means of horizontal cords that run over pulleys on the rim of a graduated circle and carry adjustable loads. A pin holds the junction of the cords in position at the center. When a test for equilibrium is to be made, the pin is removed.

a. Imagine that two of the pulleys on the force table are set at the angles marked 30° and 150° and that weights of 100 gwt and

150 gwt, respectively, are placed on the weight-hangers hanging over these pulleys. Calculate with the aid of Eqs. [33] the direction and magnitude of the vector sum of the two forces. When two or more students are working together on this experiment, each should make an independent calculation, after which the results may be compared.

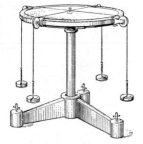

Test the foregoing calculation experimentally with the help of the force table. To do this, set two of the pulleys at the

FIG. 28. Force table

two given marks on the graduated circle and apply the given weights; remember that the weights of the weight-hangers, which will be found stamped upon them, supply part of the force applied to the ends. Then set a third pulley 180° from the *calculated* angle, apply the *calculated* weight, and remove the center pin. In order to make sure that friction is not a factor in producing equilibrium, displace the junction of the cords and tap the table; the junction should return to the center. Let the instructor see the results of this test before you begin the next part of the experiment.

1. If the weight-holders on the force table are all the same weight, is it necessary to take them into account? Explain.

2. When a particle is in equilibrium under the action of three forces, how does the vector sum of any two of the forces compare in magnitude and direction with the remaining force?

b. Choose three forces that have different magnitudes and directions and that lie in the same plane. Select a suitably orientated system of rectangular coordinates and *calculate* the x- and y-components of each force. Then *calculate* the direction and magnitude of the vector sum of the three forces by the analytic method.

Test the foregoing calculations with the aid of the force table, as in *a*.

PART II. **The Simple Crane.** Construct the model of a crane shown in Fig. 29. Before putting the spring balance S in place, obtain its "zero reading" by hanging it in the particular

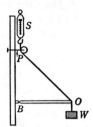

position in which it is to be used in the experiment and observing its reading when there is no load on the hook.

Hang a known weight *W*, say 1 kgwt, from the end *0* of the boom *BO*, and adjust the boom to a horizontal position with the aid of a spirit level or a steel square and plumb line.

Obtain the force in the cord *PO* by observing the reading of the spring balance S.

Fig. 29. Laboratory model of a simple crane

Measure the distances *OP* and *PB*. The point *B* may be located by dropping a plumb line from *P* to the boom. Use the known values of *OP*, *PB*, and the total load to *calculate* the force in the cord *OP*. Remember that the cord must support half the weight of the boom. (Why?) Compare the calculated result with the value obtained by observation.

3. Find by the graphical method the vector sum of the two forces specified in Part I, *a*, and compare the result with that calculated. Include the graphical construction in your report.

4. Resolve the three forces of Part I, *a*, into x- and y-components, and verify the conditions of equilibrium expressed in Eqs. [39].

5. State as clearly and concisely as possible the conclusions which you draw from this entire experiment.

○

QUESTION SUMMARY

1. What is meant by the *sum* or *resultant* of any number of forces? How is it found graphically? Describe two distinctly different methods for calculating it.

2. What is meant by the *component* of a given vector in any specified direction? How is it found graphically? How computed? Illustrate, using force vectors.

3. If the rectangular components of a vector are known, how may the magnitude and direction of the vector be obtained? Illustrate, using force vectors.

4. If a particle is in equilibrium, what must be true regarding (*a*) its momentum? (*b*) the vector sum of the forces acting upon it? (*c*) the force polygon? (*d*) the algebraic sums of the *x*-, *y*-, and *z*-components respectively?

o

PROBLEMS

1. Does the experiment with the falling airplane bomb (Sec. 2) afford any support for the principle of the independence of forces?

2. (*a*) Show that the acceleration of a body sliding without friction down an inclined plane of length *l* and height *h* is $(h/l)g$. (*b*) Show that the speed acquired in sliding down the plane is the same as that acquired in falling through the vertical height *h*. (*c*) By dividing the arc *ab*, Fig. 30, of a vertical circle into a large number of small inclined planes, show that a body sliding without friction down the arc *ab* acquires the same speed as though it fell vertically from *b* to *c*.

3. It can be shown that the time of descent down all chords which start from the top of a vertical circle is the same. After you have proved this, consult *Two New Sciences*, pp. 188–194, to see how GALILEO did it.

FIG. 30.
Problem 2 (*c*)

4. A man travels 4.0 mi south, then 6.0 mi in a direction 30° south of east, then 2.0 mi east, and finally 3.0 mi southeast. What is his final position relative to his starting point?   *Ans.* 13 mi, 44° south of east.

5. A particle of mass 300 mg is acted on by forces of magnitudes 1, 2, 3, 4, 5, and 6 dynes, all in the same plane. The angle between the first force and the *X*-axis, and between each force and the next, is 30°. Assuming these data to be correct to three significant figures, find (*a*) the sum of the forces and (*b*) the acceleration of the particle.

Ans. (*a*) 15.3 dynes at 133° 5′; (*b*) 50.9 cm · sec^{-2}, 133° 5′.

6. Three forces, of 2.0, 4.0, and 6.0 dynes respectively, act on a particle and are directed along the diagonals of the three faces of a cube meeting at the particle. Determine their sum.

Ans. 10 dynes, $\alpha = 55°\ 30′$, $\beta = 65°$, $\gamma = 45°$.

7. The top of an inclined plane is 15 m above the ground, and a car coasting down from the top traverses a distance of 1.5 m during the third second. If the friction is negligible, (*a*) what is the inclination of the plane? (*b*) how long does it take the car to coast down?   *Ans.* (*a*) 3° 30′; (*b*) 29 sec.

8. If friction is neglected, how much force would be needed to pull a load of 1500 lbwt up a 5-percent grade (percentage of horizontal distance) with constant speed? If the speed were changed uniformly from 5 mi · hr^{-1} to 15 mi · hr^{-1} in 12 sec, what total force would be needed? What if the force of friction, which can be considered to be parallel to the incline, were 10 lbwt per hundredweight of load?   *Ans.* 75 lbwt; 130 lbwt; 280 lbwt.

9. A bullet is fired from a gun with a velocity of 200 m · sec⁻¹ at an angle of 30° 0′ with the horizontal. (*a*) How high will it rise? (*b*) At what distance from the gun will the bullet strike the earth? (*c*) If the gun had been on the top of a tower 335 m high, how far from its base would the bullet have struck the ground? Disregard the resistance of the air.

Ans. (*a*) 510 m; (*b*) 3.53 × 10³ m; (*c*) 4.03 × 10³ m.

10. A man of 150 lbwt standing in the middle of a tightrope depresses the middle of the rope 5.0 ft below the ends. If the rope is 60 ft long when thus stretched, what is the force in it?   *Ans.* 4.5 × 10² lbwt.

11. A bridge span (Fig. 31) is 5 m high and 20 m long. Find the vertical force and the horizontal thrust upon each of the piers in terms of the weight W which hangs from the center of the span.   *Ans.* $W/2$; W.

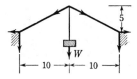
Fig. 31. Problem 11

12. A particle is in equilibrium under the action of three forces, of magnitudes 10, 25, and 30 poundals, all acting in the same plane. Find the angles between the forces.   *Ans.* 69° 30′; 162°; 128° 30′.

13. A 100-lb weight is hung from a horizontal beam by means of two strings, one of which is 60 in. long. The weight is 48 in. below the beam and the force in the 60-in. string is half that in the other string. Calculate the length of the second string and the force in each string.

Ans. 50 in.; 37 lbwt; 74 lbwt.

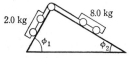
Fig. 32. Problem 15

14. How much force parallel to the plane is required to hold a 20-kg weight on a smooth inclined plane 10 m long and 6 m high? Find also the force with which the weight presses against the plane. What horizontal force would hold the weight? Find the force against the plane in this case.   *Ans.* 12 kgwt; 16 kgwt; 15 kgwt; 25 kgwt.

15. Neglecting friction, find the acceleration and the force in the cord for the system shown in Fig. 32 when (*a*) $\phi_1 = 30°$, $\phi_2 = 90°$; (*b*) $\phi_1 = 30°$, $\phi_2 = 25°$. *Ans.* (*a*) 0.70 g, 2.4 kgwt; (*b*) 2.3 × 10² cm · sec⁻², 1.5 kgwt.

16. A 1000-lb weight is supported in the manner shown in Fig. 33. Find the forces in the cable *OB* and in the boom *OA*.

Ans. 1.0 × 10³ lbwt, 1.5 × 10³ lbwt.

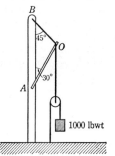
Fig. 33. Problem 16

17. A streetcar in which you are riding is traveling with a speed of 30 mi · hr⁻¹. It nearly runs over a man. The brakes are applied; the car goes 50 ft before stopping. At what angle with the vertical must you lean to keep from being thrown over? *Ans.* 31°.

18. A weight of 17.5 kgwt is supported by a pulley free to move along the string fastened at *A* and *B*, Fig. 34. Angle *BAC* = 90°, *AC* = *l*, *BC* = 2 *l*. Find (*a*) the force in the string and (*b*) the inclinations of the two parts of the string with the vertical. *Ans.* 10 kgwt; 30°.

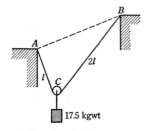

19. A stone dropped from an airplane moving horizontally is in the air 10 sec and strikes the ground in a direction making an angle of 70° with the horizontal. If air resistance can be neglected, what is the speed of the airplane? *Ans.* 1.2×10^2 ft · sec⁻¹.

FIG. 34. Problem 18

20. If a man standing on a pier 25 ft above the water pulls in a rope attached to a boat at the rate of 2.0 ft · sec⁻¹, with what speed does the boat approach the pier (*a*) when it is 100 ft from the pier? (*b*) when it is 15 ft from the pier? (*c*) Derive a general expression for the speed *v'* of the boat in terms of the speed *v* with which the rope is hauled in, the height *h* of the pier, and the distance *d* of the boat from the pier.

Ans. (*a*) 2.1 ft · sec⁻¹; (*b*) 3.9 ft · sec⁻¹; (*c*) $v' = v(h^2 + d^2)^{\frac{1}{2}}/d$.

o

STEVINUS'S deduction [Plate 8] is one of the rarest fossil indications that we possess in the primitive history of mechanics, and throws a wonderful light on the process of the formation of science generally, on its rise from instinctive knowledge. We will recall to mind that Archimedes pursued exactly the same tendency as Stevinus, only with much less good fortune. In later times, also, instinctive knowledge is very frequently taken as the starting-point of investigations. Every experimentator can daily observe in himself the guidance that instinctive knowledge furnishes him. If he succeed in abstractly formulating what is contained in it, he will as a rule have made an important advance in science.

Stevinus's procedure is no error. If an error were contained in it, we should all share it. Indeed, it is perfectly certain, that the union of the strongest instinct with the greatest power of abstract formulation alone constitutes the great natural inquirer. This by no means compels us, however, to create a new mysticism out of the instinctive in science and to regard this factor as infallible. That it is not infallible, we very easily discover. . . . The instinctive is just as fallible as the distinctly conscious. Its only value is in provinces with which we are very familiar.

E. MACH, *The Science of Mechanics* (The Open Court Publishing Company, 1893), pp. 26–27. By permission of the publishers.

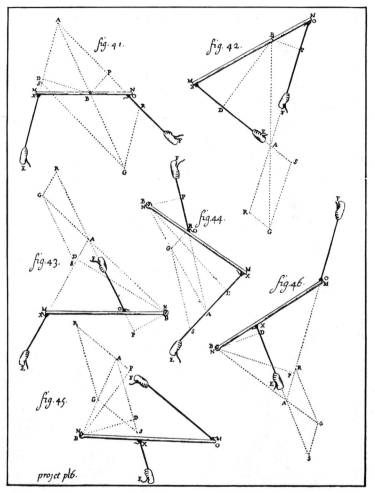

A Plate from VARIGNON'S
PROJET D'UNE NOUVELLE MECANIQUE (Paris, 1687)

VARIGNON developed the principle of the parallelogram of forces independently of
NEWTON, about 1687, and applied it to the solution of all sorts of statical problems,
some of which are illustrated on this and the following plate.

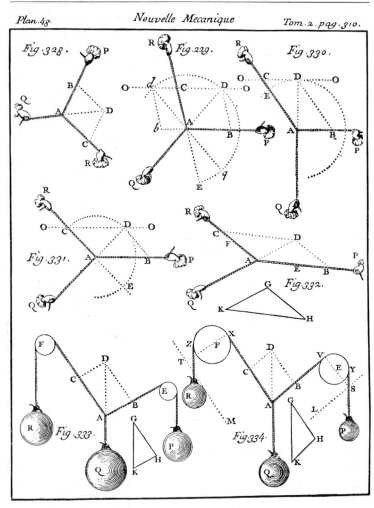

A Plate from Varignon's

NOUVELLE MECANIQUE OU STATIQUE (Paris, 1725)

There are sixty-four such plates in Varignon's book. The principle of the parallelogram of forces enabled him to treat of machines in a much simpler manner than had been the case previously. Many of the theorems and methods of presentation used in modern elementary textbooks on statics are taken from Varignon.

CHAPTER FOUR

WORK, POWER, AND ENERGY

I̶F THE Activity of an agent be measured by its amount and its velocity conjointly; and if, similarly, the Counter-activity of the resistance be measured by the velocities of its several parts and their several amounts conjointly, whether these arise from friction, cohesion, weight, or acceleration; — Activity and Counter-activity, in all combinations of machines, will be equal and opposite. From the scholium to NEWTON's third law of motion as translated from the *Principia* by WILLIAM THOMSON (LORD KELVIN) and PETER GUTHRIE TAIT in their *Treatise on Natural Philosophy*

○

The basis of our modern civilization lies in the use of machinery, and there are no notions more intimately associated with a machine than are those of work, power, and energy. The physical concepts of work and energy originated in the study of simple machines, but it has been only gradually and with great difficulty that these concepts have attained their present positions of importance in physics. This is not strange, for energy is not a notion that comes readily out of one's everyday experience, but is a scientific concept which has been created by physics for its own special purposes. It was not until the middle of the nineteenth century that the concept of energy reached the position of importance that it occupies today. This was brought about by the establishment at that time of the principle of the conservation of energy, perhaps the most important generalization that has ever been made.

○

Work

43. Definition of Work. When a force acts upon a body, the work done is defined as the product of just two factors, (1) the component of force in the direction of motion and (2) the distance through which the point of application of the force moves during its action.[1] Therefore the defining equation for work is

$$W = f \cos \phi \cdot s, \qquad [40]$$

where f and s are the magnitudes of the force and displacement, respectively, and ϕ is the angle between these two vectors. It is ob-

---

[1] This conception of *work* as the product of force and distance was introduced into physics in 1826 by the French mathematician J. V. PONCELET, at the suggestion of the French engineer and physicist G. G. CORIOLIS.

vious from this definition that the work performed by a force is also the *product of the force and the distance the body moves in the direction of the force*. Work is a scalar quantity, for it has nothing to do with direction. Consequently, to get the total work done upon a body by several different forces, the work of each may be computed separately and then the ordinary algebraic sum taken.

When the force varies from point to point on the path of the moving body, we first must divide the whole distance into parts, each so small that the force may be regarded as constant in both magnitude and direction during the motion throughout it. Eq. [40] can then be applied to each small part and the resulting increments of work added to obtain the total work done. More precisely, if the whole distance be divided into n equal parts of length Δs, and if $(f \cos \phi)_i$ be the force component while the point of application of the force moves through the ith part, then the work done in the whole distance s is accurately

$$W = \lim_{n \to \infty} [(f \cos \phi)_1 \cdot \Delta s + (f \cos \phi)_2 \cdot \Delta s + \cdots + (f \cos \phi)_n \cdot \Delta s], \quad [41]$$

or, more briefly,

$$W = \lim_{n \to \infty} \sum_{i=1}^{i=n} (f \cos \phi)_i \cdot \Delta s = \int_0^s f \cos \phi \, ds, \quad [42]$$

the last abbreviation being that used in the integral calculus.

44. Units of Work. The unit of work of course involves a unit of force and a unit of length. The cgs unit of work is the *dyne-centimeter*, more often called the *erg*; an erg of work is done when a body on which a force of one dyne acts moves one centimeter in the direction of the force. A larger unit is the *joule*, defined as 10^7 ergs.

Other units of work, used especially in mechanical and civil engineering, are the gravitational units *centimeter-gram-weight*, *meter-kilogram-weight*, *foot-pound-weight*, etc., the definitions of which are obvious from their names.[1] Since a gram of force is about 980 dynes, it is evident that 1 cm · gwt is approximately equal to 980 ergs.

45. The Work Diagram. The direct calculation of the sum in Eq. [41] obviously is tedious, if not practically impossible, when the number of terms is large. The method of the integral calculus should be employed when possible to obtain the limit of this sum,

---

[1] These gravitational units of work are ordinarily called the *centimeter-gram*, *meter-kilogram*, *foot-pound*, etc., names which the student may continue to use, provided he remembers that it is force and not mass that enters directly into the definition of work.

but it is often necessary to use a *diagram of work* for this purpose. In this diagram (Fig. 35), which is similar to the *indicator diagram* introduced by JOHN SOUTHERN [1] for use with the steam engine developed by JAMES WATT (1736–1819), the force component ($f \cos \theta$) is plotted as a function of the displacement of the body. The actual work done evidently is represented by the area under the resulting curve and can be calculated by measuring this area, due allowance being made for the scale in which the diagram is drawn.

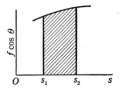

FIG. 35. The shaded area represents the work done in the displacement of magnitude $s_2 - s_1$

EXAMPLE. Experiment shows that the force needed to keep a spring stretched or compressed is proportional, within certain wide limits, to the first power of the displacement of the end of the spring from its unstrained position. When the spring of a certain safety valve is compressed from its unstrained length of 49.0 cm to a length of 45.0 cm, it then exerts a force of 232 kgwt. How much work will have to be done to compress it from 46.0 cm to 43.0 cm?

Solution. Here $\cos \phi = 1$, and $f = ks$, where k is a constant of proportionality having the value 58 kgwt · cm^{-1}. If a work diagram is drawn with the values of ks as ordinates and those of s as abscissas, the resulting curve will be a straight line (why?), and hence

$$W = 3\left(\frac{174 + 348}{2}\right) = 7.8 \times 10^2 \text{ cm · kgwt, or } 7.8 \times 10^5 \text{ cm · gwt.}$$

The same result is obtained if the method of the calculus is employed (Eq. [42]); for

$$W = \int_{s_1}^{s_2} f\, ds = k \int_{s_1}^{s_2} s\, ds = \tfrac{1}{2} k(s_2{}^2 - s_1{}^2) = \tfrac{58}{2}(36 - 9) = 7.8 \times 10^2 \text{ cm · kgwt.}$$

EXAMPLE. Prove that the work done in operating a water motor for any given time is $P\,\Delta V$, where P is the constant pressure with which the water is delivered to the motor and ΔV is the volume of water which passes through it.

Solution. Suppose that the water enters the motor through an orifice of area A. The force f driving the water forward is then PA, and if this carries the water forward a distance s during the time under consideration, the work done is PAs. But As is the volume of water which has passed through the motor. Therefore the total work done is $P\,\Delta V$. In cases where the pressure

[1] H. W. Dickinson and R. Jenkins, *James Watt and the Steam Engine* (Oxford University Press, 1927), p. 229.

varies during the operation, either the average pressure must be obtained or else the method of the calculus or the work diagram must be employed to compute the amount of work done.

46. The Work Principle. The third law of motion (Sec. 32) asserts that force is dual in nature, that in the interaction of two bodies the first body exerts a force on the second and the second exerts a force on the first, and that these two forces are equal in magnitude but opposite in direction. The reasons for this assertion and some of its consequences will be discussed in Chapter 5.

An additional and wholly distinct assertion is made in the scholium to the third law, quoted at the beginning of the present chapter. It is that in any particular machine, during any given time, *the work of the acting forces is equal to the work of the reacting forces,* whether these forces arise from friction, molecular forces, gravity, or inertia. This is a generalization from experiments upon all sorts of machines — levers, pulleys, the wheel and axle, screws, inclined planes, and all combinations of these devices. It is to be noted that in all experiments in which there is acceleration, the resistance to the acceleration which the body offers because of its inertia must be included among the reacting forces; this inertial force must be put equal in magnitude and opposite in direction to that force whose single action would produce the observed acceleration; that is, equal to $-ma$, m being the mass of the body and a its actual acceleration.

If a force applied to a body does work by moving the body in the direction of the force, this work will be called *positive.* If the motion is in the opposite direction to the force, the work will be called *negative.* With these conventions in mind, the work principle is seen to assert that, during any given time, the algebraic sum of the work done by the acting and reacting forces is zero.

○

Power

With the growing use of machines and of the steam engine in the latter part of the eighteenth century, it became important to have ways of describing them in terms of the rapidity with which they could do work. Thus JAMES WATT, wishing to compare the practical value of his engine with that of a horse as a prime mover, made some actual experiments with horses and arrived at the estimate of 33,000 ft · lbwt of work per minute for the average power of a horse.[1]

---

[1] See *H. W. Dickinson and R. Jenkins, *James Watt and the Steam Engine* (Oxford University Press, 1927), pp. 353–356.

47. Definition of Power. In ordinary language the word *power* has a variety of meanings, including especially "the ability to do some specific thing." As a physical term it means *time-rate of doing work,* and it should never be used in physics in any other sense. The defining equation for the *average power* of an agent is

$$P = \frac{W}{t},$$ [43]

where W is the work done in the interval of time t. The power developed at any *instant* is, of course, the limit of this average value when t becomes very small, or, in the notation of the calculus,

$$P = \frac{dW}{dt}.$$ [43a]

48. Units of Power. The cgs unit of power is the *erg per second.* Since this is inconveniently small, the unit usually employed is the *joule per second,* called the *watt.* The *horsepower* has, of course, lost its original significance and now means simply 33,000 ft · lbwt per minute: this is equal to 745.70 watts, or nearly three fourths of a kilowatt. The *American horsepower,* which is the unit of power now most commonly used in engineering practice in this country, is defined by the United States Bureau of Standards as exactly equivalent to 746 watts.

Most practical measurements of work involve measurements of power, and this explains why electrical energy is measured commercially in kilowatt-hours. The *kilowatt-hour* is the work done in one hour by an agent working at the rate of 1000 watts; it is a unit of *work,* not of power.

49. Relation of Power to Speed. When a body on which a force of magnitude f acts moves a small distance Δs, the force does work of amount $\Delta W = f \cos \phi \cdot \Delta s$, where ϕ is the angle between the force and the displacement. If this is accomplished in the time Δt, the power of the agent maintaining the force is

$$P = \lim_{\Delta t \to 0} \frac{\Delta W}{\Delta t} = \lim_{\Delta t \to 0} \frac{f \cos \phi \cdot \Delta s}{\Delta t} = f \cos \phi \lim_{\Delta t \to 0} \frac{\Delta s}{\Delta t}.$$

Letting v denote the linear speed of the point of application of the force, we have then, from the foregoing equation and Eq. [7], Chap. 1,

$$P = fv \cos \phi.$$ [44]

This explains, for example, why the speed attainable by an automobile increases with the horsepower.

EXAMPLE. At what rate, in kilowatts, is work done against gravity when an automobile of mass 1400 kg moves up a 5° grade with a speed of 30 km · hr⁻¹?

FOR details of the NEWCOMEN engine and other steps in the development of the steam engine see Rhys Jenkins, *Transactions of the Newcomen Society* **3**, 96–118 (1922–1923); **4**, 113–133 (1923–1924); A. P. Usher's *A History of Mechanical Inventions* (McGraw-Hill, 1929), Chap. 11; and R. H. Thurston's *A History of the Growth of the Steam Engine*. The following quotation is taken, with the permission of the Council of The Newcomen Society, from the paper by Rhys Jenkins:

"It may be asked — to what extent was Newcomen an original thinker? Did the principle of the engine — the condensation of steam under a piston — originate with him? Briefly, the sequence of ideas may be set forth thus: In 1654 Otto von Guericke, of Magdeburg, carried out an experiment to show the pressure of the atmosphere, in which he used a cylinder and piston; he connected the piston, by a rope passing over pulleys, to a heavy weight; then, by means of a small air-pump worked by hand, the air was exhausted from the cylinder, whereupon the weight was lifted by the pressure of the atmosphere upon the upper side of the piston. In 1678–79 Huygens essayed to apply this idea to a motive power engine, and constructed an apparatus in which a vacuum was produced under a piston by the explosion of gunpowder. Papin in 1687 developed further the application of gunpowder, and then, in 1690, he proposed to produce the vacuum by the condensation of steam. To Papin then belongs the distinction of first giving to the world the principle of the atmospheric engine. Whether Newcomen became acquainted with this proposal of Papin, or whether he arrived at it independently, we do not know. The apparatus devised by Papin for carrying out the idea was of a very crude character, and, while perhaps suitable for laboratory purposes, was altogether unsuitable for use as an engine. Newcomen, on the other hand, embodied the idea in a practical form, and produced a successful engine. It is by no means easy for us today to realize the difficulties which Newcomen had to surmount. Mechanical engineering as we understand it had not come into existence. Smiths had attained to a high degree of skill in their craft, but workmen for other branches were untrained; while the range of materials available for construction was quite limited. Newcomen grasped what was possible to do under the existing conditions, and he devised a machine which it was possible to build with the materials, tools, and men at his disposal; a machine which, when built, worked successfully.

" . . . In 1765 James Watt invented the separate condenser. This improvement, according to which the steam, instead of being condensed in the engine cylinder, was condensed in a separate chamber, resulted in a very considerable saving of fuel; but engines of the Newcomen form continued to be built long after this, and at least one of them is in use to the present day.

"The labours of James Watt brought the steam engine to a high pitch of perfection, both as to economy in operation, and to mechanical construction, but it would seem that his admittedly great merit has been allowed, in the public estimation, to overshadow the pioneer work done by Thomas Newcomen."

The ENGINE for Raising Water (with a power made) by Fire.

H Beighton delin 1717

HENRY BEIGHTON's Drawing
of NEWCOMEN's Steam Engine (1717)

Reduced from 7⅛ × 7⅛ in.

THE engine of THOMAS NEWCOMEN, erected at Griff in Warwickshire, England, in 1712, was the first cylinder-and-piston steam engine of which there is any record, and this rare print, which is reproduced through the courtesy of The Science Museum, London, and The Newcomen Society, is the earliest known document of any kind dealing with its construction. The operation of the NEWCOMEN engine, which was used for pumping water from mines, was briefly as follows : Steam from the boiler B entered the cylinder C and lifted the piston. A jet of cold water then condensed the steam, leaving a partial vacuum in the cylinder, and the pressure of the air forced the piston down again. It was through studies of a model of this engine at Glasgow University that JAMES WATT was led to his important discoveries and to the improvements which made the steam engine an efficient machine suitable for driving factories as well as for pumping water from mines.

Solution. Here $f = mg = 14 \cdot 10^5 \cdot 980$ dynes; $v = 833$ cm $\cdot$ sec^{-1}; cos ϕ = sin $5° = 0.0872$. Therefore, by Eq. [44],

$$P = 1.0 \times 10^{11} \text{ erg} \cdot \text{sec}^{-1} = 10 \text{ kw.}$$

EXAMPLE. If, because of friction, the force required to keep the automobile in the preceding example moving 30 km $\cdot$ hr^{-1} on a straight, level road is 5 percent of its weight, what power must the engine develop in going up the grade? How many kilowatt-hours of work does the engine do in 5.0 min?   *Ans.* 16 kw, 1.3 kw $\cdot$ hr.

○

Energy

50. Kinetic Energy of Translation. It was stated in Sec. 46 that NEWTON's assertion made in the scholium to the third law of motion is equivalent to the statement that in all mechanical operations the work done by the acting forces is equal to the work done against the resisting forces. If all these resisting forces except inertia are absent in a given case, then the only effect of the forces acting on a particle is to change the speed of the particle. Now a particle that has acquired speed itself becomes possessed of the capacity for doing work, for it can now move itself against frictional resistance, compress a spring, raise itself against gravity, or by impact overcome the inertia of some other body. This capacity for doing work that a body possesses in virtue of the speed which has been communicated to it is called *kinetic energy.*

As the simplest case, let a particle of mass m acquire a speed v under the action of a constant force of magnitude f which is applied for a time t and in that time moves the particle a distance s in the direction of the force. Next let the particle be brought to rest by the action of an oppositely directed constant force f' which requires a time t' and a distance s' in order to destroy the speed v. The work W done in setting the particle in motion is fs, and the kinetic energy E, or work which the particle is capable of doing because of its speed, is by definition $f's'$. Now f is measured by the rate at which it imparts momentum and f' by the rate at which it destroys momentum or, by Eq. [22], Chap. 2,

$$f = Kma \quad \text{and} \quad f' = Kma'.$$

Also, since the forces and hence the accelerations are assumed to be constant, the distances s and s' are, by Eq. [17], Chap. 1,

$$s = \frac{v^2}{2\,a} \quad \text{and} \quad s' = \frac{v'^2}{2\,a'}.$$

Hence
$$W = fs = Kma \frac{v^2}{2\,a} = K \cdot \frac{1}{2} mv^2,$$

and
$$E = f's' = Kma' \frac{v'^2}{2\,a'} = K \cdot \frac{1}{2} mv'^2.$$

But, by hypothesis, the speed v imparted by f and the speed v' destroyed by f' are the same; therefore

$$W = E = K \cdot \tfrac{1}{2} mv^2. \tag{45}$$

Thus, if the only resistance is that due to inertia, the kinetic energy of translation imparted to a body by the action of a force is equal to the work done upon it. This conclusion holds equally well when f and f' are variable, for it is only necessary to conceive of them as made up each of the same number of very small elements, each element being a constant force; then v will be the speed gained under the action of one of these constant elements of f and destroyed under the action of the corresponding element of f'.

According to Eq. [45], kinetic energy [1] E is proportional to $\tfrac{1}{2} mv^2$. This expression enables us to calculate the kinetic energy of a particle at any instant without having any knowledge of the forces which produced the motion or of the previous history of the particle, provided we know its mass and speed at the instant in question. Since the kinetic energy depends on the *square* of the magnitude of the velocity and hence not on direction, it is a scalar quantity. It is a quantity equivalent to work and hence is measured in the same units as work. If cgs units are used, the constant K in Eq. [45] obviously is unity and

$$E = \tfrac{1}{2} mv^2 \text{ dyne} \cdot \text{cm, or ergs.} \tag{46}$$

If m is in kilograms, and v in meters per second, E in Eq. [46] is in joules, not kilogram-meters. (Why?)

51. Comparison of Kinetic Energy and Momentum. For over half a century a controversy raged between the followers of RENÉ DESCARTES (1596–1650) and of GOTTFRIED WILHELM LEIBNIZ (1646–1716) as to whether the ability of a moving object to overcome opposing forces and to produce changes in other bodies was proportional

---

[1] The term *energy* was first used to denote mv^2 by THOMAS YOUNG in his course of lectures on natural philosophy given in 1801–1803 at the Royal Institution in London; the passage is reproduced in *A Source Book in Physics (1935), pp. 59–60. It was used in its present sense of $mv^2/2$ by LORD KELVIN [*Mathematical and Physical Papers* (Cambridge University Press, 1884), Vol. 2, p. 34, footnote; also, *Treatise on Natural Philosophy* (Cambridge University Press, 1912), p. 222]. Still earlier (1695) LEIBNIZ had called mv^2 the *vis viva*, or "living force," a term which was later applied by CORIOLIS to $mv^2/2$.

to the velocity or to the square of the velocity.[1] It turns out that both of these views are correct; every moving body has both momentum mv and kinetic energy $mv^2/2$, and which we should choose in a given case depends on the purpose in view. Thus, for the particle considered in Sec. 50, we may write

$$fs = \tfrac{1}{2}\,mv^2 = E. \qquad [47]$$

But by NEWTON'S second law we have also $f = ma = mv/t$, or

$$ft = mv = p. \qquad [48]$$

Eq. [47] tells us *how far* and Eq. [48] *how long* the particle will continue to move against a given force f before it can be brought to rest. Hence the time required for a bullet to be brought to rest by a wooden target depends on its momentum, whereas the distance it will penetrate the target depends on its kinetic energy.

The quantity ft in Eq. [48] is called the impulse [2] of the force. Thus change of momentum depends on the impulse of the force, whereas change of kinetic energy depends on the work done by the force.

The Cartesian point of view, adopted by NEWTON and followed in the present text, makes *force, mass,* and *momentum* the basic mechanical concepts. The fundamental equation in this system is Eq. [48], and force ($= p/t$) is measured by the time-rate of change of momentum or, in the calculus notation (Eq. [21], Chap. 2), by

$$f = \frac{dp}{dt}. \qquad [48a]$$

The Leibnizian view, followed by HUYGENS, J. V. PONCELET, and others, makes *work, mass,* and *energy* the basic concepts. The fundamental equation then is Eq. [47], and force ($= E/s$) is measured by the space-rate of change of energy or, in the notation of the calculus, by

$$f = \frac{dE}{ds}. \qquad [47a]$$

Many modern thinkers, arguing that kinetic energy has objective reality whereas force does not, regard the latter view as the sounder philosophically.

52. Potential Energy. Let us consider a case in which the resistance experienced by the working force is gravity alone, as when an object of mass m is pulled from the floor to the ceiling. The work

---

[1] See *A Source Book in Physics* (1935), pp. 50–60, for excerpts from some of the more important papers bearing on this controversy.

[2] This use of the term *impulse* was first proposed in 1847 by the Frenchman J. B. BÉLANGER; it was later adopted by J. CLERK MAXWELL in his *Matter and Motion* (London, 1876), Art. XLIX.

done on the body is the force of the pull times the distance to the ceiling. The pull is a variable force, being a little greater than the weight of the body when the motion is starting and a little less when it is stopping. The kinetic energy imparted to the body during the initial instants, when the speed is being acquired, is all given back during the final instants, when the speed is being lost. Taking the operation as a whole, no speed is imparted; hence, neglecting air friction, the only resistance is gravity. Then, by the work principle stated in Sec. **46**, the work which has been done upon the body is equal to the work against gravity, the latter being the weight mg of the body times the distance h to the ceiling, or mgh. But in this case, as in that of Sec. **50**, in which the resistance was inertia alone, the work may all be regained without the expenditure of any more work on the part of the agent. For, if the body be dropped from the height h, its speed just before hitting the floor is given by $v^2 = 2\,gh$, and thus its kinetic energy is $mv^2/2 = m \cdot 2\,gh/2 = mgh$. In other words, if the body be attached over a light, frictionless pulley to another body of mass m, it is capable when it descends of lifting the latter through the height h. This ability to do work which a body possesses in virtue of its position is called *potential energy*.[1] If work be done against molecular force alone, as when a spring is compressed, the work can be regained by releasing the spring, which, when compressed, is possessed of potential energy. Potential energy is, then, in general, *any capacity for doing work that is put into a system by a change in the position of its parts against the forces which hold them together.*

From our discussion of kinetic and potential energy it is evident that we may define *energy* as the capacity for doing work; or, better still, since we shall see later that a body or system may possess an enormous store of energy that is not available for work, *the energy of any system is that property which diminishes when the system does work on any other system by an amount equal to the work so done.*

53. Conservative Forces. We have seen that when the resisting force is gravity or molecular force alone, an amount of potential energy equal to the work done is stored up; likewise, when the resistance is inertia alone, kinetic energy equal to the amount of work done appears. But when the resistance is friction alone, the work done *cannot* be entirely regained; a moving body that is opposed by fric-

---

[1] This term was first used by the Scottish engineer W. J. M. RANKINE in a paper read before the Philosophical Society of Glasgow in 1853. See Rankine's *Scientific Papers* (Griffin, 1881), pp. 203, 229.

tion loses kinetic energy as its velocity decreases, but it does not at the same time gain an equivalent amount of potential energy.

When forces like that of gravity or of elasticity act, the total kinetic and potential energy of the bodies is conserved; that is, $E + V = $ constant, where V denotes the potential energy. Hence such forces are called *conservative forces*, and any system of bodies between which the forces are wholly conservative is called a *conservative system*. The fundamental characteristic of a conservative force is that the work which it does during a displacement of a body from one point to another depends only upon the initial and final positions, not upon the path followed; for example, when an object of mass m falls in vacuum from any point in one horizontal plane to any point in a lower parallel plane at a vertical distance h, the work done is mgh, regardless of the path. Friction, on the other hand, is a *nonconservative* force, for it causes a permanent decrease of the kinetic and potential energy of a system, regardless of the direction in which the motion takes place. The work done in moving a body against a nonconservative force depends on the path followed and not alone on the initial and final positions.

54. Heat as a Form of Energy. We have seen that GALILEO, HUYGENS, and NEWTON divined the principle of conservation of energy so far as isolated conservative systems were concerned. But it remained obscure as to what became of the energy expended against friction. It is true that the suggestion had been made, even before NEWTON'S time, that heat is simply a mode of motion. "The very essence of heat . . . is motion and nothing else," said FRANCIS BACON in his *Novum Organum* (1620), and a similar view was later held also by ROBERT BOYLE [1] (1627–1691), ROBERT HOOKE [2] (1635–1703), and others. Toward the end of the eighteenth century, however, the doctrine that heat is an indestructible fluid came into favor, and this led to the idea that the energy expended against friction is lost.

---

[1] BOYLE wrote: "If a somewhat large nail be driven by a hammer into a plank, or a piece of wood, it will receive divers strokes on the head before it grows hot; but when it is driven to the head, so that it can go no farther, a few strokes will suffice to give it a considerable heat; for whilst at every blow of the hammer, the nail enters farther and farther into the wood, the motion, that is produced, is chiefly progressive, and is of the whole nail tending one way; whereas, when that motion is stopped, then the impulse given by the stroke, being unable either to drive the nail further on, or destroy its entireness, must be spent in making a various vehement and intestine commotion of the parts themselves, and in such an one we formerly observed the nature of heat to consist" [*Of the Mechanical Origin of Heat and Cold* (1675), Sec. II, Exp. VI, pp. 59–62; *Works*, ed. by T. Birch (London, 1772), Vol. IV, pp. 249–250].

[2] "Observation VIII," *Micrographia* (1665), p. 45.

The first to investigate the nature of heat without prejudice in favor of any particular theory was the versatile COUNT RUMFORD (BENJAMIN THOMPSON), an American Tory, who, after having fled the United States during the Revolution, enlisted himself in the service of the Elector of Bavaria and was put in charge of an arsenal. From qualitative experiments on the production of heat in the boring of cannon, made in 1798, he was led to conclude that " anything which any insulated body, or system of bodies, can continue to furnish without limitation, cannot possibly be a material substance." [1] Heat must be motion, RUMFORD decided, and this view received complete quantitative confirmation when JAMES PRESCOTT JOULE, in a series of famous experiments extending from 1840 to 1878, demonstrated the equivalence of heat and work by showing that for every definite amount of work done against friction there always appears a definite quantity of heat.[2]

55. Joule's Equivalent. The experiments of JOULE consisted in transforming work into heat in as large a variety of ways as possible, by causing various substances to rub against one another, by percussion, by compression, by the generation of electric currents, the energy of which was finally dissipated in heat, and so forth. The experiments usually were so arranged that the heat generated was taken up by a given quantity of water, and it was observed that a given expenditure of mechanical energy always produced the same rise of temperature in the water.

The quantity of mechanical energy which, if entirely converted into heat, is capable of raising the temperature of a unit mass of water one degree is called *Joule's equivalent* or the *mechanical equivalent of heat*. This important experimental constant is designated by the symbol J. It was first calculated by J. R. MAYER in 1842 from assumptions which were at that time very uncertain. The first direct experimental determinations were those of JOULE, but the best modern value is that obtained by LABY and HERCUS[3] in 1927.

---

[1] For an account of some of these experiments see * *The Life and Works of Count Rumford* (Macmillan, 1876), Vol. II, pp. 471–493; also * *A Source Book in Physics* (1935), pp. 151–161.

[2] *Philosophical Magazine* **23**, 263, 347, 435 (1843), et seq.; *Philosophical Transactions* **140**, 61 (1850). JOULE'S experiments on the mechanical equivalent of heat covered a period of about forty years. Reference to the papers themselves is necessary if one is to gain a correct idea of the enormous experimental labor they represent. See *The Scientific Papers of James Prescott Joule* (London, 1884, 1887). Excerpts are given in * *A Source Book in Physics* (1935), pp. 203–211. See also *A. Wood, *Joule and the Study of Energy* (Bell, 1925).

[3] *Philosophical Transactions* **227**, 62 (1927).

FIGURE 1 shews the cannon used in the foregoing experiments in the state it was in when it came from the foundry.

" Fig. 2 shews the machinery used in the experiments No. 1 and No. 2. The cannon is seen fixed in the machine used for boring cannon. *w* is a strong iron bar (which, to save room in the drawing, is represented as broken off), which bar, being united with machinery (not expressed in the figure) that is carried round by horses, causes the cannon to turn round its axis.

" *m* is a strong iron bar, to the end of which the blunt borer is fixed ; which, by being forced against the bottom of the bore of the short hollow cylinder that remains connected by a small cylindrical neck to the end of the cannon, is used in generating Heat by friction.

" Fig. 3 shews, on an enlarged scale, the same hollow cylinder that is represented on a smaller scale in the foregoing figure. It is here seen connected with the wooden box (*g, h, i, k*) used in the experiments No. 3 and No. 4, when this hollow cylinder was immersed in water.

" *p*, which is marked by dotted lines, is the piston which closed the end of the bore of the cylinder.

" *n* is the blunt borer seen sidewise.

" *d, e*, is the small hole by which the thermometer was introduced that was used for ascertaining the Heat of the cylinder. To save room in the drawing, the cannon is represented broken off near its muzzle ; and the iron bar to which the blunt borer is fixed is represented broken off at *m*.

" Fig. 4 is a perspective view of the wooden box, a section of which is seen in the foregoing figure. (See *g, h, i, k*, Fig. 3.)

" Fig. 5 and 6 represent the blunt borer *n*, joined to the iron bar *m*, to which it was fastened.

" Fig. 7 and 8 represent the same borer, with its iron bar, together with the piston which, in the experiments No. 2 and No. 3, was used to close the mouth of the hollow cylinder."

Plate 12 and the foregoing description were taken from RUMFORD's paper, "An Inquiry Concerning the Source of the Heat which is Excited by Friction," read before the Royal Society on January 25, 1798, and published in the *Philosophical Transactions* 88, 80 (1798). It is reproduced in *The Life and Works of Count Rumford* (Macmillan, 1876), Vol. II, pp. 471–493. Excerpts will be found in W. F. Magie's *A Source Book in Physics* (1935), pp. 151–161, and A. Wood's *Joule and the Study of Energy* (Bell, 1925).

O

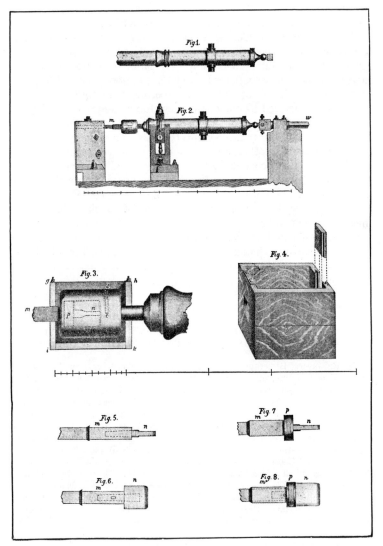

Apparatus used by RUMFORD
to Adapt his Cannon-Boring Operations
to the Study of the Source of Frictional Heat

These investigators used a continuous-flow calorimeter and obtained the value[1]
$$J = 4.1852 \pm 0.0006 \text{ joules}$$
per gram of water heated from 14.5° C to 15.5° C.

56. The Principle of Conservation of Energy. The clear and definite establishment of the principle of conservation of energy[2] is due to JOULE, J. R. MAYER, HERMANN VON HELMHOLTZ, RUDOLF CLAUSIUS, and WILLIAM THOMSON (later LORD KELVIN), about the middle of the nineteenth century[3]; and after the experiments of JOULE, it became generally recognized as one of the most fundamental and fruitful of all our physical laws. The principle asserts that *the sum total of the energy in any isolated system remains the same.* This energy can change from one form to another, and, as we shall see as we pursue our study of physics, it tends constantly to become less and less available; but it was not observed to change in amount in any case in which it was absolutely certain that no energy had either entered or left the system.

Thus every physical or chemical change of condition has a fixed mechanical equivalent; that is, it can be equated, under all circumstances, to one and the same amount of mechanical work. In other words, whenever a change takes place in the condition of a body be-

---

[1] R. T. Birge, *Reviews of Modern Physics* **1**, 30 (1929).

[2] For an account of the history of the principle of conservation of energy and an objective appraisal of the merits of the various claims put forward for MAYER, JOULE, CARNOT, SÉGUIN, and COLDING, see G. Sarton, *Isis* **13**, 18 (1929). An account of the interesting anticipations of the principle by the famous Swiss anatomist and physiologist ALBRECHT VON HALLER (1708–1777) is given by P. S. Epstein, *Thermodynamics* (Wiley, 1937).

[3] J. R. Mayer, "Bemerkungen über die Kräfte der unbelebten Natur" ("Remarks on the Forces of Inorganic Nature"), *Annalen der Chemie und Pharmacie* **42**, 233 (1842); a translation by G. C. Foster (1862) is reproduced in *A Source Book in Physics* (1935), pp. 197–203. MAYER published elaborations of his views in many other papers. See collection of J. J. Weyrauch (Stuttgart, 1893); translations of some of these will be found in the *Philosophical Magazine* (4) **28**, 25 (1864); **25**, 241, 387, 417, 493 (1863).

H. Helmholtz, "Ueber die Erhaltung der Kraft" ("On the Conservation of Energy") (Berlin, 1847). See *A Source Book in Physics* (1935), pp. 212–220.

R. Clausius, "Ueber die bewegende Kraft der Wärme" ("On the Motive Power of Heat"), Poggendorff's *Annalen der Physik und Chemie* **79**, 368, 500 (1850). Translations are given in *The Second Law of Thermodynamics*, ed. by W. F. Magie (Harper, 1899), pp. 65–107, and in *A Source Book in Physics* (1935), pp. 228–236. CLAUSIUS'S papers on the mechanical theory of heat are collected in his *Die Mechanische Wärmetheorie* (Wieweg, 1867–1879); there is a translation by W. R. Browne (Macmillan, 1879).

W. Thomson, "On the Dynamical Theory of Heat," *Philosophical Magazine* (4) **4** (1852); also in *Mathematical and Physical Papers* (Cambridge University Press, 1882–1911), Vol. 1, p. 174. It is reprinted in *The Second Law of Thermodynamics*, ed. by W. F. Magie (Harper, 1899), pp. 111–147. Excerpts are given in *A Source Book in Physics* (1935), pp. 237–247.

cause of the expenditure upon it of mechanical energy (kinetic or potential), the change is equivalent to the work done, in the sense that if the body can be brought back to its original condition, the whole of the energy expended may be regained in the form of either work or equivalent heat.

Applied to a mechanical problem, the principle asserts at once that the kinetic energy of a moving body is equal to the work done in setting it in motion. Applied to a chemical problem, it asserts, since the oxidation of 1 g of carbon generates enough heat to raise approximately 7900 g of water through 1° C, that, if it were possible to directly pull apart the united carbon and oxygen atoms, 7900 × 4.19 j of work would be required to secure 1 g of carbon from the carbon dioxide. Applied to an electrical problem, the principle asserts that if it requires 1000 m · kgwt of work per second to drive a dynamo, then the work which this dynamo does per second in the motors which it runs, plus the heat developed in all the machines and in all the connecting wires, must be exactly equal to 1000 m · kgwt.

It is an essential part of the method of physical science to search for the physical quantities that remain constant amid the variations of nature, for the laws that prevail, and for the concepts that can be retained as permanent. Nothing illustrates this better than the history of the principle of the conservation of energy. Like all the very fundamental laws of science, this principle is not capable of direct proof. It can be tested for any one kind of physical change by experiments like those of JOULE, but as a generalization it rests upon universal experience rather than upon any particular experiment. It describes nature in a way that is consistent with known facts, and has proved itself of basic importance in all the physical sciences and in engineering; yet it is to be accepted as a correct description only so long as no phenomena are discovered with which it is inconsistent.

o

Application of the Principle of Conservation of Energy and the Work Principle

The principle of conservation of energy may be regarded as comprehending the whole of abstract dynamics, because the conditions of equilibrium and of accelerated motion, in every case, may be derived from it. Together with the more restricted work principle (Sec. 46), it therefore furnishes a powerful method for the general solution of a large number of physical problems. Although NEWTON'S laws, as set forth in Chapter 2, are adequate to deal with any ordinary mechanical problem, the energy and work principles often have the

JOULE'S procedure in his earlier experiments was simply as follows: The temperature of the calorimeter containing the frictional apparatus was taken; the lead weights e, e were raised to a height measured by means of the scales k, k; the roller f was pinned to the axis of the paddle mounted inside the calorimeter; the weights were then released and the paddle revolved until the weights reached the floor. This process was repeated twenty times and the final temperature was taken.

In the final form of the apparatus the paddle was rapidly turned by hand-wheels d, e, and the calorimeter, instead of resting on a fixed stool, was suspended from the vertical shaft b which carried the paddle. The work done was measured by means of a dynamometer which balanced the torque acting on the suspended calorimeter (due to the rotation of the paddle) by a torque produced by the tension in cords wrapped around an accurately turned groove in the surface of the calorimeter. The cords passed over pulleys and supported weights k just sufficient to keep the calorimeter in equilibrium. If this torque is constant and of magnitude L, and if the paddle makes N revolutions per second, the work done per second is $2\pi LN$.

JOULE performed many experiments in which heat was produced by friction in water, mercury, and sperm oil, by friction of cast iron on cast iron, by transmission of electric currents through resistances, etc. Reference to the original papers themselves is necessary if one is to form a correct idea of the enormous experimental labor they represent and the great care with which the experiments were performed. See *Philosophical Magazine* (3) **23**, 263, 347, 435 (1843), et seq.; *Philosophical Transactions* **140**, 61 (1850); **169**, 365 (1878); also *Scientific Papers* (Taylor and Francis, 1884, 1887), pp. 149, 298, 632, etc.

o

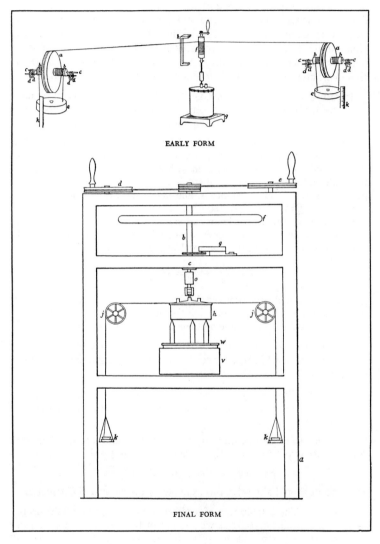

EARLY FORM

FINAL FORM

Joule's Apparatus
for Determining the Mechanical Equivalent
of Heat

advantage that they enable us to treat whole groups of problems by more or less routine forms and so to dispense with the minute discussion of them. This is an advantage, and a very important one, that every general principle has. Thus, for example, the work principle often relieves us of the necessity of troubling ourselves about the details of a machine. As MACH has pointed out, every principle of this character possesses therefore a distinctly *economical* value.

EXAMPLE. An open knife of mass m is dropped from a height h onto a wooden floor. The blade penetrates a distance s into the wood. Find the average resistance that the wood offers to the blade.

Solution. Let us first employ the detailed method of Chapter 2. The average resistance, being a force, is measured by the product of the mass stopped and the average acceleration with which it is stopped. Just before the knife touches the floor, its speed is $v = \sqrt{2\,gh}$. This speed is lost in penetrating the wood a distance s, and hence the average acceleration is $a = v^2/2\,s = gh/s$. The force required to give the knife this acceleration is $ma = mgh/s$. But if the knife had simply been placed in contact with the floor, without having any speed, it would have exerted a force mg on the wood because of its weight. The total average resistance F offered by the wood is therefore

$$F = \frac{mgh}{s} + mg = \frac{mg(h+s)}{s}.$$

This result could have been obtained at once simply by writing the equation that expresses the work principle — that the work done by the acting force must equal the work done by the resisting force; thus $mg(h+s) = Fs$.

In applying the work principle it is essential to keep in mind all four types of possible resistances: friction, molecular force, gravity, and inertia.

EXAMPLE. What average force must be applied to a kilogram weight to raise it 5 m above the earth and at the same time give it an upward speed of $10 \text{ m} \cdot \sec^{-1}$?

Solution. Here the resisting forces are gravity and inertia. The work done against gravity is mgh; that against inertia is $\frac{1}{2}\,mv^2$. Therefore

$$Fh = mgh + \tfrac{1}{2}\,mv^2, \quad \text{or} \quad F = 2 \text{ kgwt.}$$

The student should also solve this example by the method of Chapter 2.

EXAMPLE. The 1500-lb ram of a pile-driver is dropped on a 500-lb pile from a height of 10 ft above the top of the pile. The ram and pile adhere and immediately after impact have a common speed of $20 \text{ ft} \cdot \sec^{-1}$. The pile is driven a distance of 2.0 in. into the ground. By considering the potential and kinetic energies involved, find (*a*) the speed v of the ram just before impact and (*b*) the space average of the resistance F offered by the ground to the motion of the pile and ram.

Solution. *a.* The initial potential energy of the ram is $(1500 \cdot 32 \cdot 10)$ ft · poundals, and its kinetic energy at the moment before it strikes the pile is $\frac{1}{2}(1500 \cdot v^2)$ ft · poundals. By equating these two energies, one obtains

$$v = \sqrt{2 \cdot 32 \cdot 10} \text{ ft} \cdot \sec^{-1} = 25 \text{ ft} \cdot \sec^{-1}.$$

b. The kinetic energy of the system ram-pile just after impact is $\frac{1}{2}(1500 + 500)(20)^2$ ft · poundals, and the work done in stopping it is $f \cdot \frac{1}{6}$ ft · poundals. By equating these two quantities, one obtains finally

$$f = 75 \times 10^3 \text{ lbwt};$$

this is the force required to overcome the kinetic energy of the pile and driver. In order to find the total resistance F, the weight of the pile and driver must be added to f. Therefore

$$F = (75 \times 10^3) + 2000 = 77 \times 10^3 \text{ lbwt.}$$

Attention is called to the fact that the kinetic energy of the system ram-pile just after impact is less than the kinetic energy of the ram just before impact. Since there is always loss of mechanical energy on impact (see Chapter 5), one must never attempt to solve an impact problem by equating the kinetic energies *before* and *after* collision. Methods for solving such problems are given in Chapter 5.

o

Friction between Solids

The only case of friction which will be considered now is that of one solid body moving over another. We will come back in Chapters 10 and 13 to such complicated phenomena as fluid friction and the resistance offered by a fluid to a solid body passing through it.

57. Experimental Facts regarding Sliding Friction.[1] Suppose that the block in Fig. 36 is accelerated by a force of magnitude f_1, applied parallel to the surface of contact. If the magnitude of the frictional force opposing this motion be denoted by f_2, and the mass of the block by m, the acceleration is $(f_1 - f_2)/m$. Evidently, if the block is made to move with constant speed, then $f_1 = f_2$, or the applied force is exactly equal to the friction. Thus we have a

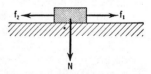

Fig. 36. Motion of a block along a rough horizontal plane

way of finding experimentally the frictional force between any two surfaces when a definite force presses them together and the speed has a definite value. Experiments of this kind were first made by

[1] Read *R. S. Ball, *Experimental Mechanics* (Macmillan, 1888), Lecture V, "The Force of Friction," pp. 65–84.

LEONARDO DA VINCI [1] (1452–1519), the celebrated artist and engineer, and, later, by the French physicist GUILLAUME AMONTONS (1663–1705). They were repeated and improved by CHARLES AUGUSTIN COULOMB [2] (1736–1806) and ARTHUR JULES MORIN [3] (1795–1880), a French artillery officer.

The frictional force f is found to be different for different substances and to vary with the condition of the rubbing surfaces. For a certain pair of surfaces moving in a definite direction it is nearly proportional to the normal force N pressing the two surfaces together, or

$$f = \mu N, \qquad [49]$$

where μ is a proportionality constant called the *coefficient of sliding* (or *kinetic*) *friction*. These are practically the only factors upon which the friction between solids depends. Within wide limits it is independent of the area of the surfaces in contact, and for moderate pressures it is nearly independent of the speed. It should be remarked that neither of these statements is true for fluid friction.

58. Sliding on an Inclined Plane. A body sliding down an inclined plane is urged downward by the component of its weight along the plane and is retarded by friction. If m is the mass of the body, and ϕ the inclination of the plane (Fig. 37), the force N with which the body is pressed against the plane is $mg \cos \phi$, and the force parallel to the plane due to gravity and friction is $mg \sin \phi - \mu mg \cos \phi$. Hence

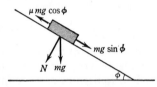

FIG. 37. Motion of a block down a rough inclined plane

$$mg \sin \phi - \mu mg \cos \phi = ma, \quad [50]$$

where a is the magnitude of the acceleration down the plane. This suggests a method for measuring μ, for if the inclination of the plane be so adjusted that the body, once in motion, moves down with constant speed, then $a = 0$ and $\mu = \tan \phi$.

If the body is initially at rest on the plane, the angle ϕ must be made greater than in the foregoing experiment before motion begins. This means that the coefficient μ is considerably greater for starting or static friction than for sliding friction. This *coefficient of static fric-*

---

[1] See *I. B. Hart, *The Mechanical Investigations of Leonardo da Vinci* (Open Court, 1925), pp. 140–141.

[2] COULOMB'S experiments on friction were published in his *Théorie des machines simples* (1779). A part of this monograph will be found in *A Source Book in Physics* (1935), pp. 103–105.

[3] See the chapter on friction in Morin's *Fundamental Ideas of Mechanics*, tr. by J. Bennett (New York, 1860).

tion is evidently equal to the tangent of the *angle of repose*, or the angle at which slipping begins. When the speed of a body is made very small, the kinetic friction increases and, at a sufficiently small speed, does not differ appreciably from the maximum static friction.

59. Rolling Friction. When one solid rolls on another without slipping, as, for example, a ball on a horizontal surface, there is no sliding friction, and yet a force is required to keep the rolling body moving with constant speed. This results from slight deformations which always occur in the surface and body, so that the latter, in effect, is always rolling uphill. It actually may be seen in the case of a heavy wheel rolling on a soft rubber mat. When the object is mounted on wheels, a part of the resistance is, of course, also due to friction in the bearings and to irregularities in the surface over which the wheels are passing. When a lubricant is used in the wheel bearings, the laws of friction between solids no longer hold, for the friction then depends in a complicated way on both the load and the speed.

60. Efficiency. In all mechanical devices the work that the machine accomplishes is inevitably less than the work put into it, simply because there is always more or less friction and hence a part of the applied work is frittered away into heat. The *efficiency* of a machine is defined as the ratio of the useful work obtained from the machine in any given time and the energy expended upon it in the same time. In a steam engine, for example, some of the energy transferred to the piston from the steam is wasted by friction in the engine itself. The rate at which energy is supplied to the piston is called the *indicated horsepower*, as contrasted with the actual horsepower available from the engine. Evidently the ratio of the actual horsepower and the indicated horsepower is equal to the efficiency of the engine.

○

EXPERIMENT IVA. EFFICIENCY OF A SYSTEM OF PULLEYS

Arrange a block and tackle as in Fig. 38. Use, preferably, three-pulley blocks of the commercial type. Regard f' as the load to be lifted and f as the force applied to the system to lift this load.

a. Hang on the movable block a weight of about 4.00 kgwt, including the weight of the weight-hanger. This, together with the weight of the movable pulley, which will usually be found stamped upon it, constitutes the load[1] f'.

---

[1] If small pulley blocks of the laboratory type are employed, make this total load f' about 400 gwt, and in *b* use, say, 500, 600, 800, 1000, 1300, 1600, 2000, 300, and 350 gwt.

Find by experiment the force f that will hoist the load f' with an *unaccelerated* motion. Because of unevenness in the bearing surfaces, it may be found impossible to adjust the weight f so that the system will move with constant speed and yet not stall at certain places; if this is found to be the case, keep the system moving with the minimum amount of assistance where it tends to stall and try to make this assistance the same throughout the experiment.

Calculate the efficiency of the pulley system. Since the efficiency is the ratio of the work done by the force f' and the work done by the force f, a determination of efficiency must involve a determination of the ratio of the distances through which the points of application of f' and f move. This can be obtained without a measurement, as a little consideration will show, from the number of strands of rope that support the weight f'.

Fig. 38. Laboratory model of a block and tackle

1. Why is it necessary to take care that the system moves without acceleration?

b. Determine the efficiencies of the pulley system for various other loads f'. Use approximately the following values for the weights, including the weight-hanger, to be hung on the movable pulley: 7.00, 10.0, 14.0, 19.0, 26.0, 34.0, 45.0, 0.500, 1.50 kgwt. Take special pains with the last two determinations. Why are they deferred until the end?

2. Plot on coordinate paper a curve having loads f' as abscissas and efficiencies as ordinates.

3. Interpret this efficiency-load curve; that is, state the physical laws or conclusions that can be drawn from it.

4. If the friction due to the bending of the cord in passing over the pulleys were negligible, so that all the friction were proportional to the load, how ought the efficiency of a system of pulleys to vary with the load? Did you find this to be true in the present experiment? Discuss.

○

EXPERIMENT IVB. EFFICIENCY AND MECHANICAL ADVANTAGE OF AN AUTOMOBILE TRANSMISSION

A *machine* may be defined as a contrivance for overcoming a resisting force at one point by the application of another force, usually at some other point. If f is the operating force required to overcome a resisting force or load f', then f'/f is called the *mechanical advantage* of the system. It is evidently a measure of force-multiplying ability. In some cases it is useful to distinguish between the *actual* mechanical advantage, as just defined, and the *ideal* mechanical advantage,

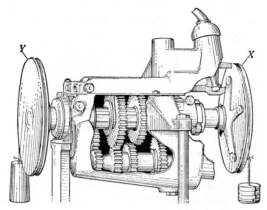

Fig. 39 A. Standard automobile transmission, with the housing partly cut away to expose the working parts. For purposes of study, two pulleys of equal diameter, X and Y, have been added to the engine and drive shafts respectively

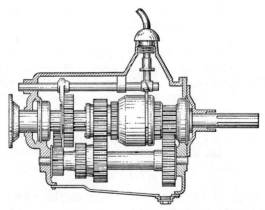

Fig. 39 B. Simplified diagram of a modern " freewheeling " gearshift. The gears are in high

which is the force-multiplying ability which the machine would have if there were no friction. It is a simple matter to prove that the ratio of the actual mechanical advantage and the ideal mechanical advantage is equal to the *efficiency* of the machine.

Since the relative speeds of two parts of a machine are often of interest, use is frequently made of the *speed-ratio* v'/v, where v' is the speed of the part of the machine to which the load f' is applied and v is the speed of the part to which the operating force is applied. Thus, in the case of an automobile transmission, the speed-ratio is the speed of the drive shaft divided by the speed of the engine shaft.

1. Show that the speed-ratio of a machine is inversely proportional to the ideal mechanical advantage, but not to the actual mechanical advantage.

2. Prove that the efficiency of a machine is equal to the product of the actual mechanical advantage and the speed-ratio.

a. Place the gearshift lever of the automobile transmission[1] (Fig. 39 *A*) in the "low" position and determine the speed-ratio of the machine by three different methods as follows:

(1) Count the number of teeth on each cogwheel and record the data on a rough sketch of the gear assembly. *Compute* the speed-ratios of the several pairs of gears from the respective numbers of teeth, and then compute from these ratios the "over-all" speed-ratio.

(2) Determine the "over-all" speed-ratio by counting the number of rotations of the engine-shaft pulley X that corresponds to one rotation of the drive-shaft pulley Y.

(3) Attach a weight f' to a cord wound on the drive-shaft pulley Y and find the weight f_1 that must be attached to a cord wound on the engine-shaft pulley X in order to cause the load f' to move *upward* with constant speed. Then find the weight f_2 which must be applied in order that the load f' shall move *downward* with constant speed. The mean f of these two weights f_1 and f_2 is the force needed to raise or lower the load f' without acceleration if there were no friction. Hence f'/f is the ideal mechanical advantage, and its reciprocal is the speed-ratio.

b. Compute the *actual* mechanical advantage f'/f_1. It will be best to make additional experimental determinations of f_1.

c. Use the values which you have determined for the speed-ratio and actual mechanical advantage to compute the efficiency of the transmission.

d. Repeat all the foregoing tests and computations for each of the other positions of the gearshift lever.

3. Summarize in a brief statement the information which you have obtained on the relative values of the mechanical advantages, speed-ratios, and efficiencies for the several positions of the gearshift lever, and discuss these results in terms of the actual performance of an automobile.

---

[1] This apparatus was described by O. H. Blackwood and E. Hutchisson in *The American Physics Teacher* **1**, 41 (1933).

EXPERIMENT IVc. EFFICIENCY OF A WATER MOTOR

The water motor is attached to the regular water supply of the room. Irregularities in the pressure of the latter are equalized by introducing, between the water outlet and the motor, an airtight tank

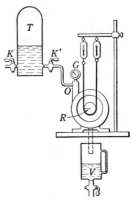

FIG. 40. Water motor with Prony brake

T (Fig. 40) of about 200-l capacity. Since the inlet and the outlet of the tank are at the bottom, it is the air in the upper part of the tank, compressed by the water-supply pressure, that is the immediate source of the pressure applied to the motor.[1]

In order to determine the energy *expended* upon the water motor in a given time it is necessary to measure both the pressure P with which the water is delivered to the motor and the volume of water ΔV that passes through it. The input of work is then the product $P \Delta V$ (Sec. 45).

The pressure at the orifice O is measured by means of a gauge [2] G. If the scale of the gauge is graduated in pounds weight per square inch, as is usually the case, it will be best to convert this reading to grams weight per square centimeter.

The volume of water is measured by deflecting the discharge water for a known time into the vessel V. At the beginning of the experiment there should be enough water in this vessel so that its initial height can be measured on the side tube, and at the end the water should be near the top of the side tube. The volume of water ΔV can then be obtained from a knowledge of the two heights of the water and the measured diameter of the vessel.[3]

---

[1] An overflow tank placed about two stories above the motor is a still better source of constant pressure.

[2] Some experimenters prefer to use a mercury manometer for measuring the pressure. If there were no water in either arm of the manometer, the pressure, measured in centimeters of mercury, would evidently be the difference between the mercury levels in the two arms, and this could be reduced to grams weight per square centimeter simply by multiplying by the density of mercury. However, since the short arm of the manometer fills with water, and since it is the pressure at the level of O which is sought, the pressure indicated by the mercury height must be diminished by that due to a water column of height equal to the difference between the level of O and the mercury level in the short arm.

[3] ΔV can also be determined simply by catching the discharge water in a jar or bucket which is weighed when empty and when nearly filled with the water.

In order to measure the *output* of work, or useful work accomplished by the motor in a given time, friction is applied to the pulley wheel of the motor by means of a Prony brake [1] (Fig. 40). This device consists of a leather belt hanging from two spring balances, [2] the friction of the belt against the pulley being adjusted by raising or lowering the clamps to which the spring balances are attached. If the radius of the pulley be R and the difference in the readings of the two spring balances be f, then f represents
the constant pull which the motor exerts at a distance R from the axis. When the pulley makes one rotation, every point on its circumference moves a distance $2 \pi R$ in a direction opposite to that of the resistance f, so that the work done by the motor during N rotations will be equal to $2 \pi RNf$.

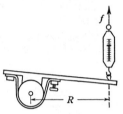

a. Starting with the valve K' closed, open valve K until a considerable pressure is produced in the tank T. Then *slowly* open K' until *nearly* the maximum available pressure is obtained. While one observer holds the gauge-reading constant by continually adjusting K, let another adjust the Prony brake so as to produce tension. Observe the force

Fig. 41. In this form of Prony brake the brake is prevented from turning by applying a known force f to the end of a lever arm of known length R. The work done by the motor is $2 \pi RNf$

f exerted by the brake. Then deflect the discharge water into the vessel V, noting accurately the time of flow and also determining with the speed-counter the number of rotations N in this time.

Measure the initial and final heights of the water in the vessel V, the diameter of this vessel, and the length R of the lever arm. Then compute the speed and the efficiency of the motor.

b. Vary the load on the motor by raising or lowering the Prony brake and determine the efficiency for this new load, the *pressure of the water being left the same as before.*

In this way make five or more different runs with loads that are evenly distributed between zero and the maximum which the machine is able to carry without stopping altogether; in other words, vary the speed of the motor between racing speed and the slowest possible speed. Be sure to adjust to the same constant water pressure during each run. Compute for each case the speed and the efficiency of the motor.

---

[1] Named for BARON G. C. F. PRONY, who was one of the first to make engineering applications of the mechanical discoveries of the second half of the eighteenth century.

[2] In the form of brake shown in Fig. 41 the force is measured by means of a single spring balance or a large-capacity beam balance placed at the end of the lever.

1. Plot a curve with speeds as abscissas and efficiencies as ordinates. Include on it the points for zero speed and for racing speed.

2. Find from your graph the speed for which the motor was most efficient at the given pressure.

3. What was the maximum efficiency of the motor?

4. Find the ratio of the speed at maximum efficiency to the speed at zero load.

○

OPTIONAL LABORATORY EXPERIMENT

Variation of the Friction in a Water Motor with the Speed. Put a constant load on a water motor and determine the force of friction for five different speeds. Plot a curve showing the variation of the force of friction with speed. Would you expect the friction in a water motor to vary with the speed?

○

QUESTION SUMMARY

1. Define *work*. In what units is it measured?

2. Define *erg*; *joule*; *centimeter-gram-weight*; *kilowatt-hour*; give the relations between them.

3. State the *work principle*.

4. Define *power*. In what units is it measured?

5. Define a *watt* and a *horsepower*, and give the relation between them.

6. Define *energy*. In what units is it measured?

7. Define, and tell how to measure, *kinetic energy* and *potential energy*.

8. What is meant by a *conservative force*? a *conservative system*?

9. Define *Joule's equivalent*.

10. State and illustrate the *principle of conservation of energy*.

11. Give the experimental facts with regard to sliding friction between solid bodies. Define the *coefficient of friction*.

12. What is meant by the *efficiency* of a mechanical system?

○

PROBLEMS

1. A force of 500 dynes is acting on a certain body. Find the work done by the force if the body moves 60 cm (*a*) in the direction of the force; (*b*) in a direction 60° away from that of the force; (*c*) in a direction at right angles to that of the force; (*d*) in a direction opposite to that of the force.

Ans. (*a*) 3.0×10^4 ergs; (*b*) 1.5×10^4 ergs; (*c*) 0; (*d*) -3.0×10^4 ergs.

2. Find the horsepower of a steam pump that lifts 10×10^4 l of water per hour from a well 30 m deep.   *Ans.* 11 hp.

3. The average flow over Niagara Falls is 270,000 ft^3 · sec^{-1}, and the height of the falls is 160 ft. (*a*) If all of this energy were utilized, what power could be developed? (*b*) If all of the energy from the waterfalls were converted into electrical energy and sold at 5 ct/kw · hr, how much would be realized in a day? Assume two-figure accuracy.
    Ans. (*a*) 4.9×10^6 hp, or 3.7×10^6 kw; (*b*) $4,400,000.

4. (*a*) Secure the necessary data for computing the horsepower which you develop when you run up a flight of stairs as fast as you can. (*b*) If you could work at this rate for 8 hr per day and were paid at the rate of 7 ct/hp · hr, what would be your daily wage?   *Ans.* 56 ct/hp.

5. Express the kilowatt-hour in (*a*) foot-pounds-weight; (*b*) horsepower-hours.   *Ans.* (*a*) 2.65×10^6 ft · lbwt; (*b*) 1.34 hp · hr.

6. How much work is done when 1.0 ft^3 of water is forced into a boiler against a pressure of 80 lbwt · in.$^{-2}$?   *Ans.* 1.2×10^4 ft · lbwt.

7. A bullet enters a target with a speed of 120 m · sec^{-1} and penetrates 10 cm. What speed should it have in order to penetrate 18 cm if the resistance offered by the target is the same as before?
    Ans. 1.6×10^2 m · sec^{-1}.

8. At what rate must energy be expended to raise 1000 kg of water per minute to a height of 22 m if the water is discharged from the top of the pipe with a speed of 4.0 m · sec^{-1}? Express the result both in kilowatts and in horsepower.   *Ans.* 3.7 kw, 5.0 hp.

9. Solve Probs. 2 (*b*) and 2 (*c*), Chap. 3, from a consideration of the potential and kinetic energies of the body at the top and bottom of the plane and of the arc.

10. A bullet of mass 12 g is fired with a muzzle speed of 30 km · min^{-1}. The gun has a smooth bore 750 mm long and 9 mm in internal diameter. (*a*) What is the energy of the bullet in joules? (*b*) in watt-seconds? (*c*) Neglecting friction, find the average pressure, expressed in grams of weight per square centimeter, inside the barrel during firing.
    Ans. (*a*) 1.5×10^3 j; (*b*) 1.5×10^3 watt · sec; (*c*) 3.2×10^5 gwt · cm^{-2}.

11. (*a*) Compare the momentums and kinetic energies of a 25-g bullet moving with a speed of 500 m · sec^{-1} and of a freight train of mass 10^6 kg moving in the same direction with a speed of 1 cm · sec^{-1}. (*b*) A constant retarding force of 10 megadynes is applied to each of them. Compare the times elapsing and the distances traversed before they are brought to rest.
    Ans. (*a*) $p_b = 1.3 \times 10^6$ g · cm · sec^{-1}, $p_t = 10^9$ g · cm · sec^{-1},
    $E_b = 3.1 \times 10^{10}$ ergs, $E_t = 5 \times 10^8$ ergs; (*b*) $t_b = 0.13$ sec,
    $t_t = 10^2$ sec, $s_b = 31$ m, $s_t = 0.5$ m.

12. Write the dimensional formulas for (*a*) work; (*b*) power; (*c*) energy; (*d*) impulse; (*e*) coefficient of friction; (*f*) Joule's equivalent.

13. (*a*) When you pay the electric-light bill are you paying for power, for energy, or for both power and energy? (*b*) Does the value of a given source of water power, say a mountain lake, depend upon the energy available, upon the power which it will develop, or upon both of these factors? (*c*) Is the term *power* used in its correct physical sense in the phrase *electric-power line*?

14. A 20-g bullet is fired from a rifle with a speed of 250 m · sec^{-1}. At a height of 2680 m it is moving at an angle of 45° with the horizontal. Find its speed at this point. *Ans.* 1.0 × 10^4 cm · sec^{-1}.

15. A train of mass 2.0 × 10^5 kg has a speed of 65 km · hr^{-1}. (*a*) What is its kinetic energy in ergs? (*b*) in joules? (*c*) in kilowatt-hours? (*d*) in meter-kilogram-weights? (*e*) If all of this energy were turned into heat, how many kilograms of water would it raise from 0° C to 100° C?
Ans. (*a*) 3.3 × 10^{14} ergs; (*b*) 3.3 × 10^7 j; (*c*) 9.2 kw · hr;
(*d*) 3.3 × 10^6 m · kgwt; (*e*) 78 kg.

16. It is found that if the clutch is disengaged from a certain automobile of weight 2200 lbwt when it is moving with a speed of 30 mi · hr^{-1} on a level pavement, the car coasts 0.50 mi before stopping. (*a*) Assuming that the frictional forces are independent of speed, compute the friction. (*b*) At what rate in horsepower must work be done to keep this car moving on a level road with a speed of 30 mi · hr^{-1}? (*c*) Considering the nature of the frictional forces involved, is it likely that these forces are independent of speed? *Ans.* (*a*) 25 lbwt; (*b*) 2 hp.

17. The top of an incline is 1 m higher than the bottom. If a body of mass 100 g sliding down this incline acquires a speed of 180 m · min^{-1}, how much work has been done against friction? *Ans.* 0.5 j.

18. A 75-kg block is drawn steadily up a 25° incline by means of a rope. The coefficient of sliding friction is known to be 0.20. (*a*) What is the force in the rope if the rope is held parallel to the incline? (*b*) if it is held horizontal? (*c*) What would the answers to these questions be if the coefficient of friction were zero? *Ans.* (*a*) 45 kgwt; (*b*) 55 kgwt; (*c*) 32 kgwt, 35 kgwt.

19. A block of weight 60 kgwt is held against a vertical wall by a horizontal force of 15 kgwt. The coefficient of sliding friction is 0.15. (*a*) What vertical force is needed to pull the block up with constant speed? (*b*) to lower it with constant speed? (*c*) If the least horizontal force that will keep the block from falling when at rest is 270 kgwt, what is the coefficient of static friction? *Ans.* (*a*) 62 kgwt; (*b*) 58 kgwt; (*c*) 0.22.

20. A particle of weight W is to be dragged along a horizontal plane by applying to it a force of magnitude F. If the coefficient of sliding friction is μ, what angle must the direction of the force make with the horizontal in order that F may be a minimum? *Ans.* tan^{-1} μ.

CHAPTER FIVE

THE LAWS OF IMPACT

Tʜᴜs ɪꜰ a sphaerical body A with two parts of velocity is triple of a sphaerical body B which follows in the same right line with ten parts of velocity; the motion of A will be to that of B as 6 to 10. Suppose then their motions to be of 6 parts and of 10 parts, and the sum will be 16 parts. Therefore upon the meeting of the bodies, if A acquire 3, 4, or 5 parts of motion, B will lose as many; and therefore after reflexion A will proceed with 9, 10, or 11 parts, and B with 7, 6, or 5 parts; the sum remaining always of 16 parts as before. If the body A acquire 9, 10, 11 or 12 parts of motion, and therefore after meeting proceed with 15, 16, 17 or 18 parts; the body B, losing so many parts as A has got, will either proceed with one part, having lost 9; or stop and remain at rest, as having lost its whole progressive motion of 10 parts; or it will go back with one part, having not only lost its whole motion, but (if I may so say) one part more; or it will go back with 2 parts, because a progressive motion of 12 parts is took off. And so the Sums of the conspiring motions 15 + 1, or 16 + 0, and the Differences of the contrary motions 17 − 1 and 18 − 2 will always be equal to 16 parts, as they were before the meeting and reflexion of the bodies.

Explanation of Corollary III to Nᴇᴡᴛᴏɴ's laws of motion
as translated from the *Principia* in 1729 by Aɴᴅʀᴇᴡ Mᴏᴛᴛᴇ

o

61. Conservation of Momentum. It was seen in Chapter 2 that the essential idea of the Gᴀʟɪʟᴇᴏ-Nᴇᴡᴛᴏɴ theory of motion is that all changes in the velocities of particles are to be regarded as the results of interaction between pairs of particles, and that no matter how such an interaction takes place, the result of it is described by the equation (Eq. [26], Chap. 2)

$$m_1 \, \Delta v_1 = - \, m_2 \, \Delta v_2. \tag{51}$$

Now $m \, \Delta v$ is *change of linear momentum*, and hence this equation is equivalent to the statement that, so long as one particle is influenced only by the other, the vector sum of their momentums does not change, however much the momentum of either one changes. In other words, the *linear momentum is conserved in all interactions*. This important principle is not stated explicitly in Nᴇᴡ-ᴛᴏɴ's laws of motion, but it follows immediately from them, and indeed was derived by Nᴇᴡᴛᴏɴ as one of several rules for expediting the treatment of the more frequently occurring problems of mechanics.

62. Conservation of Momentum in Impact. Let us consider this principle of conservation of linear momentum in connection with one particular kind of interaction between bodies, namely, that of

An Artist's Way
of Stating NEWTON's Laws of Motion

THIS illustration, taken from J. A. Paris's *Philosophy in Sport made Science in Earnest*
(London, 1827), reveals George Cruikshank (1792–1878), the famous English
artist, caricaturist, and illustrator, in the unusual role of a scientific illustrator.

An Illustration from MARCUS MARCI'S

DE PROPORTIONE MOTUS (1639)

Reproduced through the courtesy of The British Museum, London

JOANNES MARCUS MARCI (1595–1667), professor at the University of Prague, was acquainted with many of the laws of impact later discovered by HUYGENS, WREN, and WALLIS. Thus he knew that when an elastic body strikes another elastic body of the same size and material which is at rest, it loses its own motion and imparts an equal velocity to the other body.

impact. A systematic study of the various laws of impact[1] was first made in 1668 by JOHN WALLIS, CHRISTOPHER WREN, and HUYGENS, at the invitation of the Royal Society of London.[2] The solution for the general case of impact was given by NEWTON in the *Principia*.[3] It was undoubtedly this experimental and theoretical work on colliding bodies that made it possible for NEWTON to obtain his clearest and most definite conceptions of mass and momentum.

Suppose, to begin with, that the two colliding bodies are spheres which are small enough to be considered as particles and that their masses are m_1 and m_2. Let their respective velocities before impact be u_1 and u_2 and after impact, v_1 and v_2. Then, by Eq. [51],

$$m_1(v_1 - u_1) = - m_2(v_2 - u_2);$$

or
$$m_1u_1 + m_2u_2 = m_1v_1 + m_2v_2. \qquad [52]$$

The left-hand member of this equation is the vector sum of the two momentums before impact, and the right-hand member is the vector sum of the two momentums after impact; the equation then asserts that these two sums are equal. To visualize this, one may consider the impact of two billiard balls. If their total linear momentum before impact is represented by the vector $\overline{AO}$, Fig. 42, then their total linear momentum after impact is represented by the vector $\overline{OB}$, which is equal in both magnitude and direction to $\overline{AO}$.

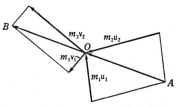

FIG. 42. The vector sum of the linear momentums does not change on impact

In making this application of the principle of conservation of linear momentum to the billiard balls, or, for that matter, to any bodies that are too large to be treated as particles, it obviously is necessary to specify what is meant by the linear momentum of an extended body. It will be shown in the next chapter that there can be associated with any body a certain point which is called the center of mass of the body, and that the *linear* momentum of the

---

[1] For the historical development of the subject, see *E. Mach, *The Science of Mechanics* (Open Court, 1893), pp. 305–330, and *A. Wolf, *A History of Science, Technology, and Philosophy in the 16th & 17th Centuries* (Allen & Unwin, 1935), pp. 231–235.

[2] Translations of these papers, together with a brief historical statement, appear in the *Abridged Philosophical Transactions* (London, 1749), Vol. I, Chap. V, pp. 457–462. Also see Bibliography.

[3] Scholium to "Axioms, or Laws of Motion"; see Cajori's revision (University of California Press, 1934), p. 22.

body is to be regarded as the product of the velocity of this point and the mass of the body. In the case of a *homogeneous* spherical body, for example, the center of mass turns out to be at the center of the sphere. Thus the linear momentum of a billiard ball is mv, where m is the mass of the whole ball and v is the velocity of the center of the ball.

63. Direct Impact. The simplest case of impact between actual bodies is that of two homogeneous spheres which are moving before impact along the line joining their centers (Fig. 43). The momentum of the system is unaltered by the impact, or

$$m_1u_1 + m_2u_2 = m_1v_1 + m_2v_2, \qquad [53]$$

it here being unnecessary to distinguish between speed and velocity, since one direction along the common line of motion can be taken as positive and the other as negative.

If the two impinging bodies have surfaces made of clay or similar soft materials, they are

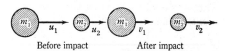

Before impact   After impact

FIG. 43. The typical case of direct impact

changed in shape by the impact, but show little or no tendency to recover their form and to thrust each other apart again; they adhere together after impact and move forward as a single body. Such a collision is said to be *inelastic*. For this special case of inelastic impact, $v_1 = v_2 = v$ and Eq. [53] reduces to

$$m_1u_1 + m_2u_2 = (m_1 + m_2)v, \qquad [53a]$$

from which the speed after impact may be calculated. This is the problem that was treated by WALLIS, and his results are all summarized in Eq. [53a].

When, on the other hand, the bodies rebound upon collision, as is usually the case, the collision is said to be *elastic*. For a complete solution of the case of *elastic* impact, Eq. [53] will not suffice, since it contains two unknown quantities v_1 and v_2. By assuming conservation of *energy* as well as conservation of *momentum*, a second equation connecting v_1 and v_2 can be obtained, namely,

$$\tfrac{1}{2}m_1u_1^2 + \tfrac{1}{2}m_2u_2^2 = \tfrac{1}{2}m_1v_1^2 + \tfrac{1}{2}m_2v_2^2,$$

and this, together with Eq. [53], suffices for the solution of those very special problems in which the energy is conserved. This was the case treated by WREN and HUYGENS. In actual impacts, however, kinetic energy almost always is lost, as we shall soon see, and without a knowledge of this loss the principle of conservation of energy is useless. In order to solve the *general* case of *actual* bodies in which the energy is not conserved, NEWTON had first to determine a new constant for colliding bodies, called the coefficient of restitution.

64. Impulse and the Coefficient of Restitution. The average force acting on either one of two impinging bodies is equal to the total change in linear momentum of the body in question divided by the time of duration of the impact. Since this time cannot in general be determined, it is convenient to confine attention to the product of the average force and the time of duration of the impact. As we already know (Sec. **51**), this quantity is called the *impulse* of the force. It is equal to the total change in momentum which the body experiences by virtue of the impact, and if its magnitude be denoted by the symbol R, then, by Eq. [53],

$$R = m_1(v_1 - u_1) = - m_2(v_2 - u_2). \qquad [54]$$

In the case of an elastic impact it is convenient to divide the impulse R into two parts, R' and R'', of which R' represents the impulse during the compression, and R'' the impulse from the instant of greatest compression to the instant of separation. Obviously, in the case of inelastic impacts, $R'' = 0$. If the colliding bodies are perfectly *resilient*, and thus recover completely their original size and shape after collision, it might be expected that R'' would be equal to R'. In point of fact this is never the case, for there are losses due to internal friction even with bodies that exhibit perfect resilience when subjected to static tests. There is, however, a definite relation between R'' and R', as was discovered experimentally by NEWTON. He found that for any two given bodies making direct impact, the ratio R''/R' is a constant, so long as the impact is not so violent as to produce permanent deformation. The names *coefficient of restitution* and *coefficient of resilience* have since been given to this ratio. Denoting it by the symbol e, we have, by definition,

$$e = \frac{R''}{R'}. \qquad [55]$$

The coefficient of restitution is a *kinetical* [1] measure of the elasticity of bodies and is used only in dealing with the phenomenon of collision. It is found by experiment that e is always less than unity.

At the instant of greatest compression, two impinging bodies evidently are moving with the same velocity. Consider again the

---

[1] It is often convenient to think of the science of dynamics as being divided into two parts, *kinetics* and *statics*, the former dealing with bodies that have accelerations and the latter with bodies that have no accelerations and hence are in equilibrium.

simple case of the direct impact of two spheres, and let their common speed at the moment of greatest compression be V; then, from the definitions of R' and R'',

$$R' = m_1(V - u_1) = - m_2(V - u_2),$$
$$R'' = m_1(v_1 - V) = - m_2(v_2 - V).$$

Division of the second of these equations by the first yields

$$\frac{R''}{R'} = e = \frac{v_1 - V}{V - u_1} = \frac{v_2 - V}{V - u_2}.$$

This gives two equations,

$$v_1 - V = eV - eu_1,$$
$$v_2 - V = eV - eu_2.$$

Elimination of V by subtraction of the second equation from the first gives

$$v_1 - v_2 = - e(u_1 - u_2). \qquad [56]$$

In words, *the relative speed after direct impact is equal to e times the relative speed before impact, and is in the opposite direction.* This is known as NEWTON'S empirical law of restitution; obviously it expresses the same property of matter as the equation $R'' = eR'$.

65. General Solution of the Problem of Direct Impact. We have seen that the principle of conservation of momentum, Eq. [53], alone cannot give a complete solution of the general problem of the direct impact of two spheres. A second relation between v_1 and v_2 is given by Eq. [56], however, and *any problem of direct impact may be solved by writing Eq. [53] and Eq. [56], and solving them simultaneously.*

Thus, by combining Eqs. [53] and [56], there result the following expressions for the final speeds in the line joining the centers of the spheres:

$$v_1 = \frac{m_1 u_1 + m_2 u_2 - e m_2(u_1 - u_2)}{m_1 + m_2},$$
$$v_2 = \frac{m_1 u_1 + m_2 u_2 + e m_1(u_1 - u_2)}{m_1 + m_2}.$$

For the sake of simplicity we have chosen the case where all the speeds were in the positive direction (Fig. 43). In applying the equations to other cases, care must be taken to give proper signs to the speeds.

EXAMPLE. (*a*) Show that when two spheres of equal masses, and for which $e = 1$, make direct impact, they simply exchange velocities, so that the result is the same as though one had passed through the other without in any way influencing it. (*b*) Explain why the only effect of a direct impact of one marble upon a row of equal marbles is to drive off the end marble. (*c*) What will happen if two marbles make impact

An Illustration from CHRISTIAAN HUYGENS's

DE MOTU CORPORUM EX PERCUSSIONE (1703)

ALTHOUGH the correct laws of impact for elastic bodies apparently were discovered by HUYGENS before 1656, his novel and original proofs were not published in detail until 1703, eight years after his death. Starting from the assumption that two equal elastic masses that collide with equal and opposite speeds v rebound after impact with exactly the same speeds, he imagined such an impact to take place in a boat moving with the speed v. For the man in the boat the speeds of the two colliding masses were still equal, but for a spectator on shore the speeds before impact were $2v$ and 0 respectively and after impact 0 and $2v$. Then by letting the boat have any speed whatever, he showed that equal elastic masses simply exchange speeds on impact. In a similar manner he deduced the laws of impact for unequal elastic masses.

See E. Mach's *The Science of Mechanics* (Open Court, 1893), pp. 314–326; or, better still, read HUYGENS's paper, which is reprinted in the original Latin and in French translation in *Œuvres complètes de Christiaan Huygens* (Société Hollandaise des Sciences, 1929), Vol. XVI, and in German translation in Ostwald's *Klassiker der Exakten Wissenschaften*, No. 138 (Engelmann, 1903).

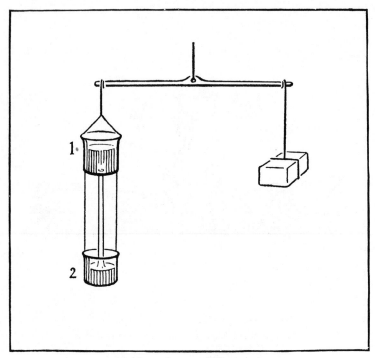

An Experiment of GALILEO's
on the "Force of Percussion"

GALILEO made a very ingenious attempt to measure the force exerted by the impact of a jet of water. He hung two buckets, one above the other, from one arm of a beam balance and balanced them by means of a weight hung from the other arm. The upper bucket (1) was filled with water, the lower one (2) was empty. A plugged orifice in the upper vessel was opened and the water flowed from the upper into the lower bucket. GALILEO expected the force of the impact of the water on the lower bucket to cause a deflection of the balance and planned to measure this force by means of a counterweight. Much to his surprise, however, except for a slight *rise* of the buckets at the *instant* the plug in the upper one was removed, there was no effect. GALILEO apparently was unable to clear up the matter satisfactorily in his own mind. See if you can do so; then consult E. Mach's *The Science of Mechanics* (Open Court, 1893), pp. 308–313, or A. Wolf's *A History of Science, Technology, and Philosophy in the 16th & 17th Centuries* (Allen & Unwin, 1935), pp. 47–48.

with the row of marbles? (*d*) What will happen if a larger or smaller marble makes impact with the row? [These considerations have numerous applications in the kinetic theory (Chap. 10) and in wave motion (Chap. 15).]

66. Oblique Impact. When the centers of two impinging spherical bodies are not moving in the same line before impact, the impact is described as *oblique*. In this case the velocity of each body before impact may be resolved into two components, one along the line through the centers at the moment of impact, the other in a direction perpendicular to that line. If the spheres are perfectly smooth, only the first component will be affected by the impact, since the whole force in this case is in the direction of the line of centers. The change in this component may be calculated as in the case of direct impact. If the spheres are not frictionless, there will be a tangential action between the bodies giving rise to a motion of rotation and the problem becomes too complicated for present consideration.

EXAMPLE. A smooth ball of mass 100 g is rolling on a table with a velocity of 100 cm · sec^{-1} east. It strikes a second smooth ball of mass 200 g, which is moving with a velocity of 100 cm · sec^{-1} northeast. If the coefficient of restitution is 0.750, what are the velocities of the two balls after impact?

Solution. Assume that the line of centers of the two balls is in the east-west direction during impact, and let this be the X-axis. Let u_x, u_y, u'_x, u'_y represent the x- and y-components of the velocities before impact and v_x, v_y, v'_x, v'_y those after impact, the primes referring to the 200-g ball. Then $u_x = 100$ cm · sec^{-1}, $u_y = 0$, $u'_x = 70.7$ cm · sec^{-1}, and $u'_y = 70.7$ cm · sec^{-1}. The x-components of the velocities after impact can be computed by the method of Sec. **65**; thus, by Eq. [53],

$$100 \cdot 100 + 200 \cdot 70.7 = 100\, v_x + 200\, v'_x,$$

and by Eq. [56],

$$v_x - v'_x = -0.750(100 - 70.7).$$

By solving these two equations simultaneously, we obtain $v_x = 65.8$ cm · sec^{-1} and $v'_x = 87.8$ cm · sec^{-1}. Since the balls are perfectly smooth, the y-components of the velocities will be unchanged by the impact; hence $v_y = 0$ and $v'_y = 70.7$ cm · sec^{-1}. The velocity of the first ball after impact must therefore be 65.8 cm · sec^{-1} east and that of the second ball $\sqrt{(87.8)^2 + (70.7)^2} = 113$ cm · sec^{-1} at an angle 38° 50′ north of east ($= \tan^{-1} 70.7/87.8$).

67. Loss of Kinetic Energy in Direct Impact. The fact that there is never any loss of momentum in an impact by no means implies that the kinetic energy is also conserved. The total kinetic energy is always less after impact than before, unless e is unity.

EXAMPLE. A putty ball of mass 50 g strikes a billiard ball of mass 500 g with a speed of 40 cm · sec⁻¹ along their line of centers. Calculate the common speed v after impact and the kinetic energies before and after impact.

Solution. The principle of conservation of momentum gives for this case $50 \cdot 40 = 550\,v$, or $v = 3.6\ \text{cm} \cdot \text{sec}^{-1}$. The total kinetic energy before impact is $\frac{1}{2} \cdot 50 \cdot (40)^2$, or 4.0×10^4 ergs, and that after impact is $\frac{1}{2} \cdot 550 \cdot (3.6)^2$, or 3.6×10^3 ergs. The loss in kinetic energy is therefore 3.6×10^4 ergs. Thus, even though the momentum is conserved, 90 percent of the kinetic energy is lost in this particular case of impact. (What has become of it?)

It is of interest to derive the expression for the loss ΔE of kinetic energy for the general case of direct impact. For this case, evidently,

$$\Delta E = (\tfrac{1}{2}\,m_1 u_1{}^2 + \tfrac{1}{2}\,m_2 u_2{}^2) - (\tfrac{1}{2}\,m_1 v_1{}^2 + \tfrac{1}{2}\,m_2 v_2{}^2), \qquad [57]$$

or $\qquad \Delta E = \tfrac{1}{2}\,m_1(u_1 - v_1)(u_1 + v_1) + \tfrac{1}{2}\,m_2(u_2 - v_2)(u_2 + v_2).$

In view of Eq. [54] this may be written

$$\Delta E = -\frac{R}{2}(u_1 + v_1) + \frac{R}{2}(u_2 + v_2) = -\frac{R}{2}\left[(u_1 - u_2) + (v_1 - v_2)\right]. \quad [57a]$$

It is desirable, however, to express ΔE in terms of the coefficient of restitution, the masses of the two bodies, and their initial speeds. To do this, first combine Eqs. [57a] and [56], thus obtaining

$$\Delta E = -\frac{R}{2}(u_1 - u_2)(1 - e). \qquad [57b]$$

Next solve Eq. [54] for v_1 and v_2 and substitute these values in Eq. [56]; this gives

$$R = -\frac{m_1 m_2}{m_1 + m_2}\,(1 + e)(u_1 - u_2). \qquad [58]$$

Substitution of this value for R in Eq. [57b] gives finally

$$\Delta E = \frac{1}{2}(1 - e^2)(u_1 - u_2)^2\,\frac{m_1 m_2}{m_1 + m_2}. \qquad [59]$$

It appears from this equation that the sole condition for no loss of kinetic energy in an impact is that $e = 1$; this is the case of so-called *perfectly elastic impact*, for which the impulse of restitution is equal to the impulse of compression. The fact that e is always somewhat less than unity then means that in all impacts there is some transformation of mechanical energy into energy of heat and sound. (Why cannot the value of e exceed unity? What is the meaning of zero value for e?)

EXAMPLE. By dividing Eq. [59] by the expression for the initial kinetic energy, obtain the expression for the fractional loss l of kinetic energy in impact; then show that for the special case where m_2 is initially at rest

$$l = (1 - e^2) \frac{m_2}{m_1 + m_2}.$$ [60]

This equation shows that the fraction of the initial kinetic energy lost in a direct impact is independent of the speed of the striking body, provided the struck body is initially at rest. The fraction is large or small according as m_1 is large or small in comparison with m_2. In driving a nail, much of the energy is spent in deforming the nail; since this diminishes the supply available for driving the nail, it is clearly advantageous to use a hammer of large mass. If, on the other hand, a hammer is to be used to shape an object, as in forging, the work of deformation should be made large by repeated blows of a hammer of small mass compared to the anvil.

EXAMPLE. Suppose that a small steel ball is dropped vertically upon a horizontal slab of stone. Show that in this case

$$e = -\frac{v}{u} \quad \text{and} \quad l = 1 - e^2.$$ [61]

68. Effect of a Series of Rapid Impulses. The effect of a series of rapid impulsive forces is the same as that of a constant force, and this force, by NEWTON'S second law, is measured by the change of momentum per unit time.

EXAMPLE. A machine gun of mass 15 kg fires 30-g bullets at the rate of 5.0 per second with a speed of 500 m · sec^{-1}. What force must be applied to the gun to hold it in place?

Solution. The bullets are discharged at intervals of 0.2 sec. During the short time of one discharge the forces between the gun and bullet are equal in magnitude and opposite in direction, and, acting for the same time, produce momentums that are also equal in magnitude and opposite in direction. At each discharge, therefore, the magnitude of the momentum acquired by the gun is $30 \cdot 50,000$ g · cm · sec^{-1} and the speed imparted to it is $30 \cdot 50,000/15,000$, or 100 cm · sec^{-1}. Suppose now that a constant force of f dynes be applied to the gun in a forward direction so as to oppose this motion and that this force starts to act 0.1 sec before the first discharge. In 0.1 sec the gun will gain a forward speed of $v = 0.1 f/15,000$ cm · sec^{-1}. If f be chosen so that this speed is just one half the backward speed imparted to the gun by each discharge, then the first discharge will send the gun backward with a speed of 50 cm · sec^{-1}. In order that this may be the case, $0.1 f/15,000$ must equal 50, or $f = 7.5 \times 10^6$ dynes. Since this applied force is constant, it will continue to act on the gun during the 0.2 sec before the next discharge. In 0.1 sec it will have overcome the backward speed of 50 cm · sec^{-1} which the first discharge imparted to the gun, and in another 0.1 sec (or by the time of the second discharge) will have imparted a forward speed of the same amount, which in turn will be reversed by the next discharge, and the whole process will be repeated. If, then, a constant force of 7.5×10^6 dynes is applied to the gun, the discharge of 30-g bullets at the rate of 5 per second will merely cause the gun to jerk back and forth over a distance of 2.5 cm with a speed that varies from 0 to 50 cm · sec^{-1}. Since, on the average, the gun stays in place, the series of rapid discharges is exactly balanced by the force of 7.5×10^6 dynes, which is therefore the force required to hold the gun in place.

In the foregoing analysis the effect has been treated as discontinuous, as it really is. In order to show, however, that it has the same average effect as has a steady force, consider that the momentum is imparted to the gun by the discharges at the average rate of 30 · 50,000/0.2, or 7.5×10^6 g · cm · sec^{-2}, and that this, by definition, is equal to the force; that is, $f = 7.5$ megadynes, as before.

It is not easy to make *direct* measurements of the speeds in an impact experiment. If, however, the bodies whose interactions are to be studied are made the bobs of pendulums, their speeds before and after impact can be calculated from easily observable quantities. The following experiments illustrate the laws of direct impact for the cases of both inelastic and elastic impact.

○

EXPERIMENT VA. SPEED OF A BULLET BY THE BALLISTIC PENDULUM[1]

If a bullet of mass m and unknown speed u is fired into a heavy pendulum bob of mass M, so that the bullet buries itself in the bob, the impact is then inelastic and the two colliding bodies move together after impact with some common speed v. Since no external forces are called into play during the interaction, the total linear momentum after impact is the same as that before, and therefore, by Eq. [53a],

$$u = \frac{(m + M)v}{m}.$$

Thus, determinations of m, $m + M$, and v are all that are needed to calculate u, the speed of the bullet.

The speed after impact v can be calculated if one knows the length r of the pendulum and the horizontal distance s traversed by the bob after the impact (Fig. 44). To see this, consider that $(m + M)v^2/2$ is the kinetic energy of the pendulum bob immediately

---

[1] The ballistic pendulum was invented by BENJAMIN ROBINS and was described by him in his famous treatise on *New Principles of Gunnery* (1742).

after the collision, and consider that, if the friction is negligibly small, all this kinetic energy is used to lift the pendulum through the height h, so that
$$(M + m)v^2/2 = (M + m)gh.$$

The vertical height h is best obtained as follows. By the theorem of Pythagoras,
$$r^2 = s^2 + (r - h)^2 = s^2 + r^2 - 2rh + h^2. \qquad [62]$$
Since h is small compared to s and r, h^2 may be neglected. Hence
$$2rh = s^2,$$
or
$$h = \frac{s^2}{2r}. \qquad [63]$$

For a pendulum bob one may use a piece of brass pipe about 20 cm long and 10 cm in diameter. The pipe is fitted with four small eyebolts placed two at each end, at opposite ends of diameters, and to these are fastened four suspending cords of about 2 m length each. The bottom of the pipe is equipped with a pin P; this pin makes contact with a light aluminum rider R, which slides along the horizontal meter stick. A softwood cylinder is turned to fit inside the pipe. Whenever this core becomes badly splintered, it should be removed from the pipe and a new one inserted.

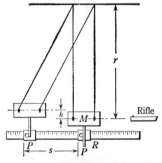

Fig. 44. Ballistic pendulum for measuring the speed of a bullet. Only the front pair of suspensions is shown

To reduce the danger of accident to a minimum, fasten the rifle in a clamp in one end of a rectangular wooden box having sides about 2 cm thick and the end opposite the barrel about 10 cm thick. Cover the part of the trough not occupied by the rifle with a sliding wooden top; before attempting to remove the pendulum bob from the trough for the purposes of weighing, first slide this top over so that it covers the rifle, thus making it impossible to fire the latter. Openings in the sides of the box should be provided for the escape of gas; these may be covered with metal screen.

Adjust the bob M until it is horizontal, and then measure the length r of the suspending cords

See that the meter stick is parallel to M and that the rider R is barely in contact with the pin P in the bob. Take the reading of one end of the rider on the scale.

For firing the bullet a 22-caliber rifle may be employed. To prevent the pendulum bob from receiving momentum from the exploding gases, place a sheet of paper between the end of the barrel and the bob. The rifle is to be fired only when it is fixed firmly in the clamp provided for the purpose and aimed at the center of the pendulum, and then only with the approval of the instructor. Secure a cartridge from the instructor, and, after he has inspected the apparatus and has given his approval, fire a bullet into the center of the wooden core in the bob. Record the new position to which the rider R is moved. The difference of the two readings is the horizontal distance s moved through by the bob as a result of the impact.

Clean the gun after each shot. Weigh the bob after each shot to obtain the mass $m + M$. Obtain the mass m of the bullet by drawing similar bullets from three cartridges and finding their average mass. To draw a bullet, clamp the body of the cartridge between two blocks of softwood in a vise and extract the bullet with pliers. *Do not* clamp the cartridge between metal jaws.

Repeat the entire procedure to secure three values of u. Calculate (*a*) the mean speed of the bullet; (*b*) the percentage of deviation of each value of u from the mean; (*c*) the total energy of the bullet; (*d*) the loss in energy in the impact, by Eq. [57]; and (*e*) l, by Eq. [60].

1. Would it make any difference in your determination of u if M, rather than $m + M$, were taken as the mass of the system after impact?

2. How much error was introduced into your result by neglecting h^2 in Eq. [62]?

3. What becomes of the kinetic energy that is lost in an inelastic impact?

4. Should you expect the fractional loss of kinetic energy to be greater in the present case than in the case of elastic impact? Explain.

○

EXPERIMENT VB. COEFFICIENT OF RESTITUTION

Make a test of NEWTON's law as to the constancy of the coefficient of restitution e for two given bodies. This may be accomplished by dropping a spherical body of some suitable material through a vertical distance h_1 onto the smooth horizontal surface of some hard, massive body, and noting the height h_2 of rebound. Since in general, for bodies falling freely from rest, $v = \sqrt{2\,gh}$, one has, in view of Eq. [61],

$$e = -\frac{v}{u} = \sqrt{\frac{h_2}{h_1}}.$$

Use first a steel ball. Drop it from the clamp *c*, Fig. 45, through the hole in the metal sheet *R* and onto the smooth top of, say, a heavy steel plate or a slab of slate. By successive trials adjust the height of *R* until, in the first rebound from the surface *S*, the bottom of the ball becomes just visible above *R*. Measure both h_1 and h_2 from the surface *S* to the bottom of the ball. Only the maximum value of h_2 that can be obtained consistently should be recorded and used in the calculations.

Make observations for at least three different heights of fall h_1 which vary between, say, 30 cm and 100 cm. For each case calculate both *e* and the fraction *l* of kinetic energy lost in the impact.

Repeat the experiment, this time with a glass ball.

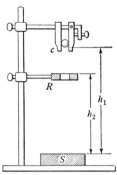

Fig. 45. A method for measuring the coefficient of restitution

1. Would the value of *e* which you obtained have been different if the balls had been dropped upon a different material, say a large slab of glass?

2. Is the momentum conserved in this case of impact — that is, is the momentum after impact equal to the momentum before impact? Explain.

○

EXPERIMENT Vc. ELASTIC IMPACT

Test the law of conservation of momentum as applied to direct *elastic* impact by swinging two steel balls of known masses as pendulums and finding their speeds before and after impact through observations of their motions over a graduated arc (Fig. 46). It will be found most convenient to cause the ball of larger mass m_1 to swing down and make impact with the one of smaller mass m_2, when the latter is

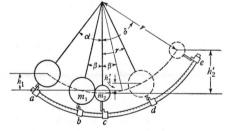

Fig. 46. A method of studying the elastic impact of two spherical bodies

hanging freely at the bottom of its arc. The momentum of the system immediately before impact is then m_1u_1, and that immediately after impact is $m_1v_1 + m_2v_2$. The object is to see if these two momentums are equal, within the limits of the experimental error.

The ball m_1 acquires its speed u_1 by swinging from rest down the arc ab (Fig. 46). Since the amount of friction is negligibly small, the speed thus acquired is the same as though the ball had fallen freely through the vertical distance h_1; that is, $u_1 = \sqrt{2\,gh_1}$. After the impact, m_1 and m_2 lose their speeds by moving along the arcs bd and ce respectively, and thus are lifted by the impact through the respective vertical distances h'_1 and h'_2; hence their respective speeds at the beginning of this motion, or immediately after impact, must be $v_1 = \sqrt{2\,gh'_1}$ and $v_2 = \sqrt{2\,gh'_2}$. Now it is evident from Fig. 46 that

$$h_1 = r(\cos \beta - \cos \alpha),$$
$$h'_1 = r(\cos \beta - \cos \gamma),$$
$$h'_2 = r(1 - \cos \delta),$$

where r is the vertical distance from the point of support to the center of either ball. Thus the equation to be tested in this experiment, namely, $m_1u_1 = m_1v_1 + m_2v_2$, may be written in the form

$$m_1\sqrt{\cos \beta - \cos \alpha} = m_1\sqrt{\cos \beta - \cos \gamma} + m_2\sqrt{1 - \cos \delta}. \qquad [64]$$

A similar substitution in Eq. [56] gives for the coefficient of restitution

$$e = \frac{\sqrt{1 - \cos \delta} - \sqrt{\cos \beta - \cos \gamma}}{\sqrt{\cos \beta - \cos \alpha}}. \qquad [65]$$

The fraction l of the kinetic energy lost in the impact can then be obtained from Eq. [60]. Evidently the problem of testing the momentum equation and of calculating the quantities e and l has been reduced to one of measuring the angles α, β, γ, δ and the two masses m_1 and m_2.

In the apparatus used, the two balls are swung from a rigid support by adjustable bifilar suspensions. The circular scale, which is graduated in degrees, has its center at the axis of suspension. The angles moved through by the balls are measured by means of two light aluminum riders which slide along this scale and give the scale-readings of the centers of the respective balls. The riders are so designed that one will be caught only by the index on the bottom of the large ball, the other only by the index on the bottom of the small one.

To insure direct impact, adjust the leveling screws and the lengths of the supporting cords until the centers of the balls are in the plane of the graduated arc and equidistant from the axis of suspension. Then draw m_1 back through some angle α, tie it in this position by means of a thread, and observe this initial position a of m_1 by bringing the large-ball rider into contact with the index of m_1.

Now slide the large-ball rider down to a point on the scale just short of its expected final position d, as approximately located by a preliminary trial. Similarly, place the small-ball rider in the neighborhood of e. Then burn the thread holding m_1 at a point near the ball. Catch m_1 with the hand as it swings back from d after the impact. Take the readings at d and e.

With the small ball hanging freely, bring its rider into contact with its index and take the reading c. In a similar way find the position b of the center of the large ball at the moment of impact, that is, the position of m_1 when m_2 hangs freely and m_1 is brought down so as just to touch m_2.

Measure r and, if the masses of the two balls are not stamped upon them, find m_1 and m_2 by weighing.

Calculate (a) the angles α, β, γ, δ; (b) the two members of Eq. [64] and the percentage of difference between them; (c) the momentums immediately before and after impact; (d) e by Eq. [65]; and (e) l.

1. Do the terms in Eq. [64] have the dimensions of momentum? Why, then, does this equation provide a test of the equalities of momentums before and after impact?

2. Analyze all the forces acting on the two balls at the moment of impact. Are any of them external forces? If so, why is the total momentum not changed by the impact?

3. Show that the end points of the arc cd differ in height by the amount $r(1 - \cos \gamma)$ and explain why it is not this quantity, but rather $r(\cos \beta - \cos \gamma)$, which is taken as the vertical distance h'_1 moved through by the large ball after impact.

4. How does the value of e obtained in this experiment on elastic impact compare with that obtained in Exp. Vb? Explain.

5. How should you expect the fractional loss of kinetic energy in an inelastic impact to compare with that found in the present case? Why?

o

EXPERIMENT Vd. INELASTIC IMPACT

The form of apparatus for studying inelastic impact is the same as that used in Exp. Vc except that a lead ball and a brass cylinder filled with lead replace the two steel balls. The index on the bottom of the cylinder is so placed that it indicates the position of the center of mass of the cylinder and ball *taken together*. Complete inelasticity of impact is secured by interposing pads of soft wax at the contact surfaces, thereby causing the ball and cylinder to stick together after impact.

The momentum equation which it is sought to verify is $mu = (m + M)v$, where m and M are the masses of the ball and cylinder respectively. If h be the vertical distance through which the ball m falls before impact, and if h' be the vertical distance through which the center of mass of $m + M$ is raised by the impact, then, in view of Fig. 47,

$$h = r(\cos \beta - \cos \alpha),$$
$$h' = r(\cos \gamma - \cos \delta),$$
[66]

and the momentum equation takes the form

$$m\sqrt{\cos \beta - \cos \alpha} = (m + M)\sqrt{\cos \gamma - \cos \delta}.$$
[67]

Note that h' is the vertical distance between the points d and e, the arc de being the path traversed by the center of mass of $m + M$ *because of the impact alone.*

The point d is the point up to which the center of mass of the system ball-cylinder would move if the ball were placed initially at b instead of at a, and were then released. Hence the motions over the arc bc, which subtends angle β, and over the arc cd, which subtends angle γ, are not to be included in obtaining the expression for h and for h'.

Adjust the apparatus as in Exp. Vc, making sure by preliminary trials that when the ball is released by burning the thread,

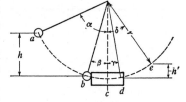

Fɪɢ. 47. Apparatus arranged for inelastic impact

the cylinder and ball swing smoothly up the arc without wobbling. Move the rider up the circular scale nearly to the point e which will be reached by the center of mass of $m + M$. Observe the initial scale-reading a of the center of the ball and, after the impact, the final scale-reading e of the center of mass of the *ball and cylinder together.*

Observe the scale-reading b of the center of the ball at the moment of impact and the scale-reading c of the center of the ball when the ball is hanging freely. Observe also the scale-reading c of the center of mass of the ball and cylinder together. If this reading does not agree with that of the ball alone, correct the cylinder index until it does.

Then, with the cylinder hanging freely and the ball held barely in contact with it, place the rider in contact with the cylinder index and release the ball. The first swing will move the center of mass of the whole system to the point d, which can thus be observed.

Measure r and also m_1 and m_2, if the latter are not stamped on the ball and cylinder.

Calculate (*a*) the angles α, β, γ, δ ; (*b*) the two members of Eq. [67] and the percentage of difference between them ; (*c*) the momentums immediately before and after impact ; and (*d*) l, by Eq. [60].

Repeat all of the foregoing procedure and thus obtain a second complete set of the required quantities.

1. With the aid of a diagram, explain fully why Eqs. [66] give the correct values of h and h'. In Fig. 47, why is the arc de, and not the arc ce, taken as the path traversed by the center of mass of $m + M$ because of the impact alone?

2. What becomes of the kinetic energy that is lost in an inelastic impact?

3. How does the fractional loss of kinetic energy in the present case compare with that in the case of elastic impact? Explain.

o

OPTIONAL LABORATORY PROBLEMS

1. Find by experiment whether the speed of a bullet is perceptibly less with each successive shot when the rifle is not cleaned between shots.

2. Measure and compare the bullet speeds of various types of cartridges.

3. Vary the kind of wood used for the core of the ballistic pendulum and find the average resisting force that each kind of wood offers to a bullet.

o

QUESTION SUMMARY

1. State clearly the *principle of conservation of momentum* and illustrate this principle by applying it to the case of the impact of two spheres. For what particular case does it alone give a complete solution to the problem of the impact of two bodies?

2. Define, and tell how to measure, the *coefficient of restitution*.

3. Show how the principle of conservation of momentum, together with either the value of the coefficient of restitution or a knowledge of the loss in kinetic energy, suffices to solve completely any problem of direct impact of two bodies. What is the value of e for inelastic impact? for perfectly elastic impact?

4. State the procedure for solving the problem of the oblique impact of two spheres.

5. Is mechanical energy conserved in an impact? Reconcile your answer with the principle of conservation of energy. If mechanical energy were conserved in an impact, what would be the value of the coefficient of restitution and what would be the nature of the impact? How large a loss of mechanical energy is it possible to have in an impact without having a loss of momentum?

PROBLEMS

1. A projectile of mass 4.0 kg is shot with a muzzle speed of 350 m · sec^{-1} from a gun of mass 3000 kg. What is the initial speed of recoil?

Ans. 47 cm · sec^{-1}.

2. A truck of mass 10 tons moving with a speed of 14 ft · sec^{-1} is stopped by buffers in 0.30 sec. What is the magnitude of the total impulse? of the average force? *Ans.* 2.8 × 10^5 poundals · sec; 15 tons.

3. Explain the rise of a rocket. Would the rocket rise in a vacuum?

4. Is it true that, at the start, the wagon pulls back with the same force with which the horse pulls forward? If so, how is any motion produced? If not, reconcile your answer with the third law of motion.

5. What becomes of the linear momentum of a meteorite when it collides with the earth? What becomes of its energy?

6. A 300-g baseball approaches a bat with a velocity cf 50 m · sec^{-1}; it leaves with an oppositely directed velocity of 100 m · sec^{-1}. Find the average force of the blow if the impact lasts 0.020 sec. *Ans.* 2.3 × 10^2 kgwt.

7. Two perfectly elastic balls having masses in the ratio of 3 to 1 are suspended by threads so as to swing with pendular motions in a plane and make direct impacts (Fig. 46). (*a*) Prove that if the initial speeds of the balls are equal, on the first impact the more massive ball will come to rest whereas the less massive one will rebound with twice its initial speed, on the second impact the two balls will rebound with equal speeds, and hence a definite cycle of impacts consisting of two different types of motion will be established. (*b*) Show that a different cycle of impacts occurs if the less massive ball is initially at rest and the more massive one swings down upon it.

8. Two spheres of masses 50 g and 100 g, respectively, make direct impact. Their velocities before impact are in the same directions and have magnitudes of 600 and 350 cm · sec^{-1}. Find the velocities after impact (*a*) if $e = 1.00$; (*b*) if $e = 0.900$.

Ans. (*a*) 267 cm · sec^{-1}, 517 cm · sec^{-1}; (*b*) 283 cm · sec^{-1}, 508 cm · sec^{-1}.

9. Suppose that one particle is initially at rest and that a second particle makes inelastic impact with it. What relation exists between the masses of the particles when the fractional part of the kinetic energy transformed into heat is (*a*) $\frac{1}{2}$? (*b*) $\frac{1}{4}$? (*c*) $\frac{3}{4}$? *Ans.* (*a*) $m_1 = m_2$; (*b*) $m_1 = 3 m_2$; (*c*) $m_2 = 3 m_1$.

10. A rapid-fire gun projects 300 20-g bullets per minute with a velocity of 400 m · sec^{-1} directly against a steel plate. If the coefficient of restitution between the steel plate and the bullets is 0.25, what average force does the steel plate exert against the wall which supports it? *Ans.* 5.1 kgwt.

11. A 100-g billiard ball moving east with a speed of 200 cm · sec^{-1} was struck by a 4.00-g putty ball moving south with a speed of 2000 cm · sec^{-1}. Find the velocity of the balls the instant after impact.

Ans. 2.07 m · sec^{-1}, 21° 50′ south of east.

12. A bullet of mass 10.0 g is shot horizontally and with a speed of 2400 m · min^{-1} into the center of a wooden ball of mass 500 g which is rolling along a horizontal plane with a speed of $400 \text{ cm} \cdot \text{sec}^{-1}$. If the directions of the bullet and ball make an angle of 45° before impact and if it be assumed that the bullet remains in the ball, (a) how much is the direction in which the ball is traveling changed by the impact and (b) what is the speed of the ball after impact? *Ans.* (a) 7° 5′; (b) $453 \text{ cm} \cdot \text{sec}^{-1}$.

13. A fire engine throws 16 l of water per second from a hose furnished with a nozzle 3.0 cm in diameter. (a) If inelastic impact is assumed, what force does a wall experience against which the jet is directed normally and at short range? (b) If each particle of water rebounded with the speed of approach, what then would be the value of the force? (c) What normal force would the wall experience if the jet were directed at short range and at an angle of 30° with the wall, the coefficient of restitution being 0.50?
Ans. (a) 37 kgwt; (b) 74 kgwt; (c) 28 kgwt.

14. A smooth ball moving with a speed of $600 \text{ cm} \cdot \text{sec}^{-1}$ makes impact with a second smooth ball of twice its mass, the latter being initially at rest. If the velocity of the smaller ball before impact makes an angle of 30° with the line of centers at the instant of impact, and if e is 0.5, find the velocity of the smaller ball after impact. *Ans.* $3 \text{ m} \cdot \text{sec}^{-1}$, ⊥ line of centers.

15. A ball falls from a height of 6 m onto a smooth floor. If the coefficient of restitution is 0.7, how high will the ball rise after striking the floor the third time? *Ans.* 0.7 m.

16. A 50-g bullet is fired into a block of mass 125 g. (a) Find the fractional loss of kinetic energy. (b) What if the bullet and block had been elastic bodies for which e is 1? *Ans.* (a) 71 percent; (b) 0.

17. A 200-g billiard ball rolls on a smooth floor with a speed of $100 \text{ cm} \cdot \text{sec}^{-1}$. It strikes a smooth wall at an angle of 45°. If e is 0.500, find (a) the direction of motion after impact; (b) the loss in kinetic energy.
Ans. (a) 26° 35′; (b) 3.75×10^5 ergs.

18. A mass of 5.00 kg moving with a speed of $120 \text{ cm} \cdot \text{sec}^{-1}$ impinges directly upon a mass of 20.0 kg which is at rest. The former is observed to rebound with a speed of $60.0 \text{ cm} \cdot \text{sec}^{-1}$. Find (a) the energy lost in the impact; (b) the coefficient of restitution. *Ans.* (a) 0.675 j; (b) 0.875.

19. One of the heaviest rainfalls on record anywhere in the United States was that reported for a storm occurring in the mountains of California, when 1.02 in. of rain fell in 1 min. Assuming that the speed of the raindrops was $300 \text{ ft} \cdot \text{sec}^{-1}$, that they struck a 60×150 ft horizontal roof at an angle of 75° with the roof, and that the coefficient of restitution was 0.2, find the total force exerted on the roof. *Ans.* 4 tons.

20. The coefficient of restitution is very nearly unity for spheres of the same size consisting of materials like steel, glass, and ivory, but it is not unity for unequal spheres of these same substances. It may be as low as 0.75 for two steel balls of greatly different sizes. Thus e is a constant of the colliding bodies, as well as of the material. By considering vibration losses and the conditions under which vibrations will persist in bodies after impact, explain why this is true.

CHAPTER SIX

RIGID BODIES AND EQUILIBRIUM

I POSTULATE the following:

1. Equal weights at equal distances are in equilibrium, and equal weights at unequal distances are not in equilibrium but incline towards the weight which is at the greater distance.

2. If, when weights at certain distances are in equilibrium, something be added to one of the weights, they are not in equilibrium but incline towards that weight to which the addition was made.

ARCHIMEDES, " On the Equilibrium of Planes or the Centres of Gravity of Planes." Translation by T. L. HEATH [1]

o

Although the science of dynamics in its broadest sense had its beginning with GALILEO, a scientific treatment of the branch of the science called statics (Sec. 41) had been begun by the Greeks and constituted a considerable body of knowledge at the time of GALILEO. Statics developed before kinetics (Sec. 64) because it is much the simpler and is more directly concerned with the implements and simple machines known to the people of antiquity. These ancient inventions of course arose largely by chance to meet immediate needs, and previous to the time of the Greeks no generalized theory, such as that of the equilibrium of a machine, had been developed to explain them or to suggest further discovery. Genuine physical theory, treated mathematically, began with the clearing up of the theory of the lever and of the center of gravity by ARCHIMEDES (*c.* 287–212 B.C.), who was the greatest mathematician, physicist, and engineer of antiquity, and one of the greatest men of all times. ARCHIMEDES also laid the foundations for the science of hydrostatics (Chap. 13), "but above all he introduced into science what might be called the Archimedian spirit, a new way of submitting things to scientific analyses, the method (and point of view) of mathematical physics." [2] Very little advance in statics beyond ARCHIMEDES occurred for the next eighteen hundred years, until STEVIN again dealt with the lever and attacked such problems as that of the inclined

---

[1] * T. L. Heath, *The Works of Archimedes* (1897), p. 189. By permission of The Macmillan Company, publishers. ARCHIMEDES stated, in all, seven postulates which he regarded as self-evident, and on the basis of them derived the general law of the lever and proved propositions for finding the centers of gravity of various plane figures.

[2] G. Sarton, *Introduction to the History of Science* (Williams & Wilkins, 1927–1931), Vol. I, p. 166.

plane (Chap. 3). Thus we see that the sixteenth century witnessed both the revival of statics and its absorption into the more fundamental and powerful dynamics of GALILEO and NEWTON. In this text we will not develop statics as an independent science, as did the ancients, but will follow the modern method of treating it as a special case of dynamics.

All the deductions made in the previous chapters for the mutual actions of two particles can be applied at once to a system of particles, and hence to bodies of finite volume, provided we assume that the latter can be regarded as made up of particles. Now when a rigid body of finite volume has translation without rotation, its motion is the same as that of a single particle (Sec. 5). But when such a body is in rotation, then its particles no longer all move in exactly the same way and there is no longer any one particle having a motion that is completely representative of the whole body. The motions of the individual particles are still described adequately by NEWTON'S three laws, but there is an immense practical advantage in having special rules, derived from these laws, that apply to the motion of the body as a whole. Two of these rules were suggested to NEWTON by his third law, when he regarded the latter as an extension of the first law to a system of particles. For it immediately became apparent that there is one particular point in any body, called the *center of mass*, which for certain purposes may be taken as representing the body, so that, for these purposes, the mass of the body may be conceived as concentrated in a particle at this point.

○

The Center of Mass and Its Properties

69. Definition of Center of Mass. The center of mass of two particles is defined as a point that divides the line joining them inversely as their masses. Thus if c be this point for two particles at A_1 and A_2, Fig. 48, then

$$\frac{A_1 c}{c A_2} = \frac{m_2}{m_1}, \qquad [68]$$

by definition.

If the rectangular coordinates of the two particles are (x_1, y_1, z_1) and (x_2, y_2, z_2), and if the coordinates of their center of mass are (x_c, y_c, z_c), then, from Eq. [68],

$$\frac{x_c - x_1}{x_2 - x_c} = \frac{m_2}{m_1},$$

THIS mosaic is the only ancient pictorial representation of the death of ARCHI-MEDES which we possess. For its history, see F. Winter's *Der Tod des Archimedes* (De Gruyter, 1924), from which the reproduction here given, as well as much of the following discussion, was taken with the permission of the publisher.

Since most mosaics are copies of actual paintings, it is possible that this one is a reproduction of a painting made shortly after ARCHIMEDES' death, when the recollection of it and of ARCHIMEDES himself was still fresh. If this should be the case, the painting, though no longer in existence, would antedate any of the historical accounts of ARCHIMEDES' death, and the stories, such as those given below, may have been inspired by it. Since writers on iconography apparently do not recognize an ARCHIMEDES among existing portraits, it is to be hoped that this mosaic, which has only recently been called to their attention, may fill the need.

ARCHIMEDES was, as this mosaic so well portrays, and as we know from his life and work, very different from the secluded bookworm most modern paintings depict.

" He was the type of man who, with all his science, leads a full life in the world of practical affairs. After he had participated under Hiero II in great and novel undertakings such as the preparation of a magnificent ship built for Ptolemy IV and had played a most important part in them, he it was who, when Rome in 214 B.C. laid siege to Syracuse, enabled his native city to make the celebrated defence and bid defiance for two years to all the efforts of the hostile superior power. On this account he found an admirer in the Roman general Marcellus who regretted extremely that his orders for the protection of Archimedes issued upon the taking of the city, should have been brought to naught by the violent act of an ignorant soldier. This Archimedes, the last representative of the great tradition of the city, the mosaic brings before our eyes. He is an elegantly dressed, solid, thickset figure sitting there. Rudely disturbed in his studies, he rises suddenly pointing to the intruder with furious glances and words. He spreads his arms over the [sand] table containing his diagrams with a movement which shows him dominated entirely by the problem which he has just been studying. He betrays, however, not a trace of pedantic sentimentality. As characteristic as the whole figure is the head with its broad forehead over which the somewhat unkempt tufts of hair stand up. His eyes blaze. The exceedingly energetic mouth is closely framed by the short-cropped beard. One has the impression of a portrait gained from personal contact with the man himself."

Accounts of ARCHIMEDES' death, differing somewhat as to details, have been given by Livy, Plutarch, Valerius Maximus, and others. Plutarch gives three different versions in his "Life of Marcellus":

" Syracusa beinge taken, nothinge greved *Marcellus* more, then the losse of *Archimedes*. Who beinge in his studie when the citie was taken, busily seekinge out by him selfe the demonstracion of some Geometricall proposition which he hadde drawen in figure, and so earnestly occupied therein, as he neither sawe nor hearde any noyse of enemies that ranne uppe and downe the citie, and much lesse knewe it was taken: He wondered when he sawe a souldier by him, that bad him

The Death of ARCHIMEDES

PHOTOGRAPH of a mosaic which probably dates back to before the time of Julius
Caesar. The original, which is in color and is 51 cm long and 43 cm high, is
owned by the Scabell family of Wiesbaden, Germany

○

go with him to *Marcellus*. Notwithstandinge, he spake to the souldier, and bad
him tary untill he had done his conclusion, and brought it to demonstracion: but the
souldier being angry with his aunswer, drew out his sword, and killed him. Other
say, that the Romaine souldier when he came, offered the swords poynt to him, to
kill him: and that *Archimedes* when he saw him, prayed him to hold his hand a litle,
that he might not leave the matter he looked for unperfect, without demonstracion.
But the souldier makinge no reckening of his speculation, killed him presently. It is
reported a third way also, sayinge, that certeine souldiers met him in the streetes
going to *Marcellus*, carying certeine Mathematicall instrumentes in a litle pretie
coffer, as dialles for the sunne, Sphaeres and Angles, wherewith they measure the
greatnesse of the body of the sunne by viewe: and they supposing he hadde caried
some golde or silver, or other pretious Iuells in that litle coffer, slue him for it."

and therefore
$$x_c = \frac{m_1 x_1 + m_2 x_2}{m_1 + m_2},$$
[69]

with two analogous equations for y_c and z_c. Thus if the rectangular coordinates and masses of two particles are known, the rectangular coordinates of their center of mass can be calculated.

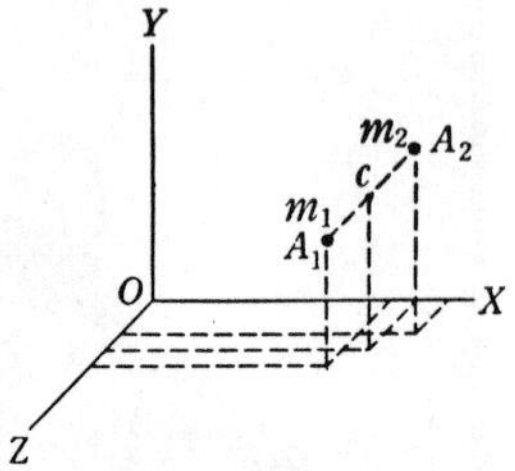

Imagine the particles m_1 and m_2 replaced at their center of mass by a single particle of mass $m_1 + m_2$. The center of mass of *three* particles m_1, m_2, and m_3 is then defined as that of $m_1 + m_2$ and m_3. By proceeding in this way to a system of n particles, we obtain for the x-coordinate of the center of mass of the system

$$x_c = \frac{m_1 x_1 + m_2 x_2 + \cdots + m_n x_n}{m_1 + m_2 + \cdots + m_n},$$

Fig. 48. A_1 and A_2 are two particles of a body, and c is their center of mass

and for y_c and z_c two analogous equations. These three equations for the system of n particles may be written in the briefer form

$$x_c = \frac{\sum_{i=1}^{i=n} m_i x_i}{M}, \qquad y_c = \frac{\sum_{i=1}^{i=n} m_i y_i}{M}, \qquad z_c = \frac{\sum_{i=1}^{i=n} m_i z_i}{M},$$
[70]

where M is equal to $m_1 + m_2 + \cdots + m_n$, the total mass of the system.

When the n particles and their center of mass are located by means of directed lines r_1, r_2, $\cdots$, r_n, r_c, drawn from the origin O, the three scalar equations [70] may be written as a single vector equation,

$$r_c = \frac{\sum_{i=1}^{i=n} m_i r_i}{M}.$$
[71]

In many problems involving a body it proves convenient to take the origin of coordinates at the center of mass of the body, in which case $x_c = y_c = z_c = 0$ and Eqs. [70] become

$$\sum_{i=1}^{i=n} m_i x_i = 0, \qquad \sum_{i=1}^{i=n} m_i y_i = 0, \qquad \sum_{i=1}^{i=n} m_i z_i = 0.$$
[72]

(What does Eq. [71] become in this case?)

Eqs. [70] are perfectly general and can easily be put into the particular form required for any given case.

EXAMPLE. Two homogeneous solid cylinders of lengths l_1 and l_2, radii R_1 and R_2, and common density ρ are joined end to end, so that their axes coincide (Fig. 49). Locate the center of mass of the combination.

Solution. By symmetry, the center of mass lies somewhere on the common axis. If we take this axis as the X-axis and take the origin O at the free end of the larger cylinder, we need solve only for x_c. Since each cylinder is homogeneous, its center of mass lies at its geometrical center, and hence the problem reduces to one of finding the center of mass of two particles. By Eq. [70] or [69], then,[1]

FIG. 49. Find the center of mass of this combination of two cylinders

$$x_c = \frac{\pi R_1{}^2 l_1 \rho \, \frac{l_1}{2} + \pi R_2{}^2 l_2 \rho \left(l_1 + \frac{l_2}{2}\right)}{\pi R_1{}^2 l_1 \rho + \pi R_2{}^2 l_2 \rho}.$$

EXAMPLE. A circular disk of diameter 16 cm contains a hole 12 cm in diameter which is tangent to the circumference, as shown in Fig. 50. Locate the center of mass.

Solution. It is interesting to note that Eq. [69] can be extended to this case of a body containing a cavity merely by first treating the body as if it had no cavity and then considering as a "negative mass" the imaginary body which would just fill the cavity. Let σ be the so-called surface density of the disk, or mass per square centimeter. Take the origin of coordinates at the center of the larger circle and the X-axis through the centers of both circles. Then

$$x_c = \frac{64\,\pi\sigma \cdot 0 - 36\,\pi\sigma\,(-2)}{64\,\pi\sigma - 36\,\pi\sigma} = 2.6 \text{ cm.}$$

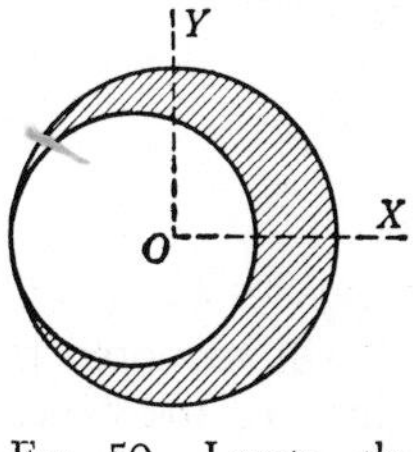

The center of mass is therefore 2.6 cm to the right of the center of the larger circle.

FIG. 50. Locate the center of mass of this ·disk

The positions of the centers of mass of many bodies are evident from inspection. In the preceding problems, for example, use was made of the obvious fact that if a homogeneous body is symmetrical about a point or an axis, the center of mass lies in that point or axis. When a body can be divided into parts such that the center of mass of each is known, the center of mass of the whole can usually be found. Thus, if a triangular area be divided into narrow strips

---

[1] The advantage of the implicit method of solving problems is apparent here, for $\pi\rho$ cancels out, thus saving much tedious numerical work.

parallel to one side, as in Fig. 51, it is seen that the center of mass of each strip lies on the line joining the middle of the side to the opposite vertex; hence the center of mass of the triangle is at the intersection of the three lines that join the vertices to the middles of the opposite sides. The most general methods of finding the centers of mass of regular bodies involve the use of the integral calculus. Such methods are illustrated in Appendix 7, which also contains the formulas that may be used for bodies of various common shapes.

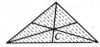

Fig. 51. Method of finding the center of mass of a triangular area

70. Displacement, Velocity, and Acceleration of the Center of Mass. Suppose that a body is in motion relative to some point O, considered as fixed, and that the positions of its particles and their center of mass at a certain moment are given by the vectors $r_1, r_2, \cdots, r_n, r_c$, drawn from O. After an interval of time Δt these positions will in general have changed to new ones given, say, by $r'_1, r'_2, \cdots, r'_n, r'_c$, also drawn from O. By using Eq. [71], one obtains for the *displacement* of the center of mass in the time Δt the vector equation

$$r'_c - r_c = \frac{\displaystyle\sum_{i=1}^{i=n} m_i r'_i}{M} - \frac{\displaystyle\sum_{i=1}^{i=n} m_i r_i}{M},$$

or

$$r'_c - r_c = \frac{\displaystyle\sum_{i=1}^{i=n} m_i (r'_i - r_i)}{M}. \qquad [73]$$

It will be noted that in obtaining this equation the masses of the particles and hence the mass of the body have been regarded as not varying with the time (Sec. 25).

To obtain an expression for the *velocity* v_c of the center of mass, divide both members of Eq. [73] by Δt and let Δt decrease without limit; this gives, according to Eq. [10], Chap. 1,

$$v_c = \frac{\displaystyle\sum_{i=1}^{i=n} m_i v_i}{M}, \qquad [74]$$

where $v_1, v_2, \cdots, v_n$ are the velocities of the particles. The terms $m_i v_i$ are the momentums of the individual particles, and hence Eq. [74] states that the total linear momentum of a body of finite volume is equal to $M v_c$, the product of the total mass and the velocity of the center of mass.

EXAMPLE. Show that the acceleration of the center of mass is given by the expression

$$a_c = \frac{\sum_{i=1}^{i=n} m_i a_i}{M}.$$ [75]

Solution. Begin with Eq. [74] and apply the same reasoning as was used in obtaining it; then make use of Eq. [12], Chap. 1.

It remained for JEAN LE ROND D'ALEMBERT (1717–1783) to see that Eq. [75] has a very important interpretation. The term $m_i a_i$ is, by the definition of force, the total force acting on the ith particle to produce the acceleration a_i. Now there are in general two classes of forces acting on a *system* of particles: the forces that the individual particles exert on one another, called *internal* forces; and the forces applied to the system from the outside, called *external* forces. But NEWTON'S third law asserts that the internal forces occur in pairs that are equal in magnitude but oppositely directed, and D'ALEMBERT[1] saw that if this be true, *the sum of the internal forces is zero.* Hence $\sum_{i=1}^{i=n} m_i a_i$ is simply the sum F of all the external forces, and Eq. [75] becomes

$$F = M a_c.$$ [76]

Here we have the great physical property of the center of mass, the one that makes the dynamics of a particle so useful. For Eq. [76] tells us that, *no matter what the shape of the body, how it may move, or where the external forces are applied to it, the center of mass moves precisely as if the total mass were concentrated there and all the external forces were transferred, with their directions unchanged, to the center of mass.* This statement is true no matter whether the particles are rigidly connected or are moving relative to one another, as is the case with the molecules of a mass of gas. Thus, when gravity is the only external force acting on a bomb dropped from an airplane, the center of mass of the bomb describes the same parabolic path as would a particle under the same circumstances; if the bomb explodes during its descent, the center of mass of the fragments continues along the same path as if no explosion had occurred, for all the forces brought into play by the explosion are internal forces and thus add up to zero.

It is important to note, however, that the external forces applied at the center of mass are those actually operating on the various particles of the body and not necessarily those that would exist if these particles were transferred to the center of mass.

---

[1] J. d'Alembert, *Traité de Dynamique* (1743).

71. Equilibrium as Regards Translation. A body is said to be in equilibrium *as regards translation* when its linear momentum $M\mathbf{v}$ is constant, that is, when its center of mass has no acceleration. According to Eq. [76], this will be true when the sum $\mathbf{F}$ of the external forces acting on the body is zero. Therefore the condition for equilibrium as regards translation is $\mathbf{F} = 0$, or, what is the same thing,

$$F_x = 0, \quad F_y = 0, \quad F_z = 0, \qquad [77]$$

where F_x, F_y, and F_z are the algebraic sums of the force components along the *X*-, *Y*-, and *Z*-axes respectively.

○

Moments of Force

72. The Principle of Moments. When one attempts to set down the conditions that govern the *rotation* of a body, it is found to make a great difference where the force is applied. We will solve this problem with the help of the scholium to NEWTON's third law, the essence of which is that in any machine, during any given time, the work of the acting forces is equal to the work of the resisting forces (Sec. **46**). This generalization must, of course, be applied to some particular case, and one of the simplest is the wheel and axle. This machine (Fig. 52) is merely a kind of continuous lever, which has for its lever arms the two rigid radii l_1 and l_2 leading out to the tangential points of the cords. A weight f_1 is attached to the cord wound around the axle. A second cord is wound in the opposite direction about the rim of the wheel, and a spring balance attached to this cord can be used to measure the

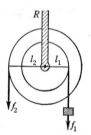

FIG. 52. The wheel and axle, and the principle of moments

force needed to raise the weight f_1 at constant speed and also to lower it steadily. In the first case the weight must be lifted and the friction in the bearings overcome, whereas in the second case the friction aids the applied force in supporting the weight, so that the average of the two balance readings is the force f_2 that would be needed to raise or lower the weight f_1 without acceleration if there were no friction. Now if the weight be raised or lowered through some distance s_1 by a steady rotation of the axle through an angle of, say, θ radians, the work of the resisting force f_1 is $f_1 s_1$, or $f_1 l_1 \theta$, where l_1 is the radius of the axle. The applied force f_2 in the same time moves through some distance s_2 and its work is $f_2 s_2$, or $f_2 l_2 \theta$, where l_2 is the radius

of the wheel. No other work is done, for there is neither acceleration nor friction, and the wheel is regarded as rigid and the support
rod R as fixed. Therefore, by the work principle, $f_1 l_1 \theta = f_2 l_2 \theta$, or

$$f_1 l_1 = f_2 l_2. \qquad [78]$$

This equation is true only for steady rotations or for rest, that is,
for rotation without angular acceleration. It is therefore the condition for equilibrium as regards rotation of a rigid, frictionless wheel
and axle.

It appears from Eq. [78] that two forces will have equal turning
effects about a given axis if the product obtained by multiplying
the magnitude f of either force by the perpendicular distance l from
the line in which it acts to the axis is equal to the corresponding
product for the other force. This product fl bears the same relation
to rotation that force does to translation. It is a measure of the
importance of a force in producing rotation about a given axis and
hence is quite appropriately called the *moment* or *torque* of the force.
We will denote its magnitude by the symbol L.

LEONARDO DA VINCI seems to have been the first to perceive that
the two factors which properly measure the tendency of any force to
produce rotation are the magnitude of the force f and what we now
call the *lever arm l*, or *perpendicular* distance from the axis of rotation
to the direction of the force.[1] It is improbable, however, that he
ever thought of the product fl as constituting a single quantity L
which is the proper measure of the importance of the force in producing rotation, or that he considered it to be a directed quantity.

As another simple example of the fact that the torque L of a
force is the proper measure of its tendency to produce rotation,
imagine a thin flat sheet of metal, free to rotate in a horizontal
plane about a fixed vertical axis A, Fig. 53, and in steady rotation
under the action of the forces f_1, f_2, f_3, and f_4, all of which lie in the
plane of rotation. The force f_1 is producing clockwise rotation,
f_2 and f_3 are resisting this rotation, and f_4 has no influence on the
rotation, since it passes through the axis A. Now the direction in
which f_1 acts is not parallel to the direction in which its point of
application moves, and thus the force that is effective in producing
the motion is not f_1 but the component of f_1 in the direction of
motion; this component has the magnitude $f_1 \sin \epsilon_1$, where ϵ_1 is the
angle between the force and the line r_1 drawn from the axis A to the
point of application of the force. Similarly, the effective resistances,

---

[1] See *I. B. Hart, *The Mechanical Investigations of Leonardo da Vinci* (Open
Court, 1925), pp. 109–113.

or resistances in the directions in which the points of application of f_2 and f_3 must move, have the magnitudes $f_2 \sin \epsilon_2$ and $f_3 \sin \epsilon_3$. Provided there are no additional resistances due to friction or molecular forces, the work principle asserts that, for a steady rotation of the body through an angle of θ radians,

$$(f_1 \sin \epsilon_1)r_1\theta = (f_2 \sin \epsilon_2)r_2\theta + (f_3 \sin \epsilon_3)r_3\theta,$$

or
$$f_1 r_1 \sin \epsilon_1 = f_2 r_2 \sin \epsilon_2 + f_3 r_3 \sin \epsilon_3. \qquad [79]$$

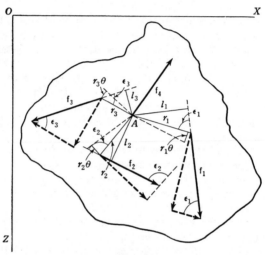

Fig. 53. The work principle applied to the rotation of a body acted upon by several forces in the plane of rotation

If perpendiculars l_1, l_2, and l_3, Fig. 53, be dropped from A upon the lines of action of the several forces, then $r_1 \sin \epsilon_1 = l_1$, etc., and Eq. [79] becomes

$$f_1 l_1 = f_2 l_2 + f_3 l_3, \qquad [80]$$

or
$$L_1 = L_2 + L_3, \qquad [81]$$

where L_1, L_2, and L_3 are the torques of the forces f_1, f_2, and f_3 with respect to the axis A. Thus if any number of forces are acting in the plane of rotation, the condition for rotational equilibrium is that *the sum of the torques producing clockwise rotation must be equal to the sum of the torques producing counterclockwise rotation*; or, if clockwise rotations be called negative, and counterclockwise positive, the algebraic sum L of the torques must be zero. In symbols, this condition is $L = \Sigma(fr \sin \epsilon) = 0$, or, what is the same thing

$$L = \Sigma fl = 0. \qquad [82]$$

Although Eq. [82] has been developed only with reference to torques about the *fixed* axis A, it holds, nevertheless, for torques taken about any *imaginary* axis conceived as passing parallel to the original axis through any point whatever in the plane of the forces, provided the body is in equilibrium as regards translation as well as regards rotation, and *provided the force f_4 be numbered among the acting forces.* Thus, if Eq. [82] be true for a body pivoted at A, Fig. 53, then it is also true that, if O be any point whatever in the plane,

$$f_1 l'_1 + f_2 l'_2 + f_3 l'_3 + f_4 l'_4 = 0, \qquad [83]$$

where $l'_1, \cdots, l'_4$ are the respective perpendiculars let fall from O to the lines of action of the forces. This conclusion follows from the consideration that, since the body is also in equilibrium as regards translation, $f_1 + f_2 + f_3 = -f_4$; that is, so far as any effects produced by f_1, f_2, and f_3 are concerned, these three forces may be replaced by the single force $-f_4$. But the torque of $-f_4$ about any axis whatever differs only in sign from the torque of f_4; hence the sum of the torques of the forces which $-f_4$ replaces, namely, f_1, f_2, and f_3, must differ only in sign from the torque of f_4.

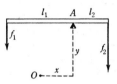

Fig. 54. The selection of the reference point for torques is wholly arbitrary

EXAMPLE. Two forces f_1 and f_2 act on a rod of negligible weight which is pivoted at A as indicated in Fig. 54. Given that the sum of the torques of these forces about the fixed axis A is zero, show that the sum of the torques about any parallel axis through any point O in the plane of the figure also is zero.

If a number of torques are applied simultaneously to a body and these torques act about different axes, they may, with certain restrictions, be added like ordinary vectors (Sec. **80**). In the special case where the torques all act about the *same* axis, they may then be added algebraically, for the vectors representing them are all localized in the same straight line. This is the case in Fig. 53, where the common axis is parallel to the Y-axis of coordinates.

73. Conditions for Complete Equilibrium. A body is in equilibrium *as regards rotation* if the vector sum L of all the torques acting on it is zero. This condition may be expressed analytically by the statement that the algebraic sum of the torques tending to produce rotation about each of three rectangular axes must be zero.

By combining this condition for rotation with that given in Sec. **71** for translation, we have as the conditions for the *complete* equilibrium of a body the two vector equations $\mathsf{F} = 0$ and $\mathsf{L} = 0$, or, what is the same thing, the six independent relations

$$
\begin{aligned}
F_x &= 0, & L_x &= 0, \\
F_y &= 0, & L_y &= 0, \\
F_z &= 0, & L_z &= 0,
\end{aligned}
\qquad [84]
$$

where L_x, L_y, and L_z are the sums of the torques tending to cause rotation about the X-, Y-, and Z-axes respectively.

74. Center of Gravity. In his theory of the lever,[1] ARCHIMEDES made use of the fact that there is a point in any body, called its *center of gravity*, such that the body will balance in all positions when supported at this point. Suppose, for example, that the center of gravity C of a flat piece of tin has been located by experiment and that the piece of tin is then supported on a horizontal axis through C so as to be free to rotate in a vertical plane (Fig. 55). Let C be the origin of rectangular coordinates, the Z-axis the axis of rotation, and the XY-plane the plane of rotation. Since the body is near the surface of the earth, not only is each of its n particles subject to the force of gravity but the directions of all these n forces may be considered as sensibly parallel. The only

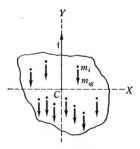

FIG. 55. When a small body is located near the earth, its weight may be conceived as concentrated at the center of mass

other force acting on the body is f, the upward thrust of the axis C. Since the body is supported at the center of gravity, it is in complete equilibrium in all positions, and hence, by Eqs. [84],

$$
\begin{aligned}
F_x &= 0, & L_x &= 0, \\
F_y &= f - \sum_{i=1}^{i=n} m_i g = 0, & L_y &= 0, \\
F_z &= 0, & L_z &= \sum_{i=1}^{i=n} m_i g x_i = 0;
\end{aligned}
$$

or
$$
f = g \sum_{i=1}^{i=n} m_i = Mg, \qquad \sum_{i=1}^{i=n} m_i x_i = 0. \qquad [85]
$$

---

[1] The first mathematical treatment of the lever, in a literal translation of ARCHIMEDES' own words, and also a less cumbrous proof due to STEVIN, are given in *J. Cox, *Mechanics* (Cambridge University Press, 1919), pp. 4–6. See also T. L. Heath, *The Works of Archimedes* (Cambridge University Press, 1897), pp. 189–194, and E. Mach, *The Science of Mechanics* (Open Court, 1893), pp. 8–19.

The first of these two equations tells us that the line of action of the total weight Mg of the body passes through the center of gravity, however the body is situated with reference to the earth, and this explains why the center of gravity is sometimes defined as that point at which the weight of the body may be conceived as concentrated. The second equation, when interpreted in the light of Eqs. [72], enables us to conclude that *the center of gravity coincides with the center of mass.*

One speaks of the center of gravity only when the body is situated in a *uniform* gravitational field; that is, when the body is in a region in which the gravitational force acting on a particle does not change either in magnitude or direction as the particle moves about in it. Thus the term *center of gravity* is appropriate only when applied to a small body at the earth's surface. If a body has any center of gravity at all, this point must always coincide with its center of mass. The center of mass is, however, a perfectly definite point which depends in no way upon the location of the body and which has remarkable properties quite independent of weight (Sec. 70).

ARCHIMEDES' definition of the center of gravity as the point such that the body will balance in all positions when supported thereat affords an excellent method of finding either the center of gravity or the center of mass experimentally. It also enables one to calculate the position of the center of mass from the principle of moments. The student should solve the illustrative examples given in Sec. 69 from this point of view.

75. Solution of a Typical Problem in Statics. Although a good part of the value to be gained from solving problems consists in developing the ability to analyze any new physical situation with which one is confronted and to plan a mode of attack, it is still useful to outline the general procedure which experience has shown to be the most effective in attacking a given type of problem. The solution of a problem in statics is usually best carried out in the following steps:

1. Isolate the body (or the parts of the body one at a time) to which the problem applies. The selection of the body to be isolated is made in such a manner that all external forces are either known or can be calculated by the method given below and so that the internal forces between the various parts of the body play no part in the problem. This is the step that differentiates one problem from another and that will test the student's ingenuity and physical insight. Once it is successfully carried out, the solution of the problem becomes largely a matter of routine.

2. Make a diagram in which the known and unknown forces involved in the problem are represented by vectors. The forces which are considered must all act on the *same* body, and *all* the forces that act on the body must be taken into account.

3. Choose three mutually perpendicular coordinate axes, compute and tabulate the components of the forces with reference to them, and then write the three equations that express the conditions for equilibrium as regards translation.

4. Compute and tabulate the torques, about the chosen axes, of the components of the various forces; then write the three equations that express the conditions for rotational equilibrium. If there is any force with which you do not wish to deal, choose the origin of your set of coordinate axes at the point of application of this force.

5. Solve the resulting equations for the required unknown quantities.

EXAMPLE. A uniform ladder of length S and weight f_1 stands on a horizontal floor and leans against a vertical wall, the coefficients of static friction at the floor and wall being μ and μ' respectively. The ladder makes an angle ϕ with the floor. If a man of weight f_2 starts to climb the ladder, how far can he ascend the ladder before it starts to slip?

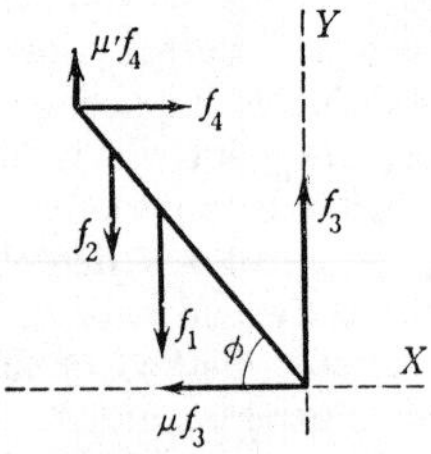

Solution. The ladder is chosen as the body to be isolated in this problem. A force diagram is then drawn, with the forces acting *on the ladder* represented by arrows. An origin and set of axes are next chosen, as shown in Fig. 56, this particular choice of origin being suggested by the fact that two forces act through this point. The axes are orientated in such a way as to make the determination of the force components as simple as possible. Let s be the unknown distance up the ladder to the point where the man is standing and let f_3 be the vertical component of the floor reaction and f_4 the horizontal component of the wall reaction. Since the ladder is on the point of slipping, the frictional forces are μf_3 and $\mu' f_4$ in the directions shown. Applying the conditions for complete equilibrium, Eqs. [84],

FIG. 56. The ladder is the only body shown in the force diagram, for the floor, the wall, and the man have been replaced by the forces which they exert

$$F_x = -\mu f_3 + f_4 = 0,$$
$$F_y = f_3 - f_1 - f_2 + \mu' f_4 = 0,$$
$$F_z = 0,$$
$$L_x = 0,$$
$$L_y = 0,$$
$$L_z = f_1 \frac{S}{2} \cos \phi + f_2 s \cos \phi - f_4 S \sin \phi - \mu' f_4 S \cos \phi = 0.$$

Eliminating the unknown forces f_3 and f_4 and solving for s,

$$s = \frac{S}{f_2} \left[\frac{\mu(f_1 + f_2)}{1 + \mu\mu'} (\tan \phi + \mu') - \frac{f_1}{2} \right].$$

(What is the effect on s of placing the ladder closer to the wall?)

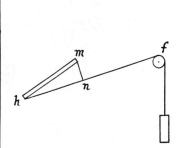

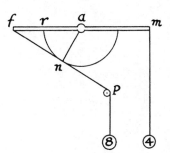

If the bar is pivoted at *m*, then *nm*, which is perpendicular to *hf*, is the lever arm

If $an = ar = af/2 = am/2$, then, for equilibrium, the weights are in the ratio 8 to 4, as shown

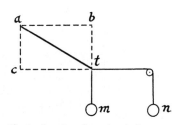

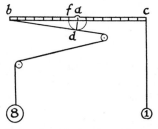

If *at* is a bar pivoted at *a*, then the lever arm of *m* is *ab*, not *at*, and the lever arm of *n* is *ac*, not *at*; hence *m* and *n* are in the ratio of *ac* to *ab*, when the bar is in equilibrium

If $ad = af = ab/8$, then the weights are in the ratio 8 to 1, when the bar is in equilibrium

Copies of Illustrations
from the Manuscripts of LEONARDO DA VINCI

THESE diagrams show LEONARDO DA VINCI's mastery of the principles of the lever, even for oblique forces, and his appreciation of the significance of the perpendicular from the axis of rotation to the line of action of the force.

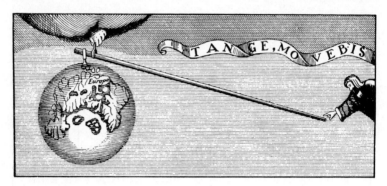

An Illustration from Varignon's

PROJET D'UNE NOUVELLE MECANIQUE (Paris, 1687)

Archimedes' appreciation of the advantages of the lever and the other "mechanical powers" is clearly indicated by the words attributed to him: "Give me a place to stand on, and I can move the earth." The vignette above was drawn to illustrate this saying. The Latin motto may be translated thus: "Touch it and you will move it."

The account given in Plutarch's "The Life of Marcellus" of the extent to which Archimedes was able to make good his boast is worth quoting:

"But *Archimedes* havinge tolde king *Hieron*, his kinseman and very frende, that it was possible to remove as great a weight as he would, with as little strength as he listed to put to it: and boasting him selfe thus (as they reporte of him) and trusting to the force of his reasons, wherewith he proved this conclusion, that if there were an other globe of earth, he was able to remove this of ours, and passe it over to the other: kinge *Hieron* wondering to heare him, required him to put this devise in execution, and to make him see by experience, some great or heavy weight removed, by little force. So *Archimedes* caught hold with a hooke of one of the greatest carects, or hulkes of the king (that to draw it to the shore out of the water, required a marvelous number of people to go about it, and was hardly to be done so) and put a great number of men more into her, than her ordinary burden: and he himselfe sitting alone at his ease farre of, without any straining at all, drawing the ende of an engine with many wheeles and pullyes, fayer and softly with his hande, made it come as gently and smoothly to him, as it had floted in the sea." — Translation by Thomas North (1579).

76. The Couple. An interesting case arises when a body is acted upon by two forces that are equal in magnitude and opposite in direction, but that do not act along the same line (Fig. 57). An example is the pair of forces applied by the thumb and finger in winding a watch. Such a pair of parallel forces was given the name *couple* in 1803 by the French mathematician LOUIS POINSOT.[1]

The two forces constituting a couple add up to zero, and there-- fore a couple has no tendency to produce translation of the body on which it acts. But it does tend to pro- duce rotation. Its torque about any axis is the algebraic sum of the torques of its two forces about the same axis. The value of this torque depends in general upon the axis about which the rotation is supposed to take place, but it may easily be shown that for an axis perpendicular to the plane in which the two forces act, the torque of a couple is equal to the magnitude of either force multiplied by the distance between the lines of action of the forces. The stu- dent should prove this, first for the case when the normal axis is at any point O

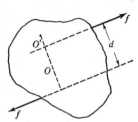

FIG. 57. A rigid body under the action of a couple. The magnitude of the couple is fd for any axis normal to the plane of the forces

between the forces and then when it is at O', outside their lines of action (Fig. 57). It may also be proved that if the axis makes an angle ϕ with the normal to the plane of the forces, the magnitude of the torque is $fd \cos \phi$. We thus arrive at the important result that *the magnitude of a couple is the same for all axes that are parallel*; in other words, a couple, unlike either a force or the torque of a force, is a *nonlocalized*, or *free*, vector.

Since a couple acting on a body has no effect on the velocity of the center of mass, the change in rotation produced by the couple must be about some axis through the center of mass. A couple evidently cannot be balanced by a single force but only by another couple of equal magnitude fd but opposite sign, lying in the same plane or in a parallel plane.[2] The theory of couples, developed by POINSOT, provides a method for simplifying complicated systems of forces, for it shows that the whole of the forces acting on a rigid body are always reducible to a single force, acting at a given point, and a couple.

---

[1] *Éléments de statique* (1803). An excerpt from a translation by T. Sutton (1848) is given in *A *Source Book in Physics* (1935), pp. 65–68.

[2] For an elementary discussion of other properties of the couple, see *J. Cox, *Mechanics* (Cambridge University Press, 1919), pp. 179–185.

EXPERIMENT VI. ROTATIONAL EQUILIBRIUM AND
THE BEAM BALANCE

The object is to become familiar with the conditions for rotational equilibrium and with the principles involved in the beam balance. The whole experiment is based upon the principle of the equality of opposing torques when no rotation exists. The beam of the model balance (Fig. 58) is a meter stick provided with a sliding frame which may be clamped at any desired point on the stick. This frame carries a knife-edge c which is fixed and a knife-edge b which is adjustable ver-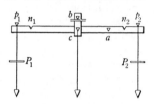tically. The frame and beam are supported by resting either of these knife-edges in an adjustable groove clamped to a tripod support. The beam is also provided with a knife-edge a at the 75-cm mark, and with two notches n_1 and n_2.

Fig. 58. Model of a balance

The pans P_1 and P_2 are hung from the upper edge of the beam on knife-edges p_1 and p_2. Plumb lines are dropped from p_1, p_2, and b to facilitate the measurement of the lever arms.

Part I. The Beam Balance. *a. Weights of the Pans.* Support the beam from the knife-edge b, which should be set about 1 cm above c, and slip the beam through the frame until it balances in an approximately horizontal position. Then hang the pans P_1 and P_2 from the beam, well out toward the ends, and slide them along the beam until the latter is again in horizontal equilibrium. Read off the lever arms of the two pans on the graduated beam. Now place a 100-g weight in the pan P_1 and move this pan in toward the fulcrum until the horizontal equilibrium is restored. Record the new lever arms. For each case write down the condition for rotational equilibrium, namely, Eq. [82], and solve these two equations for the weights P_1 and P_2. If the weights of the pans, with attached plumb bobs, are not stamped upon them, weigh each pan on the laboratory scales and thus check your results.

b. Bringing the Center of Gravity of the System into Coincidence with the Knife-edge c. Remove the pans and raise the upper knife-edge as high as is possible. Support the beam this time from the knife-edge c, and adjust the positions of the movable knife-edge and of the beam until the latter shows no tendency to move out of any position in which it is placed. The system consisting of the beam and its fittings is then in *neutral equilibrium* about c, and hence its

center of gravity has been brought into coincidence with the point *c*.
The foregoing adjustment will be greatly facilitated by the use of
rubber bands or a small bit of soft wax, which may be attached to
the beam at any desired point. Also the
final adjustments can be made more accurately if the knife-edge *c* is made to rest
on the smooth portion of its support
rather than in the notch.

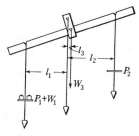

c. Weight of the Beam. Having brought
the center of gravity of the system into
coincidence with *c*, again support the
beam from *b*, hang the pans this time in
the notches, and place known weights
W_1 in the pan P_1 until the beam assumes
a position inclined about 45° to the hori-

FIG. 59. One method for
weighing the beam

zontal (Fig. 59). The equation for rotational equilibrium now involves the three forces $P_1 + W_1$, P_2, and the weight W_3 of the beam,
and the three corresponding lever arms l_1, l_2, l_3. The lever arms
are to be measured by bringing a meter stick, which is held in a
horizontal position in a meter-stick clamp, up to the plane of the
plumb lines hanging from the knife-edges. It may prove better to
measure l_3 with the help of a vernier caliper.[1] (Why?) Write the
equation for rotational equilibrium and solve for W_3.

Obtain the weight W_3 of the beam in another way, by supporting
it upon the knife-edge *a* and producing horizontal equilibrium by
means of a known weight *W* (Fig. 60).

Compare the two values of W_3 thus obtained with that found
stamped on the beam or determined by weighing the beam on the
laboratory scales.

PART II. Double Weighing. All the
errors of a balance are eliminated by a
double weighing, a method devised by
KARL FRIEDRICH GAUSS (1777–1855).
To see this, hang *unequal* pans P_1 and
P_2 from the notches and slip the beam
through the frame until a balance is

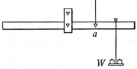

FIG. 60. A second method for
finding the weight of the beam

obtained. The arms of the balance are now unequal, the pans are
of unequal weight, and the center of gravity of the beam is not
beneath the point of support. Put an unknown weight *x* into P_1
and find by trial the known weight W_1 that balances it. Next
place *x* in P_2 and balance it by means of a known weight W_2 placed

---

[1] See Appendix 5, *Vernier Caliper.*

in P_1. If l_1 and l_2 are the lever arms of P_1 and P_2 respectively, the equation for rotational equilibrium gives $xl_1 = W_1l_2$ and $xl_2 = W_2l_1$. The solution of these two equations for x gives $x = \sqrt{W_1W_2}$. As can easily be shown,[1] if W_1 and W_2 have nearly the same value, it is sufficiently correct to write $x = (W_1 + W_2)/2$.

As a check, find x by weighing on the laboratory scales. Compute the percentage of difference.

1. Explain why a balance beam returns to a horizontal position when displaced therefrom.

2. Why must the center of gravity of the beam be below the knife-edge? What would happen if it were exactly at the knife-edge? above the knife-edge?

3. In Part I, c, since the arms l_1 and l_2 are measured with an ordinary meter stick, what is to be gained by using a vernier caliper for measuring l_3?

4. Which of the three determinations of the weight of the beam do you consider the most trustworthy? the least trustworthy? Why?

5. Prove that if W_1 and W_2 have nearly the same value, then $\sqrt{W_1W_2}$ is approximately equal to $(W_1 + W_2)/2$.

○

OPTIONAL LABORATORY PROBLEM

Sensitivity of a Balance for Different Loads. A balance is said to have great sensitivity when a very small difference of weights causes a large deflection. The *sensitivity* may be defined as the displacement produced when some arbitrarily chosen small weight w is added to one pan.

a. Arrange a laboratory model of a balance in such a way that the knife-edges p_1, p_2, b, Fig. 58, are not in the same straight line, and determine whether the sensitivity depends upon the load. To do this, set the knife-edge b about 4 cm above the line connecting p_1 and p_2 and support the beam from b. Hang the pans from the beam, but not in the notches, since a little friction at the knife-edges completely vitiates the results. Bring the beam into a horizontal position and then read off on a vertically placed meter stick the deflection of the beam produced by adding 1 gwt to one of the pans. Add, say, 300 gwt to each pan and, if necessary, again bring the beam to a horizontal position by changing the position of the pan to which the small weight was *not* added. Find the sensitivity for this new load. How does the sensitivity change with the load? What if b were below the line p_1p_2?

b. Arrange the balance so that the knife-edges p_1, p_2, and b are in the same straight line, and find how the sensitivity varies with the load in this case. If a slight bending of the beam occurred with the increased load, how would this affect your results?

---

[1] See Appendix 3, *Errors Introduced by Common Approximations.*

QUESTION SUMMARY

1. Define *center of mass*. What is its physical importance? How is its position calculated?

2. Define *center of gravity*. Distinguish between center of gravity and center of mass.

3. Define *torque*. Distinguish between torque and work. Do they have the same dimensions?

4. State and illustrate the *principles of moments*.

5. State the conditions for rotational equilibrium.

6. What are the conditions for complete equilibrium of a body?

7. What is meant by a *couple*? How may one compute the magnitude of the torque produced by a couple? What is required to balance a couple?

o

PROBLEMS

1. At the corners of a square of area 4 ft² are placed particles of masses 1, 2, 3, and 4 lb respectively. Find the position of the center of mass of this system.

> *Ans.* If the 1-lb and 2-lb masses lie on the X-axis and the 1-lb and 4-lb masses on the Y-axis, then $x_c = 1$ ft, $y_c = 1.4$ ft.

2. A thin iron rod 63 cm long is bent so that the two parts, of lengths 27 cm and 36 cm, are at right angles to each other. Find the position of the center of mass.

Ans. If the shorter arm is made the X-axis, then $x_c = 5.8$ cm, $y_c = 10$ cm.

3. A cubical block having 1-ft edges has a cylindrical cavity 10 in. in diameter and 8 in. deep cut centrally in the top. Show that if this cavity is half filled with a liquid one fourth as dense as the material composing the vessel, the center of mass of the whole is 7 in. below the top of the vessel.

4. The line of action of a 1000-dyne force lies in the XZ-plane and cuts the Z-axis at a point 6 cm from the origin. What is the torque about the Y-axis if the angle between the line of action of the force and the Z-axis is (*a*) 60°? (*b*) 180°? (*c*) 330°?

> *Ans.* (*a*) $3\sqrt{3} \times 10^3$ dyne · cm; (*b*) 0; (*c*) -3×10^3 dyne · cm.

5. Two men carry between them a load of 50 kgwt supported upon a uniform pole weighing 10 kgwt. Where must the load be placed in order that one man may carry twice as much of the whole weight as the other?

> *Ans.* 0.3 l.

6. A man wishes to overturn a hollow cubical block which has a 5.0-ft edge and a mass of 100 kg. (*a*) In what direction and with what force must he push in order that he may do this most easily? (*b*) After the block has once been set in motion will the required force increase or decrease? Why?

(c) If the coefficient of friction is 0.4, will the block slide or tip when the man pushes horizontally? when he pushes at the best angle for overturning the block? *Ans.* (a) 45°, $50/\sqrt{2}$ kgwt; (b) decrease; (c) slide; not slide.

Fig. 61. Problem 7

7. A circular ring of weight 5 lbwt rests horizontally upon three points of support 120° apart (Fig. 61). What is the least downward force, applied to the ring in a direction perpendicular to its plane, that will cause it to leave one of the points of support? *Ans.* 5 lbwt.

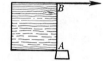

Fig. 62. Problem 8

8. A uniform board 3.00 ft square and weighing 25.0 lbwt rests on a block at A, Fig. 62, and is kept from rotating by a horizontal force at B. Find the force at B and the vertical and horizontal forces upon the block at A. *Ans.* 12.5 lbwt, 25.0 lbwt, 12.5 lbwt.

9. A 100-lb bench is dragged steadily along the floor by the force F, Fig. 63. The center of mass of the bench is at c and the coefficient of sliding friction between the bench legs and the floor is 0.10. Find the magnitude of F and of the downward push of each leg on the floor.
Ans. $F = 11$ lbwt, $F_A = 49$ lbwt, $F_B = 45$ lbwt.

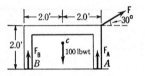

Fig. 63. Problem 9

10. Given the five-panel truss shown in Fig. 64, determine the magnitude of the force exerted by each member and also whether it produces a tensile or a compressive force. The lengths of the members are in the ratio 3 : 4 : 5.

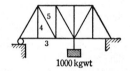

Fig. 64. Problem 10

11. The *sensitivity* of a balance may be defined as the displacement produced in the beam when some arbitrarily chosen small weight w is added to one pan. Given a balance having arms of equal length l and the knife-edges p_1, p_2, b in the same straight line (Fig. 65), (a) show that for such an ideal balance the addition of w to one pan will cause the beam to come to rest at an angle θ with the horizontal, such that $\tan \theta = wl/Wd$, where W is the weight of the beam and d the distance of its center of gravity from the point of support b. (b) Does the sensitivity depend upon the load in the pans? (c) What are some of the points to be observed in designing a satisfactory balance of high sensitivity? (d) Is it necessary that the center of gravity of the beam be below the point of support?

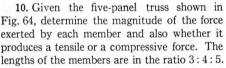

Fig. 65. Sensitivity of an ideal balance

12. The middle knife-edge b of a certain balance is at a perpendicular distance h above the middle point of the straight line p_1p_2 connecting the two pan knife-edges, and the center of gravity is distant d below the knife-edge b. (*a*) Show that the beam will come to rest when p_1p_2 makes an angle θ with the horizontal such that

$$\tan \theta = w \frac{l}{Wd + (P_1 + P_2)h},$$

where P_1 and P_2 are the unequal loads on the pans, w is the difference between P_1 and P_2, and $2\,l$ is the length of the line p_1p_2. (*b*) How does the sensitivity change as the load is increased? (*c*) Write and interpret the expression for $\tan \theta$ for the case where the knife-edge b is below the line p_1p_2; (*d*) for the case where $h = 0$.

13. A uniform rod 10.0 m long and of mass 20.0 kg rests with the upper end against a smooth vertical wall, with which it makes an angle of 60°. The lower end is prevented from slipping by a peg in the floor. Find the forces exerted on the rod at the wall and at the peg. Is the resultant force on the peg in the direction of the rod? *Ans.* Wall, 17.3 kgwt, horizontal; peg, 17.3 kgwt, horizontal, 20.0 kgwt, vertical; no.

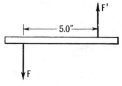

Fig. 66. Problem 14

14. (*a*) Calculate the magnitude, direction, and point of application of the third force required to balance the two parallel forces F and F′ shown in Fig. 66 if the magnitudes of these forces are 20 and 15 lbwt respectively, and the distance between them is 5.0 in. (*b*) What single force is equivalent to F and F′? (*c*) What single force would balance the two forces F and F′ if they were equal in magnitude?
Ans. (*a*) 5.0 lbwt up, 20 in. from F′.

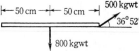

Fig. 67. Problem 15

15. Find the magnitude, direction, and point of application of the single force that will prevent the bar shown in Fig. 67 from moving. Consider the weight of the bar to be negligible. *Ans.* 640 kgwt, 129°, 20 cm from left end.

CHAPTER SEVEN

THE DYNAMICS OF RIGID BODIES

IF EACH element of a body be multiplied into the square of its distance from the axis OA and all these products be collected into one sum and if this sum is put $= Mkk$, which I call the moment of inertia of the body with respect to the axis OA, then the moment of force required to produce acceleration a will be $Mkk \cdot a$.

Translated from a paper by LEONHARD EULER
which is reproduced in his *Opera Postuma* [1]

○

As has already been remarked in the introduction to the preceding chapter, the general problem of rotations may be approached by regarding a rotating body as a system of particles and applying to these particles NEWTON'S three laws of motion. Since the linear speeds of the particles vary with their distances from the axis of rotation, this procedure is laborious and complicated, but fortunately we do not have to employ it every time a problem involving rotation is solved. The fact that, in equilibrium, torques bear the same relation to rotation that forces bear to translation (Sec. 72) suggests the practicality of defining still other rotational analogues of the various translational quantities. For the case of rigid bodies it turns out that these new rotational quantities can be so defined that they reduce the equations for rotation to forms which are not only exceedingly simple but are the exact analogues of the familiar equations for translation. No really new mechanical principles are revealed by such transformations, but they save us the individual consideration of the separate particles and afford simple rules for investigating the rotational motion of a body as a whole with a minimum expenditure of thought. This economy of thought is one of the necessities for accomplishment in any field of knowledge.

○

The Kinematics of Rotation

Consider the simple though important case of a rigid body rotating about a *fixed axis*, for example, the flywheel of a stationary engine.

---

[1] EULER'S notation has been partly modernized in this translation. The use of the notation kk (introduced by THOMAS HARRIOT in 1631), instead of the modern k^2, persisted even beyond the time of EULER, although exponential notation was used by RENÉ DESCARTES in 1637. The earliest known attempts to frame such symbolic notations were those of RAFAELLO BOMBELLI in 1572 and SIMON STEVIN in 1586. See W. W. Rouse Ball, *A Short Account of the History of Mathematics* (Macmillan, 1927), p. 242.

The particles of the body move in circles about the axis and turn in equal times through equal angles (Sec. **5**). Consequently, if polar coordinates (Fig. 68) be employed to describe the positions of the particles, the angle θ is the only space-coordinate that changes during the rotation, and its change is the same for every particle of the body.

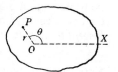

77. Angular Position and Angular Speed. Let r in Fig. 68 be the perpendicular distance of a particle P of the body from the axis of rotation O, and let OX be a fixed reference line.

Fig. 68. A rigid body rotating about a fixed axis O that is perpendicular to the plane of the paper

If $\theta_0 (= XOP)$ be the *angular position*, or *phase* as it is usually called, of the particle P at the time t_0, and if θ_1 be its angular position, or phase, at a later time t_1, then the body has turned through the angle $\theta_1 - \theta_0$ in the interval of time $t_1 - t_0$. The ratio of the angle turned through to the corresponding time interval is called the *average angular speed* ω_{av} during that time interval. By passing to the limit as the time interval approaches zero, one has the defining equation for *instantaneous angular speed*, namely,

$$\omega_t = \lim_{(t_1 - t_0) \to 0} \frac{\theta_1 - \theta_0}{t_1 - t_0} = \frac{d\theta}{dt}. \qquad [86]$$

It will be observed that the definition for angular speed is precisely analogous to the definition for linear speed given in Sec. **12**. If ω and n denote angular speeds expressed in radians per unit time and in number of complete rotations per unit time respectively, the two are evidently connected by the relation $\omega = 2\pi n$.

When the angles are expressed in *radians*, as is usually the case in theoretical discussions, the simple relation

$$s = r\theta \qquad [87]$$

exists between the angle θ turned through by the body and the linear distance s traversed by a particle of the body distant r from the axis; this relation follows immediately from the definition of a radian. By dividing both members of Eq. [87] by Δt and passing to the limit, there results

$$v_t = r\omega_t, \qquad [88]$$

which is an equally simple relation existing between the angular speed of a body and the linear speeds of its particles.

78. Angular Acceleration. When a body is rotating about a fixed axis with varying speed, the magnitude α_t of its *angular acceleration* is given by the equation

$$\alpha_t = \lim_{(t_1 - t_0) \to 0} \frac{\omega_1 - \omega_0}{t_1 - t_0} = \frac{d\omega}{dt}, \qquad [89]$$

where ω_0 and ω_1 are the angular speeds of the body at the times t_0 and t_1 respectively. It is left to the student to prove that, for a body having an angular acceleration of magnitude α_t, any particle distant r from the axis of rotation has a *linear* acceleration *along the tangent to its path* of amount

$$a_t = r\alpha_t. \qquad [90]$$

Note carefully that a_t in this equation is only *one component* of the *total* linear acceleration; the particle is moving in a circle and hence must also have a component of acceleration in a direction normal to its path (see Chapter 14).

79. Constant Angular Acceleration about a Fixed Axis. If the axis of rotation is fixed and the angular speed is changing by equal amounts in equal intervals of time, we have a case of constant angular acceleration which is precisely analogous to that of the constant linear acceleration treated in Sec. **19**. Consequently the student should experience no difficulty in deriving the following three equations:

$$\omega_t = \omega_0 + \alpha t, \qquad [91]$$

$$\theta = \omega_0 t + \tfrac{1}{2}\alpha t^2, \qquad [92]$$

$$\omega_t{}^2 = \omega_0{}^2 + 2\,\alpha\theta. \qquad [93]$$

These equations, like their linear analogues (Secs. **19** and **20**), represent a very restricted type of acceleration, but they will often be of use in connection with the motion of rigid bodies turning on a fixed axis and subject to constant external influences such as friction or driving weights. Obviously one must never attempt to apply them to an actual motion of rotation until he has made certain that the angular acceleration is constant.

80. Vectorial Nature of the Angular Quantities. As in the case of linear motion, the description of an angular motion is not complete unless both the direction and the magnitude of the motion are specified. For this reason it is found advantageous to define three new directed quantities which, for obvious reasons, are given the names *angular displacement*, *angular velocity*, and *angular acceleration*.

The two specifications that describe an angular displacement $\boldsymbol{\theta}$ are (*a*) the angle θ through which a body has rotated and (*b*) the direction of the axis of rotation, together with the sense of rotation about this axis. Evidently these two specifications can be represented graphically by laying off *along the axis of rotation* a directed line of length numerically equal to the number of radians in θ. There must also be some agreement as to which direction along the axis shall represent a given sense of rotation about the axis; the adopted rule is to let the sense of the directed line along the axis be related to the sense of rotation as the direction of advance of an ordinary right-handed screw is related to its direction of rotation (Fig. 69).

A *finite* angular displacement is not properly a vector quantity, since the addition of two *successive finite* angular displacements about different axes is not commutative. To see this, hold a book in the hand in any fixed position and imagine a set of rectangular coordinate axes drawn in it. Rotate the book 90° in a clockwise sense about the *X*-axis and then, from this position, 90° clockwise about the *Y*-axis; note the final position of the book. Now reverse the order of these two rotations, starting with the book in the same initial position as before; its final position is quite different from what it was in the first case. If the experiment with the book be repeated, but this time with the two displacements made very small, it will become clear that the sum of two *infinitesimal* angular displacements *is* independent of the order in which the quantities are added. Thus an angular displacement that is infinitesimal may be regarded as a vector.

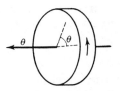

Fig. 69. Illustration of the right-handed-screw rule. Notice that this is also the relation which exists between the directions of an electric current and the corresponding magnetic field

Angular velocity ω_t is the analogue of linear velocity, and it bears the same relation to angular speed that linear velocity bears to linear speed. The two specifications that describe an angular velocity are the angular speed ω_t in Eq. [86], and the direction of the axis of rotation together with the sense of rotation about the axis. As in the case of angular displacement, angular velocity may be represented graphically by a directed line drawn along the axis of rotation in accordance with the right-handed-screw rule. For example, since the angular speed of the earth about its axis is 2π radians per day, a directed line to represent the angular velocity of the earth would be made 2π units long and would be drawn on the earth's axis, toward the north pole. The addition of angular velocities is commutative, since angular velocity is, by definition, an infinitesimal angular displacement $d\theta$ divided by the corresponding infinitesimal time dt.

Similar considerations hold for *angular acceleration* α_t, which is the analogue of linear acceleration. Evidently the statement that a body has an angular acceleration may imply (*a*) a change $\Delta\omega$ in the angular speed, while the direction of the axis of rotation remains fixed, this being the case already discussed in Secs. **78** and **79**; (*b*) a change in the direction of the axis while the speed about the axis remains constant; (*c*) a simultaneous change in both of these quantities, this being, of course, the most general case.

The Dynamics of Rotation about a Fixed Axis

81. Kinetic Energy of Rotation. The kinetic energy of any body in motion is equal to the sum of the kinetic energies of its n particles (Sec. 50); that is,

$$E = K \cdot \tfrac{1}{2}(m_1 v_1{}^2 + m_2 v_2{}^2 + \cdots + m_n v_n{}^2). \qquad [94]$$

If the motion of the body is one of pure translation, so that all the particles have the same linear speed v, Eq. [94] reduces to $E = K \cdot \tfrac{1}{2} M v^2$, where M is the total mass of the body. This is an equation that we have made use of many times. On the other hand, if the motion is one of pure rotation, Eq. [94] cannot immediately be simplified, since the linear speeds of the particles then differ. The angular speeds are the same, however, provided the body is rigid, and hence we may write, in view of Eq. [88],

$$E = K \cdot \tfrac{1}{2} \, \omega^2 (m_1 r_1{}^2 + m_2 r_2{}^2 + \cdots + m_n r_n{}^2). \qquad [95]$$

To simplify the notation let us denote the quantity in parentheses by the symbol I; that is, let

$$I = m_1 r_1{}^2 + m_2 r_2{}^2 + \cdots + m_n r_n{}^2 = \sum_{i=1}^{i=n} m_i r_i{}^2. \qquad [96]$$

The equation for the kinetic energy of a rotating body then becomes

$$E = K \cdot \tfrac{1}{2} I \omega^2. \qquad [97]$$

One will note the similarity of this expression to that for kinetic energy of translation, namely, $K \cdot \tfrac{1}{2} M v^2$, the quantity I corresponding to mass and the angular speed to linear speed. We shall find the quantity I appearing in all kinetical problems that involve rotation. It was familiar to HUYGENS through his study of the physical pendulum, but was first given a name by LEONHARD EULER [1] (1707–1783), who called it the *moment of inertia of the body about the axis of rotation*. It is evident from the definition, Eq. [96], that I may be found by imagining the rotating body divided into minute particles, multiplying the mass of each particle by the square of its distance from the axis, and then adding these products.

82. Energy Equation for a Rotating Body. Consider a body mounted on a fixed axis and acted upon by a constant torque of magnitude L. Let the moment of inertia of the body about the given axis be denoted by I. While the body is turning through an angle θ, the

---

[1] L. Euler, *Theoria Motus Corporum Solidorum seu Rigidorum* (1765), p. 166. See also the quotation at the beginning of this chapter.

torque L does work of amount $L\theta$ (Fig. 70). If the body be rigid and the friction of the bearings be negligible, this work will equal the increase in the kinetic energy, or

$$L\theta = K \cdot \tfrac{1}{2} I(\omega_1{}^2 - \omega_0{}^2). \qquad [98]$$

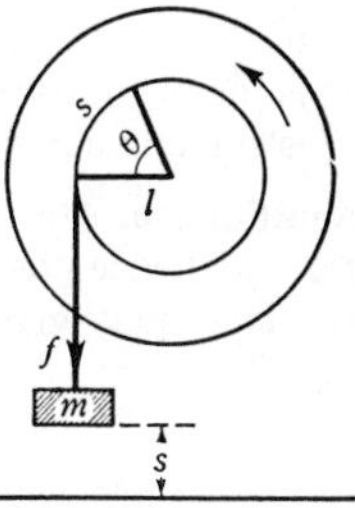

83. Rotational Analogues of Newton's Laws of Motion. In the foregoing case, let t denote the time of rotation through the angle θ. The average angular speed during this time is $\frac{1}{2}(\omega_1 + \omega_0)$, and therefore $\theta = \frac{1}{2}(\omega_1 + \omega_0)t$. By substituting this expression for θ in Eq. [98] and solving for L, there results

$$L = K \cdot \frac{I\omega_1 - I\omega_0}{t}. \qquad [99]$$

This equation for the rotation of a body which is acted upon by a constant torque is merely a special case of the general vector

Fig. 70. Work done by a torque. While the wheel rotates through the angular distance θ, the point of application of the force moves through the linear distance s; hence $W = f \cdot l\theta = L\theta$

equation for the rotation of a rigid system about any fixed axis, namely,

$$\mathsf{L} = K \cdot \lim_{\Delta t \to 0} \frac{I\omega_1 - I\omega_0}{\Delta t} = K \frac{d}{dt}(I\omega) = K \frac{d\mathsf{H}}{dt}, \qquad [100]$$

where $\mathsf{H} = I\omega$. From the analogy of linear momentum (Sec. **25**), the vector quantity $I\omega$, the product of moment of inertia and angular velocity, is called the *angular momentum* or *moment of momentum* of the body about the axis in question. Eq. [100] then asserts that the *time-rate of change of the angular momentum of a body is proportional to the impressed torque, and takes place about any axis having the direction of the torque.* This principle is the rotational analogue of NEWTON's second law of motion (Sec. **28**).

If $\mathsf{L} = 0$, then $I\omega_1 = I\omega_0$; that is, the *angular momentum of a body is constant if the vector sum of the external torques is zero.* This is the rotational analogue of NEWTON's first law (Sec. **27**) and is the condition for equilibrium as regards rotation which we obtained in Sec. **73**. Sometimes the moment of inertia changes while the body is in free rotation, and the angular velocity must then change in an inverse ratio. Thus an acrobat turning in the air can regulate the rate at which he turns by thrusting out or drawing in his legs or arms.

In view of the definition of angular acceleration, Eq. [100] may also be written in the form

$$\mathsf{L} = K \cdot I\alpha, \qquad [101]$$

provided I does not change with the time.

EXAMPLE. A flywheel mounted on a shaft of radius l is set in rotation by a driving weight of mass m (Fig. 70). If I be the total moment of inertia of the wheel and shaft about the axle and if L' be the magnitude of the torque due to the frictional forces at the axle bearing, (*a*) what is the magnitude of the angular acceleration of the wheel, and (*b*) what is the angular speed when m has descended a distance s from rest?

Solution. *a*. The force due to the driving weight is $m(g - a)$ absolute units, and hence the total accelerating torque is $m(g - a)l - L'$ absolute units. The equation of motion of the system is, therefore, $m(g - a)l - L' = I\alpha$, by Eq. [101], or

$$\alpha = \frac{mgl - L'}{I + ml^2}.$$

b. Since α is constant, we may employ Eq. [93]. Remembering that $\omega_0 = 0$ and $s = r\theta$, we finally obtain

$$\omega = \left(2\,\frac{mgl - L'}{I + ml^2} \cdot \frac{s}{l}\right)^{\frac{1}{2}}.$$

A second and more direct way of getting this result is to equate the initial potential energy of the system, mgs, to the sum of the work done against friction, $L'\theta$, and the final kinetic energy, $\frac{1}{2}mv^2 + \frac{1}{2}I\omega^2$.

As for the rotational analogue of NEWTON'S third law (Sec. 32), both experiment and theory lead to the conclusion that for every torque twisting a body about any axis there always exists a corresponding torque exerted *by the body* about the same axis; and these two torques are equal in magnitude but opposite in direction. In other words, when one of the two interacting bodies exerts a torque

TABLE II · *The Important Linear Expressions and Their Angular Analogues*

	Linear	Angular
Displacement	$\mathbf{s}$	$\boldsymbol{\theta}$
Velocity	$\mathbf{v} = \dfrac{d\mathbf{s}}{dt}$	$\boldsymbol{\omega} = \dfrac{d\boldsymbol{\theta}}{dt}$
Acceleration	$\mathbf{a} = \dfrac{d\mathbf{v}}{dt}$	$\boldsymbol{\alpha} = \dfrac{d\boldsymbol{\omega}}{dt}$
Constant acceleration	$v_t = v_0 + at$, etc.	$\omega_t = \omega_0 + \alpha t$, etc.
Inertia	M	I
Momentum	$\mathbf{p} = M\mathbf{v}$	$\mathbf{H} = I\boldsymbol{\omega}$
Impulse	$\mathbf{R} = \mathbf{f}t$	$\mathfrak{R} = \mathbf{L}t$
NEWTON'S second law	$\mathbf{f} = \dfrac{d\mathbf{p}}{dt} = M\mathbf{a}$	$\mathbf{L} = \dfrac{d\mathbf{H}}{dt} = I\boldsymbol{\alpha}$
Work	$W = fs$	$W = L\theta$
Kinetic energy	$E = \frac{1}{2}Mv^2$	$E = \frac{1}{2}I\omega^2$
Power	$P = fv$	$P = L\omega$
$s = r\theta \qquad v = r\omega$	$a = r\alpha \qquad \omega = 2\pi n$	$I = \Sigma mr^2$

on the other, the second body exerts a torque on the first; these torques are about the same axis and are equal in magnitude but opposite in sense.

Table II will be useful to the beginner. If the various angular analogues are borne in mind, problems involving rotation about a fixed axis can be solved as readily and in the same manner as the corresponding translational problems with which the student has become familiar.

o

Moment of Inertia

Moment of inertia I might also be called *rotational inertia*, for it plays the same part in rotation that mass plays in translatory motion. Just as the inertia or mass of a body is a measure of the resistance that the body offers to linear acceleration, the moment of inertia of a body is a measure of the resistance that it offers to angular acceleration. It is evident, however, that I is not proportional to mass alone, for everyday experience teaches that two rotating bodies may have the same mass and yet offer widely different amounts of resistance to the operation of starting and stopping, as in the case of two wheels one of which has its mass concentrated near the axle, the other on the circumference. Nor is I constant for any one body, as is mass under ordinary circumstances, but varies in passing from one axis of rotation to another. In other words, moment of inertia is a function both of mass and of the *distribution* of mass, that is, of the distances of the elements of mass from the axis. All of this is evident from Eq. [96], the defining equation for moment of inertia, namely,

$$I = \sum_{i=1}^{i=n} m_i r_i^2.$$

Moment of inertia is not a vector quantity, for if the direction of the axis of rotation is reversed, the moment of inertia is unchanged. Nor is moment of inertia a scalar quantity, because its value does change in general when the direction of the axis of rotation is changed. Quantities whose dependence upon direction is of this type are called *tensors*.

84. Calculation of Moments of Inertia. In order to calculate I, one evidently must multiply the mass of each particle in the body by the square of its distance from the axis of rotation and then find the sum of all these products. In the case of continuous bodies it is generally necessary to accomplish this summation with the aid

THE following graphic picture of LORD KELVIN as he appeared to his students has been given by Andrew Gray in his *Lord Kelvin, an Account of his Scientific Life and Work* (Dent, 1908), pp. 280–285. It is reproduced by permission of the publisher.

"The writer will never forget the lecture-room when he first beheld it, from his place on Bench VIII, a few days after the beginning of session 1874–1875. Sir William Thomson [Lord Kelvin], with activity emphasised rather than otherwise by his lameness, came in with the students, passed behind the table, and, putting up his eye-glass, surveyed the apparatus set out. Then, as the students poured in, an increasing stream, the alarm weight was released by the bell-ringer, and fell slowly some four or five feet, from the top of the clock to a platform below. By the time the weight had descended the students were in their places, and then, as Thomson advanced to the table, all rose to their feet, and he recited the third Collect from the Morning Service of the Church of England. It was the custom then, and it is still one better honoured in the observance than in the breach (which has become rather common) to open all the first and second classes of the day with prayer; and the selection of the prayers was left to the discretion of the professors. Next came the roll-call by the assistant; each name was called in its English, or Scottish (for the clans were always well represented) form, and the answer 'adsum' was returned. . . .

"The vivacity and enthusiasm of the Professor at that time was very great. The animation of his countenance as he looked at a gyrostat spinning, standing on a knife-edge on a glass plate in front of him, and leaning over so that its centre of gravity was on one side of the point of support; the delight with which he showed that hurrying of the precessional motion caused the gyrostat to rise, and retarding the precessional motion caused the gyrostat to fall, so that the freedom to 'precess' was the secret of its not falling; the immediate application of the study of the gyrostat to the explanation of the precession of the equinoxes, and illustration by a model . . . — all these delighted his hearers, and made the lecture memorable.

"Then the gyrostat, mounted with its axis vertical on trunnions on a level with the fly-wheel, and resting on a wooden frame carried about by the Professor! The delight of the students with the quiescence of the gyrostat when the frame, gyrostat and all, was carried round in the direction of the spin of the fly-wheel, and its sudden turning upside down when the frame was carried round the other way, was extreme, and when he suggested that a gyrostat might be concealed on a tray of glasses carried by a waiter, their appreciation of what would happen was shown by laughter and a tumult of applause."

Much of KELVIN's scientific work had to do with the mechanics of rotating bodies, and his clear ideas on, and physical insight into, this subject were undoutedly due largely to the practical knowledge which he gained as a boy spinning tops and rounded stones and to his many experiments with gyrostatic models. See S. P. Thompson's *Life of William Thomson, Baron Kelvin of Largs* (Macmillan, 1910), Vol. II, pp. 736–745. Plate 21 was taken from this volume by permission of the publisher.

Lord Kelvin's Lecture-Room
in the University of Glasgow
as Arranged for a Lecture on Gyroscopes

of the integral calculus, and for this reason the expression which usually must be evaluated in order to obtain the moment of inertia in a particular case is [1]

$$I = \lim_{\Delta m \to 0} \Sigma r^2 \cdot \Delta m;$$

that is,

$$I = \int_0^M r^2 \cdot dm.$$ [102]

For example, by employing Eq. [102] one obtains for the moment of inertia of a solid homogeneous cylinder of radius R and mass M rotating about its geometrical axis,

$$I = \tfrac{1}{2} MR^2.$$ [102a]

The moment of inertia of a homogeneous solid sphere of radius R and mass M rotating about any diameter is found in like manner to be

$$I = \tfrac{2}{5} MR^2.$$ [102b]

In the case of a homogeneous rectangular bar of length a, width b, thickness c, and mass M, rotating about an axis parallel to c and through the center of mass,

$$I = \tfrac{1}{12} M(a^2 + b^2).$$ [102c]

85. Radius of Gyration. If M is the total mass of a body and k is a quantity such that Mk^2 equals the body's moment of inertia about a given axis, then k is called the *radius of gyration* of the body about that axis. Thus the equation

$$I = \Sigma mr^2 = Mk^2$$ [103]

defines the radius of gyration k. It often will be found convenient to write the expression for the moment of inertia in terms of the radius of gyration instead of the more complicated expressions involving geometrical form to which Eqs. [96] and [102] lead. The actual evaluation of k, however, is best carried out by means of Eq. [103] after I has been determined experimentally or has been calculated by means of Eq. [102].

EXAMPLE. (a) Show that the radius of gyration about any given axis may be defined as the distance from that axis at which the whole mass M of any rotating body might be concentrated without changing the value of the moment of inertia about that axis. (b) Show that the radius of gyration of a solid homogeneous cylinder of radius R about its own axis is 0.707 R. (c) Show that the expression for the kinetic energy of rotation of any rigid body may be written in the form $E = K \cdot \tfrac{1}{2} Mk^2\omega^2$.

---

[1] Derivations of the expressions for I for various bodies will be found in Appendix 8. The student who is unfamiliar with the integral calculus may make use of the results given in this appendix even though he is unable to follow the derivations.

86. Moments of Inertia about Parallel Axes. Fortunately, if one knows the moment of inertia I_c about an axis passing through the center of mass, the moment of inertia about any other *parallel* axis A can be obtained simply by adding to I_c the mass of the body multiplied by the square of the distance between the two axes. To establish this theorem, which was enunciated by JOSEPH LOUIS LAGRANGE [1] in 1783, let the axis of rotation A be perpendicular to the plane of Fig. 71, and let m be the mass of any particle distant r from this axis. Then, by Eq. [96], $I = \Sigma m r^2$. Next imagine an axis drawn through the center of mass c parallel to the axis A. Let d

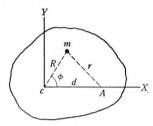

FIG. 71. LAGRANGE's theorem of parallel axes

denote the distance between these two axes, R the distance of the particle m from the axis through c, and ϕ the angle between d and R. Then $r^2 = R^2 + d^2 - 2\,dR\cos\phi$, and therefore

$$I = \Sigma m R^2 + d^2 \Sigma m - 2\,d\,\Sigma m R \cos\phi = I_c + M d^2 - 2\,d\,\Sigma mx,$$

since Σm is the total mass M of the body. The last term of the foregoing equation is zero, since Σmx is zero when the origin of coordinates is taken at the center of mass (Eq. [72], Chap. 6). Accordingly

$$I = I_c + M d^2. \tag{104}$$

The whole problem of moments of inertia is thus reduced to finding moments of inertia about axes passing in different directions through the center of mass.

87. Experimental Determination of *I*. If a body is irregular in form or variable in density, the calculation of its moment of inertia is generally an impossible undertaking and one must resort to experiment. The experiment consists of applying to the body a known torque and observing the angular acceleration thus produced (Eq. [101]), or of some procedure that is equivalent to this.

o

The Dynamics of Rotation about Movable Axes

If a body is not mounted on a fixed axis, but is free, the applied forces will in general produce both a change in the linear momentum of the center of mass (Sec. **70**) and a change in the angular momentum about some axis through the center of mass, in a direction

[1] Œuvres (Gauthier-Villars. 1870). Vol. V. p. 535.

that depends on the nature of the body and the forces. These two motions of translation and rotation are independent of each other and are produced independently by the acting forces. Hence in considering the motion of any rigid body we may discuss the translation as if the body were a single particle with its mass concentrated at the center of mass, and the rotation as if the center of mass were fixed.[1] For describing the translation we make use of Eq. [76], Chap. 6, and for describing the rotation, Eq. [100] or Eq. [101].

In impact problems it is often better to employ the equations for conservation of linear momentum and conservation of angular momentum, which really are generalizations of the foregoing equations. It should be noted, however, that *both* the linear and the angular momentum must be considered. In Chapter 5 only the linear momentum was considered. (Why?)

An illustration of the independence of translation and rotation can be obtained by suspending an ordinary meter stick loosely from the fingers of one hand and giving a blow to the stick. Whenever the blow is struck so as to pass through the center of mass, the stick flies out of the hand with a motion of translation only. When, on the other hand, the blow is struck through any other point, not only is a translatory motion imparted to the center of mass but the stick also rotates about this moving center of mass. It is left to the student to explain why, if a chair be tilted back and then let fall forward again, it not only rotates into its original position, but also translates forward along the floor.

Fig. 72. The dotted lines show the positions assumed by various bodies during rotation about a vertical axis

The general case of rotation of a free body is much more involved than the foregoing considerations might indicate. For, although the motion does consist of a simultaneous translation of the center of mass and a rotation about an axis through this point, the direction of the axis through the center of mass will in general change with time. In other words, the body tends to shift its position with respect to an axis fixed in space and, as a result, I will change with time as well as ω, so that the angular momentum will not always have the same direction as the angular velocity. This case is too involved

---

[1] This was first shown in 1834 by LOUIS POINSOT in a paper in Liouville's *Journal de Mathématiques*, entitled "Théorie Nouvelle de la Rotation des Corps." A translation by C. Whitley was published at Cambridge in 1843 under the title "Outlines of a New Theory of Rotary Motion."

for discussion here. Suffice it to say that, owing to the inertia which it possesses, every particle in the body tends to get as far as possible from the axis of rotation and so increase the moment of inertia about that axis. Hence the body will not spin stably until the moment of inertia is as large as possible (Fig. 72).

88. Rotation of a Body with One Point Fixed. The Gyroscope. We will consider briefly one interesting case of a rotating body that has only one point fixed, namely, the *gyroscope*, the principle of which has many important applications. A simple form of gyroscope is shown in Fig. 73. A bicycle wheel supported by a cord attached to one end O of its axle (Fig. 74) will serve as another simple example.[1] When the wheel is set spinning on its axle OY and released, its weight exerts a torque

Fig. 73. A simple form of gyroscope

about the horizontal axis OZ at right angles to OY. The wheel does not fall, however, but begins to revolve, or *precess*, about the vertical axis OX.

In order to understand this apparently extraordinary result, let Fig. 75 represent the face of the wheel as viewed from the positive end of the OY-axis, and let the wheel be spinning about OY with constant

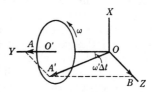

Fig. 74. Theory of the simple gyroscope

angular speed ω in the direction of the arrow. If the wheel were not spinning, the torque due to its weight would cause it to fall and, in so doing, to rotate about the axis MN in such a way that the quadrants a and d would move out from the paper toward the observer while the quadrants b and c would move in, away from the observer. Consider now the effect of the *inertia* of a particle in each of the four quadrants separately in resisting this motion about MN. Since the wheel is spinning on the axle OY, a particle in quadrant a is being carried farther away from the axis MN and hence is being made to move faster in whatever motion there is about MN. Because of inertia it resists this increase in speed, and this resistance may be represented as a force acting on the wheel in a direction perpendicular to the plane of the paper and *away from* the observer. A particle in quadrant b is getting closer to the axis MN, and therefore its speed at right angles to

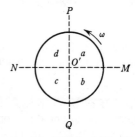

Fig. 75. Wheel of Fig. 74 viewed from the positive end of the OY-axis

---

[1] See the Optional Laboratory Problem at the end of this chapter.

the plane of the paper must diminish. This diminution results from the action of the wheel on the particle, and the particle, by virtue of its inertia, reacts on the wheel in a direction *away from* the observer. On the other hand, the reactions of particles in quadrants *c* and *d* are in the opposite direction, that is, *toward* the observer. Clearly these four reactions will cause the wheel to rotate about the axis *PQ*, if it is free to do so, instead of about *MN*. In short, if a body pivoted at a point is rotating about an axis *OY* and a torque is applied about the axis *OZ*, the inertia of the individual parts of the body causes it to turn about the third axis *OX* perpendicular to both *OY* and *OZ*.

The fundamental equation describing quantitatively the motion of a gyroscope is derived most simply by a vector treatment as follows. Let I and ω be the moment of inertia and the angular speed of the wheel about its axle, and, in Fig. 74, let the vector $\overline{OA}$, drawn according to the right-hand-screw rule, represent the angular momentum $I\omega$. Let L_z be the torque about the horizontal axis OZ caused by the weight of the wheel and let $\overline{OB}$, or its equal $\overline{AA'}$, be the change in angular momentum produced by this tipping torque in time Δt; that is, $\overline{OB} = L_z \, \Delta t$. The vectors $\overline{OA}$ and $\overline{OB}$ will automatically and instantly combine into $\overline{OA'}$, which is their vector sum, and this combination will occur long before enough tipping has occurred to be observed, and it will keep on occurring continuously. The result, then, will be that the gyroscope, instead of falling over, precesses; that is, the axle OY slowly turns about OX.

If ω' be the angular speed of precession, the axle OY will in a short time Δt turn through a small angle $\omega' \, \Delta t$, and during this time the angular momentum will have changed by an amount $\overline{OB} = \overline{AA'} = I\omega \cdot \omega' \, \Delta t$. The time-rate of change of angular momentum is therefore $I\omega\omega'$, and hence, by Eq. [100],

$$L_z = I\omega\omega'. \qquad [105]$$

This is the fundamental equation of the gyroscope. From it we see that the rate of precession ω' is directly proportional to the torque and inversely proportional to the rate of spin. If ω' be increased, L_z must increase, and the gyroscope consequently rises against gravity. If ω' be decreased, the gyroscope falls. If we try to turn the gyroscope in one direction, it turns in a direction at right angles to what we expect. If the wheel be supported at its center of gravity, so that $L_z = 0$, no precession will occur; that is, the axle will continue to point in a fixed direction in space.[1]

---

[1] For more comprehensive discussions of these phenomena and their applications see H. Crabtree, *Spinning Tops and Gyroscopic Motion*; J. Perry, *Spinning Tops*; A. Gray, *Gyrostatics and Rotational Motion*; E. S. Ferry, *Applied Gyrodynamics*; articles "Gyro-compass" and "Gyroscope," *Encyclopaedia Britannica*, ed. 14.

Illustrative Examples on the Dynamics of Rigid Bodies

In attacking problems on the rotation of rigid bodies one should first of all consider whether the axis of rotation remains fixed in direction with respect to the body. If it does, the rotational analogues of translatory motion can be applied and the problem solved in the same manner as in the corresponding translational problem. If it does not, then methods that are beyond the scope of this book must be invoked.

> EXAMPLE. Find a formula for the linear acceleration with which a homogeneous solid cylinder rolls without slipping down a plane inclined at an angle ϕ to the horizontal (Fig. 76). Neglect the rolling friction.

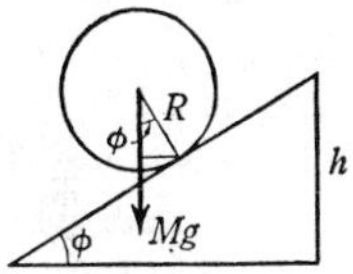

Solution. Although the axis of rotation does not remain fixed in space in this case, it does remain fixed in direction with respect to the body, and the problem can therefore be attacked by elementary methods. We will solve it in two different ways.

FIG. 76. Cylinder rolling down an incline

a. The cylinder can be considered to be rotating about the line on its surface that is in contact with the incline at any moment. The weight Mg of the cylinder acts in a vertical line through its center of gravity, and hence the torque about the momentary axis of rotation is $MgR \sin \phi$, where R is the radius of the cylinder. The moment of inertia of the cylinder about the same axis is $\frac{1}{2}MR^2 + MR^2$, by Eqs. [102a] and [104]. The angular acceleration is therefore

$$\alpha = \frac{L}{I} = \frac{MgR \sin \phi}{\frac{3}{2}MR^2} = \frac{2}{3}\frac{g \sin \phi}{R},$$

and the linear acceleration is

$$a = \alpha R = \tfrac{2}{3} g \sin \phi.$$

b. Consider the cylinder to be rotating about its geometrical axis. The potential energy of the cylinder when it is at the top of the incline is $V = Mgs \sin \phi$, where s is the length of the incline. The total kinetic energy acquired in rolling down is $E = \frac{1}{2}Mv^2 + \frac{1}{2}I\omega^2$ or, since $I = \frac{1}{2}MR^2$ and $\omega = v/R$, $E = \frac{3}{4}Mv^2$. But $V = E$, or $Mgs \sin \phi = \frac{3}{4}Mv^2$; hence $v^2 = \frac{4}{3}gs \sin \phi$. Since the acceleration is constant, $v^2 = 2\,as$. Hence, finally,

$$2\,as = \tfrac{4}{3} gs \sin \phi, \quad \text{or} \quad a = \tfrac{2}{3} g \sin \phi,$$

as before.

Does the linear acceleration depend on the size of the cylinder? How does the acceleration compare with that which the cylinder would have if it *slid* down the incline without appreciable friction? Why is it less?

IN THIS portrait MAXWELL is shown with his color-top. Even in later life he amused himself with scientific toys, and some of these led directly to scientific contributions of value. A favorite game, which he played throughout his life and which all his friends came to associate with him, was "diabolo." It is played with a sort of top in the shape of two cones joined at their vertices, which is spun, thrown, and caught by means of a string attached to two sticks. MAXWELL became very skillful with it and developed those physical intuitions which no doubt led him to construct the dynamical top with which he demonstrated the properties of rotating bodies.

MAXWELL's lifelong friend and biographer, Lewis Campbell, in his *Life of James Clerk Maxwell*, tells the following amusing story of his bringing his dynamical top with him to Cambridge in the summer of 1857 and exhibiting it at a tea party in his room in the evening: "His friends left it spinning, and next morning Maxwell, noticing one of them coming across the court, leapt out of bed, started the top, and retired between the sheets. It is needless to say that the spinning power of the top commanded as great respect as its power of illustrating Poinsot's *Théorie Nouvelle de la Rotation des Corps*."

Lewis Campbell also relates that, while at Cambridge, MAXWELL tried to find out how a cat, when dropped, always succeeds in alighting on her feet. His experiments must have made a great impression upon the minds of the other students, for he wrote to Mrs. Maxwell in 1870: "There is a tradition in Trinity that when I was here I discovered a method of throwing a cat so as not to light on its feet, and that I used to throw cats out of windows. I had to explain that the proper object of research was to find how quick a cat would turn round, and that the proper method was to let the cat drop on a table or bed from about two inches, and that even then the cat lights on her feet."

JAMES CLERK MAXWELL
as a Young Man

From a portrait at Trinity College, Cambridge

EXAMPLE. A putty ball of mass m grams moving with a speed of v cm · sec^{-1} in the direction ab, Fig. 77, makes inelastic impact with the end of a long thin bar of length $2\,l$ cm and mass M grams which is pivoted at its center of mass c. Find an expression for the angular speed communicated to the bar.

FIG. 77. Putty ball hitting a pivoted bar

Solution. The axis is fixed in this case. We will again employ two methods of solution.

    a. Let ω be the angular speed of the bar and ball after the impact and let Δt be the time of duration of the impact. The speed lost by the ball is evidently $v - \omega l$, and hence the average force acting between it and the bar during the impact is $m(v - \omega l)/\Delta t$. The average torque acting during the impact is therefore $m(v - \omega l)l/\Delta t$ and this, by Eq. [99], must equal $I_b \omega /\Delta t$, where I_b is the moment of inertia of the bar about the given axis. Thus

$$m(v - \omega l)l = I_b\omega, \quad \text{or} \quad \omega = \frac{mvl}{I_b + ml^2}.$$

    b. Regard the ball and bar as comprising a single system. Then, since there is no external torque acting about the fixed axis, the angular momentum of the system before impact must be equal to that after impact. Before impact the angular momentum about the fixed axis is mvl, and after impact it is $I\omega$, where I is the combined moment of inertia of the bar and ball about the fixed axis. Hence

$$mvl = (I_b + ml^2)\omega, \quad \text{or} \quad \omega = \frac{mvl}{I_b + ml^2},$$

as before.

How may I_b be expressed in terms of M and l? What has become of the linear momentum of the putty ball?

EXAMPLE. A uniform thin bar of length $2\,l$ rests on a smooth horizontal table. A sharp blow is delivered perpendicular to the bar at one end. How far will the bar slide before it completes one revolution?

Solution. Let $R\ (= f\,\Delta t)$ be the impulse of the blow. This impulse will impart a linear momentum of magnitude $R = Mv$ to the center of mass of the bar. At the same time there will be imparted to the bar angular momentum about an axis through the center of mass of amount $Rl = I\omega$, where I is the moment of inertia of the bar about the axis through the center of mass. Since I will be greatest for an axis perpendicular to the bar, the bar will start rotating about such an axis and will continue to do so; that is, the axis will remain fixed with respect to the body and I will remain $\frac{1}{3}Ml^2$. By eliminating R between the two equations in which it appears, we have $Mvl = Ml^2\omega/3$, or $v = l\omega/3$. If, finally, T is the time required for the bar to make one rotation, then $T = 2\,\pi/\omega$, and the distance traversed by the center of mass is given by

$$s = vT = \frac{2\,\pi l}{3}.$$

EXPERIMENT VII. MOMENT OF INERTIA

This experiment is the counterpart in rotary motion of Exps. I and II in linear motion. It includes determinations of the angular accelerations of a solid, homogeneous disk mounted on a fixed axis and acted upon by constant torques, and of calculations, by two different methods, of the moment of inertia of this disk about its fixed axis.

PART I. Some Qualitative Experiments. The apparatus for this part of the experiment is a horizontal disk mounted on ball bearings so as to rotate with almost no friction about a vertical axis. If this apparatus is not available, a rotating swing of similar construction will serve the purpose.

a. Stand on the disk, or sit on a stool placed on the disk, with arms extended and with a 1-kg or 2-kg mass in each hand. Have someone start you rotating slowly, and while you are moving lower the arms quickly and again extend them.

1. Describe and explain what you observe.

2. If your moment of inertia is halved when you lower your arms, how much will your angular speed be changed?

3. What change takes place in the kinetic energy of the system when you lower your arms? What is the source of the additional energy?

b. Stand on the disk and twist the upper portion of your body in one direction so as to rotate the disk momentarily in the opposite direction. Then resume your normal position and note whether the disk returns to its normal position.

4. Explain this effect.

PART II. Rotation of a Rigid Body about a Fixed Axis. *Adjustment and Use of Apparatus.* The rotating body employed in this part of the experiment is a solid metal disk of several kilograms mass which is mounted on precision pivot bearings so as to rotate with very little friction about a horizontal axis. A torque is applied by means of a driving weight attached to a thread wound around the circumference of the disk. To obtain a record of the successive angular positions of the disk, a series of equally timed sparks is made to jump from a spark point to a circular sheet of polar-coordinate paper fastened to one face of the moving disk. The spark point is mounted on a sliding carriage which can be moved from the periphery of the disk in toward the axis during the rotation, thus giving a spiral trace of spark perforations on the paper chart. The timing device for producing the sparks is the same as that used in Exp. I.

a. Place the apparatus on the edge of a platform which is high enough for the driving weight to have a free fall of about 2 m. Fasten the circle of polar-coordinate paper to the face of the disk by means of 6–8 drops of rubber cement placed around its rim. Adjust the spark-timer for about 3 vib · sec⁻¹ and connect it to the binding posts on the base of the rotation apparatus. Fasten one end of a thread or of a silk fishline to the circumference of the disk. If the disk is not provided with a hook for this purpose, use soft wax. Wind the thread spirally on the circumference of the disk without allowing it to cross at any point. Hang a light weight-hanger from the free end of the thread.

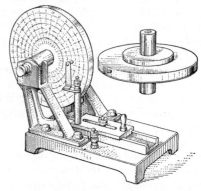

Fig. 78. Apparatus for Exp. VII. The rotating system consists of a shaft and one or more rigidly attached disks

b. Determine whether the friction in the bearings is large enough to produce a measurable acceleration of the disk and, if it is, place enough small weights on the weight-hanger so that the disk, when started, will continue to move with constant speed. Have the spark going during this test, but keep the spark point at the extreme outer edge of the disk. Observe how many times the disk rotates during the descent of the weight.

c. Rewind the thread and again start the spark. Add a mass of 100 g to the weight-hanger. Release the disk, carefully avoiding an initial impulse, and at the same time begin to move the spark point slowly in toward the axis. After the weight has fallen, open the spark switch. Before removing the paper chart from the disk, draw a small circle with a pencil around each spot made by a spark.

d. Repeat *c*, this time with, say, a 200-g mass on the weight-hanger. If the rotating system consists of several disks of different diameters, as in Fig. 78, the instructor may find it desirable to have you vary the lever arm as well as the force involved in the impressed torque.

e. It will be necessary to know the following constants of the apparatus: (1) the time interval τ between successive sparks; (2) the mass M of the rotating system, which usually will be found stamped upon the disk; (3) the radius R of the disk. If the disk is of the type shown in Fig. 78, it will be necessary to measure both the diameters and the thicknesses of its several parts.

Calculation of α. The angular accelerations are obtained in the same way as were the linear accelerations in Exp. I, Part II, *a*; that is, the data obtained from the polar-coordinate chart are substituted in the equation

$$\theta_2 - \theta_1 = \alpha, \qquad [106]$$

where θ_2 and θ_1 represent the angles through which the disk turned in successive equal intervals of time. This equation is the rotational analogue of the equation $S_2 - S_1 = a$, the derivation of which is carried out in Exp. I. It must be remembered that it gives the magnitude of α when the time interval is the time required for some arbitrarily chosen number of sparks to occur.

a. Use the data obtained with the 100-g driving weight to compute the values of α, and take their average. The method of making the chart and obtaining $\theta_2 - \theta_1$ etc. is the same as that employed to obtain $S_2 - S_1$ etc. in Exp. I, except that the points 0 and 1 on the chart need not be discarded. Make use of as many of the points on the chart as possible. Note that the angles θ can be estimated to about two tenths of a degree.

Calculate the length of time unit involved in your value of α and then express α in radians per second per second.

Calculate the magnitude L of the accelerating torque corresponding to this value of α. Note that the force in the thread is given by $m(g - a)$, where m is the mass of the driving weight and a is its linear acceleration.

b. In a similar manner calculate α and L for the motion recorded on your second polar-coordinate chart.

5. Are the angular accelerations produced in the two cases proportional to the driving weights? Are they proportional to the torques acting?

6. Derive Eq. [106]. Do its two members have the same dimensional formulas? Explain.

Moment of Inertia of the Rotating System. By substituting the corresponding values of α and L in Eq. [101], obtain two values of I and average them. Also calculate I with the aid of Eq. [102a]. Compute the percentage difference between the values of I obtained by these two methods.

7. If the rotating system used is like that of Fig. 78, find what error would have been introduced into your value of I if you had disregarded the dimensions of the shaft etc. and had treated the rotating system as a single disk of mass M and radius R.

OPTIONAL LABORATORY PROBLEM

Some Qualitative Experiments with a Gyroscope. A bicycle wheel mounted on an axis provided with handles serves admirably as a gyroscope. The moment of inertia of the wheel can be made large by winding wire around the rim. Perform each experiment while the wheel is spinning rapidly on its axle. Try to explain the effects observed; if necessary, supplement your study with references to some of the books mentioned elsewhere in this chapter.

a. Hold the gyroscope by both handles and move it in any direction with a motion of translation. Suddenly change the direction of the axle.

b. Suspend the wheel by one of the handles in a loop of cord and note the result. Push upward on the free handle. Hang various weights from the free handle. How does the rate of pre-cession (Sec. **88**) vary with the torque? with the rate of spin? Why does a bicycle bear to the left when the rider leans in that direction? Try to rotate the wheel about a vertical axis by pushing sidewise on the free handle. When an automobile turns a corner, how is its motion affected by the presence of the engine flywheel?

c. Stand on the horizontal disk of Exp. VII, Part I, hold the gyroscope with its axle parallel to a radius of the disk, and then apply a couple tending to raise one handle of the gyroscope relative to the other.

Fig. 79. A gyroscopic wheel

d. If you have access to a gyroscopic wheel that is mounted on gimbals, perform the experiments described by A. M. Worthington in *Dynamics of Rotation*, Chap. XIII, or H. Crabtree in *Spinning Tops and Gyroscopic Motion*, introductory chapter.

○

QUESTION SUMMARY

1. Define *angular displacement*; *angular velocity*; *angular acceleration*. In what different units may each be expressed? Are all of these quantities vectors? If so, how may they be represented graphically?

2. What relation exists between *angular distance* and *linear distance*, between *angular speed* and *linear speed*, and between *angular acceleration* and *linear acceleration*?

3. Define *moment of inertia* and discuss its physical importance. How is the moment of inertia of a body with reference to a given axis calculated? How can it be determined experimentally?

4. Restate all the laws and equations which have been derived in the preceding chapters for translation in terms such that they will apply to the case of rotation about a fixed axis. What changes did you have to make in order to do this?

5. If the moment of inertia of a body about an axis through the center of mass is known, how may its value about any other parallel axis be calculated?

6. Define *radius of gyration* in words and by means of an equation.

7. What is the nature of the motion of a free body under the action of forces? Why is this motion more complicated than that of rotation about a fixed axis?

8. State and discuss the fundamental equation of the gyroscope.

○

PROBLEMS

1. A bicycle having a wheel 27 in. in diameter is ridden at the rate of 11 mi · hr^{-1}. Find (*a*) the angular speed of the wheel about its axle; (*b*) the linear speed of a particle at its highest point with respect to the axle; (*c*) the linear speed of this particle with respect to the ground; (*d*) the linear speed of the particle in contact with the ground, with respect to the ground; (*e*) the linear velocity of *any* particle on the rim, with respect to the ground; (*f*) the angular speed of *any* particle on the rim, with respect to an axis passing through the point on the rim which is in contact with the ground.

Ans. (*a*) 14 rad · sec^{-1}; (*b*) 11 mi · hr^{-1}; (*c*) 22 mi · hr^{-1}; (*d*) 0; (*e*) $2\,v\cos\theta/2$; (*f*) 14 rad · sec^{-1}.

2. The flywheel of an engine runs with a constant angular speed of 150 rev · min^{-1}. When steam is shut off, the friction of the bearings and of the air brings the wheel to rest in 2.2 hr. (*a*) What is the average angular acceleration of the wheel? (*b*) How many rotations will the wheel make before coming to rest? (*c*) What is the linear acceleration, along the tangent to its path, of a particle distant 50 cm from the axis of rotation?

Ans. (*a*) -2.0×10^{-3} rad · sec^{-2}; (*b*) 9.9×10^3; (*c*) -1.0 mm · sec^{-2}.

3. A constant pull of 200 kgwt applied tangentially to the rim of a wheel of 100-cm radius changes the angular speed of the wheel from 2.00 rev · sec^{-1} to 4.00 rev · sec^{-1} in 30.0 sec. (*a*) What is the moment of inertia of the wheel? (*b*) What is the change in the angular momentum during the 30 sec? (*c*) Through what angle does the wheel turn during this change? (*d*) How much energy is expended in producing this increase?

Ans. (*a*) 4.68×10^{10} g · cm^2; (*b*) 5.88×10^{11} g · cm^2 · sec^{-1}; (*c*) $180\,\pi$ rad; (*d*) 1.11×10^{13} ergs.

4. (*a*) Find the moment of inertia of a solid copper cylinder of radius 2.0 cm, length 30 cm, about a longitudinal axis through the center of mass. (*b*) If the friction is negligible, what angular acceleration is produced about this axis by a 2.0-kg driving weight suspended by a cord wound around the cylinder? (*c*) What is the angular speed after the force has acted for

3.0 sec?   (*d*) If the driving weight were removed and a force of 2.0 kgwt were applied to the cord, what would be the resulting angular acceleration?

Ans.   (*a*) 6.7×10^3 g · cm^2;   (*b*) 2.7×10^2 rad · sec^{-2};
(*c*) 8.0×10^2 rad · sec^{-1};   (*d*) 5.8×10^2 rad · sec^{-2}.

5. (*a*) What part of the total kinetic energy is energy of translation and what part energy of rotation in the case of (1) a rolling hoop? (2) a rolling solid cylinder? (3) a rolling solid sphere? (*b*) A sphere of mass M and radius R rolls along the ground with an angular speed ω. Find its total kinetic energy by two methods: (1) by adding the kinetic energy of translation to that of rotation about the axis of the sphere; (2) by considering its motion as a pure rotation about a point in contact with the ground.

Ans.   (*a*) $\frac{1}{2}$ trans., $\frac{2}{3}$ trans., $\frac{5}{7}$ trans.;   (*b*) $\frac{7}{10} MR^2\omega^2$.

6. A merry-go-round weighs 20 tons and has a diameter of 30 ft. If its mass be considered as uniformly distributed over its surface and if friction be neglected, what speed will a 0.5-hp engine impart to it in 2 min?

Ans.   0.7 rad · sec^{-1}.

7. Write the dimensional formula for (*a*) angular distance; (*b*) angular speed; (*c*) angular acceleration; (*d*) moment of inertia; (*e*) angular momentum; (*f*) impulse of a torque.

8. The pulley of a certain Atwood's machine consists of a uniform disk of mass 150 g. The masses on the end of the cord are 200 and 250 g respectively. (*a*) Considering friction and the mass of the cord as negligible, find the linear acceleration of the moving system. (*b*) What is the equivalent mass of the pulley, that is, the mass which at its circumference would offer the same resistance to acceleration?   *Ans.*   (*a*) 93.3 cm · sec^{-2};   (*b*) 75.0 g.

9. The mass of the cylinder over which the paper passed in Exp. II, Part I (Fig. 19), is 40 g and its diameter is 3.0 cm. Assuming that the cylinder is solid, show how Eq. [30], Chap. 2, should be modified so as to take into account the inertial resistance which the cylinder offered to acceleration.

10. What relation exists between the linear speed acquired by a body in sliding without friction down an inclined plane and in rolling without slipping down the same incline, if the body is (*a*) a hoop? (*b*) a solid cylinder? (*c*) a solid sphere?

11. A hoop and a solid disk start down a hill together. Which will reach the bottom first? Find the ratio of their speeds at the bottom. If frictional effects are disregarded, why can a heavy man on a bicycle always coast faster than a light man on the same bicycle? If heavier tires were used, how would the coasting speed of a bicycle be affected? What effect will increasing the weight of the frame have upon the coasting speed?

12. A bullet of mass 5.0 g is moving with a speed of 100 m · sec^{-1} in the direction *ab*, Fig. 80, when it makes inelastic impact with the projection *b* of a wheel mounted on a fixed axis. If the wheel has a moment of inertia of 2.0×10^5 g · cm^2 and a radius of 20 cm, how many rotations per second are communicated to it? Solve by two different methods: (*a*) by means of the principle of conservation of angular momentum; (*b*) by the use of Eq. [99]. *Ans.* 0.79 rev · sec⁻¹.

13. If a thin homogeneous rod were stood vertically on end and allowed to fall over, with what angular speed would it strike the ground?
Ans. $\omega = \sqrt{3\,g/l}$.

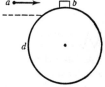

Fig. 80. Problem 12

14. Two solid wheels, each 20 in. in diameter and weighing 20 lbwt, are keyed to an axle 8 in. in diameter and weighing 40 lbwt. A string wrapped around the axle is pulled on with a force of 5 lbwt in a horizontal direction, as shown in Fig. 81. Find the linear acceleration of the center of mass of the system. *Ans.* 0.9 ft · sec⁻².

15. A thin rod 1.5 m long and of mass 5.0 kg is held in a vertical position and struck a horizontal blow with a hammer at a point 1.2 m from the lower end, the impulse of the blow being 0.20 kgwt · min south. If the stick is released at the instant of the blow, (*a*) what linear velocity is imparted to the rod and (*b*) what angular velocity is imparted about its center of mass? (*c*) Find these velocities for the case where the blow is struck at the center of mass of the rod. (*d*) Compare the kinetic energies of translation and of rotation acquired by the rod in the two cases.
Ans. (*a*) 24 m · sec⁻¹ south; (*b*) 56 rad · sec⁻¹ east.

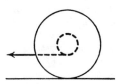

Fig. 81. Will the spool move to the left or to the right?

○

To those who study the progress of exact science, the common spinning-top is a symbol of the labours and the perplexities of men who had successfully threaded the mazes of the planetary motions. The mathematicians of the last age, searching through nature for problems worthy of their analysis, found in this toy of their youth, ample occupation for their highest mathematical powers.

No illustration of astronomical precession can be devised more perfect than that presented by a properly balanced top, but yet the motion of rotation has intricacies far exceeding those of the theory of precession.

Accordingly, we find Euler and D'Alembert devoting their talent and their patience to the establishment of the laws of rotation of solid bodies. Lagrange has incorporated his own analysis of the problem with his general treatment of mechanics, and since his time M. Poinsôt has brought the subject under the power of a more searching analysis than that of the calculus, in which ideas take the place of symbols, and intelligible propositions supersede equations.

J. Clerk Maxwell, "On a Dynamical Top", *Transactions of the Royal Society of Edinburgh*, 21, Part 4 (1857)

CHAPTER EIGHT

ELASTIC BODIES

WE TOOK then a long Glass-Tube, which by a dexterous hand and the help of Lamp was in such a manner crooked at the bottom, that the part turned up was almost parallel to the rest of the Tube, and the Orifice of this shorter leg of the Siphon (if I may so call the whole Instrument) being Hermetically seal'd, the length of it was divided into Inches, (each of which was subdivided into eight parts) by a straight list of paper, which containing those Divisions was carefully pasted all along it: then putting in as much Quicksilver as served to fill the Arch or bended part of the Siphon, that the Mercury standing in a level might reach in the one leg to the bottom of the divided paper, and just to the same height or Horizontal line in the other; we took care, by frequently inclining the Tube, so that the Air might freely passfrom [sic] one leg into the other by the sides of the Mercury, (we took (I say) care) that the Air at last included in the shorter Cylinder should be of the same laxity with the rest of the Air about it. This done, we began to pour Quicksilver into the longer leg of the Siphon, which by its weight pressing up that in the shorter leg, did by degrees streighten [sic] the included Air: and continuing this pouring in of Quicksilver till the Air in the shorter leg was by condensation reduced to take up but half the space it possess'd (I say, possess'd, not fill'd) before; we cast our eyes upon the longer leg of the Glass, on which was likewise pasted a list of Paper carefully divided into Inches and parts, and we observed, not without delight and satisfaction, that the Quicksilver in that longer part of the Tube was 29. Inches higher than the other.

ROBERT BOYLE, "Two new Experiments touching the measure of the Force of the Spring of Air compress'd and dilated"[1]

○

In the preceding chapters we have, for the most part, treated all bodies as if they remained unaltered in size and shape when acted upon by forces. This conception of bodies as perfectly rigid is of course an idealization, like that of the particle, but it is an exceedingly useful one for simplifying the initial approach to the study of the mechanics of bodies. Indeed, for certain practical purposes many bodies may be treated as if they were perfectly rigid. Thus the steel parts of such structures as bridges and buildings ordinarily may be regarded as perfectly rigid by everyone except the architects and engineers who design them. On the other hand, in dealing with fluids the concept of a perfectly rigid body is entirely insufficient.

A body is said to be *elastic* if it tends to return to its original size or shape when the forces which deform it are removed. It is

---

[1] *Part II, Chap. V, p. 58, of *A Defence of the Doctrine touching the Spring and Weight of the Air*, published by BOYLE in his *New Experiments Physico-Mechanical*, ed. 2 (Oxford, 1662).

" GALILEO's Problem," an Illustration
from *Two New Sciences*

IN HIS *Two New Sciences* GALILEO not only solved the problem of falling bodies and in so doing laid the foundations of the science of kinetics, but he also began the study of the strength of materials (see " Second Day "). As A. E. H. LOVE says in the " Historical Introduction " to *A Treatise on the Mathematical Theory of Elasticity*:

" The first mathematician to consider the nature of the resistance of solids to rupture was Galileo. Although he treated solids as inelastic, not being in possession of any law connecting the displacements produced with the forces producing them, or of any physical hypothesis capable of yielding such a law, yet his enquiries gave the direction which was subsequently followed by many investigators. He endeavored to determine the resistance of a beam, one end of which is built into a wall, when the tendency to break it arises from its own or an applied weight; and he concluded that the beam tends to turn about an axis perpendicular to its length, and in the plane of the wall."

This problem, through its association with GALILEO's name, has come to be called " GALILEO's Problem."

Ex rerum Causis Supremam noscere Causam.

H. Crivelot inv et delin .

ROBERT BOYLE, 1627–1691

THIS vignette, which is reproduced through the courtesy of the Mount Wilson Observatory of the Carnegie Corporation of Washington, appears on the title-page of *The Works of the Honorable Robert Boyle*, edited by T. Birch (London, 1772). In the preface Dr. Birch says of this likeness of BOYLE:

" There being only two original pictures of Mr. *Boyle* now known to be extant, it was thought proper to have them both engraved. One, which represents him in the 38th year of his age, is placed in the title-page of each volume, copied from a drawing of Mr. *Faithorne*, communicated by *Sir Hans Sloane*, from which likewise Mr. *Faithorne* himself engraved his print, with the instruments accompanying the head, according to the design of Dr. *Robert Hooke*, who thought the *face very carefully and well done, and very like . . .*"

The air pump shown at the right is the early form described by BOYLE in his *New Experiments Physico-Mechanicall*. In designing and making this pump, and in carrying out the experiments performed with it, BOYLE was assisted by ROBERT HOOKE, who was the best mechanician of his day. It is almost certain that the "we" in the quotation given at the beginning of Chapter 8 refers to BOYLE and HOOKE.

said to be *perfectly elastic* if the magnitude of the restoring force depends only on the extent and character, and not at all on the rate, of the deformation. Thus a wire would show perfect elasticity if the successive removal of a number of stretching weights caused it to resume exactly the lengths which it had during the successive addition of the weights, no matter how long or how short a time the weights had been in place. The imposing body of knowledge now known as the *theory of elasticity* had its beginning in certain problems proposed by GALILEO[1] in 1638 in connection with his endeavor to determine the resistance of beams to rupture. In the history of the theory[2] undoubtedly the two great landmarks are the discovery of HOOKE's law in 1660 and the formulation of the general differential equations of elasticity by LOUIS MARIE HENRI NAVIER in 1821. The use of the term *elasticity* in its present sense is due largely to ROBERT BOYLE (1627–1691), who wrote of "the elastic forces" exerted by deformed solid bodies and of "the spring or elastic power in the air in which we live."[3] BOYLE was the first to make a systematic investigation of the elastic properties of gases, and the law that bears his name was the first law of elasticity to be published.

Our very brief discussion of elasticity will be confined to materials that are *homogeneous* and *isotropic* as regards their elastic properties, that is, to materials that have the same elastic properties at every point in them and that have, at any point, the same elastic properties in all directions.

89. Young's Modulus. Suppose that a wire or rod is clamped at one end and that it is stretched or compressed in the direction of its length by a force applied at the other end (Fig. 82). Now any change occurring in the relative positions of the parts of a body is called a *strain*; thus we speak of the wire or rod as having under-

---

[1] *Two New Sciences*, dialogues of the "First and Second Day."

[2] For a brief historical account of the theory see the Introduction to A. E. H. Love, *A Treatise on the Mathematical Theory of Elasticity*, one of the good texts on elasticity. The standard history is I. Todhunter and K. Pearson, *A History of the Theory of Elasticity and of the Strength of Materials.*

[3] *New Experiments Physico-Mechanicall, touching the Spring of the Air, and its Effects, (Made, for the most part, in a New Pneumatical Engine)* (Oxford, 1660). In the second edition (1662), BOYLE added a defense against certain criticisms which had been made of the first edition. This contains a description of experiments establishing the relation between pressure and volume in a gas which we now call "BOYLE's law." Because of its interest and importance this description has been reproduced in *The Laws of Gases*, ed. by C. Barus (Harper, 1899), pp. 3–10, and in *A Source Book in Physics* (1935), pp. 84–87. A portion of it is quoted at the beginning of this chapter.

gone longitudinal strain. Since any part of the wire is lengthened or shortened in the same proportion as the whole wire, we take as the measure of a *longitudinal strain* the change in length per unit length; or, denoting the original length of the wire by l and the change in length by Δl, the longitudinal strain is $\Delta l/l$.

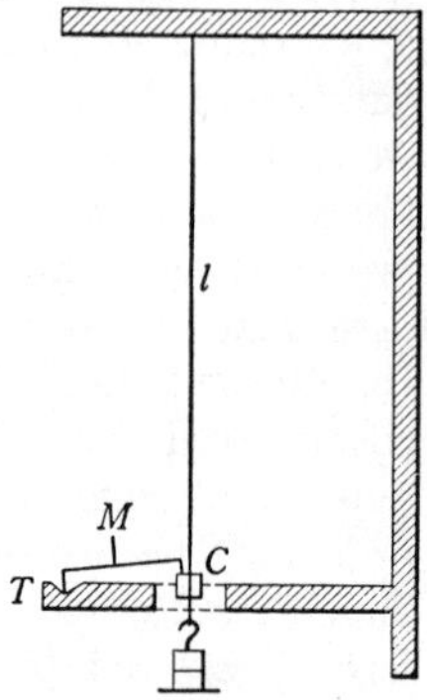

If the material in the wire be elastic, the strain will be accompanied by internal restoring forces arising between the contiguous parts of the wire. If the cross section at any point is considered as a dividing plane, the part of the wire on one side will act with a certain force on the part on the other side, and the latter will react with a force that is equal in magnitude but opposite in direction. When such forces arise in an elastic body, the body is said to be under *stress*, the measure of a stress being the ratio of the internal force to the area across which it is transmitted. In the case of a longitudinal strain in an elastic body, the accompanying longitudinal stress is equal in magnitude to the *external* force per unit of cross-sectional area that acts on the body to produce the strain. Hence, if f be the external force applied to one end of a wire or rod and A be the cross-sectional area, the *longitudinal stress* anywhere in the body is f/A.

By increasing progressively the load carried by a wire and observing the corresponding dimensions of the wire, the data for a curve like Fig. 83 may be obtained. If the body be maintained at a constant temperature, the first part of this curve will be found to be practically a straight line. For small strains, then, the longitudinal stress is proportional to the longitudinal strain; that is,

Fig. 82. An apparatus for measuring the change in the length of a wire as weights are added or subtracted. By means of a small lever, one end of which rests on C, the other end on the fixed table T, the mirror M is rotated through an angle as the wire moves up or down relative to T. The motion of a beam of light reflected from the mirror is observed through a telescope. Can you prove that the beam reflected by the mirror turns through twice the angle turned through by the mirror?

$$\frac{f}{A} = Y \frac{\Delta l}{l}. \qquad [107]$$

The constant of proportionality Y, given by

$$Y = \frac{fl}{A\Delta l}, \qquad [107a]$$

is usually called *Young's modulus,* in honor of THOMAS YOUNG (1773–1829), who was the first to give the constant a physical meaning.[1] It is also called the *stretch modulus,* although its value for a given material is found to be the same whether the stress f/A is *tensile* (Fig. 82) or *compressive* (Fig. 84).

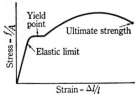

For the small strains that correspond to the straight-line portion of the curve in Fig. 83 it is found that a decrease in the load is accompanied by a shortening of the stretched wire until the length of the wire is the same as it was before the load was put on. When the range of perfect elasticity represented by the straight line is exceeded, in which case the wire is said to have passed the *elastic limit,* removal of the load results in the wire's contracting somewhat, but a permanent set will remain; the wire from then on acts like a different wire with a new elastic limit.

FIG. 83. Stress-strain diagram for a steel wire. The *yield point* is the stress at which the strain begins to increase rapidly, so that the wire begins to flow much like a very viscous liquid. The *ultimate strength* is the stress at rupture. This wire will not recover its original length after removal of the load because the stress corresponding to the *elastic limit* has been exceeded. What does the *slope* of the straight-line portion of this diagram represent?

> EXAMPLE. If the stress existing in a stretched elastic wire does not exceed the elastic limit; how much energy is stored per unit volume in the wire?

Solution. Employing the method of Sec. **45,** one finds that the total work done by the stretching force while it increases from zero to f is $W = f \, \Delta l/2$. By dividing both members of this expression by Al, which is the volume V of the wire, one obtains finally

$$\frac{W}{V} = \frac{1}{2} \text{ stress} \cdot \text{strain}. \qquad [108]$$

90. Hooke's Law. Eq. [107] expresses the famous law of proportionality of stress and strain that ROBERT HOOKE discovered in 1660 but did not publish until 1676, and then only in the form of an anagram meant to represent the words *Ut tensio sic vis* (Plate 25). Although approximately true for a wide range of deformations,

---

[1] On page 46, Vol. II, of his *A Course of Lectures on Natural Philosophy and the Mechanical Arts* (London, 1807), YOUNG states: "The modulus of the elasticity of any substance is a column of the same substance, capable of producing a pressure on its base which is to the weight causing a certain degree of compression, as the length of the substance is to the diminution of its length." For illustrations of this definition see Lecture XIII, Vol. I of the same work, or *A Source Book in Physics* (1935), pp. 95–97, in which a portion of this lecture is reproduced. It should be noted, however, that the modulus defined by YOUNG differs from the present one. In fact, what we now call YOUNG'S modulus is the *weight* of the column per unit of area of its base.

THIS photostat was made from a copy of HOOKE's A DESCRIPTION OF HELIO-SCOPES, And some other INSTRUMENTS which is in the private library of DR. GEORGE E. HALE. HOOKE announced his law of elasticity in 1676 in the anagram given in Invention 3. In his LECTURES De Potentia Restitutiva, or of SPRING Explaining the Power of Springing Bodies, published two years later, he stated that he had first found out the theory of springs eighteen years before, but had not published his discovery because he was anxious to obtain a patent for a particular application of it. He then continued, "About two years since I printed this Theory in an Anagram at the end of my Book of the Descriptions of Helioscopes, viz. c e i i i n o s s s t t u u, id est, Ut tensio sic vis; That is, The Power of any Spring is in the same proportion with the Tension thereof."

Facsimile reproductions of both the Description of Helioscopes and the Potentia Restitutiva will be found in R. T. Gunther's Early Science in Oxford (Oxford, 1931), Vol. VIII, pp. 119–152, 331–356. A portion of the latter is also reproduced in W. F. Magie's A Source Book in Physics (1935), pp. 93–95.

To fill the vacancy of the enfuing page, I have here added a *decimate* of the *centefme* of the Inventions I intend to
publifh, though poffibly not in the fame order, but as I can
get opportunity and leafure; moft of which, I hope,
will be as ufeful to Mankind, as they are yet unknown and
new.

1. *A way of Regulating all forts of* Watches *or* Time-
keepers, *fo as to make any way to equalize , if not exceed the*
Pendulum-Clocks *now ufed.*

2. *The true Mathematical and Mechanichal form of all
manner of* Arches *for Building, with the true butment neceffary
to each of them.* A Problem which no *Architectonick* Writer hath ever yet attempted , much lefs performed. abccc
ddeeeee f gg iiiiiiii llmmmmnnnnnnooprr ssstttttttuuuuuuuux.

3. *The true Theory of* Elafticity *or* Springinefs, *and a particular Explication thereof in feveral Subjects in which it is to
be found: And the way of computing the velocity of Bodies
moved by them.* ceiiinosssttuu.

4. *A very plain and practical way of counterpoifing* Li-
quors, *of great ufe in* Hydraulicks. Difcovered.

5. *A new fort of Object-Glaffes for* Telefcopes *and* Mi-
crofcopes, *much outdoing any yet ufed.* Difcovered.

6. *A new* Selenofcope, *eafie enough to be made and ufed,
whereby the fmalleft inequality of the Moons furface and limb
may be moft plainly diftinguifhed.* Difcovered.

7. *A new fort of* Horizontal-Sayls *for a* Mill , *performing
the moft that any Horizontal-Sayls of that bignefs are capable of;
and the various ufe of that principle on divers other occafions.*
Difcovered.

8. *A new way of* Poft-Charriot *for travelling far, without
much wearying Horfe or Rider.* Difcovered.

9. *A new fort of* Philofophical-Scales, *of great ufe in Experimental Philofophy.* cdeiinnoopsssttuu.

10. *A new Invention in* Mechanicks *of prodigious ufe, exceeding the* chimera's *of perpetual motions for feveral ufes.*
aaaæbccddeeeeeegiiilmmmnnooppqrrrrs
tttuuuuuu.
aaeff hiiiiillnrrsstuu.

Photostat of the Concluding Paragraph (pages 31–32)
of A DESCRIPTION OF HELIOSCOPES, AND SOME OTHER INSTRUMENTS,
by ROBERT HOOKE (London, 1676)

the law holds precisely only for very small strains taking place under isothermal conditions. The first use of the law was made by EDME MARIOTTE[1] (1620–1684), who stated it independently and applied it to GALILEO'S problem of the resistance of beams to rupture.

Experience has shown that HOOKE'S law may be generalized for various kinds of elastic changes into the statement

Stress = constant · strain.   [109]

The *constant* thus defined is a characteristic of the elastic material and is generally referred to as a *modulus*, or measure, of elasticity. There are as many kinds of elastic moduluses for a given material as there are kinds of strains. While these moduluses must be determined from measurements made upon a particular piece of material, they really are properties of the *substance* of which the specimen is made and not of the dimensions of the particular sample. They do vary, however, with the temperature, diminishing, in the case of solids, as the temperature rises; thus, a spring balance stretches a little farther for the same force when hot than when cold. There is no connection between these elastic constants of a material and its "degree" of elasticity. The former measure the magnitudes of the stresses required to produce given strains, the latter the perfectness of the return to the initial conditions. In popular usage a

FIG. 84. A column under compressive stress

body is said to be "very elastic" if it possesses nearly perfect elasticity through wide limits. But in technology it is regarded as *highly elastic* only if it has large elastic constants; this is because the ability of a material to withstand large forces is generally of more importance in engineering than its ability to withstand large deformations. In the technical sense, then, lead and steel are highly elastic, since their elastic constants are large, whereas rubber is not very elastic, for its elastic constants are small.

o

Elasticity of Size

91. Volume Elasticity. A strain that consists of change in size of every volume element of a body without change in shape, is called a *volume*, or *isotropic*, strain. It is measured by the proportion

---

[1] *Traité du Mouvement des Eaux* (Paris, 1686) [see *Œuvres* (Leiden, 1717; The Hague, 1740)]. A translation by J. T. Desaguliers was published in London in 1718 under the title *The Motion of Water and other Fluids.*

$\Delta V/V$, in which the volume of the body is changed. In elastic bodies such a strain is accompanied by a stress which, if the body be isotropic, will be a pressure equal in all directions. Such a stress is called a *hydrostatic pressure* because, as BLAISE PASCAL[1] (1623–1662) clearly showed, it is the only kind of stress that can exist in fluids at rest. The term *pressure* is used here in its correct physical sense to mean a normal force per unit area; that is, if pressure be denoted by P, normal force by f, and area by A, then the defining equation for the *average pressure*[2] over the area A is $P = f/A$.

Experiments show that for small volume strains, Eq. [109] is applicable. Now, to produce a volume strain, a uniform pressure must be applied at all points on the surface of the body, and this external pressure will be equal in magnitude to the resulting stress in the body. Hence, if the external pressure for a volume V be P, and that for a volume V' be P', then

$$P' - P = - k \frac{V' - V}{V},$$

or
$$\Delta P = - k \frac{\Delta V}{V}, \qquad\qquad [110]$$

the negative sign merely indicating that the volume gets smaller as the pressure increases. The modulus k is called the *volume*, or *bulk*, modulus.

All materials, whether their state of aggregation be solid, liquid, or gaseous, possess elasticity of volume. Liquids offer less resistance to compression than solids, as a rule, but this resistance is so large that even liquids may for many purposes be treated as if incompressible. However, BRIDGMAN[3] has succeeded in compressing water into three fourths its ordinary volume by employing a pressure[4] of 20,000 atmospheres.

EXAMPLE. The reciprocal of the volume modulus is called the *compressibility*. Show that the change of volume of a body due to a given change of pressure is equal to the product of the change of pressure, the original volume, and the compressibility.

---

[1] *Traitez de l'équilibre des liqueurs et de la pesanteur de la masse de l'air* (Paris, 1663).

[2] If the force is not uniform over the surface, one must then use the more general concept of the *pressure at a point*, for which the defining equation is

$$P = \lim_{\Delta A \to 0} \frac{\Delta f}{\Delta A},$$

where Δf is the normal force on a small portion ΔA of the surface.

[3] P. W. BRIDGMAN, *The Physics of High Pressures* (Bell, 1931).

[4] The *standard atmosphere* (abbreviation A_s) is defined as the pressure exerted by a column of mercury 76 cm high, at 0° C and subject to standard acceleration due to gravity g_s. It is equivalent to 1.01325×10^6 dyne · cm^{-2}, or 1.01325 bars. A *bar* is a pressure of 10^6 dyne · cm^{-2}

92. Boyle's Law. When BOYLE published his *New Experiments*, the book met with a certain amount of opposition, although it was on the whole a sound record of experimental work. The criticism was for the most part unwarranted, but it served to incite BOYLE to renewed research, and in 1662 he brought to light the famous law that describes the relation between the volume occupied by a confined gas and the pressure exerted by the gas upon the walls of the containing vessel. BOYLE found that *so long as the temperature of a given mass of gas remained unchanged, the product of the pressure and volume of the gas was constant* (Fig. 85). In 1676 MARIOTTE [1] independently discovered the same law and carefully tested it. The *constant* in BOYLE's law evidently is proportional to the mass of gas used; for, if the pressure and temperature be kept the same, one must take twice the original volume in order to double the mass of gas. We may therefore express BOYLE's law by means of the equation

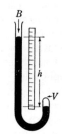

FIG. 85. BOYLE'S apparatus for investigating pressure greater than that of the atmosphere. Under isothermal conditions $(B + h)V$ is constant, where B is the atmospheric pressure expressed in centimeters of mercury, h is the difference in height of the two mercury columns, and V is the volume of the gas under test

$$PV = MC,   [111]$$

where P is the pressure and V is the volume of a mass M of gas. The constant C depends on the kind of gas and its temperature. In view of Eq. [111] and the definition of density (namely, $\rho = M/V$), BOYLE's law may be stated in the alternative form: Under isothermal conditions the pressure in a gas is proportional to the density and is independent of the mass of the gas.

Measurements made with even an ordinary meter stick are accurate enough to reveal the fact that the simple law enunciated by BOYLE represents the behavior of many gases only approximately. Since we shall have occasion to consider this matter again in chapters that follow, it will suffice to say here that for gases like hydrogen and air the departures from the law are less than 1 percent up to 10 atmospheres of pressure, but for substances like carbon dioxide and water vapor the law is inadequate except at temperatures well above their liquefaction points. We shall often find it convenient to speak of an *ideal* or *perfect* gas, which is an imaginary gas whose behavior under isothermal conditions is described exactly by BOYLE's law. This imaginary gas is

---

[1] *Discours de la Nature de l'Air* [see *Œuvres* (Leiden, 1717; The Hague, 1740)]. A portion of the paper appears in *A Source Book in Physics* (1935), pp. 88–92.

called *ideal* because of its simple properties and because it represents the behavior of all actual gases to a first approximation. Indeed, experiment shows that as the pressure decreases the behavior of all gases becomes more and more like that of the ideal gas. At low enough pressures, therefore, any gas will act like an ideal gas to any desired degree of approximation.

93. Isothermal Elasticity of an Ideal Gas. Suppose that the initial pressure and volume of a given mass of an ideal gas are P and V, respectively, and that the application of an additional small pressure ΔP changes the volume by the small quantity ΔV, the temperature being kept constant by abstracting heat from the gas during the compression. Then, by BOYLE'S law, $PV = (P + \Delta P)(V + \Delta V)$. By performing the indicated multiplication and neglecting a term which is the product of the two small quantities ΔP and ΔV, we obtain $P \Delta V + V \Delta P = 0$. If this equation be solved for P, there results, in view of Eq. [110],

$$k = P. \qquad [112]$$

The obvious interpretation of this result [1] is that, for small strains, the volume modulus of an ideal gas is equal to the initial pressure provided the process is isothermal. If, however, a gas is compressed so quickly that its temperature rises, the resistance to compression is much larger than is predicted by Eq. [112]. Liquids also resist a quick compression a little more than a slow one, but the difference in their case is so small as usually to be negligible.

○

Elasticity of Shape

In the long interval between the discovery of HOOKE'S law and that of NAVIER'S equations, work on the elasticity of solids was confined largely to the solution and extension of GALILEO'S problem by such eminent mathematical physicists as JACQUES BERNOULLI (1654–1705), DANIEL BERNOULLI (1700–1782), EULER, and LAGRANGE. The theory of the flexure of beams was considered by COULOMB,[2] who

---

[1] This result might have been obtained in a more elegant manner by the use of the differential calculus. For infinitesimal changes, by definition (Eq. [110]),

$$k = -V \frac{dP}{dV}.$$

But, by BOYLE'S law, $P = MC/V$, from which $dP/dV = -MC/V^2 = -P/V$. Hence

$$k = -V \frac{dP}{dV} = P.$$

[2] *Théorie des machines simples* (1779). See **A Source Book in Physics* (1935), pp. 98–103.

also investigated the resistance of thin wires to torsion. COULOMB
was also the first to give attention to the kind of strain that we now
call a shear, though he considered it in connection with ruptures
only. It was not until 1807 that the conception of a shear as an
elastic strain was introduced by YOUNG.

94. Shearing Strain and Stress. A *shearing strain*, or *shear*, is a
strain that consists in a change of shape only, without any change of
volume. Since our everyday ideas of change of shape are essentially
qualitative, we must first agree on a way of measuring the amount
of shear existing at any point in a distorted body. Suppose that
abcd, Fig. 86, represents a little rectangular element of the body.
If the base *ab* be firmly fixed, a force *f*
applied tangentially to the upper face
will strain the element into the shape
abc'd', just as a thick book lying on the
table would be distorted by applying a
tangential force to the cover. Such a
deformation in which the adjacent layers of the body merely slip past one
another is evidently a shear, for configuration alone has been changed, the
volume remaining the same as at first.
As the measure of this shearing strain

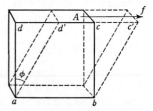

FIG. 86. The shearing strain is
dd'/da or, if the strain be small, ϕ.
The shearing stress is *f/A*

we shall agree to take the ratio *dd'/da*. When the shear is small,
as is usually the case in practice, this ratio is equal to the angle ϕ,
expressed in radians.

If the shear tends to disappear when the forces that produce
it cease to act, the material is said to possess *shear elasticity* or *rigidity*. In Fig. 86 imagine a plane drawn anywhere in the element,
parallel to the top; the part on one side of this plane will exert a
tangential force on the part on the other side, and this force will
equal the force applied to the top. The magnitude of this force
per unit area of the plane (namely, *f/A*) is the measure of the
shearing stress. If HOOKE's law holds, this stress should be proportional to the shearing strain; that is,

$$\frac{f}{A} = n\phi, \qquad [113]$$

where *n* is the *shear modulus* or *modulus of rigidity*.

Only solid bodies possess elasticity of shape. In fact, it is upon
this difference in the elastic behavior of different bodies that the
definitions of solids and fluids are usually based. A body is called a

solid[1] if it possesses elasticity of both size and shape; it is called a *fluid* if it exhibits elasticity of size but not of shape.

95. Torsion of Hollow Tubes. When a thin hollow cylinder (Fig. 87) is twisted about its geometrical axis by holding one end and applying a torque to the other end, a pure shear results. Let us imagine the whole length of the cylinder divided into thin rings of equal thickness by planes perpendicular to the axis. If a series of imaginary planes passing through the axis now be drawn before the cylinder is twisted, they will divide each ring into very small elements,

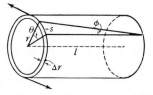

Fig. 87. Torsion of a thin-walled tube, an example of pure shearing action

as in Fig. 88. After the torsion each of these elements will have a strain like that of the element in Fig. 86.

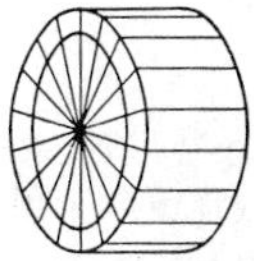

In the present case it is not so easy to measure ϕ directly as it is to measure θ, the angle through which the end of the cylinder is twisted (Fig. 87). Now $\phi = r\theta/l$, where r is the mean radius and l is the length of the cylinder. Therefore Eq. [113] may be written

Fig. 88. Each ring of the hollow tube is divided into small elements

$$\frac{f}{2\pi r \Delta r} = n\frac{r\theta}{l},$$

[114]

where f is the tangential shearing force applied to the end of the cylinder and Δr is the thickness of the cylinder wall. Suppose that the force f is brought into play in the manner shown in Fig. 89. By the principle of moments, $fr = FR$, or $f = FR/r$. In view of this relation and Eq. [114], one obtains finally

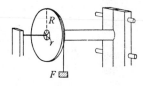

$$n = \frac{FRl}{2\pi r^3 \cdot \Delta r \cdot \theta}.$$

[115]

This equation expresses the shear modulus in terms of quantities all of which are easily measurable.

96. Torsion of Rods and Wires. The

Fig. 89. The shear modulus of the material in a tube may be investigated by clamping one end of the tube, fastening the other end rigidly to a grooved circular disk of radius R, and applying a twisting force F to the rim of the disk by means of weights

problem of a *solid* cylinder which is twisted about its axis may be approached by imagining the cylinder to be made up of a large

---

[1] For a review of modern work on the solid state, see *R. E. Gibbs, *Science Progress* **29**, 661 (1935).

number of concentric hollow cylinders of thicknesses Δr_i and of mean radii $r_1, r_2, \cdots, r_i, \cdots, r_s$ (Fig. 90). The torque, say L_i, needed to twist any particular hollow cylinder of radius r_i through the angle θ is, from Eq. [115],

$$L_i = \frac{2\,\pi n r_i{}^3 \Delta r_i \theta}{l}.$$ [116]

The total torque L which must be applied in order to twist the solid cylinder through the angle θ is evidently the sum of all the torques L_i, or

$$L = \frac{2\,\pi n\theta}{l} \lim_{s\to\infty} \sum_{i=1}^{i=s} r_i{}^3 \Delta r_i = \frac{2\,\pi n\theta}{l} \int_0^r r^3\,dr = \frac{2\,\pi n\theta}{l} \cdot \frac{r^4}{4}.$$

Hence, finally, for a solid cylinder of radius r,

$$n = \frac{2\,lL}{\pi r^4 \theta}.$$ [117]

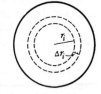

Eq. [117] shows that for a given cylinder the torque is proportional to θ. The proportionality factor $\pi r^4 n/2\,l$ is called the *constant of torsion* of the wire or rod; denoting this constant of the cylinder by L_o, we may write Eq. [117] in the form

Fig. 90. Cross section of a solid cylinder

$$L = L_o \theta.$$ [118]

Obviously L_o might also be defined as the *torque per unit twist*. Eqs. [117] and [118] permit us to express the shear modulus in terms of the constant of torsion and dimensions of the cylinder; thus

$$n = \frac{2\,lL_o}{\pi r^4}.$$ [119]

○

Relations between the Elastic Constants

Of all possible elastic changes, the only two that are independent are *volume strain* and *shearing strain*. Other changes involve both of these fundamental kinds. For instance, the stretching of a wire (Sec. **89**) involves a change in form as well as in volume, for instead of increasing or decreasing all dimensions, it increases the length and decreases the diameter; this lateral contraction is strikingly shown by a stretched rubber band. The various elastic changes have moduluses of their own, but these are always expressible in terms of the two fundamental ones, k and n. For example, it can be shown that YOUNG'S modulus is given by

$$Y = \frac{9\,kn}{3\,k + n}.$$ [120]

Since the direct determination of k is not easy, this equation enables us to find k if we know Y and n, both of which can be easily determined experimentally.[1] Of the various coefficients of elasticity, YOUNG'S modulus is the one most used in practical work.[2]

○

EXPERIMENT VIIIA. DENSITY OF AIR

Suppose that a glass bulb of known volume V is weighed when it is filled with dry air at atmospheric pressure P_1 and that the bulb is again weighed after the pressure in it has been reduced to P_2 by pumping out some of the air. If ρ_1 and ρ_2 be the densities of air corresponding to the pressures P_1 and P_2 and if m be the mass of air pumped out, as revealed by the difference of the two weighings, it is evident that $V\rho_1 - V\rho_2 = m$. But, by BOYLE'S law (Sec. 92), $P_1/\rho_1 = P_2/\rho_2$; hence

$$\rho_1 = \frac{P_1 m}{V(P_1 - P_2)}. \qquad [121]$$

In making the weighings, a counterpoise, which consists of a closed bulb having the same external volume as the bulb containing the air, is placed on the pan of the balance with the weights. This device was first used by the French chemist and physicist HENRI VICTOR REGNAULT (1810–1878), known for his exceedingly careful and exact measurements of many physical constants, in his classical determination of the density of air. It minimizes any errors which would arise from changes in temperatures or atmospheric pressure during the experiment, provided only that the evacuation of the bulb from the pressure P_1 to pressure P_2 is accomplished isothermally; for, with this arrangement, the buoyant effect of the air upon both sides of the balance is the same no matter how rapidly the barometric pressure or the temperature may change.

a. To Fill the Bulb with Dry Air at Atmospheric Pressure. See that the stopcocks on the bulb are well greased and that the surface of the bulb is clean and dry. By means of a vacuum pump or compressed-air system, pass a gentle current of air first through a calcium chloride drying tube and thence through the bulb for several minutes.

---

[1] See Appendix 13, Table F, for values of the moduluses of elasticity for various substances.

[2] *For more comprehensive elementary treatments of the topics discussed in this chapter the student is referred to H. Bateman, article "Elasticity," *Encyclopaedia Britannica*, ed. 14; E. Edser, *General Physics for Students*, Chaps. VII–VIII; J. H. Poynting and J. J. Thomson, *Properties of Matter*, Chaps. IV–IX; P. G. Tait, *Properties of Matter*, Chaps. VIII–XI.

Then turn off the air at its source and carefully close the stopcock in the outlet tube of the bulb. Allow the bulb to remain connected to the drying tube for several minutes, so that it may assume the temperature of the room. Observe the temperature by means of a thermometer hung near the bulb and observe the pressure P_1, in centimeters of mercury, by means of a mercurial barometer.[1] Then carefully close the inlet stopcock, disconnect the bulb from the drying tube, and weigh the bulb.

b. The First Weighing; Use of the Analytical Balance. The analytical balance (Fig. 91) is controlled by means of the *pan-arrest p* and the *beam-arrest b.* The pans or their contents should never be touched except when the beam is raised and clamped by means of the beam-arrest. Remove all dust from the bulb and the pans of the balance by means of a camel's-hair brush. Then very carefully suspend the bulb from the hook c of the left-hand pan and place upon the right-hand pan the counterpoise and a few gram weights from the box of weights. *Handle the weights only with the forceps.* Next release the pans by pushing in and fastening the button p. Then

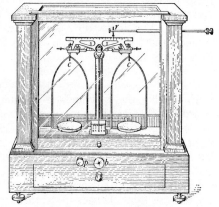

FIG. 91. An analytical balance

slowly release the beam-arrest b just enough to see whether the chosen weight is too heavy or too light; this done, raise the beam-arrest immediately but so slowly as not to endanger the knife-edges by the slightest jar. By proceeding thus, make a systematic trial of the gram weights until you know between what two consecutive numbers of grams the condition of balance must lie. Then try in the same way the milligram weights in order of magnitude until a weight is found such that, when the beam-arrest is completely lowered, the pointer oscillates *near* the middle of the scale over a distance of from three to six divisions; a larger swing than this indicates insufficient care in lowering the beam-arrest. The rider r may be used in place of the small milligram weights, if desired, but it is best not to attempt to add fractional portions of one milligram by means of the rider.

---

[1] Follow the directions given in Appendix 6.

The next step in weighing is to take the resting point. Before doing this, raise the beam-arrest again, taking pains to avoid a jar by raising it only at a time when the pointer is at the center of the scale. Close the door of the balance so as to shut out air currents, and stop all swinging of the pans by gently manipulating the pan-arrest p. Then carefully lower first the pan-arrest and then the beam-arrest and take the *resting point* R_1. This is done by averaging the mean of three successive turning points of the pointer on one side with the mean of the two intervening turning points on the other side. This use of an odd instead of an even number of turning points eliminates completely the effect of damping. The following example, taken in connection with Fig. 92, will make clear the method of procedure:

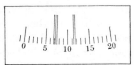

Fig. 92. The scale of the balance

TURNING POINTS

Left	Right
6.8	11.7
7.1	11.2
7.4	
Means 7.10	11.5

$$\therefore R = 9.3$$

Having determined R_1, slowly raise the beam-arrest when the pointer is in the middle of a swing. Count up the weights on the scale pan, calling this sum M_1. To avoid error, first record in your notebook the values of the weights as obtained by counting the vacant spaces in the box and then check this by counting again as the weights are returned to the box. Remember exactly which of the larger weights were used, for you will want to use these same weights again later.

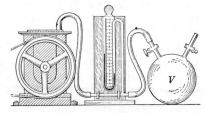

Fig. 93. Apparatus for evacuating the bulb

c. Evacuation of the Bulb. Attach the bulb V to a vacuum pump and open-arm mercury manometer (Fig. 93) and pump as much of the air as possible out of the bulb. Wait several minutes for the evacuated bulb to regain the temperature of the room. Then read the two arms of the manometer and at the same time close the stopcock. Since one arm of the manometer is open to the outside air and the other to the exhausted bulb, the difference of the heights of the mercury in the two arms gives directly the difference $P_1 - P_2$ between the barometric pressure and the pressure in the bulb. Observe the temperature of the bulb; if it differs by more than half a degree from that previously taken, the first weighing must be discarded and another taken for the present temperature. This need not be done until after the completion of *d*.

d. The Second Weighing. Place the bulb V upon the same scale pan as before and, proceeding exactly as in *b*, balance it by means of such weights of mass M_2 as will cause the pointer again to oscillate near the middle of the scale. In doing this, choose the same larger weights as were used in *b*. Then take the new resting point R_2. If R_2 coincides exactly with R_1, then evidently $M_1 - M_2$ would be the difference in the masses of the bulb in the two cases. But, in general, R_2 will not coincide with R_1, and it is not best to make it do so by repeatedly shifting the rider. The most rapid and the only correct method of making an accurate weighing is to determine the *sensitivity*,[1] or number of scale divisions that the rest point is shifted by the addition of 1 mg, and then to calculate by interpolation the exact correction which must be applied to M_2 in order to bring R_2 precisely into coincidence with R_1. This is done as follows: Immediately after finding R_2, add to the *lighter* side a small weight, say 2 mg, or 1 mg if the balance is very sensitive, and take the corresponding rest point R_2'. This procedure simply determines the value in milligrams of the scale divisions. Thus, if $R_2 = 10.6$ and if, upon the addition of 2 mg, $R_2' = 7.0$, then the sensitivity is 1.8 scale divisions per millimeter. From a knowledge of R_1, R_2, and the sensitivity, it is an easy matter to calculate the mass which must be added to or subtracted from M_2 in order that the pointer may be brought exactly to the original resting point R_1. Thus, in this case, the number of milligrams which would be required to move the pointer from R_2 (= 10.6) back to R_1 (= 9.3) is 0.72. This number of milligrams must be added to or subtracted from M_2, according as the point 10.6 is farther from or nearer to the object being weighed than the point 9.3. Let M_2' represent the corrected value of M_2. Then $M_1 - M_2'$ is the mass m of Eq. [121].

e. Volume of the Bulb. If the volume V of the bulb is not marked upon it, fill the bulb with water and weigh upon a trip scales. Ob-

---

[1] Owing to the bending of the beam, the sensitivity varies with the load and therefore should ordinarily be determined at each weighing (Chap. 6). Since, however, it generally requires a considerable change in load to produce an appreciable change in the sensitivity of a good balance, one determination is usually sufficient so long as the loads involved are of about the same magnitude.

serve the temperature of the liquid and obtain its density from tables. Thoroughly dry the bulb before using it again.

f. Calculations. Calculate the density ρ_1 of dry air for the conditions of temperature and pressure that exist in the room. The pressures which appear in Eq. [121] — namely, P_1 and $(P_1 - P_2)$ — must of course be expressed in the same units. If they are expressed in centimeters of mercury, the two columns of mercury which they represent must have the same temperature; a difference not exceeding 5° C will lead to an inappreciable error, however, as an inspection of a table of mercury densities [1] will show.

Compare your value of the density of dry air with the values given in Appendix 13, Table D. Since the pressures in the tables represent heights of mercury columns at 0° C, it will be necessary to reduce your observed value of the barometric height P_1 to 0° C and then to find in the tables the value of the density corresponding to this reduced value of the barometric height. This reduction is made by multiplying the observed height by the ratio of the densities of mercury at the room temperature and at 0° C; this correction, together with a slight correction for the brass barometric scale, will be found already worked out in Appendix 13, Table G.

1. Why was it necessary in the two weighings which you made to have the air in the bulb initially at the same temperature?

2. Derive Eq. [121].

3. ARISTOTLE attempted to determine whether air has weight by weighing a flexible bladder, first when it was deflated and afterwards when it was inflated with air. He found both weighings to be equal, and concluded that air has no weight. Criticize this conclusion.

4. Show why the true zero of a vibration which is gradually dying down because of damping is not obtained by taking the mean of two successive turning points, one on the right and one on the left.

5. Show why the true zero is obtained by averaging the mean of two successive turning points on one side with the intervening turning point on the other side.

o

OPTIONAL LABORATORY PROBLEMS

1. Variation of the Sensitivity of an Analytical Balance with Load. Determine the sensitivity of an analytical balance for zero load in the pans and then for each of several loads up to the capacity of the balance. Find the capacity from the manufacturer's catalogue. Plot a curve showing the variation of the sensitivity with the load and keep it for use with the balance. (See the Optional Laboratory Problem in Chapter 6.)

---

[1] Appendix 13, Table C.

2. Double Weighing. Determine the weight of a piece of brass on an analytical balance by the method of double weighing (see Exp. VI, Part II). Prove that the formula to be used in making a double weighing is

$$W = W' + \frac{1}{2}\frac{R_1 - R_2}{s},$$

where W' is the weight that balances the object to within 1 mg, R_1 and R_2 are the resting points for the load on the left-hand pan and on the right-hand pan respectively, and s is the sensitivity with the load in either pan. Do you find that a double weighing requires a larger number of observations than a single weighing?

3. Ratio of the Lengths of the Arms of an Analytical Balance. Weigh an object on an analytical balance by the method of single weighing, first when the object is in the right-hand pan and then when it is in the left-hand pan, and from these two data compute the ratio of the lengths of the arms of the balance. How much error does this inequality of the lengths of the arms introduce into the determination of the weight of the object by a single weighing?

o

EXPERIMENT VIIIB. TORSIONAL PROPERTIES OF RODS

A statical method is employed. The apparatus (Fig. 94) consists of two heavy table-clamps, one of them carrying a wheel about 15 cm in diameter. In the hub of the wheel is a socket in which the rod to be tested is rigidly fastened. The other end of the rod is held in a similar socket mounted in the other clamp. The torque for twisting the rod about its axis is applied by adding weights to the weight-hanger attached to the rim of the torsion wheel. Angles of twist are measured by means of a graduated scale on the rim of the wheel and a vernier.

Four rods are to be used, of which three are of steel of the same length but of different diameters; the fourth is of brass. One of the steel rods has a third bushing near its middle, which allows it to be

Fig. 94. Apparatus for studying the torsion of a rod

used in measurements at two different lengths. Numbers should be assigned to the rods for identification. Assign No. 1 to the largest steel rod, No. 2 to the next largest, No. 3 to the smallest, No. 4 to the half-length of the same rod, and No. 5 to the brass rod. Record also the set-number of the rods which you test, for in a later experiment (Exp. XIVB) you will study the elastic properties of these rods by a kinetical method.

a. Place rod No. 1 in position. The rod must be initially straight, and the clamps must be so aligned that the rod remains straight and free from longitudinal stress when both it and the clamps are fastened firmly in position.

b. With the weight-hanger in position, place enough weights on it to take up any slack or lost motion in the apparatus; the weight-hanger may be sufficient to do this. Carefully set the vernier at zero and take this "no-load" reading. Then gently add weights, one at a time, to the weight-hanger, and record at each step the load and the corresponding scale-reading.[1] Also take readings as these weights are successively removed. Take two distinct readings with the maximum load.

c. Measure the distance between the inside faces of the bushings on the ends of the rod and record this as the length l of the rod; if the rod is equipped with a third bushing near its middle point, the length of this bushing as determined with a vernier caliper must be deducted from the length as found above.

Measure the diameter of the torsion wheel and record its radius R.

Measure the diameter of the rod in at least ten places equally distributed along its length and circumference. Be sure to check the zero reading of the micrometer caliper.[2]

d. Repeat the preceding observations with rods Nos. 2, 3, 4, and 5.

e. For each rod calculate the mean twist θ for 100 gwt added to or subtracted from the weight-hanger. The following example illustrates the method of averaging to be employed. Suppose that, in the case of one of the rods, five weights were successively added and then subtracted from the weight-hanger, so that twelve readings were made. In the list of recorded scale-readings, subtract reading 1 from reading 6, 2 from 5, and 3 from 4; likewise subtract 12 from 7, 11 from 8, and 10 from 9. Do the same with the list of loads. Then add up the resulting list of load differences and the list of angle differences, and divide the former sum into the latter.

---

[1] If the diameter of the rod is about 2 mm, add 100 gwt at a time up to 500 gwt; if about 3 mm, add 200 gwt at a time up to 1000 gwt; if about 4 mm, add 500 gwt at a time up to 2500 gwt. The maximum weight should not be so great as to exceed the elastic limit of the steel. (How can you tell whether you are still within this limit?)

[2] See Appendix 4.

Calculate by means of Eq. [118] the constant of torsion of each rod. Then use Eq. [119] to compute n for each rod. Average the values of n found for the steel rods. Compare (*a*) the ratio of the lengths of rods Nos. 3 and 4 with the ratio of the mean twists produced in each by unit torque; (*b*) the ratio of the fourth powers of the radii of rods Nos. 1 and 2 with the ratio of the mean twists per unit torque; (*c*) the ratio of the maximum and minimum twists of rod No. 2 with the ratio of the maximum and minimum loads.

1. State in words the relations shown by your experiment to exist between the angle of twist and the dimensions of the rod, and between the angle of twist and the torque producing the twist. Are these relations in accord with Eq. [117]?

2. Suppose that a, b, $\cdots$, h are the values of the successive scale-readings obtained as weights are added to and then subtracted from the weight-hanger. Show that if the mean twist is calculated simply by averaging the differences $(b - a)$, $(c - b)$, $\cdots$, $(g - h)$, the readings b, c, f, and g are wasted. Show that if the readings are averaged by the method which you actually employed in this experiment, no observation is used more than once, so that all of them are utilized in obtaining the mean. Also show that the latter method of averaging gives to each observation precisely the amount of consideration that it deserves; that is, it gives a weight of 3 to the observed twist for 300 gwt, etc.

3. One observer finds the diameter of a rod to be 2.513 mm, and another finds it to be 2.501 mm. What is the percentage difference of their two values of the diameter? What would be the resulting percentage difference in the two values of n which they obtained?

4. If the radius of the rod were measured to 0.001 cm, with what accuracy should the length be measured in order that the value of n may be affected to the same extent by both?

5. Decide from a study of your data which one of the quantities involved in the determination of n introduces the largest error.

○

OPTIONAL LABORATORY PROBLEMS

1. Young's Modulus and the Elastic Limit for an Annealed Copper Wire. Determine YOUNG'S modulus for an annealed copper wire. Between successive loadings of the wire return to the initial "no-load" condition, so that you will know when the elastic limit is reached. Try to verify the assertion that if a wire is strained beyond its elastic limit, the yield point is raised and a permanent elongation is produced. Does the wire which you are testing become more or less elastic after it has been stretched beyond its elastic limit? What, approximately, is its ultimate strength?

2. Indirect Determination of the Volume Modulus. Find the volume modulus of some given material by determining YOUNG's modulus and the shear modulus for a specimen of the material that is in the form of a rod or wire.

○

QUESTION SUMMARY

1. Define *stress* and *strain* for each of the following cases: stretching, or the application of forces of equal magnitude in opposite directions along the same line; volume compression or extension; shear.

2. State HOOKE's law.

3. What is meant by a *modulus of elasticity*? Define in words and by means of an equation: *Young's modulus*; the *volume modulus*; the *shear modulus*. How is each determined experimentally?

4. When is a body said to be *perfectly elastic*? Distinguish between an elastic constant of a material and its degree of elasticity.

5. What is meant by the *constant of torsion* of a wire or rod? Of what is it a constant? How does it depend upon the length of the rod? upon its radius?

6. State BOYLE's law and specify the conditions under which it applies.

○

PROBLEMS

1. It is found that when a cast-steel rod of length 5.00 ft and diameter 0.280 in. is subjected to a stretching force of 300 lbwt, the resulting elongation is 0.016 in. Compute (*a*) the longitudinal stress; (*b*) the longitudinal strain; (*c*) YOUNG's modulus for cast steel.
   Ans. (*a*) 4.9×10^3 lbwt · in.$^{-2}$; (*b*) 2.7×10^{-4}; (*c*) 1.8×10^7 lbwt · in.$^{-2}$.

2. A wire 80 cm long and 0.30 cm in diameter is stretched 0.30 mm by a force of 2.0 kgwt. If another wire having the same YOUNG's modulus is 180 cm long and 8.0 mm in diameter, what force is required to stretch it to a length of 180.1 cm?   *Ans.* 21 kgwt.

3. A copper wire 31 cm long and 0.50 mm in diameter is joined to a drawn brass wire 108 cm long and 1.0 mm in diameter. If a certain stretching force produces an elongation of 0.50 mm in the whole wire, what is the elongation of each part?   *Ans.* Brass, 0.27 mm; copper, 0.23 mm.

4. The work done per unit volume in straining a material to its elastic limit is called the *resilience* of the material. (*a*) If YOUNG's modulus for a certain piece of steel is 2.98×10^7 lbwt · in.$^{-2}$ and its elastic limit is 6.20×10^4 lbwt · in.$^{-2}$, what is the resilience? (*b*) If 1.0 ft^3 of this steel were in the form of a spring so designed as to be strained in every part to its elastic limit when wound up, how much energy could be stored in the spring?   *Ans.* (*a*) 9.3×10^3 ft · lbwt · ft^{-3}; (*b*) 17 hp · sec.

5. A given mass of an ideal gas is compressed isothermally in a manometer (Fig. 85), and the successive volumes and corresponding pressures of the gas are observed. (*a*) If the pressures be plotted as ordinates and the corresponding volumes as abscissas, what kind of curve will be obtained? (*b*) What is the physical significance of the area under this curve? (*c*) If the product PV is plotted as a function of P, what will be the nature of the resulting curve?

6. The space above the mercury in a barometer tube contained some air. When the volume of this space was 10.0 cm³, the barometer indicated 70.0 cm of pressure. When the volume of the space was reduced to 5.00 cm³ by pushing the barometer tube down into the cistern, it indicated 69.5 cm. What was the true barometric height?   *Ans.* 70.5 cm.

7. When the level of the mercury in a McLeod gauge is at b, Fig. 95, let the pressure of the air in bc be P_1 and let the volume between b and c be V_1. Suppose that the reservoir R is now raised, causing the mercury to rise into the upper part of the small closed tube. Let this be continued until the mercury in D comes to a level with the top c of the closed tube. If A be the cross-sectional area of the closed tube and h the final difference in heights of the mercury columns, show that $P_1 = (A/V_1)h^2$, approximately.

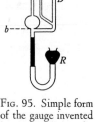

Fig. 95. Simple form of the gauge invented by H. McLeod for measuring low gas pressures. [*Philosophical Magazine* (4) **48**, 110 (1874)]

8. At the greatest oceanic depths, about 10 km, the pressure is approximately 1.0 kilobar, the *bar* being a unit of pressure equal to 10^6 dyne · cm⁻². (*a*) If a piece of ordinary mild steel sinks to this depth, how much is its density changed? (*b*) If the density of sea water at the surface is 64.9 lb · ft⁻³, what is its density at this depth?   *Ans.* (*a*) 0.05 percent; (*b*) 68 lb · ft⁻³.

9. To the opposite faces of a cubical block of gelatin, 30 cm on each edge, parallel and opposite tangential forces of 100 gwt are applied. If the edge of one face is displaced 1.0 cm, what is (*a*) the shearing stress? (*b*) the shearing strain? (*c*) the shear modulus?
    Ans. (*a*) 1.1×10^2 dyne · cm⁻²; (*b*) 0.033 rad; (*c*) 3.3×10^3 dyne · cm⁻².

10. Write the dimensional formulas for (*a*) strain; (*b*) stress; (*c*) YOUNG's modulus; (*d*) the volume modulus; (*e*) the constant C in BOYLE's law; (*f*) the shear modulus; (*g*) the constant of torsion of a cylinder.

11. A man grips the circumference of a bar which is 100 in. long and 1 in. in diameter and twists it through 1°. He applies the same *force* to the circumference of a similar bar 80 in. long and 2 in. in diameter. Find the resulting twist.   *Ans.* 0.1°.

12. The steel propeller shaft of a ship is designed to be 30 ft long and to be driven by a 1200-hp engine. What must be the diameter of the shaft if the twist is not to be greater than 1° 0′ when the speed of the shaft is 200 rev · min⁻¹?   *Ans.* 9.0 in.

13. The constant of torsion of a particular wire is 7.21×10^6 absolute units; the diameter of the wire is 2.732 mm, and its length is 50.1 cm. Find the torque required to twist a wire of the same material, of diameter 1.00 mm and length 4.00 cm, through 90.0°. *Ans.* 2.55×10^6 dyne · cm.

14. A glass tube open at one end is 60 cm long. The inside is covered with a soluble pigment. After a sea sounding, in which the tube is lowered vertically, open end down, the pigment was found to be dissolved to within 5.0 cm of the top. The average density of sea water being $1.03 \text{ g} \cdot \text{cm}^{-3}$, find the depth of the sea at the place in question. *Ans.* 1.1×10^2 m.

15. Torques of the same magnitude are applied to a solid glass rod of length 100 cm and radius 1.0 cm, and to a hollow glass tube having the same length, a mean radius of 1.0 cm, and a wall thickness of 0.10 cm. Compare the twists produced. *Ans.* $\theta_t = 2.5 \ \theta_r$.

○

D R. ROBERT HOOKE was Born at *Freshwater*, a Peninsula on the West side of the Isle of *Wight*, on the eighteenth of *July*, being *Saturday*, 1635, at twelve a Clock at Noon. . .

For his Age he was very sprightly and active in Running, Leaping, &c. tho' very weak as to any robust Exercise : Was very apt to learn any thing, and after his English soon learnt his Grammar by Heart; but, as he says, with but little understanding, till his Father designing him for the Ministry, took some pains to instruct him. But he still being often subject to the Head-ach which hindered his Learning, his Father laid aside all Thoughts of breeding him a Scholar, and finding himself also grow very infirm through Age and Sickness, wholly neglected his farther Education, who being thus left to himself spent his time in making little mechanical ' Toys, (as he says) in ' which he was very intent, and for the Tools he had successful ; so that there was nothing he saw ' done by any Mechanick, but he endeavoured to imitate, and in some particulars could exceed, (which are his own words.) His Father observing by these Indications, his great inclination to Mechanicks, thought to put him Apprentice to some easy Trade (as a Watchmakers or Limners) he shewing most inclinations to those or the like curious Mechanical Performances; for making use of such Tools as he could procure, ' seeing an old Brass Clock taken to pieces, he attemted to ' imitate it, and made a wooden one that would go: Much about the same time he made a small ' Ship about a Yard long, fitly shaping it, adding its Rigging of Ropes, Pullies, Masts, &c. with ' a contrivance to make it fire off some small Guns, as it was Sailing cross a Haven of a pretty ' breadth: He had also a great fancy for drawing . . .'

RICHARD WALLER, " The Life of Dr. Robert Hooke " in the Introduction to *The Posthumous Works of Robert Hooke* (London, 1705)

CHAPTER NINE

TEMPERATURE AND SOME OF ITS EFFECTS

L ET THE First Instrument be that represented by *Fig.* 1. [Plate 26] which may serve, (as likewise several others) to shew the changes of the *Air*, in reference to *Heat* and *Cold*, and is commonly call'd a *Thermometer*: 'tis made of Cristal-glass, after this manner. The *Artificer* by blowing with his own Mouth (instead of Bellows) through a Glass-Pipe upon the flame of a Lamp, forces it in one continued Stream, or several, at pleasure, from one place to another, where it is requisite; and by this means, shapes most curious, and admirable Works of Glass. Such an *Artificer* we call a *Lamp blower*. Let him then make the Ball of this Instrument of such a Capacity, and joyn thereto a Cane of such a bore, that by filling it to a certain mark in the Neck with Spirit of Wine, the simple cold of Snow or Ice *Externally Applyed*, may not be able to condense it below the 20 *deg.* of the Cane; nor on the contrary, the greatest vigour of the Sun's Rays at *Midsummer*, to Rarifie it above 80 *deg.* which Instrument may be thus fill'd, *viz.* by heating the Ball very hot, and suddenly plunging the open end of the Cane in the Spirit of Wine, which will gradually mount up, being suck'd in as the Vessel Cools. But because 'tis hard, if not altogether impossible to evacuate the Ball of all the *Air* by Rarefaction; and the Ball will want so much of being fill'd as there was *Air* left in it; we may thus quite fill it with a Glass Funnel, having a very slender shank, which may easily be made when the Glass is red hot, and ready to run; for then it may be drawn into exceeding small hollow Threads, as is well known to those that work in Glass. Put the small shank of this Funnel into the Cane to be fill'd, and by forcing the Spirit of Wine through the Funnel with ones Breath, or sucking it back again when there is too much; you may fill the *Instrument* up to what mark in the Neck you please. The next thing is to divide the Neck of the *Instrument* or *Tube* into Degrees exactly; therefore first, divide the whole *Tube* into Ten equal Parts with Compasses, marking each of them with a knob of white *Enamel*, and you may mark the intermediate Divisions with green Glass, or black *Enamel*: these lesser Divisions are best made by the Eye, which Practice will render easie. This done, and with the proof of *Sun* and *Ice*, the proportion of the Spirit of Wine found; the Mouth of the *Tube* must be closed with *Hermes* Seal at the flame of a Lamp, and the *Thermometer* is finish'd.

> *Essayes of Natural Experiments Made in the Academie del Cimento,*
> pp. 2–3. Translated by RICHARD WALLER (London, 1684)

○

By the *temperature* of a body is meant the number which expresses, on some definite scale, how "hot" or how "cold" the body is. The general notions which we have of temperature are gained through our thermal senses, but these yield neither quantitative results nor objective means of making measurements. In fact, temperature cannot be measured directly, for it is impossible to express one "hotness" in terms of another in the same sense that one length or mass may be expressed in terms of another. Temperature, like force, can be measured only in terms of its effects.

97. Early Thermometers.[1] Although the ancients were familiar with some of the effects that heating produces in a body, there was no application of this knowledge to the estimation of temperatures until GALILEO invented the thermoscope (Plate 2). This instrument was merely a glass bulb containing air and having a long stem which extended downward into a vessel of water. As the temperature changed, the air in the bulb expanded or contracted and the water in the stem rose or fell. That changes in the atmospheric pressure also affected the height of the water column was clearly recognized by PASCAL[2] and by BOYLE,[3] but thermometers of this kind continued to be used throughout the first half of the seventeenth century.

The first use of the expansion of a liquid for the estimation of temperature was made in 1631 by a French physician, JEAN REY,[4] who determined the body temperatures of his fever patients by means of a glass bulb similar to that used by GALILEO except that it was inverted and filled with water. Important improvements, which consisted in sealing the end of the tube to prevent evaporation and in substituting other liquids, such as alcohol, for water, were made about 1657 by the Accademia del Cimento at Florence (Plate 26). All that was needed to turn these Florentine thermometers into modern instruments was a scale that could be accurately duplicated, so that measurements of a given temperature made at various places and times with different instruments would be the same.

98. Development of a Temperature Scale. In setting up a temperature scale it eventually became clear that the first step is always to choose at least two temperatures which can be accurately and easily reproduced. The two that are now ordinarily chosen are the melting point of pure ice and the temperature of condensing steam, under a pressure of one standard atmosphere ($1 A_s$). These two particular *fixed points* are termed the *ice point* and the *steam point* respectively. HOOKE[5] suggested the first of these points in 1664, and

---

[1] For interesting accounts of the development of thermometers see *H. C. Bolton, *The Evolution of the Thermometer, 1592–1743* (Chemical Publishing Co., 1900), or *A. Wolf, *A History of Science, Technology, and Philosophy in the 16th & 17th Centuries* (Allen & Unwin, 1935), pp. 82–92.

[2] *Traitez de l'équilibre des liqueurs et de la pesanteur de la masse de l'air* (Paris, 1663); also *Récit de la grande expérience de l'équilibre des liqueurs* (Paris, 1648). Portions of these papers appear in **A Source Book in Physics* (1935), pp. 73–80.

[3] *New Experiments and Observations touching Cold* (London, 1665), p. 71; *Works*, ed. by T. Birch (London, 1772), Vol. II, p. 498.

[4] In a letter to MARIN MERSENNE, dated January 1, 1632 [Rey's *Essays* (1777), p. 136].

[5] *Micrographia* (1665), pp. 38–39.

THE Accademia del Cimento was founded at Florence, Italy, in 1657, as a direct result of GALILEO's teaching and to carry on his method of investigating truth by experiment alone. Among its moving spirits was VIVIANI, one of GALILEO's most distinguished disciples, and its interested and active patrons were the GRAND DUKE FERDINAND II of Tuscany and his brother LEOPOLD, in whose apartments the first meetings were held. In spite of its brief existence (only ten years), the Academy achieved remarkable results and exerted an enormous influence in spreading the experimental method all over Europe. It made noteworthy improvements in the thermometer. Florentine thermometers soon became world-famous and contributed powerfully towards the advance of science; some of them remain to this day and are marvels of glass blowing and veritable *objets d'art*. Its members published a joint account of their experiments and discoveries in 1667. At the request of the Royal Society this account was translated into English by Richard Waller under the title ESSAYES OF NATURAL EXPERIMENTS Made in the ACADEMIE DEL CIMENTO (London, 1684). It makes most interesting reading. In it are described an improved barometer, classical experiments on air pressure, experiments on the speed of sound, radiant heat, phosphorescence, the compressibility of water and its expansion on freezing, and the discovery of the rotation of the plane of oscillation of a pendulum which was used later by FOUCAULT to prove the earth's rotation. For further details consult A. Wolf's *A History of Science, Technology, and Philosophy in the 16th & 17th Centuries* (Allen & Unwin, 1935), pp. 54–59, 87, 307, and M. Ornstein's *The Rôle of the Scientific Societies in the Seventeenth Century* (University of Chicago Press, 1928), pp. 73–90, from which the following quotation has been taken:

" Italy was the home of the first organized scientific academy, the Accademia del Cimento of Florence (1657–67). It illustrates more perfectly than any other the functions of such societies as centers of the cultivation of experiment. Here nine scientists, supplied with the means of scientific research, gave ten years of united effort to the elaboration of instruments, the acquisition of experimental skill, and the determination of fundamental truths : so completely were their efforts welded together that their work was sent into the world like that of a single individual ; so exhaustive were their labors that the book they published became the 'Laboratory Manual', so to speak, of the eighteenth century, and their own work and methods the model and inspiration of other learned societies."

Figs. 1 and 2 of the plate represent the usual type of thermometer made by the Florentine Academy. They had either 50, 100, or 300 divisions marked by minute glass beads (white enamel for the larger divisions, green or black for the smaller).

Fig. 3 shows a thermometer with 300 divisions, which, in order to make it more compact, was ingeniously coiled into a spiral, " This Instrument being made rather for *fancy* and *curiosity* to see the Liquor run the Decimals of Degrees by the onely impulse of a warm breath, &c. than for any accurate Deduction, or Infallible Proportion of *Heat*, and *Cold* to be learnt thereby."

Fig. 4 is a thermometer of quite a different type. A number of blown glass bubbles were suspended in alcohol, the weights being adjusted so that first one and then another would sink as the temperature rose and the density of the alcohol became less.

Fig. 5 is a hygrometer for measuring the humidity of the air. It consisted of a hollow cone of cork with an outer cover of tin. To the bottom of the cork cone was attached a glass cone. When the instrument was filled with ice or snow, moisture from the air was deposited on the glass cone and ran into the measuring vessel. Relative humidities at different places and times could be compared by comparing the quantities of water collected in a given time.

Fig. 6 is the timing device used by the Academicians. In order to increase the accuracy of their measurements they introduced the bifilar suspension. The period of the pendulum was varied by means of the sliding clamp shown.

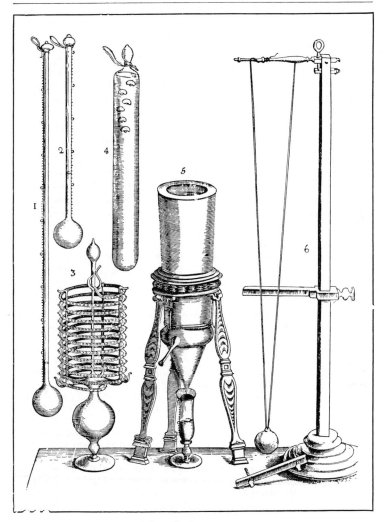

Scientific Instruments
of the Accademia del Cimento,
Florence, 1657–1667

Plate 1 of ESSAYES OF NATURAL EXPERIMENTS Made in the ACADEMIE DEL CIMENTO
(London, 1684), somewhat reduced in size

HUYGENS,[1] the second a few months later; but neither of these men appears to have realized that at least two points are needed if readings from different thermometers are to be made comparable with one another. The first thermometer graduated by the use of two fixed points with which useful observations were made was probably that of NEWTON,[2] although JOACHIM D'ALENCÉ [3] had in 1688 suggested the same method of graduation. The modern mercury-in-glass thermometer with a scale depending upon the ice point and the steam point was introduced in 1714 by GABRIEL DANIEL FAHRENHEIT.[4]

After the selection of the fixed points the next problems to be considered are the subdivision of the interval between them and the measurements of temperatures outside this interval. Here use is made of the principle that as a body grows hotter to the touch many of its physical properties change. Selecting a particular substance and a suitable property, such as volume, pressure, or electrical resistance, whose variations can be accurately and conveniently measured, we use the changes in this property to indicate changes in the temperature of the body, or of any other body with which it is in thermal equilibrium. For example, we could observe the *volume* of a given piece of *iron* first at the ice point and afterwards at the steam point, and then divide the computed increase in volume into any convenient number of equal parts. Such a change in temperature as will produce a volume change equal to one of these parts is then defined as the *degree*. On the *centigrade system*, which was proposed in 1742 by the Swedish astronomer ANDERS CELSIUS,[5] the number

---

[1] In a letter of January 2, 1665, to R. MORAY; see *Œuvres complètes de Christiaan Huygens*, Société Hollandaise des Sciences, Vol. V, p. 188.

[2] *Philosophical Transactions* 22, 824 (1701); 5th abridged ed. (London, 1749), Vol. IV, Part II, p. 1. This interesting paper is reproduced in *A Source Book in Physics* (1935), pp. 125–128.

[3] *Traittez des Baromètres, Thermomètres, et Notiomètres, ou Hygromètres* (Amsterdam, 1688). This is the first work which lays down rules for graduating thermometers. It contains many interesting illustrations of early thermometers.

[4] FAHRENHEIT described his process of making thermometers in five short papers in Latin in the *Philosophical Transactions* (1724–1726). German translations of these papers, together with papers by RÉAUMUR and CELSIUS, will be found in Ostwald's *Klassiker der Exakten Wissenschaften*, No. 57 (Engelmann, 1894). One of them appears in *A Source Book in Physics* (1935), pp. 131–133. See also *F. Cajori, A History of Physics* (Macmillan, 1929), pp. 114–116; *Isis* 4, 17 (1921).

[5] Published in the transactions of the Swedish Academy, Stockholm 4, 197 (1742), A German translation appears in Ostwald's *Klassiker der Exakten Wissenschaften*. No. 57 (Engelmann, 1894), pp. 117–124. CELSIUS actually marked the steam point 0° and the ice point 100°. The inverted scale with the ice point marked 0° is due to CHRISTIN of Lyons (1743) and STRÖMER of Uppsala (1749). CHRISTIN, however, disregarded the variation in the steam point with barometric pressure.

of equal parts is made 100, and the ice point is taken as the zero of the scale (0° C).

It is to be noted, however, that the choice of the zero and of the size of the degree is entirely independent of the choice of a *thermometric substance* and of a *thermometric property*. It is found that the properties of different substances are not generally the same functions of temperature, and therefore thermometers constructed from different substances do not agree exactly with one another at temperatures other than the fixed points. Hence it becomes necessary to choose some particular thermometric substance and some particular property of this substance, and to agree that the changes in the latter shall be taken as the measure of temperature. We shall see that certain of the gases possess peculiar advantages for this purpose.

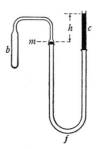

FIG. 96. A simple constant-volume gas thermometer, due to P. von Jolly and first described in Poggendorff's *Jubelband* (1874), p. 82

99. The Constant-Volume Hydrogen Thermometer. The *constant-volume* gas thermometer is based on the principle that a given mass of gas contained in a closed vessel of constant volume assumes a pressure which is entirely determined by the temperature. The bulb b, Fig. 96, which contains the gas, is connected by a bent capillary tube to the tube c through a flexible tube f containing mercury. By raising or lowering the tube c, the surface of the mercury in the other tube is always brought to a certain fiducial mark m. This is done first when the bulb b is in melting ice, then when it is in the vapor arising from boiling water, and finally when it is at the unknown temperature t which is to be determined. The pressure of the confined gas in each case can be measured by reading the difference in level, h, of the two mercury columns and adding to this the atmospheric pressure as determined by a barometer.

The *centigrade degree* on the *constant-volume hydrogen* thermometer is defined as any temperature change which, starting from any temperature whatever, will produce in the confined hydrogen a change in pressure amounting to 1/100 of that observed when the gas is heated from the ice point to the steam point. Accordingly we *define* any centigrade temperature t, measured on the constant-volume hydrogen scale, as

$$t = \frac{P_t - P_0}{\frac{1}{100}(P_{100} - P_0)}, \qquad [122]$$

where P_t, P_0, and P_{100} are the pressures of the gas at $t°$, $0°$, and $100°$ C respectively.

The expression $(P_{100} - P_0)/100\, P_0$, applied to any constant-volume apparatus, is called the *coefficient of increase of pressure* or simply the *pressure coefficient* of the gas. Evidently it represents the ratio of the increase in pressure to the pressure at 0° C for 1° change in temperature. If this coefficient be denoted by β, we may write Eq. [122] in the form

$$t = \frac{P_t - P_0}{\beta P_0},$$ [123]

where β is the pressure coefficient for hydrogen.

All gases increase in pressure if heated at constant volume. Experiments show that the pressure of a gas at any temperature t, when t is measured with a constant-volume hydrogen thermometer, is given approximately by the equation

$$P_t = P_0(1 + \beta t).$$ [124]

The pressure coefficient β is found to have somewhat different values for different gases. It will be noted that Eq. [124] can be put into the same form as Eq. [123].

100. Practical Thermometry. For reasons that will soon be apparent, H. V. REGNAULT selected the hydrogen constant-volume thermometer as his ultimate standard of reference in practical thermometry, and this instrument is now universally used for this purpose. Eq. [123], as applied to hydrogen and also sometimes to helium, is therefore to be regarded as the accepted practical definition of temperature. Experiments show that the value of β for hydrogen is approximately 1/273.04, and thus 1° C is by definition such a temperature variation as will result in a pressure change of 1/273.04 of the pressure at 0° C in a given mass of hydrogen kept at constant volume.

Because of the experimental difficulties in the use of gas thermometers and the relatively low precision attainable in a single measurement, the Seventh General Conference on Weights and Measures [1] adopted provisionally in 1927 a standard working scale designated as the *international temperature scale*. This scale is defined by a series of fixed points, the temperatures of which have been determined by gas-thermometer measurements, and by the specification of suitable thermometers for interpolation between the fixed points and for extrapolation to higher temperatures. The platinum electrical-resistance thermometer is employed in the range – 190° to 660° C; with this instrument temperatures can be determined with ease to 0.0001° C. For higher temperatures thermoelectric and optical pyrometers are used.[2]

---

[1] See *Journal of Research of the National Bureau of Standards* 1, 635 (1928) for a complete text of the decisions of the Conference.

[2] *Consult H. L. Callendar, article "Thermometry," *Encyclopaedia Britannica*, ed. 14; G. K. Burgess and H. Le Chatelier, *The Measurement of High Temperatures*; E. Griffiths, *Methods of Measuring Temperature*.

TABLE III · *Comparison of Hydrogen, Air, and Mercury-in-Glass Thermometer Temperatures. (The temperatures are reckoned from the ice point)*

Hydrogen	Air	Mercury-in-Jena-normal-glass
0°	0°	0°
20	20.008	20.091
40	40.001	40.111
60	59.990	60.086
80	79.987	80.041
100	100	100

Among liquids, mercury is found to agree fairly closely with the gas scale, and mercury-in-glass thermometers doubtless will continue to be employed in cases where facility of observation is more important than the highest obtainable degree of precision. As will be seen from Table III, the variations between corresponding readings on the hydrogen, air, and mercury-in-glass thermometers due to irregularities of expansion are for many purposes negligible.

○

The Expansion of Gases with Temperature

Because of the very great effect of pressure on the volume of a gas, one must be careful to specify the pressure conditions that are to hold during an investigation of expansion. The knowledge which we have already gained of controlled quantitative studies as they are made in physical science would suggest that the simplest way to approach the problem of expansion is to maintain the pressure constant and to measure the changes in volume of a given mass of the gas that occur with variations in the temperature alone.

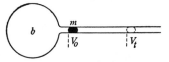

FIG. 97. A simple device for studying the thermal expansion of a gas under constant pressure. Evidently it may also be used as a constant-pressure gas thermometer

101. Expansivities of Gases. Consider, then, the bulb *b* in Fig. 97, in which a given mass of gas is confined by means of an indicating globule of mercury *m*, which moves with little friction forward or backward in the stem as the temperature rises or falls. The end of the stem is open so that the *constant pressure* used is that of the atmosphere. Experiments made with this apparatus show that the increase in volume is approximately proportional to the original

THE first human being to ascend in a balloon was JEAN FRANÇOIS PILÂTRE DE ROZIER. On November 21, 1783, after having made several ascents in a captive balloon, he and the MARQUIS D'ARLANDES rose from the Jardin du Château de la Muette, in the Bois de Boulogne, Paris, in a large hot-air balloon to a height of about 500 ft, and after remaining in the air for 20 to 25 min descended about 5 mi from the starting point.

Only ten days later, JACQUES ALEXANDRE CÉSAR CHARLES (1746–1823), the discoverer of the gas law known as the law of CHARLES and GAY-LUSSAC, ascended from the Tuileries in a balloon inflated with hydrogen gas. The balloon, which was constructed by the brothers ROBERT, one of whom took part in the ascent, was made of lutestring coated with gum elastic and had a diameter of 27 ft. The car was suspended from a hoop surrounding the middle of the balloon and fastened to a net, which covered the upper hemisphere. After ascending to a height of about 2000 ft and covering a distance of 27 mi in about 2 hr, CHARLES and ROBERT descended near the small town of Nesle (see the small map above), where ROBERT left the car and CHARLES reascended alone for a journey lasting a further 35 min, during which he reached a height estimated at 2 mi.

Most of the features of modern balloons are due to CHARLES. Thus he was the first to use hydrogen successfully for inflating balloons, and he invented the valve at the top of the balloon as well as the method of suspending the car which are still generally used. It is of interest that the names of CHARLES and GAY-LUSSAC, which are so intimately associated in the discovery of the gas laws, are also both important in the history of balloon flights (see Plate 28).

o

The Balloon Ascension of CHARLES and ROBERT,
Paris, December 1, 1783.
The Second Balloon Journey of Human Beings
and the First in a Hydrogen-Filled Balloon

From a wash drawing by Duperreaux in the Musée Carnavalet, Paris

volume and to the rise in temperature. If we agree always to take the volume at $0°$ C as the "original volume," it then follows that

$$V_t - V_0 = \alpha V_0 t,$$

or
$$V_t = V_0(1 + \alpha t),   [125]$$

where V_t and V_0 are the volumes at $t°$ and $0°$ C respectively, and α is a proportionality constant called the *expansivity*, or *coefficient of expansion*, of the gas.

In 1787 a Frenchman by the name of JACQUES CHARLES [1] discovered that *all gases have the same expansivities* α *when heated through the same temperature range*; and this was confirmed experimentally some fifteen years later by JOHN DALTON [2] and by LOUIS JOSEPH GAY-LUSSAC,[3] who used the apparatus of Fig. 97. This law, like that of BOYLE (Sec. **92**), has been shown by the later experiments of REGNAULT and others to be only approximately correct. However, except for the easily condensable vapors, the departures can be detected only with the most refined apparatus.[4] Since the different gases are described by BOYLE's law with different degrees of exactness, these departures from the law of CHARLES and GAY-LUSSAC were to have been expected. In fact, we shall see eventually that the kinetic theory requires the result, established by experiment, that the gases which show the largest departures from BOYLE's law, such as carbon dioxide and nitrous oxide, show also the largest variations in the expansivity α with pressure and with temperature.

The fact that α is not in general constant makes it desirable to speak of the *mean expansivity* between the temperatures t_1 and t_2, and this is defined by the equation

$$\alpha = \frac{V_2 - V_1}{V_0(t_2 - t_1)},   [126]$$

V_1 and V_2 being the volumes at $t_1°$ and $t_2°$ respectively. The limit which this quantity approaches as $t_2 - t_1$ approaches zero is defined

---

[1] CHARLES did not publish his results. See the paper of GAY-LUSSAC referred to in footnote 3.

[2] *Memoirs of the Manchester Literary and Philosophical Society* **5**, Part 2, 595 (1802). This paper is reproduced in part in *The Expansion of Gases by Heat*, ed. by W. W. Randall (American Book Co., 1902), pp. 19–22.

[3] *Annales de Chimie et de Physique* (1) **43**, 137 (1802). A translation of this paper, together with papers by BIOT, REGNAULT, and others, is available in *The Expansion of Gases by Heat*, ed. by W. W. Randall (American Book Co., 1902), pp. 27–48. Excerpts will be found in *A Source Book in Physics* (1935), pp. 166–170.

[4] See, for example, *T. Preston, The Theory of Heat*, ed. 4 (Macmillan, 1929), pp. 196–210.

as the *true expansivity* at the temperature t in the interval $t_2 - t_1$; that is,

$$\alpha = \lim_{(t_2 - t_1) \to 0} \frac{V_2 - V_1}{V_0(t_2 - t_1)} = \frac{1}{V_0} \frac{dV}{dt}, \qquad [127]$$

the pressure remaining constant.

○

The Laws of Ideal Gases

102. The Equation of State of an Ideal Gas. Any equation that expresses the relation between the pressure, volume, and temperature of a substance is called the *equation of state* of the substance. In the case of a given mass of an *ideal* gas (Sec. **92**), the question of how the pressure varies with simultaneous changes in the volume and temperature can be answered by combining BOYLE'S law with either Eq. [124] or Eq. [125]. Let us suppose that the ideal gas is initially in the state described by the three *gas coordinates* P_0, V_0, $0°$, and that it is heated at constant volume V_0 until it arrives at the state P', V_0, t. Then, by Eq. [124],

$$P' = P_0(1 + \beta t).$$

Next assume that the volume is changed, without changing the temperature, until the gas arrives at a third and final state, which is characterized by the coordinates P, V, t. For this isothermal change, by BOYLE'S law,

$$PV = P'V_0.$$

By eliminating the intermediate pressure P' which occurs in both of these equations, we obtain, finally,

$$PV = P_0V_0(1 + \beta t). \qquad [128]$$

EXAMPLE. Prove that, for an ideal gas, the expansivity α and the pressure coefficient β are identical quantities.

Solution. By combining BOYLE'S law with Eq. [125], one obtains, finally, $PV = P_0V_0(1 + \alpha t)$. A comparison of this equation with Eq. [128] yields $\alpha = \beta$.

Experiments show that for any *actual* gas, α and β are never exactly equal; nor should one expect them to be, as an actual gas is never described exactly by BOYLE'S law.

EXAMPLE. Show that, if an ideal gas is used as the thermometric substance, the equation $t = 100(V_t - V_0)/(V_{100} - V_0)$, when applied to a constant-pressure thermometer, gives precisely the same definition of temperature as does Eq. [122] applied to a constant-volume thermometer.

IN 1804, the French Academy of Science, desirous of ascertaining whether the earth's magnetic field ceased at a distance above its surface, obtained the use of a balloon which had been employed in Napoleon's campaign in Egypt, and selected GAY-LUSSAC and J. B. BIOT for the task. On August 23 they ascended from the garden of the Conservatoire des Arts et Métiers in Paris and rose to a height of 13,000 ft. GAY-LUSSAC was not satisfied with the altitude reached, and so on September 16 he made a second ascent, this time alone, and succeeded in reaching an elevation of 23,000 ft above sea level, the greatest height to which man had attained up to that time. Even though the temperature at this height was nearly 10° below freezing, he remained in the air for some time, making magnetic observations, collecting samples of air at different heights, and measuring the variations in the temperature, pressure, and humidity of the air with altitude. He concluded that the earth's magnetic field remained sensibly constant even up to 23,000 ft, and on analyzing his samples of air could find no difference in their composition at different heights.

The following description of the ascent of GAY-LUSSAC and BIOT, the first ever undertaken for strictly scientific purposes, is quoted from the *Proceedings of the American Academy of Arts and Sciences* **6**, 20 (1862):

" Supplied with a full complement of barometers, thermometers, hygrometers, electrometer, and instruments for measuring magnetic force and dip, as well as frogs, insects, and birds for galvanic experiments, the scientific voyagers embarked in their aerial car on the 23d of August, 1804. They began their experiments at the altitude of 6,500 feet, and continued them up to the altitude of 13,000 feet, and with a success commensurate with their wishes. The last part of the excursion, and especially the landing which they made, was so difficult, and even dangerous, that, according to the statement of Sir John Leslie, 'Biot, though a man of activity, and not deficient in personal courage, was so much overpowered by the alarms of their descent, as to lose for the time the entire possession of himself.' "

It is interesting to compare this description with accounts of more recent ascents, such as those of A. PICCARD in 1931 and 1932 [*National Geographic Magazine* **63**, 353 (1933)], that of T. G. W. SETTLE and C. L. FORDNEY in 1933 [*Proceedings of the National Academy of Science* **20**, 79 (1934)] or that of A. W. STEVENS and O. A. ANDERSON in 1935 [*National Geographic Magazine* **69**, 59 (1936)].

◉

Gay-Lussac and Biot Making a Balloon Ascension
for Scientific Observations in 1804

From J. H. Appleton's *Beginners' Hand-Book of Chemistry* (Chautauqua Press, 1888)

103. The Absolute Thermodynamic Scale. In 1824 N. L. SADI CARNOT published his remarkable memoir [1] on the theory of heat engines, in which he showed that the maximum efficiency of an engine working through any fixed small range of temperature is independent of the nature of the working substance used in the engine and is a function of the temperature alone. Twenty-four years later LORD KELVIN [2] made the brilliant suggestion that a temperature scale based on CARNOT'S engine would rid the definition of temperature of the peculiarities characteristic of any one real substance. This new scale is *absolute* in the sense of being independent of any particular property of any particular thermometric substance and is therefore called the *absolute thermodynamic scale*. Now it can be shown that an *ideal gas* would give this thermodynamic scale precisely if it were used in a gas thermometer. Thus, either Eq. [123] or Eq. [125], when applied to a thermometer containing an ideal gas, may for our purposes be regarded as the definition of temperature on the thermodynamic scale.

Since an ideal gas thermometer cannot be realized in practice, it becomes of importance to see how much the temperatures determined by a thermometer containing a real gas differ from absolute temperatures. With this object in mind, KELVIN devised his famous porous-plug experiment (Chap. 11). The work was carried out in conjunction with JOULE and finally resulted in showing that hydrogen behaves so nearly like an ideal gas that the hydrogen scale may be taken for all practical purposes as agreeing exactly with the absolute scale at ordinary temperatures.[3]

---

[1] N. L. S. Carnot, *Réflexions sur la Puissance Motrice du Feu* (1824). A translation of this memoir, together with a biography of CARNOT and an account of his theory, the latter by LORD KELVIN, has been published under the title *Reflections on the Motive Power of Heat*, ed. by R. H. Thurston (Wiley, 1897). A translation also appears in *The Second Law of Thermodynamics*, ed. by W. F. Magie (Harper's Scientific Memoirs, 1899), and in *A Source Book in Physics* (1935), pp. 221–228. Although so little known that his name appears in few biographical dictionaries, and although he died before he was forty and has only this one paper to his credit, CARNOT really inaugurated the development of the modern science of thermodynamics. As P. G. TAIT says in his *Thermodynamics*, "Without this work of CARNOT, the modern theory of energy, and especially that branch of it which is at present by far the most important, the dynamical theory of heat, could not have attained its now enormous development."

[2] WILLIAM THOMSON, who became LORD KELVIN in 1892, was one of the great English physicists of the nineteenth century. His first ideas regarding the thermodynamical scales were published in 1848 but were vastly improved upon in 1851, in the memoir referred to in footnote 3, p. 76.

[3] Data on the departure of various gas scales from the thermodynamic scale are given in the *International Critical Tables* (1926), Vol. I, p. 53.

Thermodynamics deals with the relations of heat and work. Like mechanics, it is a science of great power and universality because it employs a small number of primary postulates from which may be drawn a variety of far-reaching deductions. The science is based on two very general principles, which are known as the *first* and the *second laws of thermodynamics*, the first of these principles being the law of conservation of energy, with which we are already familiar. Thermodynamics differs from the kinetic-molecular interpretation of heat phenomena (Chap. 10) in that it is independent of any special theory of the mechanism of heat transfer and of the structure of the substances involved; it deals with the equilibrium conditions that exist at the beginning and at the end of a process, and not with the forces and the conditions that intervene, and for this reason requires no model of the structure of matter. Indeed, we have already seen (p. 78) how such a method affords a powerful means of attacking many mechanical problems without the necessity of making the minute analysis that the direct application of NEWTON'S laws requires. Thermodynamics achieved its greatest usefulness, however, when it was properly correlated with a kinetic-theory point of view, for it then furnished a general theory which is far in advance of any that could have been reached by either point of view taken alone.

104. The Absolute Zero and Absolute Temperature. Let us rewrite Eq. [128] in the form

$$PV = \frac{P_0 V_0 \left(\frac{1}{\beta} + t\right)}{\frac{1}{\beta}}. \tag{129}$$

By putting $(1/\beta) + t = T$, we introduce a temperature T, corresponding to t but referred to a zero point lying $1/\beta$ below the ice point. This zero point is called the *absolute zero*,[1] and temperatures reckoned from it are termed *absolute temperatures*. The centigrade absolute scale is commonly called the *Kelvin scale* and is referred to by the abbreviation $T°$ K. On the ordinary centigrade scale the temperature of the ice point is $t_0 = 0°$ C, but on the KELVIN scale it is found to be[2] $T_0 = 1/\beta = 273.18°$ K; in other words, it turns out that the pressure coefficient β for an ideal gas is $1/273.18$ and hence that the absolute zero corresponds to $-273.18°$ C.

---

[1] For an interesting history of the notion of the absolute zero see the *Collected Papers of Sir James Dewar* (Cambridge University Press, 1927), Vol. II, pp. 768–775. There due credit is given to the French physicist GUILLAUME AMONTONS for having had the conception of an absolute zero as early as 1703, and for having calculated its value as $-240°$ C. Note, however, that the concept of an absolute zero arises merely from the way in which temperature has been defined and from the behavior of an ideal gas, and that it can be entirely avoided simply by defining temperature in a different way.

[2] The latest values reported from Leiden and from Berlin are 273.144° and 273.16° K, respectively.

The absolute zero and absolute temperature are concepts that greatly facilitate the mathematical expression and application of the gas laws and should be treated as such and nothing else. If we attempt to apply the gas laws at the absolute zero, we are led to the meaningless equation $PV = 0$. from which we deduce that at the absolute zero a gas either has no volume or exerts no pressure. We know, however, that all gases liquefy before this point is reached, so that at the absolute zero the gaseous state probably no longer exists.

In practice, many attempts have been made to attain temperatures close to the absolute zero. Up to 1877, the lowest temperature attained was $-110°$ C, produced by MICHAEL FARADAY [1] in 1844, by the rapid evaporation in vacuum of ether and solid carbon dioxide, a mixture that ordinarily has a temperature of about $-80°$ C. In 1877 a Swiss, RAOUL PICTET, and a Frenchman, LOUIS PAUL CAILLETET, independently liquefied oxygen,[2] which has a boiling point of $-182.97°$ C. But as neither of these experimenters obtained the liquid oxygen in a static condition, they could make no observation of its temperature. The lowest measured temperature was $-140°$ C, obtained by PICTET by the rapid evaporation of nitrous oxide. Following these, the two Poles SIGMUND v. WROBLEWSKI and KARL OLSZEWSKI and the Britisher JAMES DEWAR (1842–1923) accomplished the liquefaction of oxygen, nitrogen, and air in quantity, and by their evaporation in vacuum produced and measured temperatures as low as $-210°$ C. In 1885 OLSZEWSKI liquefied hydrogen and located its boiling point at $-243.5°$ C. Hydrogen has since not only been liquefied in quantity (1898) but also solidified by DEWAR (1900)[3]; the boiling and melting points of hydrogen are $-252.7°$ and $-259.1°$ C, respectively, according to recent determinations. Of all the gases, helium was the last to resist liquefaction and solidification. In 1908 KAMERLINGH ONNES [4] (1853–1926), working in his celebrated cryogenic laboratory at Leiden, accomplished its liquefaction at a temperature of only 4.3° above the absolute zero. ONNES made an unsuccessful attempt in 1921 to solidify helium but did succeed in reaching a temperature of 0.82° K by evaporating the liquid in high vacuum.[5] Four months after the death of ONNES, in 1926, helium was successfully solidified, in the same laboratory, by placing it in a brass tube in a helium bath and

---

[1] *Faraday's Diary*, ed. by T. Martin (Bell, 1932–1936), Vol. IV, pp. 166, 191–192. See also *Philosophical Transactions* **135**, 155 (1845). This paper is reproduced in Faraday's *Experimental Researches in Chemistry and Physics* (Taylor and Francis, 1859), pp. 98–100, and in *The Liquefaction of Gases*, Alembic Club Reprint No. 12 (Edinburgh, 1912), pp. 37–39.

[2] The papers of PICTET and CAILLETET appear in *A Source Book in Physics* (1935), pp. 193–196.

[3] All of DEWAR's papers on the liquefaction of gases will be found in the *Collected Papers of Sir James Dewar* (Cambridge University Press, 1927).

[4] See *Nature* **78**, 370 (1908); *Communications from the Physical Laboratory of the University of Leiden*, No. 108.

[5] K. Onnes, *Transactions of the Faraday Society* **18** (Dec., 1922); *Communications from the Physical Laboratory of the University of Leiden*, No. 159.

Professor KAMERLINGH ONNES (right)
Working at the Helium-Liquefying Apparatus
in the Cryogenic Laboratory
of the University of Leiden

Photograph kindness of Professor C. A. Crommelin

The Physical Laboratory, University of Leiden, Holland, 1922

Photograph kindness of Professor C. A. Crommelin

THIS laboratory is famous for the cryogenic work done under the late KAMERLINGH ONNES. A temperature of less than 0.005° K has been attained.

The Cavendish Laboratory at Cambridge University, England.
Original Wing Built by MAXWELL and Opened in 1874

THIS laboratory is still among the most productive in the world. The high ideals and standards of accomplishment set by MAXWELL have been carried on by a re-markable succession of directors — J. CLERK MAXWELL (1871–1879), LORD RAYLEIGH (1879–1884), SIR J. J. THOMSON (1884–1919), and LORD RUTHERFORD (1919–1937) — all men whose names will ever be landmarks in the history of physics. An interesting history of this laboratory was published by Longmans, Green, and Company in 1910.

subjecting it to pressure.[1] In this work temperatures as low as 0.71° K were reached. Recently P. DEBYE and W. F. GIAUQUE have suggested a method of reaching very low temperatures by means of the adiabatic demagnetization of certain magnetic substances kept at the temperature of liquid helium. The essential idea is that if heat is prevented from entering the substance while it is being demagnetized, the work of demagnetization must be furnished by the substance itself, with a consequent lowering of its temperature. This method has been applied with success [2] at the University of California and at Leiden in 1933, and at Oxford in 1934. At Leiden, temperatures of less than 0.005° K actually have been reached.

105. The Gas Constants. In view of the definition of absolute temperature, we may write Eq. [129] in the form

$$\frac{PV}{T} = \frac{P_0 V_0}{T_0}.$$ [130]

But V_0 is equal to m/ρ_0, where m is the mass of the gas and ρ_0 is its density at the ice point T_0; hence

$$\frac{PV}{T} = \frac{P_0 m}{\rho_0 T_0},$$

or

$$\frac{PV}{T} = mR'.$$ [131]

The constant R', which has made its appearance here in the equation of state of an ideal gas, obviously can be interpreted as the value of PV/T for unit mass, or of $P/\rho T$, of the particular gas considered. It is therefore called the *gas constant for unit mass*. The fact that its value is different for different gases should be noted.

EXAMPLE. Compute the value of the gas constant R' for air, expressing it in cgs units.

Solution. Tables show that the density of dry air at T_0 and 1 A_s is 0.0012930 g · cm^{-3}.
Then, since $T_0 = 273.2°$ K, and 1 $A_s = 1.0132 \times 10^6$ dyne · cm^{-2},

$$R'_{air} = \frac{P_0}{\rho_0 T_0} = \frac{1.0132 \times 10^6}{1.2930 \times 10^{-3} \times 273.2} = 2.869 \times 10^6 \text{ erg} \cdot \text{deg}^{-1} \cdot \text{g}^{-1}.$$

In order to obtain a final form for the equation of state of an ideal gas, one must take into account a thoughtful guess made in 1811 by the Italian physicist AMEDEO AVOGADRO, to the effect that *equal volumes of different gases at the same temperature and pressure contain*

---

[1] W. H. Keeson, *Nature* **118**, 81 (1926); see also *Science* **64**, 132 (1926) and *Communications from the Physical Laboratory of the University of Leiden*, No. 184 (*b*).

[2] *Physical Review* **43**, 768 (1933). *Physica* **1**, 1 (1933); **2**, 81, 335 (1935). *Proceedings of the Royal Society* **149**, 152 (1935).

equal numbers of molecules. This important law, now known to be exact only for ideal gases, although approximately true for real ones, leads immediately to the conclusion that *moles*, or *gram-molecular weights*, of the various gases will occupy the same volume at the same pressure and temperature. Consequently, if M_1 be the number of grams in a mole of one kind of gas, M_2 that of another kind, etc., one has, from Eq. [131],

$$\frac{PV}{T} = M_1 R'_1 = M_2 R'_2 = \cdots = R, \qquad [132]$$

where V is the volume of a mole and R is a new constant called the *gas constant per mole*. Evidently R is a universal constant; that is, it is the same for all gases, its numerical value depending only upon the units in which P, V, and T are expressed. Its value in cgs units, as computed from experimental data, is $R = (8.3136 \pm 0.0010) \times 10^7$ erg · deg^{-1} · mole^{-1}. When the pressure is expressed in atmospheres, R has the easily remembered numerical value of 82 (approximately).

If, in Eq. [131], we put $R' = R/M$, from Eq. [132], and $m = NM$, where N is the number of moles of gas used, then the equation of state of an ideal gas attains the final form that is commonly used in physical science and in engineering, namely,

$$PV = NRT, \qquad [133]$$

where V is the volume of N moles.

o

The Expansion of Liquids and Solids with Temperature

106. Expansivities of Liquids and Solids. That the expansion of a liquid or solid is not exactly proportional to the change in its temperature is evident from the fact that thermometers made from such bodies by dividing the increase in volume between the ice point and the steam point into 100 equal parts do not agree at intermediate temperatures with the gas thermometer (Sec. **98**). The volume V of a liquid or solid at any temperature $t°$ C can in general be represented by an equation of the form

$$V_t = V_0(1 + at + bt^2 + ct^3 + \cdots), \qquad [134]$$

where V_0 is the volume of the specimen at 0° C and a, b, c, $\cdots$ are constants characteristic of the substance. If this entirely empirical expression be put into the form $V_t = V_0 [1 + (a + bt + ct^2 + \cdots)t]$, and this in turn be compared with Eq. [125], it is seen that the *mean* volume expansivity of a liquid or solid between the temperatures 0° and $t°$ C is

$$\alpha = a + bt + ct^2 + \cdots. \quad [135]$$

EXAMPLE. Show that the *true* volume expansivity of a liquid or solid is $\alpha = a + 2\,bt + 3\,ct^2 + \cdots$.

Solution. Substitute the empirical expression for V_t in Eq. [127] and perform the indicated differentiation.

The number of constants a, b, c, $\cdots$ that must be used in any particular case depends upon the accuracy desired, the temperature range involved, and the solid or liquid under investigation. For some substances, b, c, etc. are so much smaller than a that this one constant is sufficient. This is true for most solids and for mercury at temperatures below 100° C. For example, a mercury-in-glass thermometer graduated by dividing the increase in volume between 0° and 100° into 100 equal parts differs from a hydrogen thermometer at no point in this interval by more than 0.2°. Hence, in ordinary work with these substances, α is usually considered to be constant and equal to a, and the equation

$$V_t = V_0(1 + \alpha t) \quad [136]$$

is applied in the same way as in the case of a gas.

If a body be heated from $t_1°$ to $t_2°$, the increase in volume will be $V_2 - V_1 = V_0\alpha(t_2 - t_1)$. For most liquids and solids, but not for gases, α is so small that the error introduced by replacing V_0 by V_1 in this equation is usually less than the unavoidable errors of observation which occur in measuring the quantities involved.

107. Measuring the Expansion of a Liquid. The expansion coefficients of most liquids other than mercury increase rapidly with the temperature. For example, in passing from 0° to 40° C the mean expansivity of ethyl alcohol increases about 6 percent, and that of turpentine, about 8.5 percent. Between 0° and 4° C water possesses the peculiar property, which is also shown by bismuth and by certain alloys, of contracting as the temperature rises; that is, between 0° and 4° C it has a negative coefficient.

An accurate and convenient method of studying the expansion of a liquid at various temperatures is to place it in a glass vessel having a large bulb and a graduated capillary stem (Fig. 98), to heat it from $t_1°$ to $t_2°$, and then to observe the resulting rise l of the liquid in the stem. If the volume corresponding to one scale division on the stem be v, the apparent increase in the volume is vl. This increase is, in reality, the difference between the expansion of the glass and of the liquid. Now, the increase in the volume of the dilatometer is precisely the same as would be the increase in the volume of a solid piece of glass of the same volume as the interior of the vessel, for the solid piece may be conceived as made up of a series of concentric hollow vessels, each of which expands independently of all the rest. This means that, if we let V_0 represent the volume occupied by the liquid at $0°$ C and let α_g represent the volume expansivity of the glass, then the increase in the volume of that part of the dilatometer which is occupied by the liquid is $V_0\alpha_g(t_2 - t_1)$. Similarly, if α_l be the expansivity of the liquid, the true increase in the volume of the liquid is $V_0\alpha_l(t_2 - t_1)$. Since the apparent expansion vl is the difference between these quantities, there results

$$\alpha_l - \alpha_g = \frac{vl}{V_0(t_2 - t_1)}. \qquad [137]$$

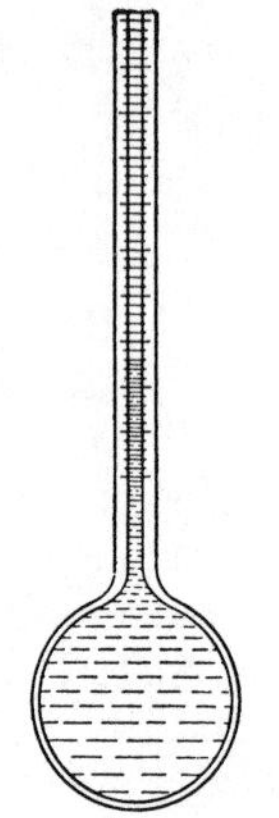

This equation shows that it is impossible to obtain α_l from measurements made with a dilatometer unless α_g is also known. The right-hand member of Eq. [137] is referred to as the *apparent expansivity* of the liquid in glass.

108. The Absolute Expansion of Mercury. If the expansion coefficient of some one liquid were determined by a method that is independent of the expansion of glass or of any other substance, this liquid could be used for determining α_g for a particular dilatometer, and the latter could thereafter be used for investigating other liquids. Early in the last century P. L. DULONG and A. T. PETIT,[1] and later REGNAULT[2] and others, studied mercury by a very elegant method of this kind. The principle is simple, though its application is less so. The vertical tubes in Fig. 99 are connected at the bottom by a capillary tube and contain mercury. One is surrounded by melting ice and the other by a water jacket of known temperature t. According to the laws of hydrostatics (Chap. 13), the levels of the liquids in the two tubes must adjust themselves so that the pressures are equal at any two points A and B in

FIG. 98. The dilatometer is simply a thermometer with a very large bulb. It has been used in many of the classical researches made on the expansion of liquids

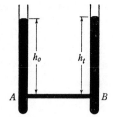

FIG. 99. Apparatus for determining the expansivity of a liquid

[1] *Annales de Chimie et de Physique* (2) **7**, 124 (1818).

[2] *Relation des Expériences* (Paris, 1847), Vol. 1, p. 271; *Mémoires de l'Académie des Sciences* **21** (1847)

the same horizontal plane. The pressure at A is $h_0\rho_0 g$ and that at B is $h_t\rho_t g$, where ρ_0 and ρ_t are the densities of the liquid at $0°$ and $t°$ C respectively; hence

$$h_0\rho_0 = h_t\rho_t. \qquad [138]$$

Now, by combining Eq. [136] with the defining equation for density, there results

$$\rho_t = \frac{\rho_0}{1 + \alpha t}. \qquad [139]$$

Therefore, from this equation and Eq. [138], one obtains for the mean expansivity of the liquid between $0°$ and $t°$ C

$$\alpha = \frac{h_t - h_0}{h_0 t}. \qquad [140]$$

By using a very much improved form of the DULONG and PETIT apparatus, CALLENDAR and MOSS[1] found the mean expansivity of mercury between $0°$ and $100°$ C to be 1.8205×10^{-4} per deg C. A similar apparatus has been used recently at the Reichsanstalt[2] to make a careful study of the expansivity of water.

109. Linear Expansivities of Solids. The volume expansivity which thus far has been discussed is the only thermal expansion coefficient that must be defined for fluids. For solids, on the other hand, it is advantageous to consider also the expansion in any one direction. The *mean linear expansivity*, or *mean coefficient of linear expansion*, λ, for the temperature range t_1 to t_2 is defined as the expansion per degree per unit length reckoned from zero; that is,

$$\lambda = \frac{l_2 - l_1}{l_0(t_2 - t_1)}, \qquad [141]$$

where l_0, l_1, and l_2 are the lengths of any given line of the body at $0°$, $t_1°$, and $t_2°$ respectively. Evidently the quantity $(1/l_0)\, dl/dt$ is the *true* linear expansivity at the temperature t.

It is found that solids generally expand in such a way that the length l_t at any temperature may be represented as a function of t by an empirical equation similar to the one used for volume expansion, namely by $l_t = l_0(1 + a't + b't^2 + \cdots)$. As before, the constants b', c', $\cdots$ are much smaller than a', so that for small temperature changes the linear expansivity λ may be regarded as constant.

It is the coefficient λ that is usually made the subject of measurement in the case of solids; for, if λ is known, then α, the volume

---

[1] *Philosophical Transactions* **211**, 1 (1911).

[2] The Reichsanstalt is the national physical laboratory of Germany.

expansivity, can be computed. In order to see this, consider a cube of the material whose side is l_0 at 0° C and whose volume at 0° C is therefore $V_0 = l_0{}^3$. When heated to $t°$ the volume of the cube becomes V_t, where

$$V_t = [l_0(1 + \lambda t)]^3 = V_0(1 + 3\,\lambda t + 3\,\lambda^2 t^2 + \lambda^3 t^3).$$

Since λ is always very small, scarcely ever greater than 3×10^{-5} per deg C, the terms containing λ^2 and λ^3 may be neglected in comparison with the term containing λ; hence $V_t = V_0(1 + 3\,\lambda t)$, and therefore

$$\alpha = 3\,\lambda. \qquad [142]$$

The volume expansivity is, then, to a first approximation, equal to three times the linear expansivity. This is true for an isotropic solid of any form whatever, since a body can always be considered as made up of infinitesimal cubes. But it is not true for anisotropic bodies, since the linear expansivity is then not the same for all directions. Substances that crystallize in the cubic system are truly isotropic. Metals are quasi-isotropic; they have the same *average* properties in all directions, for they are composed of a very large number of small crystals orientated entirely at random.

In the case of solids, different specimens of the same material do not expand alike, and thus the data on expansivities given in tables[1] must be regarded merely as mean values. The determination of linear expansivity can be accomplished by various optical methods or by the use of a comparator. The optical method illustrated in Fig. 100 was introduced by LAVOISIER and LAPLACE toward the end of the eighteenth century. When only small specimens of a substance are available, another optical method can be employed which involves the use of an interferometer. In the comparator method, two separate fixed microscopes are focused upon fine scratches near the ends of the bar to be investigated; these microscopes are provided with micrometer eyepieces by means of which a direct measurement is made of the elongation produced by a given rise in temperature.

FIG. 100. Measuring the expansion of a bar with an optical lever

---

[1] Appendix 13, Table J.

[2] See the Optional Laboratory Problem which accompanies Exp. IXB.

EXPERIMENT IXA. THE PRESSURE COEFFICIENT OF AIR

The constant-volume gas thermometer which is to be used is shown in Fig. 101. The volume is kept constant by bringing the mercury in the closed arm, before each reading, exactly into coincidence with the fiducial mark m, which consists of a line etched upon the short section of glass tubing connecting the bulb to the manometer system. This adjustment is made by turning the screw S at the base of the instrument. The copper jacket J surrounding the bulb serves both as a steam bath and to hold shaved ice and water.

Let P_0 represent the observed pressure of the confined gas when the bulb is at the ice point (0° C). Let P_t represent its observed pressure when the bulb is at the temperature t of condensing steam; this temperature is usually appreciably lower than 100° C. If it could be assumed that the volume of the bulb remains constant and that all of the confined gas undergoes the same change of temperature, then the expression for the pressure coefficient β would be simply

Fig. 101. The gas thermometer

$$\frac{P_t - P_0}{P_0 t}, \qquad [143]$$

from Sec. 99. But, in point of fact, the *observed* pressures are all slightly different from the pressures appearing in the foregoing expression, for the latter correspond to an absolutely constant volume and to a condition in which all of the confined gas undergoes the heating or cooling operation. Since neither of these conditions is realized in practice, two corrections must be applied to the observed pressures.

Consider first the correction for the expansion of the bulb. If V_0 be the initial volume of the bulb and if α_i be the volume expansivity of the cast iron of which the bulb is made, then the volume of the bulb at $t°$ C will be $V_0(1 + \alpha_i t)$. If this volume of the gas in the hot bulb were reduced back to V_0 without changing its temperature, the resulting pressure x of the confined gas would be such that, by Boyle's law,

$$P_t V_0 (1 + \alpha_i t) = x V_0,$$

or

$$x = P_t (1 + \alpha_i t).$$

When this is substituted for P_t in expression [143], the expression for β becomes

$$\frac{P_t - P_0}{P_0 t} + \alpha_i \frac{P_t}{P_0}. \qquad [144]$$

Consider next the correction that must be applied to all of the observed pressures in order to make allowance for that portion of the confined gas which escapes the changes in temperature taking place within the bulb. Let v be the volume of the gas in the dead space between the bulb and the fiducial mark m, let t_1 be the temperature of the room near this dead space, and let ρ be the density of the gas at $0°$ C and pressure P_0. Since no gas enters or leaves the apparatus during any of the operations, the total mass of gas in the bulb and dead space when the bulb is at $0°$ C is equal to that in the bulb and dead space when the temperature of the bulb is $t°$ C. It can be shown that this results in the equation

$$V_0\rho + \frac{v\rho}{1 + \beta t_1} = \frac{P_t V_0 \rho}{P_0(1 + \beta t)} + \frac{P_t v\rho}{P_0(1 + \beta t_1)}. \qquad [145]$$

By dividing all the terms of this equation by $V_0\rho$ and neglecting the small quantity βt_1 whenever it occurs in a term that is multiplied by the very small fraction v/V_0, one obtains

$$1 + \frac{v}{V_0} = \frac{P_t}{P_0}\frac{1}{1 + \beta t} + \frac{P_t}{P_0}\frac{v}{V_0}.$$

After solving this for β, there results finally

$$\beta = \frac{P_t - P_0}{P_0 t}\frac{1 + \dfrac{v}{V_0}}{1 - \dfrac{v}{V_0}\dfrac{P_t - P_0}{P_0}},$$

or

$$\beta = \frac{P_t - P_0}{P_0 t}\left(1 + \frac{v}{V_0}\frac{P_t}{P_0}\cdots\right). \qquad [146]$$

The terms omitted in this expression all involve the square or higher powers of the very small quantity v/V_0, and therefore they can be neglected. Hence the correction for the dead space is accomplished by adding [1] to expression [144] the quantity

$$\frac{P_t - P_0}{P_0 t}\frac{v}{V_0}\frac{P_t}{P_0}. \qquad [147]$$

Thus the expression for β to be used with this gas thermometer is

$$\beta = \frac{P_t - P_0}{P_0 t} + \alpha_i\frac{P_t}{P_0} + \frac{P_t - P_0}{P_0 t}\frac{v}{V_0}\frac{P_t}{P_0}. \qquad [148]$$

1. Derive Eq. [145] and also give a complete derivation of [147].

Measurements. *a.* Take the reading of the fiducial mark m on the central graduated scale with the help of a small square or draftsman's

---

[1] Strictly speaking, the two expressions are not additive, but the error introduced by so treating them is negligible.

triangle, one edge of which is held against the graduated scale while the other is brought into coincidence with the etched line. Make three independent observations and average them.

b. Screw the cap *c* to the top of the jacket *J* and pass steam into the latter from the steam-generator through the *upper* hose-nipple, all the while keeping the mercury level in the closed arm of the manometer at the fiducial mark *m*.

When the mercury has ceased to rise in the open arm, observe, by means of the sliding index *i*, the reading of the top of the mercury meniscus upon the central graduated scale. Make three independent settings and observations and average them.

c. Read the barometer.[1]

d. Disconnect the steam-generator; remove the screw-cap of the jacket *J*; close the lower hose-nipple with a short piece of rubber tubing and a pinch clamp. Pack shaved ice carefully about the bulb, again keeping the mercury level in the closed arm of the manometer at the fiducial mark *m*. Pour distilled water over the ice until the bulb is completely immersed, adding more ice if necessary; the water is added both to insure good contact and to make certain that the temperature of the ice is not below 0° C.

After the mercury in the open arm has become stationary, read, as before, the level of the meniscus.

When the readings have all been taken and checked, drain the water and ice from the jacket *J* by removing the pinch clamp from the rubber tubing on the lower hose-nipple.

e. Record the values of *v* and V_0, which will be found marked on the apparatus. The value of α_i for cast iron is 3.2×10^{-5} per deg C. The temperature of the steam bath is determined by finding from tables the boiling point of water for the observed barometric pressure.[2]

f. Calculate β by means of Eq. [148]. In computing P_0 and P_t, remember that the difference between the two mercury levels is in each case to be regarded as negative if the open-arm reading is the lower of the two and that it is the algebraic sum of this difference and the corresponding barometer reading which gives the pressure.

Taking the accepted value as 3.663×10^{-3} per deg C, calculate the percentage of error in your value of β.

2. How much error would have been introduced into your value of β if you had failed to correct for the expansion of the cast-iron bulb? if you had failed to correct for the dead space?

3. From the result of your experiment, calculate the temperature at which the pressure of the air in the bulb would be zero.

---

[1] See Appendix 6.   [2] Appendix 13, Table E.

OPTIONAL LABORATORY PROBLEM

Standardization of a Mercury-in-Glass Thermometer. Employ a water bath to compare the readings of a mercury thermometer with those of a constant-volume air thermometer at several points between the two fixed points. Then determine, in the order named, the steam point and the ice point of the mercury thermometer. The ice point thus found is termed the *depressed zero*, since it is usually lower than the value found if the two fixed points are determined in the reverse order. Why? The depressed zero point is the one to be used for calibration, and for accurate determination of temperature it always should be determined immediately after taking the temperature in question. In all the foregoing determinations, make sure that the mercury thermometer is immersed to the top of the mercury thread.

Plot a curve of corrections for the mercury thermometer. Regard the corrections as positive if the mercury thermometer reads too low, negative if it reads too high. What if the thermometer is later used with a part of the mercury thread exposed to the air of the room?

○

EXPERIMENT IXʙ   EXPANSIVITY OF A VOLATILE LIQUID

The simplest accurate method of determining the volume expansivity α_l of a nonvolatile liquid like mercury is one in which the observations are reduced to weighings. But in the case of a volatile liquid of small density, such as ethyl alcohol, the accuracy obtained by employing weighings is more than counterbalanced by the errors introduced by evaporation. For this reason the use of the dilatometer (Fig. 98) is preferable with volatile liquids. As is apparent from Eq. [137], α_l cannot be obtained by means of a dilatometer until three constants of the instrument have been determined; these are the volume v corresponding to one scale division on the stem, the volume expansivity α_g of the glass, and the volume V_0 of the bulb up to the zero mark on the stem at 0° C.

a. Determination of the Volume Corresponding to One Scale Division on the Stem.[1] Place enough pure mercury in the dilatometer to fill

---

[1] If the liquid to be tested has so small an expansivity that it is necessary to employ a dilatometer equipped with a stem of very small bore, the following more difficult method of calibration will have to be employed. Gently warm the bulb of the dilatometer and then allow it to cool with the open end of the stem under mercury. In this way almost fill the stem with a thread of mercury without allowing any of the mercury to enter the bulb. Measure the over-all length s of the mercury thread in terms of the scale divisions marked on the stem, using for this purpose either a cathetometer or a mirror-scale; if a cathetometer is used, avoid the error due to lost motion

the stem almost to the top of the scale and observe the scale-reading of the mercury meniscus. Pour nearly all of the mercury that is in the graduated part of the stem into a small weighed cup and determine its mass m. Again observe the scale-reading of the mercury meniscus in the stem. Observe the temperature t of the laboratory. The volume v corresponding to one scale division is evidently $m v_t/s$, where s is the length of the poured-out column of mercury expressed in terms of the scale divisions on the stem, and v_t is the volume of 1 g of mercury weighed in air at the temperature t; tables show that v_t is given by

$$v_t = 0.07355(1 + 0.000181\,t)\ \text{cm}^3. \qquad [150]$$

b. Expansivity of the Glass in the Bulb. If the value of α_g is not marked on the glass, determine it by means of the following procedure. Fill the dilatometer with pure mercury so that, when the bulb is immersed in an ice bath, the column of liquid in the stem rises only a small distance above the zero mark. Observe the scale-reading l_0 of the mercury meniscus when the bulb is immersed in the ice bath, and then observe the scale-reading l_t of the meniscus when the bulb is immersed in, say, a steam bath. Determine the mass M of the mercury in the dilatometer by weighing the latter first when it contains the mercury and second when it is empty. Read the barometer and find from tables the temperature t of the steam. The volumes of the mercury, and hence of the bulb, up to the zero mark on the stem are $V_0 = M v_0 - l_0 v$, at 0° C, and $V_t = M v_t - l_t v$, at t° C, where v_0 and v_t are given by Eq. [150]. Since V_0, V_t, and t are known, the mean value of α_g between the temperatures 0° and t° C can be calculated with the help of Eq. [136].

c. Volume of the Bulb at 0° C. If the observations for α_g have been carried out as described in **b**, no additional ones need be made to obtain the volume of the bulb at 0° C, since the latter is simply $M v_0 - l_0 v$. But if **b** has been omitted it will be necessary to perform an experiment of the following kind in order to obtain V_0.

---

by always bringing the cross hairs up to the mercury meniscuses from the same side. Also measure the height h of the mercury meniscus in terms of a scale division. Again warm the air in the bulb and thus expel the mercury into a small weighed cup. Determine the mass m of this mercury by weighing. Observe the temperature t of the laboratory. The ends of the mercury thread are not plane but convex, and therefore the observed length s is slightly too large. It can be shown that for a capillary tube it suffices to subtract from s the quantity 0.4 h. When this correction and Eq. [150] are taken into account, the expression for v becomes

$$v = \frac{0.07355\,m(1 + 0.000181\,t)}{s - 0.4\,h}\ \text{cm}^3. \qquad [149]$$

Determine the mass of the dilatometer, first when empty and dry, and second when nearly filled with air-free distilled water. Such water may be prepared by boiling distilled water for half an hour. Observe the temperature t of the water and the scale-reading l_t of the water meniscus in the stem. Now, tables show that 1 g of water, weighed with brass weights in air, occupies very nearly $(2.00106 - \rho_t)$ cm³, where ρ_t is the density of water at the temperature t. Therefore the volume of the bulb up to the zero mark on the stem is, at this temperature, $[M(2.00106 - \rho_t) - l_t v]$ cm³, where M is the mass of the water. The volume V_0 can then be obtained by means of Eq. [136].

d. Volume Expansivity of the Liquid. Determine the volume expansivity of, say, ethyl alcohol, turpentine, or benzol. To do this, place the dilatometer containing the liquid under investigation in a large water bath and observe the scale-readings $l_1, l_2, \cdots$ of the liquid meniscus for several bulb temperatures[1] $t_1, t_2, \cdots$. Stir the bath continually and make observations only when you are sure that the liquid in the dilatometer is in thermal equilibrium with the bath.

Employ Eq. [137] to calculate the mean expansivities of the liquid between the temperatures 0° and $t_1°$ C, 0° and $t_2°$ C, etc.

1. Calculate the *mean apparent* expansivity of the liquid in this dilatometer for the range 0° to $t_1°$ C.

2. In work of extreme precision it is necessary to calibrate the stem of the dilatometer throughout its length in order to correct for irregularities of the cross-sectional area. Explain how this could be done.

3. Derive an expression for the volume of the liquid under investigation at the temperature t, in terms of the quantities V_o, α_g, t, l, and v.

4. Devise and describe a method of obtaining α_l for a nonvolatile liquid such that all the observations, except those for the temperatures, are reduced to weighings.

○

OPTIONAL LABORATORY PROBLEM

Linear Expansivity. Determine the mean linear expansivity of a specimen of brass or of glass which is in the form of a hollow tube. Next to the difficult method involving the use of an interferometer, the following method of measuring linear expansivity is probably the most accurate. Place the tube in a heat-insulating jacket, which is mounted horizontally upon supports, and focus two micrometer microscopes upon fine scratches near the two ends of the tube. Pass a current of tap water through the tube until the temperature has become constant. Then set the movable cross

---

[1] It will probably be necessary to standardize the thermometer with which these temperatures are measured. Directions for doing this will be found in Appendix 10.

hairs of the microscopes accurately upon the scratches and take readings on the scales in the micrometer eyepiece; in each case make a number of settings. Repeat these observations while a rapid current of steam is passing through the tube. Calibrate each microscope by focusing upon a standard scale and observing how many turns of the micrometer screw correspond to 1 mm.

○

QUESTION SUMMARY

1. What is meant by *temperature*? From what does our original notion of temperature come? How is an accurate measure of temperature obtained?

2. In the development of modern thermometers, what has determined the choice of (*a*) *thermometric substance*, or *property*? (*b*) *fixed points*? (*c*) *zero* and *size of degree* to be used? Why is it necessary to have two fixed points?

3. What is the present standard thermometer for practical thermometry? Why was it selected? Give a precise definition of any temperature *t* in terms of it.

4. Why is the mercury thermometer often used to measure temperature? Define any temperature *t* in terms of it. How greatly does this temperature scale differ from the constant-volume hydrogen gas scale?

5. Discuss the possibility of giving a definition of temperature that is absolute in the sense of being independent of the properties of some particular thermometric substance. How does this scale of temperature differ from the hydrogen gas scale? from an ideal gas scale?

6. Define *absolute temperature*. To what temperature centigrade does the absolute zero correspond? How nearly has it been reached experimentally?

7. Define (*a*) *pressure coefficient of a gas*; (*b*) *expansivity of a gas, liquid, or solid*; and (*c*) *linear expansivity of a solid*. What is the relation between the linear and volume expansivities of a solid?

8. What experimental fact did CHARLES and GAY-LUSSAC discover? Is it correct to say that GAY-LUSSAC discovered that the pressures of all gases vary directly as the absolute temperatures provided the volume remains constant, and that the volumes of all gases vary directly as the absolute temperature provided the pressure remains constant? What is involved in this statement other than the experimental fact found by GAY-LUSSAC?

9. From BOYLE'S law, GAY-LUSSAC'S law, AVOGADRO'S law, and the definition of absolute temperature, derive the equation of state of an ideal gas, $PV = NRT$. What sort of constant is R?

10. How can the numerical value of R be calculated? What is its value in absolute units? What is its value when the pressure is expressed in atmospheres and the volume in cubic centimeters?

PROBLEMS

1. In a gas thermometer like that of Fig. 96, the mercury in the tube c is 15 cm above the fiducial mark when the bulb is in melting ice and the atmospheric pressure is 750 mm of mercury. Ignoring the expansion of the bulb, find the bulb temperature for which the mercury will be 5.0 cm below the fiducial mark.   *Ans.* $-61°$ C.

2. The volume of a certain mass of carbon dioxide was found to be 100.0 ml at 0° C and 1 A_s. When the temperature and pressure were raised to 100° C and 1.369 A_s respectively, the volume was found to be unchanged. What is the mean pressure coefficient of carbon dioxide between 0° C and 100° C?   *Ans.* 0.00369 per deg C.

3. The pressure in an automobile tire, as indicated by a gauge registering pressures above that of the atmosphere, was found to be 35 lbwt · in.$^{-2}$ on a cool day when the temperature was 16° C. Assuming that the volume change is negligible and that there is no leakage, find the pressure in the tire on a hot day when the temperature is 36° C.   *Ans.* 53 lbwt · in.$^{-2}$.

4. (*a*) With the help of the general gas law and the definition of density, derive a single equation that will enable one to calculate the density of an ideal gas for all temperatures and pressures if the density has once been determined for one single value of the temperature and pressure. (*b*) Given that the density of air at 16° C and under a pressure of 740 mm of mercury is 1.189×10^{-3} g · cm^{-3}, compute its density at 0° C and 760 mm of pressure. (*c*) What volume will 28.9 g of air occupy at 0° C and 760 mm pressure?   *Ans.* (*b*) 1.293×10^{-3} g · cm^{-3}; (*c*) 22.36 l.

5. How much work is done against atmospheric pressure when a quantity of air of mass 14 g is heated from 0° to 50° C under a constant pressure of 1 A_s?   *Ans.* 2.0×10^2 j.

6. If the molecular weight of a certain gas is 28, what volume will be occupied by 12 g of the gas under a pressure of 2.0 A_s and at 35° C? *Ans.* 5.4 l.

7. What is the pressure exerted by 0.50 g of argon contained in a closed vessel of capacity 5.0 l at 20° C?
Ans. 62 gwt · cm^{-2}.

8. A gas is enclosed in a tube of uniform cross section. If the conditions shown in Fig. 102 apply when the temperature is $-20°$ C and the pressure is 1 A_s,   Fig. 102. Problem 8
at what temperature will the mercury surfaces be at
the same level? Assume that the change in the density of the mercury and the change in the outside level are negligible.   *Ans.* 354° C.

9. In dealing with thermal problems by dimensional methods it is often found desirable to introduce another fundamental unit in addition to those of length, mass, and time. The unit of temperature is a convenient choice. Denote the dimension of the unit of temperature by [θ] and write the dimensional formulas for (*a*) the volume expansivity α; (*b*) the pressure coefficient β; (*c*) the product PV; (*d*) the gas constant for unit mass R'; (*e*) the universal gas constant R; (*f*) the linear expansivity λ.

10. PIERRE has found that, for chloroform between 0° and 63° C, the constants in Eq. [135] are $a = 1.1071 \times 10^{-3}$, $b = 4.6647 \times 10^{-8}$, $c = 1.7433 \times 10^{-8}$. From these data calculate the increase, in percentage, of the mean expansivity of chloroform in passing from 0° C to 40° C.   *Ans.* 2.69 percent.

11. A soft iron ball 5.000 cm in diameter is 0.01 mm too large to go through a hole in a brass plate when the ball and plate are at 20° C. At what temperature (the same for both the ball and the plate) will the ball just pass through the hole?   *Ans.* 53° C.

12. A glass bulb of capacity 10 cm³ is filled with mercury at 20° C. How much mercury will run out if the bulb is heated to 100°?   *Ans.* 1.7 g.

13. Given that α is the expansivity of mercury and that λ is the linear expansivity of a scale which is correct at the temperature t_0, derive a formula for reducing the reading B_t of a barometer at a given temperature t to what it would be if the temperature were 0° C.
          Ans. $B_0 = [1 + \lambda(t - t_0)](1 + \alpha t)^{-1}B_t$.

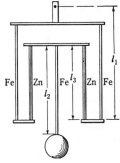

14. In one type of compensated pendulums (Fig. 103) the expansion of one set of rods lowers the bob, whereas that of another set raises it. Suppose the first set to be made of cast iron, and that $l_1 + l_2 = 90$ cm. If the material of the second set is zinc, what must be the length l_3 if the bob is to be neither raised nor lowered with changes in the temperature?   *Ans.* 36 cm.

FIG. 103.   The gridiron pendulum

15. Accurately calibrated thermometers read correctly only when the whole mercury column is immersed in the bath whose temperature is to be determined. Find the reading of a mercury-in-glass thermometer if the bulb and stem up to the zero mark are exposed to a temperature of 210° C while the remainder of the stem is at 18° C. Assume the mean linear expansivity of the glass to be 5.8×10^{-6} per deg C.   *Ans.* 203° C.

○

ACCURATE AND minute measurement seems to the non-scientific imagination a less lofty and dignified work than looking for something new. But nearly all the grandest discoveries of science have been but the rewards of accurate measurement and patient long-continued labor in the minute sifting of numerical results.
          LORD KELVIN, *Report of the British Association for the Advancement of Science* (1871), Vol. 41, p. 91.

CHAPTER TEN

THE DISCRETE NATURE OF MATTER AND THE KINETIC–MOLECULAR THEORY OF GASES

1. From nothing comes nothing. Nothing that exists can be destroyed. All changes are due to the combination and separation of molecules   2. Nothing happens by chance: every occurrence has its cause, from which it follows by necessity.   3. The only existing things are the atoms and empty space; all else is mere opinion.   4. The atoms are infinite in number, and infinitely various in form; they strike together, and the lateral motions and whirlings which thus arise are the beginnings of worlds.   5. The varieties of all things depend upon the varieties of their atoms, in number, size, and aggregation.   6. The soul consists of fine, smooth, round atoms, like those of fire. These are the most mobile of all: they interpenetrate the whole body, and in their motions the phenomena of life arise.

Democritus's principles, as quoted by John
Tyndall [1] in his Belfast Address in 1874

○

A little reflection will convince one that there are only two possible hypotheses as to the constitution of matter: either matter is continuous in structure, so that, however far division or subdivision is carried, iron is always iron and will always exhibit the characteristics of iron; or else the contrary is true and subdivision carried far enough will bring us to particles called *atoms*, meaning, literally, something that cannot be cut. It is not surprising, then, to learn that an atomistic philosophy was taught as early as the fifth century B.C. by the Greek philosopher LEUCIPPUS and his famous pupil and associate DEMOCRITUS (*c.* 460–370 B.C.). Little is known of the writings of these "atomists," but their theory of matter has been immortalized in verse by LUCRETIUS (TITUS LUCRETIUS CARUS), the famous Roman poet of the first century B.C., in "the greatest philosophical poem of all times," *De Rerum Natura*.[2] Many of the views expressed in this poem and in the principles of DEMOCRITUS which

---

[1] *Fragments of Science*, ed. 5 (Appleton, 1879), p. 472. The first few pages of this address give an excellent account of the atomistic philosophy of the ancients.

[2] This poem has been translated, under the title *On the Nature of Things*, by H. A. J. Munro (1905), Cyril Bailey (1910), and many others. There are metrical translations by John Evelyn (1656) and by William Ellery Leonard (1922) and others. "Lucretius's poem is an amazing performance; it does not contain new scientific facts or theories, but sets forth Greek views in a wonderful manner, with many flashes of genius. It is a masterly exposition of the rationality and determinism of the universe. It marks the climax of Roman scientific thought" (Sarton).

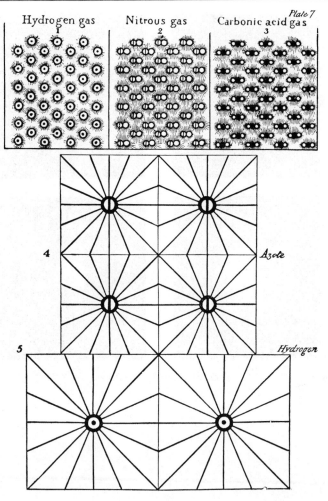

Some of DALTON's Pictures of Atoms

Reproduced from JOHN DALTON's A New System of Chemical Philosophy (Manchester, 1810),
Part II, p. 548

DALTON conceived of the atoms as being hard particles with "atmospheres of heat'
(represented by rays in his drawings) emanating from them. Note the different
volumes occupied by the atoms of hydrogen and nitrogen (azote).

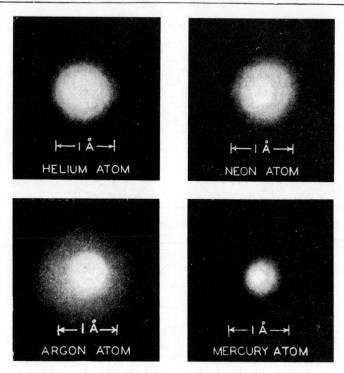

Recent Pictures of Atoms

THESE illustrations were prepared by E. O. WOLLAN and A. H. COMPTON [*Journal of the Optical Society of America* **24**, 229 (1934)] from data on x-ray scattering in gases. They represent the electron distributions in the various atoms, and therefore correspond to "photographs" of the atoms which appear, in general, as regions more or less diffusely filled with electricity. These "photographs" are in good agreement with the modern quantum theory of atomic structure. (The abbreviation Å denotes a unit of length called the Angstrom; 1 Å = 10^{-8} cm.)

are quoted at the beginning of this chapter, with a few modifications and omissions, might almost pass muster today. The fact is, nearly all the *qualitative* conceptions of the modern kinetic-atomic theory of matter were developed more than two thousand years ago, for the Greek atomists had almost as clearly developed a picture of a world made up of incessantly moving atoms as has the modern physicist. The difference is that the idea had its roots in one case in a mere speculative philosophy; in the other case, like most of our modern scientific knowledge, it rests upon direct, controlled quantitative investigations.

That matter is granular in structure is not very evident from casual observation; the human eye has never seen, and indeed can never see, an individual atom or molecule. Thus, under the opposing influence of the teachings of EMPEDOCLES (*c.* 490–435 B.C.) and ARISTOTLE (384–322 B.C.), and because the then civilized world was soon to succumb to the attacks of the barbarians, the atomic theory of the Greeks languished and died. EMPEDOCLES held that matter consists of one primordial substance which is the habitat of four elementary properties — earth, air, fire, and water. Such a view is not as fantastic as it may seem on first thought, and it at least had the merit of being productive, for it provided a theoretical basis for the idea of transmutation of base metals into precious ones, and thus helped to furnish the alchemists with the motive power for nearly all chemical experimentation down to the eighteenth century.

o

The Atomic-Molecular Theory of Matter

110. Dalton's Atomic Theory. More than twenty-two centuries elapsed before the purely speculative views of the Greek atomists were placed on the substantial foundation of experiment. This was accomplished by the English chemist JOHN DALTON [1] (1766–1844), who was led by his studies of the physical properties of gases to formulate an atomic theory having, in essence, the following two features: (*a*) each chemical element consists of identical atoms, there being as many different kinds of atoms as there are kinds of elements; (*b*) when different elements combine to form a compound, the smallest possible portion of the latter contains a definite number of atoms of each element. A study of thousands of experiments on the

---

[1] *A New System of Chemical Philosophy* (Manchester, 1808). Reprints of earlier papers will be found in *Foundations of the Atomic Theory*, Alembic Club Reprint No. 2 (Edinburgh, 1923).

proportions by weight in which various substances combine chemically reveals several very fundamental principles that find their most simple and natural interpretation in this atomic hypothesis. These principles were known to DALTON, and they have played an important part in providing tests for his theory.

111. The Laws of Chemical Combination. *a. The Law of Definite Proportions.* This law states that the proportions by weight in which the elements enter into a given compound are invariable. For example, it is found that hydrogen will unite with chlorine, so as to leave no free hydrogen and no free chlorine, only when the mass of hydrogen present bears to the mass of chlorine one definite proportion. The same may be said of the combination of hydrogen with bromine. These ratios are

$$\text{1.0000 g hydrogen to 35.183 g chlorine;} \qquad [151]$$
$$\text{1.0000 g hydrogen to 79.297 g bromine.}$$

Similarly, the chlorides of potassium and silver are always found to contain the following proportions:

$$\text{38.793 g potassium to 35.183 g chlorine;} \qquad [152]$$
$$\text{107.05 g silver to 35.183 g chlorine.}$$

Although the law of definite proportions was established by JOSEPH LOUIS PROUST (1754–1826) without reference to any particular theory as to the constitution of matter, its evident interpretation in the light of an atomic hypothesis can only be that the atoms of each element are of constant mass and that any given compound is made up of discrete portions, each of which is always of the same atomic composition.

b. The Principle of Equivalence. From the foregoing examples it is evident that 35.183 g of chlorine and 79.297 g of bromine may be called equivalent quantities in the sense that each one of them combines with exactly the same mass of hydrogen, namely 1 g. Similarly, 38.793 g of potassium and 107.05 g of silver may also be called equivalents, since each of them combines with the same mass of chlorine. These latter elements cannot be made to combine with hydrogen directly, but since the masses given combine with just the amount of chlorine that has been found to be the combining equivalent of 1 g of hydrogen, these masses may also be said to have been found in this indirect way to be the equivalents in combining ability of 1 g of hydrogen. So much for the definition of *equivalent*. Now the fact of peculiar significance is this. A quantitative analysis of the *bromides* of potassium and silver leads to precisely the same numbers for the equivalent masses of these elements as did a study of the chlorides. Thus the only proportions in which these elements will combine with bromine are

$$38.793 \text{ g potassium to } 79.297 \text{ g bromine;} \qquad [153]$$
$$107.05 \text{ g silver to } 79.297 \text{ g bromine.}$$

When, further, a study of, say, the iodides of potassium and silver leads again to the same results, it becomes certain that some very definite physical significance lies behind these numbers. The simplest possible interpretation to put upon them is, to take a particular case, that the atoms of potassium which combine with chlorine to form potassium chloride are exactly like the atoms which combine with bromine to form potassium bromide, and with iodine to form potassium iodide.

The facts of equivalence which have been presented here, and which constitute one of the strongest arguments for the atomic hypothesis, may be summarized thus: the study of many different compounds leads often to precisely the same number for the combining equivalent of a given element with reference to hydrogen.

c. The Law of Multiple Proportions. In some cases the study of different compounds leads to more than one number for the equivalent of a given element with reference to hydrogen. For example, DALTON found that ethylene gas yielded upon decomposition the two elements carbon and hydrogen in the proportions by weight of 5.96 parts of carbon to 1 part of hydrogen, whereas methane yielded the same two elements in the ratio 2.98 parts of carbon to 1 part of hydrogen. Again, in the three compounds nitrous oxide, nitric oxide, and nitrogen oxide the masses of oxygen which combine with 1 g of nitrogen were found to be in the ratio 1 : 2 : 4. Further study of other compounds revealed to DALTON the principle that whenever elements can combine in different proportions, these proportions always bear simple ratios to one another. This is known as the *law of multiple proportions*. Its evident interpretation, in the light of the atomic hypothesis, is that it is possible in some cases for two or three or some other small number of atoms of a given element to enter into the constitution of the smallest units of a compound.

Enough has now been said to show how to each element there may be assigned an experimental number which in itself, or when multiplied by some small integer, expresses the mass by which the element enters into combination with other elements. This experimental number is called the *atomic weight* of the element, although it is more properly a "combining mass." For reasons which we need not discuss, modern tables of atomic weights are not based upon the atomic weight of hydrogen as 1.0000 but upon that of oxygen taken as 16.000. On this oxygen scale, hydrogen is found to have an atomic weight of 1.0078.

112. Molecules and the Law of Avogadro. GAY-LUSSAC[1] discovered that the various gases, under conditions of equal pressure and temperature, combine in simple volumetric proportions. He found, for instance, that two volumes of hydrogen combine with one volume of oxygen to form two volumes of steam. This did not fit in well with DALTON'S idea that the elementary gases were always composed of single atoms, and it led AVOGADRO[2] to modify the Daltonian idea of the ultimate particles of substances by distinguishing between two types of structures, one of which is the atom and the other that which is now called the *molecule*. Molecules are the smallest units into which either compounds or elements can be divided without chemical decomposition. The molecule is in turn made up of one or more atoms, there being always the same number of the same kind of atoms in every molecule of a given substance. In the case of an element, the molecule contains atoms of that element only, whereas the molecules of a compound must contain the atoms of at least two different elements. Today we know that compounds that are in the liquid or solid state often have for their smallest structural unit the atom and not the molecule. For this reason it is best to define the *molecule* as *the smallest unit of a substance that exists in the gaseous state.*

As we already know (Sec. **105**), AVOGADRO went farther and postulated that equal volumes of different gases at the same temperature and pressure contain equal numbers of molecules. This enabled him to determine the numbers of atoms in the molecules of the various gases. Thus he found that one volume of hydrogen and one volume of chlorine combine to form two volumes of hydrogen chloride. In accordance with AVOGADRO'S law, there must therefore be twice as many hydrogen chloride molecules present after the reaction as there were hydrogen molecules before it; since, however, each hydrogen chloride molecule must contain a hydrogen atom, the hydrogen molecules must each have consisted of two atoms before the reaction. Analogous considerations apply to the chlorine molecules.

One of the proofs of AVOGADRO'S law rests upon a remarkable relation which is found to exist between the densities of gases and the numbers representing their combining masses. Table IV brings out

---

[1] "Memoir on the Combination of Gaseous Substances with Each Other," read before the Philomathic Society, December 31, 1808, and published in the *Mémoires de la Société d'Arcueil* **2**, 207 (1809). A translation appears in *Foundations of the Molecular Theory*, Alembic Club Reprint No. 4 (Edinburgh, 1923), pp. 8–24.

[2] A. Avogadro, "Essay on a Manner of Determining the Relative Masses of the Elementary Molecules of Bodies, and the Proportions in which they Enter into Compounds," *Journal de Physique* **73**, 58 (1811). For a translation see *Foundations of the Molecular Theory*, Alembic Club Reprint No. 4 (Edinburgh, 1923), pp. 28–51.

clearly this striking relation. The densities given in this table are the results of experiments at a given temperature and pressure, the density of oxygen being taken, for convenience, as 32. It is seen that in the case of the compounds the numbers that represent the densities in terms of a gas $\frac{1}{32}$ as dense as oxygen are throughout very nearly equal to the numbers that represent the molecular weights, the latter in each case having been obtained by taking the sum of the atomic weights of the atoms in the molecule of the substance. But if the molecular weights of a number of gases bear the same ratios as the masses of the gases per unit volume, then evidently the number of molecules per unit volume must be the same in all the gases.

TABLE IV · *An Experimental Verification of Avogadro's Law*

Gas	Symbol	Density relative to O as 32	Atomic weight
Hydrogen	H	2.012	1.008
Helium	He	3.999	4.00
Nitrogen	N	28.05	14.01
Oxygen	O	32.00	16.00
Bromine	Br	159.9	79.92

Gas	Formula	Density relative to O as 32	Molecular weight
Hydrogen chloride	HCl	36.59	36.47
Methane	CH_4	16.21	16.03
Carbon monoxide	CO	28.17	28.00
Carbon dioxide	CO_2	44.75	44.00
Nitric oxide	NO	30.19	30.01
Nitrous oxide	N_2O	44.45	44.02
Water	H_2O	18.17	18.02

This remarkable conclusion, which applies necessarily to all the compounds represented in Table IV, provided only that their molecular constitutions are correctly given, is seen to apply also to all the elementary gases in the table, if only the molecular weights of hydrogen, nitrogen, oxygen, and bromine, but not helium, be assumed to be twice the values of the atomic weights of these elements, that is, if the molecules of these gases be composed each of two atoms; for then the molecular weights become numbers which are in close agreement with those given in the column of densities. In the case of helium the atomic and molecular weights are the same; helium is one of the few substances that are *monatomic*.

Now it is found that with equally simple choices as to the molecular constitution of those gases in which the combining equivalents of the constituent elements leave two or more choices open, the densities of all known gases are in close agreement with their molecular weights. This agreement is least perfect in the case of those gases that show the largest departures from BOYLE'S law. For actual gases, then, this law, like the laws of BOYLE (Sec. 92) and of CHARLES and GAY-LUSSAC (Sec. 101), is only a close approximation.

The agreement between molecular weights and gas densities is not by any means the only experimental basis for the law of AVO-GADRO. The actual determination of relative masses of hydrogen, nitrogen, and oxygen atoms in collision, as observed from the C. T. R. WILSON cloud tracks,[1] has confirmed the masses indicated by the law. More recently, accurate determinations of the relative masses of the elements in the periodic table by means of the mass-spectrograph yield values that fit in with all its predictions.[2]

o

The Kinetic Theory of Gases

Many years before the establishment of the atomic-molecular theory of chemistry, the accumulating knowledge of such purely physical phenomena as the elasticity of gases and the effects of heat on matter had led to a revival of the ancient Greek conjecture that matter is composed of particles. In order to account for the elastic properties of air, two views were advanced. The first was the repulsion theory, according to which the pressure exerted by confined air was attributed to repellent forces existing between the molecules, which were assumed to be at rest. This view was held by prominent scientists even as late as the middle of the nineteenth century. When BOYLE'S law was discovered, in 1662, it of course became necessary to reconcile the theory with it, and this could be done only by making the assumption that the molecules repel one another with forces that are inversely proportional to the distances between them. The theory has now been altogether abandoned: first, because such a law of molecular force is wholly at variance with all modern views as to the nature of molecular force; second, because it necessitates the conclusion that the pressure which a gas exerts is a function not of its density and temperature alone but also of the shape and size of the containing vessel, a conclusion which is directly contradicted by experiment; third, because the fact that a gas does not experience

---

[1] See, for example, J. A. Crowther, *Ions, Electrons and Ionizing Radiations.*
[2] F. W. Aston, *Isotopes*; also, *Mass-Spectra and Isotopes.*

a rise in temperature when it expands into a vacuum proves that no repulsion exists between its molecules. Just a few years after NEWTON's death DANIEL BERNOULLI [1] crystallized these ideas of a particle structure into the hypothesis that a gas consists of perfectly elastic particles in rapid motion of translation, and in this way made the first attempt to explain the observed properties of a gas on a simple mechanical basis. In his treatise *Hydrodynamica* (1738), he explains the pressure on the walls of a vessel enclosing a gas as due to the impacts of the particles and deduces an expression for the pressure in terms of their speeds.[2] He does not specify accurately what is meant by the speed of the gas particles, or by pressure, or by temperature, and hence this must be regarded as only the first rough quantitative sketch of what has since come to be the important branch of theoretical physics known as the *kinetic theory of gases.* It was not until after the middle of the nineteenth century, when the discovery of the mechanical nature of heat [3] gave added interest to these problems, that JOULE,[4] A. KRÖNIG,[5] and especially RUDOLF CLAUSIUS [6] (1822–1888) laid the foundations of the theory, which was then rapidly developed by JAMES CLERK MAXWELL [7] (1831–1879) and LUDWIG BOLTZMANN [8] (1844–1906).

---

[1] DANIEL BERNOULLI (1700–1782) was one of the most distinguished members of a celebrated Dutch-Swiss family. No less than fifteen members of this family earned reputations in various scholarly fields, chiefly in mathematics and physics; and of these, eight attained eminence. For an interesting account of this unusual family see *Die Naturwissenschaften* 22, 717 (1934).

[2] The *Hydrodynamica* was written in Latin, but a translation of the first six sections of the tenth chapter, which contain this first successful application of the kinetic theory, will be found in *A Source Book in Physics* (1935), pp. 247–251.

[3] See Secs. 54 to 56.

[4] *Memoirs of the Manchester Literary and Philosophical Society* 9, 107 (1851); also *Philosophical Magazine* (4) 14, 211 (1857). This paper has been reprinted in *The Scientific Papers of James Prescott Joule* (Taylor and Francis, 1884), Vol. I, pp. 290–297. An excerpt appears in *A Source Book in Physics* (1935), 255–257.

[5] *Grundzüge einer Theorie der Gase* (Berlin, 1856); Poggendorff's *Annalen der Physik und Chemie* 99, 315 (1856).

[6] Poggendorff's *Annalen der Physik und Chemie* 100, 353 (1857); 105, 239 (1858); 115, 1 (1862); et seq. Translations of these papers appeared in the *Philosophical Magazine* (4) 14, 108 (1857); 17, 81 (1859); 23, 417, 512 (1862). Most of CLAUSIUS's work on the theory of heat and of gases is collected in his *Die Mechanische Wärmetheorie.* For his contributions to the kinetic theory, see *Die Kinetische Theorie der Gase* (Wieweg, 1889–1891), Vol. III.

[7] *Philosophical Magazine* (4) 19, 19 (1860); 20, 21 (1860). *Philosophical Transactions* 156, 249 (1866) et seq. See *The Scientific Papers of James Clerk Maxwell* (Cambridge University Press, 1890), Vols. I, II.

[8] *Vorlesungen über Gastheorie*, ed. 1 (1895); ed. 2 (Barth, 1910). A French translation was published by Gauthier-Villars (1902).

113. Model of an Ideal Gas. The fact that at very low pressures the physical behavior of all gases is represented by simple laws indicates that all have a common and simple structure, and thus the construction of a *model* of an ideal gas that would yield these laws became the first objective in the kinetic theory. In the light of the atomic-molecular theory of chemistry, and the experiments on the nature of heat which showed heat to be simply the energy involved in the random motions of molecules, the following postulates came to be formulated:

1. *A chemically homogeneous gas is composed of identical molecules which are moving in so random a fashion that the number moving in any one direction is on the average the same as that moving in any other.*

2. *The actual space occupied by the molecules is negligible in comparison with the space between them.*

3. *The molecules exert no forces on one another except when they actually collide; that is, their velocities are changed only by collision with other molecules or with the walls of the containing vessel.*

4. *The impacts between molecules and with the walls must be perfectly elastic, for otherwise there would be a continual loss of energy as time progresses* (Sec. **67**). It would be better to say that, *on the average, the impacts must be perfectly elastic*; in the case of diatomic molecules, for example, an individual collision may result in some of the translational energy's being converted into rotational energy, thus apparently disappearing, whereas in other collisions the reverse effect will occur.

114. Kinetic-Theory Interpretation of Gas Pressure. In 1848 JOULE,[1] building on a sounder foundation of experimentation than had BERNOULLI and probably without knowing of the latter's work, succeeded in deducing from the foregoing postulates an expression for the pressure of a gas in terms of the number, the speed, and the mass of the molecules. Imagine the gas to be confined in a rectangular box (Fig. 104) the lengths of whose sides are x, y, and z, and let m be the mass of each molecule. We will fix our attention upon some particular molecule having at the moment a velocity v, the rectangular components of which are v_x, v_y, and v_z respectively. Suppose this molecule to impinge on, say, the top wall of the vessel. The normal force which it exerts on this wall depends upon its momentum in the Y direction only, that is, upon mv_y. Since it re-

---

[1] See footnote 4, p. 205.

bounds with a velocity the component of which perpendicular to the top surface is $-v_y$, the total momentum imparted normal to this surface in one impact is $2\,mv_y$.

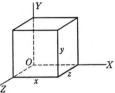

Now this particle will make against the top wall $v_y/2\,y$ impacts in unit time, for v_y is the distance that it moves in unit time in the Y direction, and $2\,y$ is the distance moved in this direction in the interval between successive impacts on the top wall. Thus the amount of momentum which it imparts to this wall in unit time is $2\,mv_y(v_y/2\,y)$, or mv_y^2/y, and this, by NEWTON'S second law, is the average normal force exerted by the

FIG. 104. Method of obtaining the kinetic-theory expression for pressure

molecule on the top surface. Since there are other molecules present in the vessel, the molecule considered will, of course, collide with them in its excursions to and fro. But, so long as the volume occupied by the molecules is negligibly small, the number of impacts on the wall will be unaltered by these collisions; for when two perfectly elastic spheres of equal masses collide, the effect is the same as though one particle had passed through the other without influencing it (Sec. 65).

The total normal force exerted on the top wall evidently is the sum obtained by adding the quantities mv_y^2/y for all the molecules, and the pressure P is this sum divided by xz, the area of the surface; thus

$$P = \frac{m}{xyz}\,\Sigma v_y^2;$$

or, letting N denote the total number of molecules in the box, V the volume xyz, and ρ the density,

$$P = \frac{mN}{V}\frac{\Sigma v_y^2}{N} = \rho\,\frac{\Sigma v_y^2}{N}. \qquad [154]$$

The quantity $(\Sigma v_y^2)/N$ is the average of v_y^2 for all the molecules, and if this average value be denoted by $(v_y^2)_{av}$, then

$$P = \rho\,(v_y^2)_{av}.$$

Now, for any one molecule, $v^2 = v_x^2 + v_y^2 + v_z^2$, from which it follows that the average of v^2 for all the molecules is given by $(v^2)_{av} = (v_x^2)_{av} + (v_y^2)_{av} + (v_z^2)_{av}$. Since the molecules are moving wholly at random, $(v_x^2)_{av} = (v_y^2)_{av} = (v_z^2)_{av}$, and therefore $(v^2)_{av} = 3\,(v_y^2)_{av}$; hence

$$P = \tfrac{1}{3}\,\rho\,(v^2)_{av}. \qquad [155]$$

As was shown by JOULE, this equation at once affords a means of calculating the quantity $(v^2)_{av}$, which is called the *mean-square* speed of the molecules. This was the first molecular magnitude to be calculated quantitatively.[1] It gave an astounding result.

By extracting the square root of the mean-square speed, there results what is known as the *root-mean-square* speed. This quantity is defined by the equation $\sqrt{(v^2)_{av}} = \sqrt{(\Sigma v^2)/N}$. It is somewhat greater than the *arithmetic-mean* speed, defined by $v_{av} = (\Sigma v)/N$, because the squares of numbers increase more rapidly than the numbers themselves.

> EXAMPLE. Find the root-mean-square speed of hydrogen molecules at 0° C.
>
> **Solution.** Since the density of hydrogen at 0° C and 1 A$_s$ is, from physical tables, 8.99×10^{-5} g · cm^{-3}, and since 1 A$_s$ = 1.0132×10^6 dyne · cm^{-2},
>
> $$\sqrt{(v^2)_{av}} = \sqrt{\frac{3\,P}{\rho}} = \sqrt{\frac{3 \times 1.0132 \times 10^6}{8.99 \times 10^{-5}}} = 1.84 \times 10^5 \text{ cm · sec}^{-1}.$$

115. Dalton's Law of Partial Pressures and Graham's Law of Diffusion. Among the various laws that any satisfactory theory of gases must be able to explain is DALTON'S experimental discovery [2] that a mixture of gases having no chemical action on one another exerts a pressure that is the sum of the pressures which would be exerted separately by the several constituents if each alone occupied the same volume at the same temperature. The law obviously follows immediately from our initial postulates (Sec. 113), inasmuch as the molecules are considered to be so small as not sensibly to obstruct one another. Thus, if oxygen and hydrogen are in the same containing vessel, the molecules of one gas will move exactly as if those of the other gas were not present and will exert the same pressures as they would exert if occupying the vessel alone. One would expect this to be true only if the pressure were low, and indeed DALTON'S law is found to be in accord with experiment only in this case.

The kinetic theory also explains a law of diffusion obtained experimentally by THOMAS GRAHAM.[3] This law states that the time-rates of flow of gases through a thin porous diaphragm, such as a

---

[1] A few attempts to estimate molecular distances had been made previously. Thus YOUNG in 1816 gave some really remarkable estimates based upon surface-tension data, in a paper on "Cohesion" written for the *Encyclopaedia Britannica* [reproduced in the *Miscellaneous Works of Thomas Young* (Murray, 1855), Vol. 1; see especially p. 461]. Later this method was made more precise by LORD RAYLEIGH [*Philosophical Magazine* (5) **30**, 285, 456 (1890); *Scientific Papers* (Cambridge University Press, 1902), Vol. III, p. 423].

[2] *Memoirs of the Manchester Literary and Philosophical Society* **5**, 535 (1802).

[3] *Philosophical Transactions* **136**, 573 (1846).

thin plate of graphite (Fig. 105), are inversely proportional to the square roots of the densities of the gases, if their pressures and temperatures are the same. Consider two gases A and B that have the same pressures and temperatures. Then, by Eq. [155],

$$\frac{\sqrt{(v_A{}^2)_{av}}}{\sqrt{(v_B{}^2)_{av}}} = \frac{\sqrt{\rho_B}}{\sqrt{\rho_A}}. \qquad [156]$$

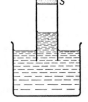

Since it can be proved that root-mean-square speeds are in the same ratio as arithmetic-mean speeds, this equation tells us that, to take a particular case, hydrogen molecules move on the average nearly six times as fast as chlorine molecules, for chlorine has a density nearly thirty-six times that of hydrogen at the same temperature and pressure. And since, at the same pressure and temperature, both gases contain the same number of molecules per unit volume, the hydrogen will diffuse six times as rapidly as the chlorine.

Fig. 105. A glass tube is stopped at one end by a porous plug S, is filled with, say, hydrogen, and then is placed with its open end under a liquid. The liquid rises in the tube, and after a time it is found that the gas in the tube is no longer hydrogen but air

116. Kinetic-Theory Interpretation of Temperature. Since $\rho = nm$, where n is the number of molecules in a unit volume of the gas, Eq. [155] may be changed to read

$$P = \tfrac{1}{3} nm(v^2)_{av}. \qquad [157]$$

The average kinetic energy E_{av} of each molecule of the gas is given by $E_{av} = \tfrac{1}{2} m(v^2)_{av}$, and hence we may also write, instead of Eq. [157],

$$P = \tfrac{2}{3} nE_{av}. \qquad [158]$$

In order to introduce the temperature, use may be made of the empirically determined equation of state $PV = RT$, where V is the volume of one mole of the gas and R is the gas constant per mole (Sec. **105**). Putting the value of P from Eq. [158] into this general gas equation gives

$$\tfrac{2}{3} nE_{av}V = RT. \qquad [159]$$

Now the product nV is equal to N_o, the number of molecules in the mole. This number, N_o, known as the *Avogadro number*,[1] is evidently the same for all substances. Its numerical value, as determined by an indirect method depending on the evaluation of the charge on the

---

[1] The number of molecules per *unit volume of an ideal gas at 0° C and pressure 1 A,* is called the *Loschmidt number* (symbol n_o). Its value can be computed easily; or see Appendix 13, Table K.

electron, is $(6.064 \pm 0.006) \times 10^{23}$ per mole. Putting the AVOGADRO number N_o into Eq. [159], we have

$$\tfrac{2}{3} N_o E_{av} = RT. \qquad [160]$$

This important deduction not only tells us that the *average* kinetic energy of the molecules is a function of the temperature alone but is still more specific in that it shows it to be *directly proportional to the absolute temperature*. The kinetic theory thus makes it easy to understand how the temperature of a body can be raised by doing work upon it, as when a gas is compressed, or how a heated body can itself do work. (What is the physical meaning given to the quantity RT by Eq. [160]?)

> EXAMPLE. (*a*) Show that Eq. [160] renders possible the calculation of the root-mean-square speed of the molecules of any ideal gas at a given temperature merely from a knowledge of R, T, and the molecular weight of the gas. (*b*) Given that the atomic weight of helium is 4.00, calculate the root-mean-square speed of its molecules at 0° C; (*c*) at 820° C.

A form of Eq. [160] that will often be found in the literature is obtained by solving for E_{av}; thus

$$E_{av} = \tfrac{3}{2} k_o T, \qquad [161]$$

where

$$k_o = \frac{R}{N_o}, \qquad [162]$$

k_o being another universal constant, which is known as the *Boltzmann constant* or as the *gas constant per molecule*. It will be noted that, from the point of view of the kinetic theory, the absolute zero on the thermodynamic scale is the temperature at which the gas molecules have no kinetic energy at all and therefore come completely to rest. Although actual gases liquefy, and even solidify, before reaching this temperature, the gas helium can be used in a gas thermometer to within a few degrees of the absolute zero.

117. Equipartition of Energy. If two gases A and B are contained in different vessels but have the same temperature, it follows immediately from Eq. [161] that the average kinetic energy of the molecules, namely E_{av}, is the same for both gases; that is,

$$\tfrac{1}{2} m_A (v_A{}^2)_{av} = \tfrac{1}{2} m_B (v_B{}^2)_{av}. \qquad [163]$$

If the gases were mixed in the same vessel, there is no reason to suppose that the same condition would not hold, and in fact direct experiment shows that it does. It is concluded, therefore, that the molecules in a mixture of any gases have the same *average* kinetic energy of translation. Eq. [163] thus represents a particular case of the very important law called the *law of equipartition of energy* or equal distribution of energy among the molecules. This law is applicable only when dealing with a large number of molecules.

118. The Brownian Movement. One of the proofs of the law of equipartition of energy comes out of the study of Brownian movements. JEAN PERRIN,[1] for example, studied under the microscope the motions of various kinds of colloidal particles suspended in water, and found that the mean kinetic energies of these particles were the same, and equal to that of a gas at the same temperature; this was true for particles varying in mass from 60,000 to 1.

The highly interesting phenomenon known as the Brownian movement was first observed in 1827 by ROBERT BROWN,[2] a botanist, who found, to use his own words, that "Extremely minute particles of solid matter, whether obtained from organic or inorganic substances, when suspended in pure water, or in some other aqueous fluids, exhibit motions for which I am unable to account and which, from their irregularity and seeming independence, resemble in a remarkable degree the less rapid motions of some of the simplest animalcules of infusions." The cause of the Brownian movement was long in doubt, but with the development of the kinetic theory came the realization that the particles move because they are bombarded unequally on different sides by the rapidly moving molecules of the fluid in which they are suspended. The Brownian movement never ceases. "It can be seen in liquid occlusions in quartz, which have been sealed up for thousands of years. It is inherent and eternal."[3] Here, then, is direct experimental evidence of the same sort of perpetual motion that we have assumed the molecules themselves to have.

119. The Maxwell-Boltzmann Law and the Introduction of Statistics into Physics. In our discussion of the motions of molecules we have dealt with *mean* speeds and therefore have tacitly assumed that the molecules do not all move alike. The laws of elastic impact tell us that such an assumption is warranted, whether or not the gas as a whole is in a state of equilibrium; for, even if the speeds originally were equal, the encounters between the molecules would produce an

---

[1] *Brownian Movement and Molecular Reality*, tr. by F. Soddy (Taylor and Francis, 1910); also, *Atoms*, tr. by D. L. Hammick (Constable, 1923).

[2] BROWN'S original paper makes most interesting reading. It was published as a pamphlet with the title *A Brief Account of Microscopical Observations made in the Months of June, July and August, 1827, on the Particles Contained in the Pollen of Plants; and on the General Existence of Active Molecules in Organic and Inorganic Bodies.* A portion of it is reproduced in *A Source Book in Physics* (1935), pp. 251–255.

[3] *J. Perrin, article "Brownian Movement," *Encyclopaedia Britannica*, ed. 14.

inequality. Since the velocities of the molecules are constantly changing in both direction and magnitude, it is obviously hopeless to attempt to follow their individual behaviors. Curiously enough, however, it is just this complete disorderliness of the motions, combined with the assumption that the number of molecules is enormously large, that makes it possible to describe the behavior of a gas as a whole in terms of the motions of its molecules; for it is under such circumstances that one can apply the well-known laws of probability.

The problem of how the velocities are distributed among the various molecules was solved by the joint efforts of MAXWELL[1] and BOLTZMANN,[2] and the famous law developed by them is known as the *Maxwell-Boltzmann law of the distribution of molecular velocities* (Figs. 106 and 107). Its mathematical treatment is beyond the scope of this text; but, to quote MAXWELL[3] himself, "the distribution of the molecules according to their velocities is found to be of exactly the same mathematical form as the distributions of observations ac-

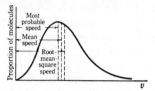

FIG. 106. The MAXWELL-BOLTZMANN distribution for a given temperature. Speeds, v, are represented by the abscissas, and the proportions of molecules having these speeds are represented by the ordinates. The curve has zero ordinate at the origin, and after passing through its maximum it comes down asymptotically to the v-axis. The abscissa value corresponding to the peak of the curve is called the *most probable speed* for the temperature represented. The *arithmetic-mean speed* comes a little farther out on the abscissa, and the *root-mean-square speed* still farther out. It can be shown that these three speeds are in the ratio $0.8165 : 0.9213 : 1$. Calculations show that about 0.4 of all the molecules have speeds that do not exceed the most probable speed. About one molecule in 10^{10} has a speed that is five times the most probable speed, and about one in 2×10^{42} has a speed that is ten times that value

cording to the magnitude of their errors, as described in the theory of errors of observation. The distribution of bullet holes in a target according to their distances from the point aimed at is found to be

[1] *Philosophical Magazine* (4) **19**, 19 (1860); also *The Scientific Papers of James Clerk Maxwell* (Cambridge University Press, 1890), Vol. I, p. 377. The portion of this paper which deals with the distribution of molecular velocities is reproduced in *A Source Book in Physics* (1935), pp. 258–261.

[2] *Wiener Sitzungsberichte* **58**, 517 (1868); **66**, 275 (1872). *Vorlesungen über Gastheorie*, ed. 2 (Barth, 1910), Vol. I, p. 15; *Leçons sur la Théorie des Gaz* (Gauthier-Villars, 1902), p. 15.

[3] *Theory of Heat*, ed. 10 (Longmans, Green, 1891), p. 316. Chapter 22, on molecular theory, and also the first three chapters of the book are especially valuable for beginners.

of the same form, provided a great many shots are fired by persons of the same degree of skill." This introduction of statistical methods into physics is historically a step of the greatest importance and significance. In the hands of MAXWELL, BOLTZMANN, and WILLARD GIBBS[1] (1839–1903), it has led to the development of *statistical mechanics.* Today statistical methods are proving to be a powerful means of attacking many of the most fundamental problems of physical science.

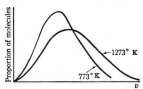

FIG. 107. The MAXWELL-BOLTZMANN distribution for each of two temperatures. The areas under the two curves are equal; that is, both curves refer to the same total number of particles. It will be noticed that even at 773° K an appreciable proportion of the molecules have speeds greater than the most probable speed for 1273° K

120. The Mean Free Path. In Sec. 114 we saw how the kinetic theory leads to the result that the speeds of gas molecules are of the order of magnitude of a kilometer per second. This was very puzzling at first in view of the fact that the time required for the odor from, say, an open bottle of ammonia to travel a few meters across a room free from air currents is a matter of hours rather than fractions of a second. CLAUSIUS saw that the explanation lay in the collisions of the molecules: a molecule of ammonia cannot move very far in one direction without striking a molecule of air, and the succession of unequal zigzag paths which it thus traverses is such that it might take several hours to travel a few meters in any given direction. CLAUSIUS proceeded to deduce a formula for the *mean* distance that a molecule travels between collisions. This distance he called the *mean free path,* and it has come to be an important quantity in molecular theory. Its meaning is illustrated by the following considerations.

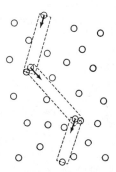

FIG. 108. Method of determining the mean free path of a molecule that has been projected into a group of stationary molecules distributed at random

Assume a molecule to be a sphere of diameter d, and imagine it to be projected into a group of similar molecules which are at rest and distributed at random (Fig. 108). The projected molecule will

---

collide with any other molecule whose center lies within the distance d of the line drawn through the center of the projected molecule in the direction of its motion. In a time t the molecule travels a distance $v_{av}t$, where v_{av} is the arithmetic-mean speed appropriate to the temperature of the gas; and, since the linear dimensions of the molecules are assumed to be negligibly small compared with the distances between them, the number of collisions in the time t is equal to the number of molecules whose centers lie in a cylinder of volume $\pi d^2 v_{av}t$. If n is the number of molecules in unit volume, the number of collisions made in the time t by the projected molecule is therefore $n\pi d^2 v_{av}t$. Since the molecule traverses a total distance $v_{av}t$ in this time, its mean free path λ is given by

$$\frac{1}{n\pi d^2}.$$

It is easy to see that this simple calculation cannot be exact, for it has been made on the assumption that all the molecules but one are at rest. By taking into account the fact that all the molecules are moving, thus increasing the chance of collisions, and also that their speeds are unequal, MAXWELL found the mean free path to be given by the expression

$$\lambda = \frac{1}{n\pi d^2 \sqrt{2}}. \tag{164}$$

Thus the mean free path is inversely proportional to n and therefore to the density ρ of the gas. Obviously, if both λ and n are determined experimentally, Eq. [164] permits the calculation of molecular diameters. If, however, n is not known (and of course in the early days of the kinetic theory it was not), then a second equation connecting n and d must be obtained before either can be calculated.[1]

As a result of extensive experiments on the motions of electrons through gases, the predictions of kinetic theory as to mean free paths and the distribution of molecular velocities have been verified to an accuracy of better than one part in a thousand. In passing, it should be remarked that, according to more modern theory, each atom of a gas molecule consists of a positively charged nucleus surrounded by negative electrons, and it is the forces proceeding from these electrical charges that may be regarded as acting between the molecules to divert them from their rectilinear paths. The deviations become larger the closer the mutual approach of two molecules, and hence the molecular paths, although still zigzag in nature, are without any sharp corners. Moreover, it is now improper to speak of the diam-

---

[1] See Prob. 15 at the end of this chapter.

eter of a molecule as though it were a rigid sphere. The quantity d in Eq. [164] is therefore better defined as *the distance of nearest approach of the centers of two molecules in an impact*; this obviously would equal the diameter of one of the molecules if the molecules were rigid spheres and came into actual contact. The term *diameter of a molecule* is, however, a convenient one, and for this reason we shall continue to use it.

Methods have recently been developed for making direct measurements of mean free paths,[1] but originally definite values for this quantity, and therefore for the diameters of the molecules, had to be obtained indirectly from studies of the phenomena of viscosity, of diffusion, and of the conduction of heat. These phenomena provide three distinct means of evaluating λ, and the comparatively good agreement of the results obtained was regarded as strong evidence for the validity and accuracy of the methods of the kinetic theory (Table V).

TABLE V · *Molecular Radii Calculated from the Kinetic Theory of Gases* [2]

Gas	From the coefficient of viscosity	From the coefficient of diffusion	From conduction of heat	From deviations from BOYLE's law
Hydrogen . .	1.36×10^{-8}	1.36×10^{-8}	1.36×10^{-8}	1.27×10^{-8}
Helium . . .	1.09		1.10	0.99
Nitrogen . .	1.89	1.92	1.89	0.78
Air	1.87	1.87	1.87	1.66
Oxygen . . .	1.81	1.82	1.81	
Carbon dioxide	2.31	2.19	2.41	1.71

121. Viscosity. By *viscosity* is meant the internal friction that arises when contiguous layers of a fluid are in bodily motion relative to one another. Except for the fact that it is a nonconservative force (Sec. 53), so that energy spent in doing work against it is changed into heat, this viscous resistance to fluid motion in no way resembles friction between solids (Sec. 57).

Let Fig. 109 represent the traces of two imaginary parallel planes, Δs apart, described in a fluid that is flowing in parallel layers from

---

[1] For a description of the elegant method afforded by the use of molecular beams, see, for example, R. G. J. Fraser, *Molecular Rays* (Cambridge University Press, 1931), Chap. 2.

[2] Extracted from J. H. Jeans, *The Dynamical Theory of Gases*, ed. 4 (Cambridge University Press, 1925), Chap. 14. The radii are given in centimeters. This table, however, somewhat overdoes the consistency of the observational data.

left to right. Let the layer of fluid in the lower plane be moving with a speed u while that in the upper plane is moving with a larger speed $u + \Delta u$. The quantity $\Delta u/\Delta s$ is evidently the average change in the speed of flow per unit distance in a direction perpendicular to the layers, and the limit du/ds which this approaches as the interval Δs approaches zero is called the *velocity gradient* at a point in this interval.

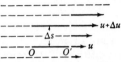

The fluid above the plane OO' in Fig. 109 is flowing more rapidly than the fluid below it and therefore exerts on the latter a tangential force f tending to speed it up; at the same time the fluid above the plane OO' is acted upon by a retarding force of equal

FIG. 109. Velocity gradient in a fluid

magnitude. This force f is evidently proportional to the area A of the plane OO'; and, since it is called into play by the relative motion of the parts of the fluid above and below the plane, it is natural to assume, as did NEWTON, that it is also proportional to the velocity gradient du/ds at OO'. This assumption cannot be justified by direct experiments, but very extensive experimental tests of predictions based upon it afford indirect proof of its validity. It follows that

$$f = \eta A \frac{du}{ds},\qquad [165]$$

where η is a proportionality constant, called the *coefficient of viscosity*. The value of η is found to depend on the nature of the fluid and on its temperature. For gases its value increases with temperature, whereas for most liquids it decreases very rapidly.

It is to be recalled that friction between solids is, within wide limits, independent of the speed with which one body slides over the other (Sec. **57**). Since this is not at all true in the case of viscosity, there must be a fundamental difference in the nature of these two kinds of resistance.

> EXAMPLE. Show that the definition of η given by Eq. [165] is in accord with the following one, which appears in MAXWELL'S *Theory of Heat*: "The coefficient of viscosity of a substance is measured by the tangential force on the unit area of either of two horizontal planes at the unit of distance apart, one of which is fixed, while the other moves with the unit of velocity, the space between them being filled with the viscous substance."

122. Kinetic-Theory Explanation of Viscosity in Gases. In Fig. 110 let OO' again represent the trace of an imaginary plane in a *gas* which is flowing from left to right and let the vectors which are drawn parallel to this plane represent the velocities of flow of the layers *relative* to the velocity of the layer OO'. The molecules above OO' have a greater velocity of flow than those below it. Because they also possess a random heat motion, some of these molecules will cross OO' from the upper to the lower side, and an equal number will pass upward to replace them. Thus the layer above OO' is continually losing the momentum associated with its flow, and that below is continually gaining it. This will tend to stop the relative motion of adjacent layers of the gas; in other words, as a result of this transfer of momentum there exists a tangential force on the plane OO', and it is this which constitutes the viscous drag. Since the energy of flow of the gas as a whole is being changed continually into the energy of random motion of the molecules, the process is accompanied by the appearance of heat.

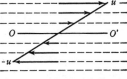

Fig. 110. Theory of the viscosity of a gas

It will be supposed that the bodily speed of flow u is very small in comparison with the random speed v_{av} of the molecules. Owing to the latter the molecules are of course moving in all possible directions, but the effects in the gas are the same as if, in each unit volume, $n/3$ of them were moving with a speed v_{av} normally to OO' while the other two groups of $n/3$ each were moving in mutually perpendicular directions parallel to this plane. Only the first group of $n/3$ molecules per unit volume crosses OO', and of these the number per unit volume moving downward through the plane OO' is $n/6$ and the number moving upward is also $n/6$. The total number of particles that, in unit time, cross the plane from the upper to the lower side is equal to the number moving in this direction that are contained in a column of height v_{av} and cross-sectional area A, where A is the area of the plane; this number is $\frac{1}{6} nv_{av}A$. An equal number are moving upward through the plane in unit time. Now these molecules thus crossing the plane because of their heat motion began their paths at different depths, but on the average they come from a distance from OO' which is equal to the mean free path λ without suffering an encounter on the way. Hence their mean horizontal speed of flow upon arrival at the plane is $\lambda \, du/ds$, where du/ds is the velocity gradient perpendicular to OO'. Therefore each molecule carries with it momentum $m\lambda \, du/ds$, and the total momentum carried over in unit time is $\frac{1}{6} nv_{av}A \cdot m\lambda \, du/ds$. Simultaneously there pass through the plane in the opposite direction $\frac{1}{6} nv_{av}A$ molecules, each of which also comes from an average distance λ from the plane, and the total momentum carried over in this direction is $-\frac{1}{6} nv_{av}A \cdot m\lambda \, du/ds$. Therefore the layer of gas above OO' loses in unit time the momentum

$\frac{1}{3} nv_{av}A \cdot m\lambda \, du/ds$, and this is the tangential force f exerted on the plane. By solving for f/A and substituting in Eq. [165], one obtains finally

$$\eta = \frac{1}{3} \rho v_{av} \lambda, \tag{166}$$

since nm is the density ρ of the gas.

A more exact calculation,[1] made by BOLTZMANN and by TAIT on the basis of the MAXWELL-BOLTZMANN law of distribution of molecular velocities, gives

$$\eta = 0.35 \, \rho v_{av} \lambda, \tag{167}$$

where v_{av} is the arithmetic-mean speed of the molecules. The constant appearing in this equation is not accurately known, however.

Since λ is inversely proportional to ρ (Sec. 120), it follows that $\rho\lambda$ in Eq. [167] is independent of the density, and hence that the coefficient of viscosity η for a given gas is independent of the pressure to which the gas is subjected. Thus, concluded MAXWELL, the oscillations of a torsion-pendulum in a gas should be equally damped by gaseous friction, however low the pressure might be. This conclusion was so contrary to the general opinion of physicists at the time that its experimental confirmation, which took place soon afterward, formed one of the most brilliant successes of the kinetic theory in its earlier history.[2] Later experiments have shown that at very low pressures, when the mean free path is comparable with the dimensions of the vessel enclosing the gas, the viscosity is no longer independent of pressure. That there is also an upper pressure limit beyond which the law no longer holds may also be inferred from the method used in deducing the law.

It is left to the student to show that Eq. [167] also leads to the conclusion that the coefficient of viscosity of a gas increases with rising temperature in linear ratio with the square root of the absolute temperature. Actually, experiment shows that η increases more rapidly than would be the case if this law held.[3]

123. Diffusion in a Gas. The phenomenon of gaseous diffusion can be treated in a manner that is very similar to that employed with viscosity, except that we no longer investigate the transfer of mo-

---

[1] See O. E. Meyer, *The Kinetic Theory of Gases*, tr. by R. E. Baynes (Longmans, Green, 1899), p. 444.

[2] See footnotes 1 and 2, p. 225.

[3] This result is predicted by classical kinetic theory when account is taken of the forces of attraction between the molecules, which will pull in a slowly moving molecule but not a fast one [for example, see W. Sutherland, *Philosophical Magazine* (5) **36**, 507 (1893)]. It has become evident, however, that to make the most progress in accounting for the many new results of this type revealed by present experimental methods it is necessary to replace classical theory by the more fundamental methods of attack furnished by modern quantum theory.

mentum across the plane OO', Fig. 110, but rather the transfer of mass. It is a matter of experience that, if the density of a gas is not uniform, diffusion will take place from regions of greater density to those of lesser density, and will not cease until the density is everywhere the same. It follows that the time-rate of flow of the diffusing gas at any point in any direction must depend on the *density gradient* at that point and in that direction. In the case of steady diffusion in a direction perpendicular to the plane OO', it is evident that the total mass which diffuses in unit time, dM/dt, is proportional to the area A of the plane OO', and it is simplest to assume that it is also proportional to the density gradient, $-d\rho/ds$, at OO'; that is,

$$\frac{dM}{dt} = -\mu A \frac{d\rho}{ds},$$ [168]

where μ is a constant of proportionality, called the *coefficient of diffusion*.

By means of reasoning similar to that employed in obtaining the kinetic interpretation of viscosity (Sec. 122), it is possible to show that $dM/dt = -\frac{1}{3} v_{av} A m \lambda \, dn/ds$, where n is the number of molecules in a unit volume of the gas, and hence that

$$\mu = \frac{1}{3} v_{av} \lambda.$$ [169]

As in the case of Eq. [166], this result is not based on an exact calculation.

It follows from Eq. [169] that the coefficient of diffusion, unlike the coefficient of viscosity, is dependent upon the pressure of the gas. Also, at a given pressure, μ varies as $T^{\frac{3}{2}}$, a result which experiments with actual gases show to be inexact.

124. High Vacuum. Whenever the domain of experiment is extended, unexpected and often startling new facts are likely to come to light. There is no better illustration of this important truism than that offered by the extension of the experimental domain to regions of very low pressures, an advance which led to·the discovery of the electron, of x-rays, and of so many other new phenomena that at the present time experimental methods involving high-vacuum technique predominate in most physical laboratories.

The study of gases at reduced pressures has presented kinetic theory with many new and difficult problems, for it has been found that all the laws describing the behavior of gases in motion change completely at very low pressures. Indeed, we have already learned how the coefficient of viscosity varies markedly at low pressures, although ordinarily it is independent of the pressure (Sec. 122), and it is found that this is also true of the thermal conductivity of a gas.

THE five years 1860–1864, during which MAXWELL was Professor of Natural Philosophy in King's College, London, were perhaps the most fertile in his career. It was during this period that he produced his Bakerian lecture on "The Viscosity of Air at Different Temperatures and Pressures" and two other important papers on the kinetic theory of gases.

MAXWELL's experiments on the viscosity of gases were carried out in the large garret of his residence (8 Palace Gardens Terrace, Kensington). In order to raise the temperature of the room, great quantities of water were boiled in open kettles over a large fire, the exhausting work of acting as stoker being performed by Mrs. Maxwell. Later the garret was cooled by bringing in great cakes of ice.

The following amusing incident of MAXWELL's stay in London is related by Lewis Campbell in *The Life of James Clerk Maxwell*:

"On one occasion he [Maxwell] was wedged in a crowd attempting to escape from the lecture theatre of the Royal Institution, when he was perceived by Faraday, who, alluding to Maxwell's work among the molecules, accosted him in this wise — 'Ho, Maxwell, cannot you get out? If any man can find his way through a crowd it should be you.'"

BOLTZMANN compared MAXWELL's memoir "On the Dynamical Theory of Gases" to a musical drama as follows:

"A mathematician will recognize Cauchy, Gauss, Jacobi, Helmholtz, after reading a few pages, just as musicians recognize, after the first few bars, Mozart, Beethoven or Schubert. Perfect elegance of expression belongs to the French, though it is occasionally combined with some weakness in the construction of the conclusions; the greatest dramatic vigour to the English, and above all to Maxwell. Who does not know his *Dynamical Theory of Gases*? At first the Variations of the Velocities are developed majestically, then from one side enter the Equations of State, from the other the Equations of Motion in a Central Field; ever higher sweeps the chaos of Formulae; suddenly are heard the four words: 'put $n = 5$.' The evil spirit V (the relative velocity of two molecules) vanishes and the dominating figure in the bass is suddenly silent; that which had seemed insuperable being overcome as by a magic stroke. There is no time to say why this or why that substitution was made; who cannot sense this should lay the book aside, for Maxwell is no writer of program music who is obliged to set the explanation over the score. Result after result is given by the pliant formulae till, as unexpected climax, comes the Heat Equilibrium of a heavy gas; the curtain then drops."

◊

Some Classical Apparatus
at the Cavendish Laboratory,
Cambridge, England

MAXWELL's apparatus for measuring the viscosity of air at different pressures is shown on the right. A phonetic wheel of RAYLEIGH's stands to the left on the table. The coils shown are those used in some of the classical determinations of the ohm.

The physical reason for these anomalies is that in a high vacuum a gas molecule makes many more impacts with the solid walls of the containing vessel than with other gas molecules, whereas at ordinary pressures the impacts of a molecule with the walls are very infrequent compared with its impacts with other molecules. Evidently the physical criterion of whether a gas is to be described by the laws holding for low pressures or by those holding for ordinary pressures is given by the length of the mean free path as compared with the linear dimensions of the vessel in which the gas is contained.

Although the first air pumps were made about 1650 by OTTO VON GUERICKE,[1] the greatest progress in high-vacuum technique has come during the last twenty years with the investigations of electronic emission and the application of electronic devices in radio and other fields. Modern vacuum pumps may be divided into two classes: those which produce a vacuum as low as about 10^{-3} mm of mercury and which are usually called *fore pumps* because they are widely used to produce the preliminary vacuum required before pumps of the second class can come into operation; those which produce a vacuum as low as 10^{-7} mm of mercury and which are referred to as *high-vacuum* pumps. Many different types of fore pumps have been developed, but most of the modern ones are rotary pumps which work immersed in oil. The high-vacuum pumps now in use are modifications of the GAEDE diffusion pump and the LANGMUIR condensation pump, both of which operate with no moving parts larger than gas molecules.[2] In the form of high-vacuum condensation pump shown in Fig. 111, mercury is heated in the boiler b, and the stream of evaporated mercury passes down through the nozzle O, is condensed into drops in a tube surrounded by a water jacket, and then returns through f

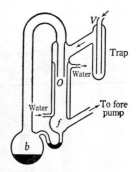

FIG. 111. The Langmuir mercury-vapor condensation pump. Original single-stage glass form

---

[1] *Experimenta Nova (ut vocantur) Magdeburgica de Vacuo Spatio* (Amsterdam, 1672). There is a German translation of Book III in Ostwald's *Klassiker der Exakten Wissenschaften*, No. 59 (Engelmann, 1894). The first published account of VON GUERICKE'S air pump, however, is by K. Schott in his *Mechanica Hydraulico-Pneumatica* (1657); it was probably this work that stimulated BOYLE to employ HOOKE to construct the pump described in *New Experiments Physico-Mechanicall* (Oxford, 1660)

[2] For detailed descriptions of various pumps and the methods of producing and measuring very low pressures, consult S. Dushman, *Production and Measurement of High Vacuum*; L. Dunoyer, *Vacuum Practice*; F. H. Newman, *Production and Measurement of Low Pressures*; G. W. C. Kaye, *High Vacua*. For good, brief accounts of vacuum technique, see *E. N. da C. Andrade, article "Vacuum," *Encyclopaedia Britannica*, ed. 14, and *S. Dushman, *Journal of the Franklin Institute* 211, 689 (1931).

to the boiler. As the high-speed mercury particles rush by O, they collide with the gas molecules diffusing in through V from the system to be exhausted and impart to them a velocity component. The gas thus removed from the vessel collects in f at a pressure much below that of the atmosphere, though this pressure is still relatively high, and from there a fore pump removes it. For rapid work, and when it is desirable to operate with a relatively high fore vacuum, use is often made of two-stage and three-stage mercury-vapor pumps, in which the stages are arranged in series.

In order that mercury vapor from the pump may not reach the exhausted system, some form of trap is inserted between the two (Fig. 111). The trap may be cooled by liquid air or solid carbon dioxide, or it may be coated on the inside with an alkali metal, which strongly absorbs the mercury vapor. Recently condensation pumps have been constructed in which the mercury is replaced by one of the phthalates, for example butyl benzyl phthalate; with such a pump it is not necessary to surround the condenser with a water jacket, and, for many purposes, the refrigerant trap can also be eliminated. To obtain the very lowest pressures, a liquid air trap, which may contain activated charcoal, must be used to remove condensable vapors. The glass parts of the whole apparatus must be subjected, during evacuation, to prolonged baking at as high a temperature as the glass will stand without collapsing; this removes water and other vapors from the walls. If metal parts are present, the gases must be eliminated from them by high-frequency heating during exhaust or by some other convenient method. If an apparatus is to be sealed off from the pumps, a small amount of some substance known as a *getter* is introduced into it; after the seal-off the getter removes residual gases either by chemical action or by adsorption. It should be pointed out that the term *adsorption* is not to be confused with *absorption*. When gases condense on a surface, the phenomenon is referred to as *adsorption*. If they actually enter the substance as a solute, the process is known as *absorption*. Gases may also, of course, combine chemically with a substance.

For measuring pressures down to 10^{-4} mm of mercury the McLeod gauge (Fig. 95) is the standard instrument. The accuracy of this type of gauge depends upon the closeness with which the gas in question is described by BOYLE'S law. (Why?) It has the great defect, as we shall presently see, of not indicating the pressures of condensable gases, such as water. In order to be able to measure very low pressures, special gauges have been devised which depend for their action on the anomalous behavior of gases at low pressures. Four different types of gauges, which depend respectively on viscosity, thermal conduction, ionization, and the radiometer effect will be mentioned. *Viscosity gauges* make use of the law that at low pressures the viscosity is proportional to the pressure. The *conductivity*, or *Pirani-Hale*, gauge is based on the similar relation existing between thermal conductivity and pressure; this gauge is much like an electric lamp, the voltage required to keep the filament at constant temperature being a measure of the change of the heat conductivity of the gas with pressure. The third type of gauge, the *ionization gauge*, is practically a thermionic valve; it utilizes the law that at low pressure the ionization produced in a gas by a definite electron current is proportional to the pressure. The fourth type, the *Knudsen absolute manometer*, is unique in that its readings do not depend on the nature of the gas; it is based on the so-called radiometer effect, or mechanical repulsion between two surfaces of unequal temperature when gases at low pressure lie between them, a phenomenon that is independent of the atomic weight of the gas.

o

EXPERIMENT XA. VAPOR DENSITY AND MOLECULAR WEIGHT

The object is to determine the ratio of the densities of carbon tetrachloride [1] and air by the method developed by J. B. A. DUMAS,[2] and thus to determine the molecular weight of carbon tetrachloride. In DUMAS's method for determining vapor densities, a glass bulb of known volume V is weighed, first when full of air at pressure P_1, temperature T_1, and then when full of vapor at pressure P_2, temperature T_2. In these weighings a closed bulb of the same volume as the density bulb is used as a counterpoise. (Why?) If the difference in mass, as determined by the difference between the first and second weighings, be represented by ΔM, this quantity being of course negative if the second weight exceeds the first, then evidently $V\rho_{a_1} - V\rho_{v_2} = \Delta M$, or

$$\frac{\rho_{v_2}}{\rho_{a_1}} = 1 - \frac{\Delta M}{V\rho_{a_1}}, \qquad [170]$$

where ρ_{a_1} is the density of air at P_1, T_1, and ρ_{v_2} is the density of the vapor at P_2, T_2. In Eq. [170] the expansion of the bulb is neglected, because in the determination that follows it will not affect the result by more than a small fraction of 1 percent.

Now, the quantity that is sought in this experiment is the density of the vapor in terms of the density of air at the same temperature and pressure, namely, the ratio ρ_{v_2}/ρ_{a_2}. This means that Eq. [170] must be modified by taking into account the equation that expresses the relation between the densities of air at P_1, T_1, and at P_2, T_2; namely,

$$\frac{\rho_{a_1}}{\rho_{a_2}} = \frac{P_1}{P_2} \cdot \frac{T_2}{T_1}. \qquad [171]$$

By combining this equation with Eq. [170], the student can easily show that the required ratio ρ_{v_2}/ρ_{a_2} is given by

$$\frac{\rho_{v_2}}{\rho_{a_2}} = \left(1 - \frac{\Delta M}{V\rho_{a_1}}\right)\frac{P_1}{P_2} \cdot \frac{T_2}{T_1}. \qquad [172]$$

---

[1] Instead of carbon tetrachloride, the student may find it desirable to determine the relative vapor density of water-free alcohol or of ether.

[2] *Annales de Chimie et de Physique* (2) **33**, 337 (1826).

All the quantities in the right-hand member of this equation, excepting ρ_{a_1}, are measured directly in the experiment. The quantity ρ_{a_1} can be obtained either from Eq. [171] and the result of Exp. VIIIA or from tables.[1]

The accepted value of the density of air relative to oxygen is 0.90469, and the molecular weight of oxygen is 32. Hence, if the ratio ρ_{v_2}/ρ_{a_2}, as determined by Eq. [172], be multiplied by $0.90469 \cdot 32$, the density of the vapor is obtained in terms of oxygen as 32. This is the quantity which, according to the law of AVOGADRO (Sec. 112), should agree with the molecular weight of the vapor.

Technique and Measurements. Tables show that the maximum pressure exerted by carbon tetrachloride vapor at ordinary temperatures is less than atmospheric pressure. Consequently it is impossible, under ordinary atmospheric conditions, wholly to replace the air in the density bulb by the vapor. This can be done easily, however, at any temperature at which the maximum pressure of the carbon tetrachloride vapor is more than atmospheric. Hence use is made of the following method, which was employed by DUMAS as early as 1826. Since parts of this method do not differ in principle and technique from Exp. VIIIA, the latter experiment should be reviewed at this time.

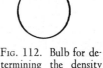

FIG. 112. Bulb for determining the density of a vapor by the method of DUMAS

a. The density bulb is a thin glass flask, furnished with a narrow stem which has been drawn to a capillary tip *o*, Fig. 112. Carefully dry the bulb by repeatedly warming and exhausting it through a calcium chloride tube. Then weigh the bulb upon an analytical balance, making use of the counterpoise and following the directions for the first weighing given in Exp. VIIIA. Read the barometer height P_1 and the temperature T_1 of the air near the bulb.

b. Introduce into the bulb about 10 ml of pure carbon tetrachloride by heating the bulb slightly and then letting it cool with its tip beneath the liquid. Heat a water bath to about 75° C, a temperature which is below the boiling point of carbon tetrachloride, and immerse the bulb in it with about 2 cm of the stem protruding. Support the bulb in this position by means of a three-pronged wire holder. Cover the bath, heat the water to boiling, and boil it steadily until the carbon tetrachloride vapor no longer escapes, as shown by its failure to deflect a flame applied at the tip. With the water boiling steadily,

---

[1] Appendix 13, Table D.

gently heat the exposed stem by means of a second burner in order to vaporize completely any liquid carbon tetrachloride that may have condensed in it. Then, with the water in the bath boiling continuously, so that no air will be forced into the bulb, quickly seal the tip; this is best done by applying a fine blowpipe flame to the stem just above the surface of the bath, heating to softness, and then drawing off the tip. Carefully preserve this drawn-off tip.

Remove the bulb from the bath, dry it thoroughly, and test for a leak by allowing the condensed carbon tetrachloride in the bulb to run down to the tip of the stem and observing whether or not fine bubbles enter the bulb. After the bulb has come to the temperature of the balance case, weigh it together with the drawn-off tip.

Read the barometer to obtain P_2, the pressure of the vapor in the bulb at the time of sealing the stem. Determine T_2, the temperature of the vapor at the time of sealing the stem, by finding from tables the boiling point of water for the observed barometric pressure.[1]

c. To obtain the volume of the bulb, weigh it upon trip scales, then crush the capillary tip with a pair of cutting pliers under water, and weigh the bulb full of water upon the trip scales; obtain the density of the water from tables [2] and compute V. If the bulb does not completely fill with water, the filling may be completed with the help of a pipette; it is true that Eq. [172] is not then rigorously correct, but unless the bubble is quite large the error introduced will be negligible.

d. Compute the ratio ρ_{v_2}/ρ_{a_2} and, from this, the density of the vapor in terms of oxygen as 32. Compare the latter value with that of the molecular weight of carbon tetrachloride as computed from the formula CCl_4.

1. State as clearly and concisely as possible the conclusions which you are able to draw from this experiment.

2. Is it reasonable to suppose that the percentage difference of your determination of the molecular weight from that corresponding to the formula CCl_4 is due mostly to deviation of the vapor from the laws of ideal gases?

3. How could Eq. [170] be modified so as to take into account the expansion of the density bulb? Can you justify the statement that neglecting the expansion of the bulb will not affect the final result by more than a small fraction of 1 percent?

4. Estimate the percentage error in the vapor density caused by leaving in the stem of the bulb 0.01 g of liquid carbon tetrachloride at the time of sealing.

---

[1] Appendix 13, Table E.   [2] Appendix 13, Table B.

EXPERIMENT XB. THE COEFFICIENT OF VISCOSITY OF AIR

The classical experiments of O. E. MEYER [1] and of MAXWELL [2] on gaseous viscosity were made by observing the damping of the torsional oscillations of a disk suspended in the gas. This method involves serious difficulties of a mathematical nature, and in recent work methods have been employed that are based on observations of the flow of a gas through a channel or tube. The idea of using tubes is not new, however, for THOMAS GRAHAM [3] used a capillary-flow method in the first experiments ever made on the viscous resistance of gases. In the present experiment on air we will employ a modification of a simple form of capillary-flow viscometer devised by A. O. RANKINE. [4] The procedure consists essentially in measuring the volume of air that is forced through the capillary tube in unit time owing to a difference in pressure of known amount between the two ends of the tube.

Experiments show that when a fluid moves at a constant rate through a narrow tube, the lines of flow are parallel to the axis of the tube, provided a certain critical speed of flow is not exceeded. Moreover, if one imagines a very large number of cylinders to be described in the fluid about the axis of the tube, the motion is found to consist in the slipping of cylinder through cylinder, like the slipping of the tubes of a pocket telescope through one another.

Let us first consider the steady flow of a *liquid* through a narrow tube; this case is simpler to treat than that of a gas, since liquids are almost incompressible, and it is also a case of great practical importance. If an *imaginary cylinder* of radius r be described about the axis of the tube, the velocity gradient du/dr (Sec. 121) obviously will be the same at every point of this cylinder, and hence the viscous drag exerted on the liquid inside the imaginary cylinder by the slower-moving liquid outside of it is given by

$$f = \eta \cdot 2 \pi r l \cdot \frac{du}{dr}, \qquad [173]$$

---

[1] Poggendorff's *Annalen der Physik und Chemie* **113**, 55, 193, 383 (1861); **125**, 177, 401, 564 (1865); **143**, 14 (1871). See also Meyer's *The Kinetic Theory of Gases*, tr. by R. E. Baynes (Longmans, Green, 1899), pp. 181–188; this book, though old, remains one of the best treatments of the kinetic theory ever written.

[2] *Proceedings of the Royal Society* **15**, 14 (1866); *Philosophical Transactions* **154**, 249 (1866); also *The Scientific Papers of James Clerk Maxwell* (Cambridge University Press, 1890), Vol. II, p. 1.

[3] *Philosophical Transactions* **136**, 573 (1846); **139**, 349 (1849).

[4] *Proceedings of the Royal Society* **83**, 265 (1910).

where l is the length of the tube. This result is derived directly from Eq. [165] by supposing that the cylindrical surface is made up of a large number of plane strips.

The pressure is greater where the liquid enters the tube than where it leaves the tube by some amount ΔP, and thus the liquid within the imaginary cylinder is subject to an accelerating force of magnitude $\pi r^2 \Delta P$. But since the flow is assumed to be steady, the total force acting upon the liquid in the cylinder must be zero, or

$$\eta \cdot 2 \pi r l \cdot \frac{du}{dr} = \pi r^2 \Delta P$$

By solving this equation for du and integrating, we obtain

$$u = \frac{\Delta P}{4 \eta l} r^2 + k, \qquad [174]$$

where k is the constant of integration. Now there is weighty evidence in favor of the assumption that very little or no slip occurs between the walls of the tube and the liquid immediately in contact with them. In other words, if R be the radius of the tube, it may be assumed that $u = 0$ when $r = R$, and Eq. [174] becomes

$$u = -\frac{\Delta P}{4 \eta l} (R^2 - r^2). \qquad [175]$$

This equation gives the speed of the liquid at a distance r from the axis of the tube.

It now becomes necessary to obtain the expression for V/t, the volume of liquid that crosses any section of the tube in unit time. Consider two of the imaginary cylinders which are of nearly equal radii, r and $r + dr$, and which therefore cut off an annular strip of area $2 \pi r \, dr$ from any cross section of the tube. The volume of liquid that crosses this strip in unit time, namely, the differential of V/t, is evidently

$$d\left(\frac{V}{t}\right) = u \cdot 2 \pi r \, dr = -\frac{\pi \Delta P}{2 \eta l} r(R^2 - r^2) \, dr;$$

therefore

$$\frac{V}{t} = -\frac{\pi \Delta P}{2 \eta l} \int_0^R r(R^2 - r^2) \, dr,$$

or

$$\frac{V}{t} = -\frac{\pi R^4}{8 \eta} \frac{\Delta P}{l}. \qquad [176]$$

This is the well-known formula that was developed by J. L. M. POISEUILLE,[1] an anatomist who was interested in the physics of

---

[1] *Annales de Chimie et de Physique* (3) **7**, 50 (1843). This is "one of the classics of experimental science and is frequently quoted as a model of careful analysis of sources of error and painstaking investigation of the effects of separate variables" (G. Barr). See Bibliography, p. 452, for complete references.

blood circulation. It is used in determining the coefficient of viscosity of a liquid flowing through a capillary tube.

POISEUILLE'S formula is not directly applicable to a gas, for a gas is highly compressible and therefore changes in density as it passes along the tube from places of higher pressure to those of lower pressure. However, in a very short element

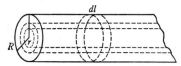

FIG. 113. Flow of a gas through a tube

of length dl of the tube (Fig. 113), between the ends of which the pressure difference dP is infinitesimally small, the density may be considered to be constant, and Eq. [176] then applies; thus

$$\frac{V}{t} = -\frac{\pi R^4}{8\eta}\frac{dP}{dl},\qquad [177]$$

where V/t is now the volume that flows in unit time through the element in question. If both members of this equation be multiplied by the density of the gas in the element, the resulting expression gives the mass of gas that flows through the element in unit time, and this quantity is the same for all elements of the tube. Let ρ' be the density of the gas under unit pressure and let P be the pressure in the element in question. Then, if it be assumed that BOYLE'S law applies, the density in the element is $\rho'P$, and the mass that flows through the element in unit time is

$$-\rho'P\frac{\pi R^4}{8\eta}\frac{dP}{dl}.$$

This mass is equal to that which enters the tube in unit time; hence, if P_1 be the pressure at the entrance to the tube and if V_1/t be the volume which enters in unit time, then

$$\rho'P_1\frac{V_1}{t} = -\rho'P\frac{\pi R^4}{8\eta}\frac{dP}{dl}.\qquad [178]$$

By multiplying both members of this equation by dl and integrating, we obtain

$$P_1\frac{V_1}{t}\int_0^l dl = -\frac{\pi R^4}{8\eta}\int_{P_1}^{P_2}P\,dP,\qquad [179]$$

where P_2 is the pressure at the end of the tube from which the gas emerges. If we carry out the indicated integration and then solve for η, there results finally

$$\eta = \frac{\pi R^4}{16\,l}\frac{(P_1{}^2 - P_2{}^2)t}{P_1V_1}.\qquad [180]$$

The Experiment. The U-tube of the viscometer (Fig. 114) is a glass tube of not more than 3.5-mm inside diameter, the two limbs being approximately 35 cm and 10 cm in length. To the shorter limb is attached, by means of rubber tubing, a piece of clean thermometer tubing of not more than 0.20-mm internal diameter and about 20-cm length. *Cleanliness is essential.* See to it that the tubes, beakers, and other glassware used are clean and use only the cleanest mercury. If the capillary tube is not clean, force through it, by means of compressed air, first an aqueous solution of potassium hydroxide, then distilled water, and finally a thread of mercury. *Do not touch the ends of the tube, as there is danger of clogging it.*

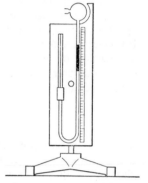

a. Rotate the apparatus *clockwise* into a horizontal position and introduce enough mercury into the bulb at the end of the long limb of the U-tube to form a pellet about 8 cm long.

Fig. 114. Rankine's simple capillary-flow viscometer

When this pellet is allowed to slide down the U-tube, it furnishes the pressure needed to force the air out through the capillary tube. Rotate the apparatus *counterclockwise* back into a vertical position, and observe the time required for the pellet to traverse some measured distance in the fall tube.

Invert the apparatus by turning it again *clockwise* and observe the time the pellet takes to return a measured distance.

Repeat these operations several times and compute the average times of fall for the erect and the inverted position.

Measure the length of the mercury pellet on the scale provided on the apparatus. Observe the barometric pressure and the temperature of the room.

b. Repeat the foregoing procedure with at least two other shorter lengths of mercury pellets, each obtained by removing some of the mercury from the tube.

c. Determine the length l and the radius R of the capillary tube. If the value of R is not marked on the tube, determine it by weighing on an analytical balance, to two significant figures at least, the thread of pure mercury of known density that fills a measured length of the tube; to obtain R accurately enough for the purposes of this experiment is evidently a very difficult matter.

d. Find the cross-sectional area of the fall tube by weighing the mercury pellet that fills a measured length of the tube.

e. Calculate the other quantities that appear in Eq. [180], namely, V_1, P_1, and P_2. The volume V_1 is obtained by multiplying the distance through which the pellet slid down the fall tube by the cross-sectional area of this tube. For the case in which the air is forced out through the capillary tube the pressure P_2 is the barometric pressure expressed in dynes per square centimeter. The pressure P_1 is obtained by adding to P_2 the pressure due to the pellet of mercury; the latter can be calculated from a knowledge of the length of the pellet, the density of mercury, and the acceleration due to gravity. (What are P_1 and P_2 for the case in which the air is drawn in through the capillary tube?)

By means of Eq. [180] calculate, in cgs units, the values of η corresponding to the six or more sets of data which you have obtained. Remember that the pressures P_1 and P_2 must also be expressed in cgs units. If two or more of the determinations were made under the same or nearly the same conditions of temperature, compute the average value of η.

1. Convert Eq. [180] into a laboratory equation, that is, into a form such that all measured and observed quantities can be substituted directly in the formula.

2. List all of the assumptions made in deriving Eq. [180] and make clear at what point each enters the derivation.

3. How much, approximately, would the temperature of the room have to change before a difference in η would be produced that could be detected with this apparatus?

4. Since to some extent the mercury sticks to the walls of the fall tube, the pressure furnished by the pellet is diminished by some amount σ; consequently the value of η given by Eq. [180] is a little too high. Assuming σ to be constant during a given run, show how the equation should be modified so as to take this effect into account.

5. Explain how the value of σ in question 4 can be determined experimentally. If the instructor so directs, correct your value of η for the sticking of the mercury pellet.

6. Employ the data of the present experiment and of Exp. VIIIᴀ to find the root-mean-square speed, the arithmetic-mean speed, and the mean free path of air molecules under standard conditions of temperature and pressure.

○

QUESTION SUMMARY

1. Contrast the ancient and modern views of the structure of matter. What is the essential difference between them? What date would you assign as the beginning of the modern period?

2. State the law of definite proportions, the principle of equivalence, and the law of multiple proportions.

3. State AVOGADRO'S law. What is its experimental basis?

4. What are the two possible explanations of the elastic properties of gases? What experiments and arguments finally settled the question?

5. Describe briefly the kinetic-molecular theory of gases, giving precisely the two fundamental hypotheses upon which it rests.

6. Deduce the expression for the pressure exerted by a gas in terms of the mass and velocity of the molecules. Point out clearly what simplifying assumptions are involved in the derivation. Are all of them necessary?

7. What was the first molecular magnitude to be determined quantitatively? How was it calculated, and what was its approximate value?

8. From the kinetic-theory point of view, what is the physical meaning of *temperature*? Give a quantitative relation connecting the average kinetic energy of the molecules and the absolute temperature.

9. What is meant by the *mean free path*? How is it related to the molecular diameter?

10. What was the second molecular magnitude to be determined quantitatively? How was it determined?

11. Define *coefficient of viscosity*. What connection exists between the coefficient of viscosity of a gas and the mean free path? What is the effect of pressure on gaseous viscosity? of temperature?

12. Define *coefficient of diffusion*. How is it related to the mean free path?

13. How would one go about determining from kinetic-theory data alone the effective diameter of a gas molecule and the number of molecules in a given volume?

○

PROBLEMS

1. Given standard conditions of temperature and pressure, calculate the volumes occupied by (*a*) 2.016 g of hydrogen; (*b*) 32.00 g of oxygen; and (*c*) 30.01 g of nitric oxide. Explain the connection between the results.

2. A spherical body of mass 2.00 mg is moving to and fro between two parallel walls which are 10.0 cm apart. If the speed of the body is 2.00×10^3 cm · sec^{-1} and its impacts with the walls are perfectly elastic, what force is needed to keep the walls from moving under the influence of the impacts, (*a*) when the diameter of the body is negligibly small, and (*b*) when the diameter is 4 mm?   *Ans.* (*a*) 800 dynes; (*b*) 833 dynes.

3. Suppose that the actual speeds of 10 molecules are found to be 1, 2, 3, 4, ···, 10 cm · sec^{-1} respectively. (*a*) Compute their root-mean-square speed. (*b*) Compute their arithmetic-mean speed. (*c*) If the mass of each molecule is *m*, what is its average kinetic energy? (*d*) Explain why the root-

mean-square speed of gas molecules will always be greater than the arithmetic-mean speed.

Ans. (*a*) 6.2 cm · sec⁻¹; (*b*) 5.5 cm · sec⁻¹; (*c*) 19 *m* ergs.

4. 1.0 g each of nitrogen and of helium are introduced together into an
evacuated bulb of radius 10 cm. Find the pressure which this mixture of
gases exerts on the walls of the bulb when the temperature is 25° C.

Ans. 1.3 × 10³ mm.

5. The density of dry air at 0° C and 1 A$_s$ is 1.2930 g · l⁻¹ and that of
oxygen under the same conditions is 1.4290 g · l⁻¹. Assuming that air is
composed entirely of nitrogen and oxygen, find the percentages by volume
of these gases in dry air.   *Ans.* 76.2 percent; 23.8 percent.

6. One cubic centimeter of a certain mixture of nitrogen and carbon
dioxide at 0° C contains 2.0 × 10²⁰ molecules of nitrogen and 5.0 × 10¹⁶
molecules of carbon dioxide. Calculate the partial pressure due to each gas
and also the total pressure.   *Ans.* 7.4 A$_s$; 0.0019 A$_s$; 7.4 A$_s$.

7. Calculate the value of the BOLTZMANN constant.

Ans. 1.37 × 10⁻¹⁶ erg · deg⁻¹ · molecule⁻¹.

8. (*a*) What is the average kinetic energy of translation of a molecule of
any ideal gas that is at 300° K? (*b*) Calculate the *total* energy of linear motion of all the molecules in 1 l of ideal gas at 0° C and 1 A$_s$.

Ans. (*a*) 6.17 × 10⁻¹⁴ erg · molecule⁻¹; (*b*) 1.52 × 10⁹ ergs.

9. (*a*) Given that the density of nitrogen at 0° C and 1 A$_s$ is 1.2505 g · l⁻¹,
find the mean-square speed of nitrogen molecules at this temperature.
(*b*) From the AVOGADRO number and the molecular weight of nitrogen, find
the mass of a nitrogen molecule. (*c*) Without using any additional data,
calculate the BOLTZMANN constant.

Ans. (*a*) 24.3 × 10⁸ cm² · sec⁻²; (*b*) 4.63 × 10⁻²³ g;
(*c*) 1.37 × 10⁻¹⁶ erg · deg⁻¹ · molecule⁻¹.

10. (*a*) Assuming the validity of the law of equipartition of energy, calculate the root-mean-square speed of a molecule of oxygen at 0° C, if the
root-mean-square speed of a molecule of hydrogen at this temperature is
1.839 × 10⁵ cm · sec⁻¹. (*b*) When the temperature is 364° K, what is the
arithmetic-mean speed and what is the most probable speed of the hydrogen
molecules?

Ans. (*a*) 461 m · sec⁻¹; (*b*) 1.96 × 10⁵ cm · sec⁻¹, 1.74 × 10⁵ cm · sec⁻¹.

11. Two trains are moving in the same direction on parallel tracks with
speeds of 10 mi · hr⁻¹ and 40 mi · hr⁻¹ respectively. If 30 sec are required
for one train to pass the other completely and if during this time 2.0 tons of
bricks are thrown from each train to the other, what average force, in addition to that necessary to overcome friction, is required to keep each train
moving at its original speed?   *Ans.* ± 1.8 × 10² lbwt.

12. Given that the coefficient of viscosity of hydrogen at 0° C and 1 A$_s$
is 8.67 × 10⁻⁵ g · cm⁻¹ · sec⁻¹, calculate (*a*) the mean free path for hydrogen
at 0° C and 1 A$_s$; (*b*) the number of collisions per second suffered by a hy-

drogen molecule at $0°$ C and 1 A$_s$;   (c) the diameter of a hydrogen molecule. (d) How does this diameter compare with that of the helium molecule, for which λ is 1.72×10^{-5} cm at $0°$ C and 1 A$_s$?

> *Ans.*   (a) 1.6×10^{-5} cm;   (b) 10^{10} sec^{-1};   (c) 2.3×10^{-8} cm;   (d) 2 percent larger.

13. (a) By employing the most refined high-vacuum technique, the air pressure in a vessel can be reduced to about 10^{-8} mm of mercury; how many gas molecules are there in 1 cm^3 at this pressure?   (b) How many are there when the pressure is 10^{-6} mm of mercury, which is the highest vacuum ordinarily used in commercial work?

> *Ans.*   (a) 4×10^8 cm^{-3};   (b) 4×10^{10} cm^{-3}.

14. (a) The mean free path in oxygen under ordinary conditions of pressure and temperature is about 10^{-5} cm. Find from this its approximate value for a pressure of 10^{-3} micron of mercury (1 micron $= 10^{-6}$ m). (b) Obtain information on the sizes of the finest capillary tubes used in gas manipulations and decide whether the flow of oxygen at atmospheric pressure through them is described by the laws holding for ordinary pressures.   (c) In the case of the connecting tubes ordinarily used in high-vacuum apparatus, how low would the pressure have to be before the laws for ordinary pressures would cease to apply?   *Ans.*   (a) 8×10^3 cm.

15. The values of the molecular diameter d and of the number of molecules n_o in 1 cm^3 of a gas at $0°$ C and 1 A$_s$ were first calculated by JOSEPH LOSCHMIDT [1] in 1865. Try making the calculation for hydrogen, given that the density of gaseous hydrogen at $0°$ C and 1 A$_s$ is 9×10^{-5} g · cm^{-3}, the density of liquid hydrogen is 0.07 g · cm^{-3}, and the coefficient of viscosity of hydrogen gas at $0°$ C and 1 A$_s$ is 8.67×10^{-5} g · cm^{-1} · sec^{-1}. [*Hint.* (a) Derive an expression for the total volume v of the molecules in 1 cm^3 of any gas in terms of n_o and d. (b) Calculate v', the volume of liquid formed by the liquefaction of 1 cm^3 of hydrogen, in terms of the density of the gas and the density of the liquid. (c) Assume that $v = v'$. (How good an approximation do you think this is? Can you make a better one?) (d) Compute the mean free path λ from the coefficient of viscosity. (e) Combine the equation obtained in (c) with Eq. [164] and solve.]

> *Ans.* $d = 2 \times 10^{-7}$ cm; $n_o = 5 \times 10^{17}$ cm^{-3}.

---

[1] *Wiener Sitzungsberichte* **52**, 395 (1865). See also O. E. Meyer, *The Kinetic Theory of Gases*, tr. by R. E. Baynes (Longmans, Green, 1899), Chap. 10.

CHAPTER ELEVEN

THE PROPERTIES OF VAPORS

THE ORDINARY gaseous and ordinary liquid states are, in short, only widely separated forms of the same condition of matter, and may be made to pass into one another by a series of gradations so gentle that the passage shall nowhere present any interruption or breach of continuity. From carbonic acid as a perfect gas to carbonic acid as a perfect liquid, the transition we have seen may be accomplished by a continuous process, and the gas and liquid are only distinct stages of a long series of continuous physical changes. Under certain conditions of temperature and pressure, carbonic acid finds itself, it is true, in what may be described as a state of instability, and suddenly passes, with the evolution of heat, and without the application of additional pressure or change of temperature, to the volume, which by the continuous process can only be reached through a long and circuitous route. In the abrupt change which here occurs, a marked difference is exhibited, while the process is going on, in the optical and other physical properties of the carbonic acid which has collapsed into the smaller volume, and of the carbonic acid not yet altered. There is no difficulty here, therefore, in distinguishing between the liquid and the gas. But in other cases the distinction cannot be made; and under many of the conditions I have described it would be vain to attempt to assign carbonic acid to the liquid rather than the gaseous state.

THOMAS ANDREWS in his 1869 Bakerian Lecture, " On the Continuity of the Gaseous and Liquid States of Matter "

o

To just what extent the ideal gas laws which we have been considering fail to describe the behavior of the gases found in nature was investigated first by that most skillful and patient experimenter REGNAULT,[1] and later by ÉMILE HILAIRE AMAGAT[2] (1841–1915). BOYLE himself had noticed small departures from constancy of the product of pressure and volume, but he ignored the failure of his law to hold accurately because of its simplicity and general usefulness. Now if BOYLE's equation, $PV = $ constant, actually did apply accurately to a real gas, the curve connecting PV and P, for a given mass of gas kept at a constant temperature, would be a *straight* line parallel to the axis of pressures. REGNAULT and AMAGAT found, however, that when a gas is tested over a wide range of pressures and at different temperatures, curves like those of Figs. 115 and 116 are obtained. In general, the value of PV diminishes at first, but when

---

[1] *Mémoires de l'Académie des Sciences* 21, 329 (1847); *Relations des Expériences* (Paris, 1847–1870). For translations of two of REGNAULT's papers on the coefficient of expansion of gases, see *The Expansion of Gases by Heat*, ed. by W. W. Randall (American Book Co., 1902), pp. 65–150.

[2] For a bibliography of AMAGAT's papers on the gas laws, and translations of two of his papers, see *The Laws of Gases*, ed. by Carl Barus (Harper, 1899), pp. 108–109.

the pressure exceeds a certain value, which depends on both the gas and its temperature, it steadily increases. It will be observed also that as the temperature increases the marked drops in the curves disappear (Fig. 116). At high enough temperatures, depending on

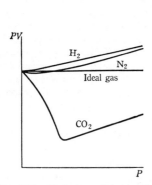

FIG. 115. Variations of the product PV with pressure for a given mass of various gases kept at constant temperature

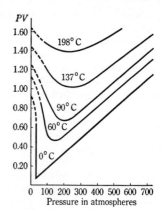

FIG. 116. Variations of PV with pressure for a given mass of carbon dioxide at various constant temperatures

the gas, the curves are found to show only an upward slope. In the case of hydrogen (Fig. 115) and helium, this is true even at room temperature.

In considering experiments such as these, there naturally arises the question of what happens to the properties of a gas when it approaches close to the liquid state. As early as 1822 CHARLES CAGNIARD DE LA TOUR [1] had observed that when a liquid was heated in a hermetically sealed tube it evaporated silently as the temperature rose up to a certain point, and then the boundary between liquid and gas grew indistinct and completely disappeared; the densities of the liquid and gas had become the same and the distinction between the two states of aggregation had vanished. A more thorough study of this effect was made later by THOMAS ANDREWS [2] (1813–1885) during the course of his classical experiments on the behavior of carbon

---

[1] *Annales de Chimie et de Physique* (2) **21**, 127, 178 (1822); **22**, 410 (1823). See *A Source Book in Physics* (1935), pp. 181–187.

[2] *Bakerian Lecture, "On the Continuity of the Gaseous and Liquid States of Matter," *Philosophical Transactions* **159**, 575 (1869); also *Scientific Papers* (Macmillan, 1889), p. 296. See *A Source Book in Physics* (1935), pp. 187–192, and also the quotation at the beginning of this chapter.

Gas Compressor Used by Thomas Andrews
in His Work on Gases
at Queen's University, Belfast

Photograph taken outside the physics laboratory by A. R. Hogg, 67 Great Victoria
Street, Belfast

A Medallion
Commemorating the Work of van der Waals.
The Model of a Surface
Represented on the Reverse Side
is a Three-Dimensional Plot
of the van der Waals Equation

Reproduced from *Isis* **9**, 258 (1927), by permission of the editor

dioxide under pressure at different temperatures. ANDREWS had begun these experiments in an attempt to solve one of the great problems of his period, the liquefaction of what were then called the "permanent gases," an extensive study of which had been made by FARADAY[1] from 1823 to 1845.

125. Andrews's Experiments. The apparatus used by ANDREWS is shown in Plate 34 and Fig. 117, and results typical of the kind which he obtained are illustrated by the isothermal curves in Fig. 118. It is seen that the curve for 48.1° C is nearly a perfect hyperbola, so that at this temperature carbon dioxide behaves almost like an ideal gas. But at lower temperatures the deviations from BOYLE'S law become greater and the curves change in character, taking a form similar to that for the isotherm marked 32.5° C. In this latter curve a jog is apparent, and for the higher pressures the curve is very steep. Indeed, in this respect the curve resembles the isotherm for a liquid, since a relatively large increase in pressure causes only a small decrease in volume. Yet no sign of liquefaction of the carbon dioxide could be noted at this temperature or above it, however much the pressure was increased.

But for temperatures below 31.1° C a radical change occurs in the nature of the isotherms. Taking the curve for 21.5° C and starting from the point A, we see that as the pressure is increased the volume diminishes rapidly until the point B is reached; between A and B we are evidently dealing with a gas. If, after reaching the point B, an attempt be made to compress the gas further, it is found that some of it liquefies and that a considerable change of volume takes place with no change of pressure; in

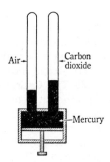

FIG. 117. Schematic diagram of ANDREWS's apparatus. The gas is compressed by forcing the mercury up into the tubes. One of the tubes contains air, which is used to measure the pressure exerted on the carbon dioxide

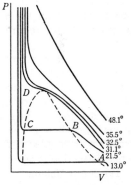

FIG. 118. Isotherms for a given mass of carbon dioxide

---

[1] *Philosophical Transactions* **113**, 160, 189 (1823); **135**, 155 (1845). These papers are reproduced in *Faraday's Experimental Researches in Chemistry and Physics* (London, 1859) and in *The Liquefaction of Gases*, Alembic Club Reprint No. 12 (Edinburgh, 1912). See also *Faraday's Diary*, ed. by T. Martin (Bell, 1932–1936).

other words, in the portion *BC* gas and liquid exist together with a visible meniscus between them, more and more liquid being formed as *C* is approached. At *C* the carbon dioxide is again homogeneous, all of it now being in the liquid state; beyond this point the gradient is very steep, since a liquid is not easily compressed.

126. The Critical Constants of a Substance. It will be seen from Fig. 118 that a smaller contraction in volume upon liquefaction takes place at 21.5° C than at 13.0° C, and thus that the difference in density between the liquid and its vapor will be smaller the higher the temperature. When a temperature of 31.1° C is reached, the horizontal part of the isotherm has disappeared altogether, and no separation into liquid and vapor can be effected, however much the pressure is increased. The temperature at which this occurs is called the *critical temperature* t_c; and the pressure that is just sufficient to liquefy the gas at its critical temperature is called the *critical pressure* P_c. Above the critical temperature a gas cannot be liquefied by pressure alone; that is to say, a separation of the gaseous and liquid states cannot be effected. Indeed, it is possible to pass from any point *A*, where the substance would undoubtedly be regarded as a gas, to a point *D*, where it is in the dense, almost incompressible condition that one would naturally call liquid, without having the liquid distinct from the vapor at any time; to do this it is only necessary to vary the pressure, volume, and temperature in such a way as not to pass through the region bounded by the dotted curve, inside which alone heterogeneity is possible. In other words, above the critical temperature the two states become identical, a property which is referred to as the *continuity of the liquid and gaseous states*.

The properties represented by the isotherms for carbon dioxide (Fig. 118) are characteristic of all substances that have been studied, but the critical constants differ widely. Thus the critical temperature of water is 374° C; of air, − 141° C; and of helium, − 268° C. It is customary to give the name *vapor* to a substance in the gaseous condition when it is below its critical temperature and to confine the term *gas* to a substance when it is above the critical temperature. This distinction is not important, however.

127. Finite Volume of the Molecules. Since the liquid and gaseous states are continuous, it should be possible to obtain a general equation of state connecting the pressure, volume, and temperature that will apply to a substance whether it be in the liquid or the gaseous state. The simple equation $PV = NRT$ obviously will not hold for this more general case, but it might be possible to modify this equation by removing some of the simplifying assumptions used in deriving it in kinetic theory (Sec. 113). A hint as to one of the ways in which this can be done is furnished by the nature of the curves in Fig. 116. It is found that for very high pressures these curves become for any given substance a system of sensibly parallel lines. The equation for any one of these lines evidently will be $PV = bP + c$, where the quantity b, which is the slope of the curves, is found to depend upon the nature of the substance, and the quantity c, which is the PV-intercept, depends upon the temperature. By writing this equation in the form

$$P(V - b) = c, \qquad\qquad [181]$$

one sees that P becomes infinite for $V = b$, and hence that b can be interpreted as the least volume into which the substance can be compressed. Thus $V - b$ is the whole space in which the gas is enclosed, diminished by the least volume of the substance. If $V - b$, rather than V, be considered as the volume of the gas, then BOYLE'S law may be said to apply to all gases even at high pressures.

From the point of view of kinetic theory, as CLAUSIUS himself recognized, $V - b$ is the volume which is available for the free motion of the molecules. In other words, the assumption that the volume occupied by the molecules themselves is negligible compared with the space between them (Sec. 113) is not exactly true for actual gases. The result is that the number of collisions, in a given time, of any molecule with the other molecules or with the walls of the containing vessel is greater than that calculated from the theory in Chapter 10, and that in the collisions with the walls the effect is the same as if the molecules themselves were negligibly small but were confined in a smaller space.

128. Influence of Intermolecular Actions; the Joule-Kelvin Experiment. In attempting to account theoretically for the departures from the laws of ideal gases exhibited in Figs. 116 and 118, one must also take into account the fact that the forces which the molecules exert on one another will not be entirely negligible. In the case of liquids and solids the existence of such forces is sufficiently evident to be familiar to everyone. The first attempts to detect their existence between gas molecules were made by GAY-LUSSAC [1] and, later, by JOULE. [2] In JOULE'S experiments (Fig. 119) a copper vessel

---

[1] *Mémoires de la Société d'Arcueil* 1, 180 (1807). A translation is given in *The Free Expansion of Gases*, ed. by J. S. Ames (Harper's, 1898), pp. 3–13, and an excerpt will be found in *A Source Book in Physics* (1935), pp. 170–172.

[2] *Philosophical Magazine* (3) **26**, 369 (1845); *Scientific Papers* (London, 1884), Vol. 1, p. 172. This paper is reproduced in *The Free Expansion of Gases*, ed. by J. S. Ames (Harper's, 1898), pp. 17–30, and an excerpt appears in *A Source Book in Physics* (1935), pp. 172–173.

containing air at a pressure of 22 atmospheres was connected with another similar vessel from which the air had been exhausted. Upon opening the stopcock between the two vessels the air expanded and filled both vessels. Now the gas expanded into a rigid container and hence did no external work during expansion; therefore, unless there were forces between the molecules, there would be no change in the potential energy of the molecules and hence none in their kinetic energy; that is, the gas *as a whole* would not change in temperature (Sec. **116**). JOULE could not detect any change in the temperature of the surrounding water bath as a result of the expansion.

JOULE varied the experiment by placing the two vessels in separate water baths, when it would be expected that the drop in temperature of the vessel initially containing the gas would be greater than the gain in the other vessel if intermolecular attraction exists. Again he found that the temperature lost by the one vessel was equal to that gained by the other. JOULE therefore concluded that if intermolecular forces

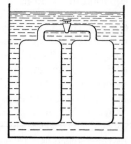

FIG. 119. JOULE's apparatus very much simplified

exist, his experiments were too insensitive to detect them. As a matter of fact, the effect sought for is very small, whereas the thermal capacity of JOULE'S apparatus was so large compared with that of the gas that it would have required a change in temperature of several degrees in the gas to produce any noticeable effect in the water.

LORD KELVIN thought that this matter should be tested by a better method, and this led him to devise his classical porous-plug experiment, which he and JOULE carried out together.[1] The improvement consisted in making the expansion of the gas continuous instead of intermittent. The gas, kept at constant pressure by a pump, was allowed to flow continuously through a porous plug of tightly packed cotton, as indicated in Fig. 120. The resistance offered by the plug to the flow was so large that the kinetic energy of the flow was negligible; the gas merely expanded through the plug. The lowering

---

[1] J. P. Joule and W. Thomson, *Philosophical Magazine* (4) **4**, 481 (1852); *Philosophical Transactions* **143**, 357 (1853); **144**, 321 (1854); **150**, 325 (1860); **152**, 579 (1862); abstracts appear in the *Proceedings of the Royal Society* for these same years. See also *Joint Scientific Papers of James Prescott Joule* (London, 1887), Vol. II, pp. 216–362; Thomson's *Mathematical and Physical Papers* (Cambridge University Press, 1882–1911), Vol. I, pp. 333–455; *The Free Expansion of Gases*, ed. by J. S. Ames (Harper, 1898), pp. 33–102.

of temperature which was actually found in this experiment for all gases except hydrogen and helium can be explained by assuming that there is a very slight attraction be-tween the gas molecules, so that there is an increase of potential energy upon expansion which takes place at the expense of the kinetic energy. The fall of temperature is found to be nearly proportional to the difference between the pressures on the two sides of the plug, and to increase as the initial tem-

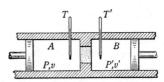

Fig. 120. The porous-plug experiment

perature of the gas decreases. These results are of great importance, because from them can be calculated the corrections necessary to re-duce the readings of a gas thermometer to absolute thermodynamic temperatures (Sec. **103**). They also form the basis of the method developed by KARL RITTER VON LINDE in 1895 for liquefying air and the other so-called "perma-nent" gases on a large scale for commercial pur-poses (Fig. 121). It has been found that even hydrogen and helium are cooled by their own expansions if their initial temperatures are below a certain value, called the *temperature of inver-sion*, which is $-80.5°$ C for hydrogen and about $-238°$ C for helium. This fact is of great impor-tance in the liquefaction of these gases.

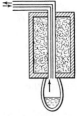

Fig. 121. In LINDE's method compressed gas passes along the inner tube, expands through a valve, and passes back through the outer tube. The cooling thus goes on progressively until some of the gas lique-fies as it escapes from the valve. Actually the tubes are arranged in the form of a spiral

In order to explain the heating which was observed during the expansion of hydrogen and helium at ordi-nary temperatures, it is necessary to analyze this ex-periment in more detail. Let $W_t + W_i$ be the total *internal* energy per unit mass of the gas (kinetic plus potential) as it enters the plug, shown diagrammati-cally in Fig. 120, and let $W_t' + W_i'$ be its total *internal* energy as it leaves the plug. Similarly, let P and v be the pressure and the volume per unit mass of gas on side A, and P' and v' the pressure and the volume per unit mass on side B. If no heat enters or leaves the gas from without and if the speeds of inflow and out-flow of the gas are equal, then the decrease in total internal energy per unit mass in passing through the plug must equal the external work per unit mass done by the gas, or

$$(W_t + W_i) - (W_t' + W_i') = P'v' - Pv.$$

If now BOYLE's law holds, so that $Pv = P'v'$, then

$$W_t + W_i = W_t' + W_i'.$$

Consequently, if there are no forces between the molecules, so that $W_i = W_i'$, then also $W_t = W_t'$ and there will be no temperature change. But if there are intermolecular forces of any sort whatever, the potential energy W_i must change on expansion, and W_i will not equal W_i'; consequently W_t cannot equal W_t' and the temperature, which is a function of W_t (Sec. 116), must change. When the intermolecular forces are attractive, the gas must cool on passing through the plug.

If, however, the product Pv increases as the pressure decreases, as AMAGAT found it did up to a certain point for all gases except hydrogen and helium (Fig. 115), then $P'v'$ is greater than Pv, and $W_t + W_i$ must be greater than $W_t' + W_i'$. In this case we should expect a cooling even though there are no intermolecular forces. On the other hand, if Pv increases with the pressure, as it does in the case of hydrogen and helium and all other gases beyond a certain point, then Pv is greater than $P'v'$, and $W_t' + W_i'$ is greater than $W_t + W_i$; there will then be a heating even though there are no intermolecular forces.

The deviations from BOYLE'S law caused by the finite size of the molecules thus produce in this experiment temperature changes which must be added (algebraically, of course) to the cooling produced by the work done against the intermolecular attractions. In the case of hydrogen and helium, at ordinary temperatures the heating due to this cause is sufficient, since the intermolecular forces are small, to produce a rise in temperature. At sufficiently low temperatures, however, the cooling effect predominates.

It having been established that intermolecular forces exist, the next question is how they affect the pressure and volume of a gas. In the interior of the gas the resultant effect is indeed negligible, for the molecules are attracted equally in all directions by the other

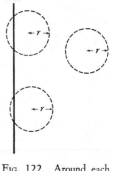

Wall

Fig. 122. Around each molecule there may be circumscribed an imaginary sphere of very small radius r, called the *sphere of molecular attraction*, outside of which the intermolecular force exerted by or upon a molecule is inappreciable. For any molecule that is at a distance less than r from the wall, only the forces tangential to the wall will be balanced, and hence there will be a resultant force acting inward toward the gas

molecules. But not so in the layers next to the walls of the containing vessel, where the resultant attraction is directed inward toward the gas (Fig. 122), thus tending to reduce the volume occupied by the gas, just as an increase of pressure would do. The effect of molecular attraction may therefore be represented by adding to the pressure P, which is applied to the gas externally, a quantity P', representing the pressure due to internal attraction. Our gas equation, Eq. [181], then becomes

$$(P + P')(V - b) = NRT. \tag{182}$$

129. Equations of State of a Fluid. Of the numerous attempts that have been made to deduce a general equation holding for any substance throughout the liquid and gaseous states, the most celebrated is that developed by JOHANNES DIDERIK VAN DER WAALS[1] (1837–1923). His equation may be derived from Eq. [182] by assuming that the quantity P' is proportional both to the number of molecules striking a unit area of the wall in unit time and to the number of molecules attracting any given molecule. Since both of these factors are proportional to the density of the fluid, P' will vary directly as the square of the density or inversely as the square of the volume; hence, denoting P' by a/V^2, Eq. [182] takes the form

$$\left(P + \frac{a}{V^2}\right)(V - b) = NRT. \qquad [183]$$

This is known as the *van der Waals equation*. The coefficients a and b depend upon the amount of gas as well as upon its nature, a being proportional to the square, and b to the first power, of the mass of gas.[2] When one considers the simplicity of this equation, its general agreement with experiment is remarkable. CLAUSIUS, C. DIETERICI, and many others [3] have proposed equations that are improvements, in one respect or another, on the VAN DER WAALS equation. Substantial advances in theory have resulted from their efforts, but the use of the equations in practical applications introduces complications that tend to offset any advantages gained.

130. The Kinetic Theory of Liquids and the Process of Evaporation. If the molecules of a gas are in rapid motion, the molecules of a liquid must be also, for we have seen that no fundamental distinction exists between the liquid and the gaseous states (Sec. **126**). In fact, at high temperatures the two states become absolutely identical. Below the critical temperature the possession of a clearly marked surface may be taken as the distinguishing feature of a liquid.

---

[1] *On the Continuity of the Liquid and Gaseous States* (Leiden, 1873). The original edition is in Dutch. A German translation by F. Roth (Barth, 1899) has been translated into English [*Physical Memoirs* (Physical Society of London, 1890), Vol. I, Part 3.]

[2] It simplifies the use of the VAN DER WAALS equation if the pressure P is expressed in atmospheres and the volume V as the ratio between the actual volume and the volume which the substance would occupy if it were an ideal gas at $0°$ C and 1 A_s. Consequently the values of a and b found in tables are usually expressed in terms of these special units. In this case NR will have the numerical value 0.00366. (Why?)

[3] See, for example, T. Preston, *The Theory of Heat*, ed. 3 (Macmillan, 1919), pp. 484–489. More than fifty such equations of state are listed by J. R. Partington and W. G. Shilling, *The Specific Heats of Gases* (Benn, 1924), pp. 29–34.

Direct experimental evidence of molecular motion in liquids is furnished by the Brownian movement (Sec. **118**). The molecules of a liquid are crowded so close together, however, that their paths between impacts are extremely minute, being of the same order of magnitude as the diameters of the molecules. At the free surface of a liquid, where there is greater freedom of motion, the paths of the molecules are influenced not only by collisions but also by the attractions of the other molecules. On account of the enormous number of molecules present in or near the surface, this downward force upon a molecule just above the surface is doubtless very large; so large, in fact, that the molecules which are moving away from the surface are, in general, unable to leave it. They simply rise to a certain distance by virtue of their velocities, after the manner of projectiles shot up from the earth, and then fall back again into the liquid (Fig. 123).

At a given temperature the *average* kinetic energy of the molecules of a given liquid is always the same. Nevertheless at any instant a large number of them will be moving with energies greater than the average and still others will be moving with less than average energy. Thus a molecule near the surface may escape from the liquid if it possesses enough kinetic energy to overcome the effect of the powerful attractive forces existing in a

FIG. 123. The kinetic picture of a liquid and its vapor. The zigzag and curved lines represent possible paths of the particles

thin layer near the surface; in this case it moves off as an independent gas molecule into the space above. (In the light of the kinetic theory, how do you account for the cooling which a liquid undergoes while it is evaporating?)

If the space above a liquid surface is enclosed, it gradually becomes filled with the gaseous form of the substance comprising the liquid. This gas becomes more and more dense as more molecules escape, but there is evidently a limit to its possible density, for many of the escaped molecules chance, in their wanderings, to return to the surface and re-enter the liquid. The number of molecules thus returning in unit time evidently increases as the number above the liquid increases; that is, it is proportional to the density of the vapor. When this density has reached a certain limit, there is set up a condition of equilibrium in which the average number returning will equal the average number escaping. A vapor thus in equilibrium with its liquid is said to be *saturated*; that is, it has the largest density which it is ever able to have at the existing temperature, and it therefore exerts the largest pressure which it ever can exert at this temperature. If the vapor is not allowed to accumulate over the liquid, it will remain unsaturated, equilibrium will not be reached, and the liquid will gradually disappear by evaporation. (Why does fanning greatly facilitate evaporation?)

131. Densities and Pressures of Saturated Vapors. If the temperature of a closed vessel containing a liquid and its vapor be raised, the vapor evidently can have a higher density; for the number of molecules escaping in unit time must be greater at the higher temperature because of the higher mean kinetic energy, and hence the density of the vapor must be greater before the condition of equilibrium is set up. Also, since the pressure exerted by the vapor is proportional both to the density and to the mean kinetic energy of each impact, and since both density and kinetic energy increase with temperature, it is evident that the pressure must rise with twofold rapidity as the temperature rises. If, on the other hand, the temperature be held constant, all attempts to increase the density or the pressure of a vapor which is in contact with its liquid in a closed vessel must be futile. To see this clearly, suppose that a few drops of some volatile liquid, such as ether, are introduced into a barometer tube so as to fill the space above the mercury with ether and saturated ether vapor (Fig. 124). As soon as the density of this vapor is momentarily increased by lowering the tube, thus compressing the vapor, the equilibrium at the surface is destroyed, and immediately more molecules begin to enter the surface in unit time than escape from it. Hence, in a very short time

Fig. 124. In *a* the space above the mercury column contains nothing but mercury vapor. In *b* a small quantity of some volatile liquid has been introduced into this space. In *c* the volume occupied by the vapor has been decreased by lowering the tube, but the height of the column of mercury has remained unchanged

enough ether condenses to restore the original condition of density and pressure. It is left to the student to explain how, when the tube is raised and the volume is thus increased, the condition of equilibrium is soon re-established at the same density and pressure.

Evidently the foregoing experimental facts may be summarized in the statement that *the pressure exerted by a saturated vapor is a function of its temperature only.* If the liquid above the barometric column in Fig. 124 be heated, the vapor pressure steadily increases, and by making a number of observations at various temperatures a *saturation vapor-pressure-temperature* curve may be constructed. Such a curve for water is shown in Fig. 125. (What is the nature of the *isotherms* for a saturated vapor?) These various properties of a saturated vapor really are already familiar to us from ANDREWS'S experiments with carbon dioxide, for in the region under the dotted line in Fig. 118 liquid and vapor exist together and the vapor is saturated. In fact, the points in the area enclosed by the dotted line represent all the physical conditions under which liquid carbon dioxide and its vapor can be in equilibrium with each other.

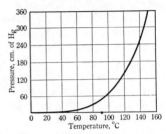

FIG. 125. The vapor-pressure-temperature curve for saturated water vapor

The fact that the VAN DER WAALS equation applies both to the gaseous and to the liquid state makes it certain that the isotherms plotted from it will not have the discontinuities that appear in the experimental curves of Fig. 118. Instead, these isotherms have the form shown in Fig. 126. Inside the dotted curve they are not only continuous but cut the horizontal lines *BC* etc. in three points; this is because the VAN DER WAALS equation is of the third degree with respect to *V*, and in this region all three roots are real. The critical point is evidently the particular point where the pressure P_c and temperature T_c are such as to make the three roots equal. Above the critical point the equation has only one real root.

It is to be noted that the part of the isotherm which lies between *B* and *C*, Fig. 126, shows a minimum and a maximum. It is, of course, impossible to realize physically the segment of the curve lying between the minimum and the maximum, for the volume would then have to increase with increasing pressure. Under certain circumstances, however, the parts of the curve extending from *C* to the minimum point and from the maximum point to *B* can be realized. The first part represents the condition of a *superheated* liquid; the second, that of an *undercooled* or *supersaturated* vapor.

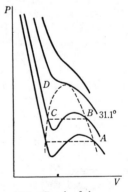

FIG. 126. Graph of the VAN DER WAALS equation for the case of carbon dioxide

132. Mixtures of Vapors and Gases. The saturated vapor pressure for a given temperature is not affected measurably by the presence of gases that do not combine chemically with the vapor. For example, in Fig. 124 the space above the mercury column contains mercury vapor as well as ether vapor, but the presence of the mercury has practically no effect on the ultimate amount of ether that will evaporate into the space. The ether vapor exerts its own pressure independently of that of the mercury, making the total pressure the sum of the two. This is also true when air, even at greater densities than that of the ether vapor, is present above the mercury. In other words, DALTON'S law of partial pressures (Sec. 115) applies to a mixture of gases and saturated vapors. It fails, of course, at high pressures.

The presence of gases does have a very marked effect, however, in *retarding* evaporation. When ether is introduced into a space containing air, the maximum density of the vapor may not be reached for several hours, whereas when it is introduced into a vacuum, the condition of saturation is reached in a few seconds.

133. Vapor Pressures of Liquids. As was shown in Sec. **131,** for a given liquid at a given temperature there is only one pressure which its vapor can have and still exist in equilibrium with the liquid. This saturation pressure ordinarily is referred to simply as the *vapor pressure of the liquid*. It is to be distinguished from the *pressure of the vapor*, which when not in contact with the liquid may have any value from zero up to one somewhat exceeding the saturation vapor pressure. The vapor pressure of a liquid increases rapidly with the temperature. No very convenient or exact formula connecting it with the temperature has been devised, however; so the relation is usually shown by means of a graph like Fig. 125 or by means of tables.[1]

From the laws of thermodynamics it is possible, however, to derive an *exact* differential expression for the *slope* of the curve that connects the vapor pressure P of a liquid with the absolute temperature T. This expression, which is known as the *Clapeyron-Clausius equation*,[2] is as follows:

$$\frac{dP}{dT} = \frac{JL}{T(v_v - v_l)},$$

where J is JOULE'S equivalent, L is the heat of vaporization (Chap. 12), and v_v and v_l are the volumes occupied by a unit mass of the substance in the saturated-vapor and liquid states respectively. The derivation of this equa-

---

[1] Appendix 13, Table E.

[2] This equation was derived from CARNOT'S work by B. P. E. Clapeyron, *Journal de l'École Polytechnique* **14,** 153 (1834) [tr. in *Scientific Memoirs,* ed. by R. Taylor (London, 1837), Vol. 1, pp. 347–376]. It was first rigorously established by CLAUSIUS in his papers on the mechanical theory of heat. A similar equation applies to the melting of solids.

tion rests upon the second law of thermodynamics and is beyond the scope of this text. There is perhaps some advantage, however, in becoming acquainted with the form of the equation, if not with its derivation. Since at the boiling temperature of a liquid the vapor pressure is equal to the external pressure upon the liquid and vapor (Sec. 135), the CLAPEYRON-CLAUSIUS equation also expresses (more clearly in inverted form) the change of boiling point with the external pressure.

The vapor pressure of a liquid with a curved surface must be different from that of the same liquid with a plane surface. The magnitude of this effect was first calculated in 1870 by LORD KELVIN.[1]

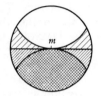

FIG. 127. Effect of surface curvature upon the saturation vapor pressure

> EXAMPLE. With the help of Fig. 127, which represents the sphere of molecular attraction outside of which the attractions of other molecules for molecule m are inappreciable, and from considerations similar to those employed in Sec. 130, show that the saturation vapor pressure over a concave surface must be smaller than it is over a plane surface, and that over a convex surface it must be larger than over a plane surface.

134. Vapor Pressures of Solutions. If a gas present in the space above a liquid is soluble in the latter, as in the case of ammonia or carbon dioxide in water, there will result a decrease in the pressure of this component of the mixture. It is found that the amount of dissolved gas is proportional to the partial pressure of that constituent and independent of the pressure due to other gases present.

When one volatile substance is dissolved in another, as when alcohol is dissolved in water, the vapor pressure of the solution is not even approximately equal to the sum of the pressures which the components would exert if each were by itself. The actual relation between the various factors involved in any given case must be determined empirically, since the relations are too complicated to be generalized easily into a single statement.

In the case of a solid dissolved in a liquid, the relation, at least for dilute solutions, is a simple one: the vapor pressure of a given solvent is lowered by an amount that is directly proportional to the number of molecules of dissolved material in unit volume of the solution. If, in the process of solution, some of the molecules of the dissolved substance dissociate into smaller electrically charged particles, called *ions*, as happens in the case of aqueous solutions of acids, bases, or

---

[1] *Proceedings of the Royal Society of Edinburgh* **7**, 63 (1870); *Popular Lectures and Addresses* (Macmillan, 1891), Vol. 1, p. 64.

salts, the lowering of the vapor pressure is then proportional to the total number of ions and un-ionized molecules present.

> EXAMPLE. Explain, from the point of view of the kinetic theory, why dissolving a solid in a liquid lowers the vapor pressure.

135. Boiling. Evaporation takes place to some extent at all temperatures whenever the space above the liquid is not saturated. If the liquid be under a constant external pressure, such as that of the atmosphere when the barometer is steady, an increase in the temperature of the liquid will result in more rapid evaporation, until finally a temperature is reached at which the evaporation begins to take place not simply at the surface but also within the body of the liquid; that is, bubbles of vapor begin to form beneath the surface upon the sides of the containing vessel, whence they rise to the top, growing rapidly as they ascend. It is evident that this condition cannot be reached until the maximum pressure exerted by the vapor formed from the liquid is at least equal to the outside pressure; for if the pressure exerted by the vapor in the bubbles were less than that outside, these bubbles, even if formed, would at once collapse.

This temperature, then, at which the pressure of the saturated vapor becomes equal to the outside pressure, is called the *boiling temperature*. The vapor-pressure-temperature curve (Fig. 125) may therefore also be called the *boiling-temperature curve*, the pressures now signifying the total pressure on the liquid. The *boiling point* of any substance, as distinguished from its boiling temperature, is defined as the temperature at which the vapor pressure of the saturated vapor is equal to 1 A_s.

> EXAMPLE. On a diagram like that of Fig. 125, indicate roughly the nature of the vapor-pressure-temperature curve for a dilute aqueous solution of a solid, and then make use of these two curves to explain what effect a dissolved solid has on the boiling temperature of a liquid.

A liquid will not always necessarily boil as soon as its temperature reaches the boiling temperature. In fact, the temperature of a boiling liquid must always be at least a trifle higher than that at which the pressure of the saturated vapor is equal to the outside pressure, for the pressure within the bubble of vapor must be sufficient to overcome not only the outside pressure but also the weight of the superimposed liquid and the surface tension of the bubble. But when the bubble rises to the surface and breaks, the pressure exerted by the vapor contained within it must fall exactly to the atmospheric condition, and the temperature of this vapor must also fall, by virtue of expansion, to that temperature at which the pressure of the saturated vapor is equal to the existing atmospheric pressure. Hence a thermometer which is to indicate the true boiling temperature must be placed not in the boiling liquid itself but in the vapor rising from it. The temperature of the liquid itself is in fact a very uncertain quantity. GAY-LUSSAC found that the temperature of boiling water in a glass vessel was usually 1° C to 3° C higher than in a metal vessel. For the reasons already mentioned, it must always be a trifle higher than the boiling temperature; but under some circumstances it may rise many degrees above this temperature. For it is by no means necessary that bubbles of vapor begin to form as soon as the temperature is reached at which they are able to exist after being formed. The presence of air in the liquid or occluded in the walls of the containing vessel or of dust particles or other impurities is found to be essential to the genesis of bubbles. A Frenchman named DONNY found, in 1844, that when he carefully removed the dissolved air he could raise the temperature of water in a clean glass vessel to 137° C before boiling began.[1] But in all such cases, since the pressure of the saturated vapor corresponding to the temperature of the water is much more than the atmospheric pressure, as soon as a bubble once starts it grows with explosive rapidity and produces the familiar phenomenon of " boiling by bumping." This usually occurs whenever a liquid has been subjected to long boiling, and has been suggested as a possible cause of boiler explosions.

136. Vapor Pressures of Solids. Solids also evaporate and may change, therefore, from the solid state directly into the vapor state, as in the evaporation of camphor, or of snow in cold dry weather. This process is called *sublimation.* Solids have definite saturation vapor pressures, and for certain substances, such as iodine, these pressures are appreciable even at room temperature. The vapor pressure increases with temperature in much the same way as does that of a liquid, and in some cases, such as that of solid carbon dioxide ("dry ice"), reaches atmospheric pressure at a temperature below the melting point. This temperature is called the *sublimation point.*

At the melting point the vapor pressures of the solid and liquid are equal, since at this temperature the solid and liquid can exist together in equilibrium; if they were not equal, the state with the greater vapor pressure would necessarily pass over gradually into the other. In Fig. 128 are shown diagrammatically the vapor-pressure curves for water. Curve A is for the liquid, curve B for ice, and curve C is the melting-temperature curve. Point p is called the *triple point*; it gives the temperature and pressure at which the three states of aggregation — solid, liquid, and vapor — can exist together in equilibrium. The vapor-pressure curve for the ice must be steeper than that for the water. This may be proved as follows.

---

[1] A more complete discussion of these questions, with references to the original literature, will be found in *T. Preston, *The Theory of Heat*, ed. 4 (Macmillan, 1929), pp. 349–353.

Consider ice, water, and water vapor in equilibrium in a heat-insulated container. If salt is added to the mixture, the vapor pressure of the water is lowered and the equilibrium is destroyed; the water vapor condenses, while the ice, having a higher vapor pressure, evaporates. Since the energy necessary to evaporate the ice must come from the system itself, the temperature falls until the vapor pressure of the ice equals that of the water, or until all the ice is melted. If, therefore, equilibrium between the ice and the solution is to be possible, the rate of change of vapor pressure with temperature must be larger for the solid than it is for the liquid; that is, the vapor-pressure curve for ice must be steeper than that for water. This conclusion was confirmed experimentally by GUS-TAV KIRCHHOFF.[1]

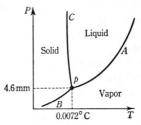

FIG. 128. Schematic triple-point diagram for water. It consists of the curve of Fig. 125 and the equilibrium curves for solid-liquid and solid-vapor, combined in the same diagram. Curve B is steeper than curve A, although the difference has been exaggerated here. (Can you show that the three curves must intersect in a common point p?)

EXAMPLE. On a diagram like that of Fig. 128 indicate roughly the nature of the vapor-pressure-temperature curve for a dilute aqueous solution of a solid and then make use of the curves to explain what effect a dissolved solid has on the freezing temperature of a liquid.

137. Hygrometry. The water vapor in the earth's atmosphere is usually unsaturated, and hence precipitation in the form of rain, fog, or dew may not occur even if the thermometer falls suddenly through many degrees. The humidity of the atmosphere at any time can be expressed exactly in terms of either the pressure P or the density ρ of the water vapor present; the latter quantity, ρ, is often called the *absolute humidity*. But it is usually preferable to describe the humidity by means of either the dew point or the relative humidity, two quantities that suggest more directly the effects due to the presence of the water vapor. The *dew point* τ is the point to which the temperature of the atmosphere must fall, without change of pressure, in order that the water vapor existing in it may be in the saturated condition; as soon as the temperature falls below this point condensation must of course ensue. The *relative humidity* r is defined as the ratio of the density of the water vapor actually in the atmosphere at any given time and the density of saturated water vapor at the

[1] Poggendorff's *Annalen der Physik und Chemie* **103**, 206 (1858).

existing temperature. The experimental determination of any one of the four quantities P, ρ, τ, and r, taken in connection with the pressure-temperature curve of a saturated vapor, suffices for the calculation of all the rest. In the air-mass analyses used in the newer methods of weather forecasting, two other hygrometric quantities are found useful. One is the *water-vapor content*, or *mixing ratio*, w, defined by the equation $w = \rho/\rho_d$, where ρ and ρ_d are the densities of the water vapor and of the dry air respectively. The other quantity is the *specific humidity* q, which measures the amount of vapor associated with the *moist* air; that is, $q = \rho/(\rho + \rho_d) = w/(1 + w)$.

The earliest known hygroscope was of the absorption type and is described in the works of NICOLAUS DE CUSA, a German cardinal of the fifteenth century.[1] An early hygrometer constructed by the Accademia del Cimento in the seventeenth century is shown in Plate 26. *Absorption hygrometers* of the type shown in Fig. 129 are still in common use, but, as was proved by REGNAULT in 1845, the only reliable method of graduating such an instrument is an empirical one, as by comparison with a dew-point hygrometer, and frequent calibration is necessary.

Accurate measurements in hygrometry began with the introduction by J. F. DANIELL, in 1823, of the *dew-point hygrometer* (Fig. 132), and this instrument in one or another of its numerous modifications has become the standard of comparison for the testing and graduation of all other hygrometers; it consists essentially of a polished metal tube, the temperature of which is in some way lowered until dew is observed to form upon its surface. The instrument which is now used most extensively in meteorological observations was first proposed by the Scottish physicist JOHN LESLIE in 1790. It is called the *wet-and-dry-bulb* hygrometer, or *psychrometer*, and consists of two similar thermometers mounted side by side, one of which

FIG. 129. An absorption hygrometer, due to DE SAUSSURE. The hair changes in length with changes in the humidity

has its bulb covered with a light wick kept moist with water (Fig. 133). Evaporation from the wet bulb will cool it as compared with the dry bulb, and the drier the air the larger will be the difference in temperature between the two. This difference, however, also depends in a complicated way upon the barometric pressure and upon the speed of the air currents near the bulb. Formulas connecting the difference of temperature with the hygrometric state

---

[1] See *F. Cajori, *A History of Physics* (Macmillan, 1929), p. 53; also *A. Wolf, *A History of Science, Technology, and Philosophy in the 16th and 17th Centuries* (Allen & Unwin, 1935), pp. 306–309.

FARADAY at Work in His Laboratory
at the Royal Institution, 1852

From a water-color drawing by Harriet Moore, in the possession of the Royal
Institution. By permission of the Managers

The Royal Institution of Great Britain

From a water color by T. Hosmer Shepherd, painted about 1840. By permission of the
Managers of the Royal Institution

THE ROYAL INSTITUTION, founded in 1799 by COUNT RUMFORD for the purpose of
"teaching by courses of philosophical lectures and experiments the application of
science to the common purposes of life," has exerted an enormous influence upon
the development of science through the researches of YOUNG, DAVY, FARADAY,
TYNDALL, DEWAR, and others, that were conducted within its walls. Its public
lectures have been equally influential in diffusing a knowledge of science and its
methods. See T. Martin's "The Professors of the Royal Institution," *Nature* **135**,
813 (1935); also F. Cajori's *A History of Physics* (Macmillan, 1929), p. 402.

of the air have been obtained both theoretically and empirically, in the latter case by means of long series of comparisons of this instrument with the dew-point hygrometer. In practice it is more convenient to use the empirical tables usually furnished with such instruments.

○

EXPERIMENT XIa. PRESSURE OF A SATURATED VAPOR BY THE STATIC METHOD

The object is to observe, by the static method, the pressures exerted by saturated water vapor at different temperatures. The apparatus is shown in Fig. 130. The bulb *B*, originally open at *c*, was first half-filled with mercury. The long arm, also originally open at the top, was then exhausted and inclined until mercury completely filled it up to a point at which it had been drawn down to capillary dimensions. The tube was then sealed off at this point, so that when the instrument was vertical the difference between the levels of the mercury in the bulb and in the tube was equal to the barometric height. Water was then inserted at *c* and boiled until the air was all driven out of the bulb, when the opening at *c* was sealed off. Since, then, only water and water vapor exist above the mercury in the bulb, the difference between the levels in the bulb and in the tube gives at once the pressure of the saturated water vapor in the bulb. It is, therefore, only necessary to vary the temperature of the bulb in order to obtain the curve which expresses the relation between the pressure and the temperature of saturated water vapor.

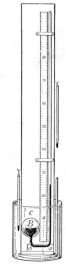

Fig. 130. Apparatus for determining the variations of the boiling temperature with the pressure by the static method

a. Immerse the bulb in cold water contained in a glass jar. Place ice or snow in the jar until the temperature falls approximately to zero. While waiting for the mercury level to cease falling, have some water heating in another vessel.

Three thermometers are needed for the experiment; the first one is placed in the water bath and should, if possible, be one which has been compared with a standard thermometer and for which a curve of corrections has been prepared [1]; the second is placed with its

---

[1] See Appendix 10.

bulb at about the middle of the exposed stem of the first one; and the third is hung near the long arm of the manometer (Fig. 130).

Read the two mercury levels and record the temperatures.

b. Remove all of the ice and part of the water from the jar, add enough warm water to raise the temperature of the bulb to 10° C approximately. Stir the water continuously and, after the mercury level has become stationary, repeat the observations described in *a.*

c. Repeat at intervals of 10° to 12° C until a temperature of 50° C has been reached.

d. Replace the jar by a metal pail and apply heat directly to the pail by means of a Bunsen flame. The level of the mercury in the bulb cannot now be read, but a method for inferring its position will be given presently.[1] Record the temperatures and the level of the mercury in the long arm at intervals of about 5° C up to the highest temperature that can be reached.

e. After obtaining the last reading, lower the pail and quickly observe the position of the mercury surface in the bulb. From a knowledge of the positions of the mercury surface in the bulb at 50° C and at the highest temperature reached, find by interpolation its positions for the intermediate temperatures.

f. Correct the observed temperatures of the vapor for the errors of the thermometer and for the error due to the exposed stem.

g. In order to compare your results with tabulated values of vapor pressures, it is necessary to reduce all pressure-readings to what their values would have been if the room temperature had been 0° C. This is effected by multiplying the observed pressure by the ratio of the densities of mercury at the mean temperature of the long arm of the manometer and at 0° C. Below 75° C this correction is so small that it may be ignored.

h. The observed pressures should also be corrected for the capillary depression of the mercury in the tube.[2]

1. Plot your observations on rectangular-coordinate paper, using temperatures as abscissas and pressures as ordinates. Employ a scale that will make the resulting curve fill the entire page.

2. On the same sheet plot the accepted values of the pressures of saturated water vapor as given by tables.[3]

3. Is the behavior of a saturated vapor correctly described by Eq. [124], Chap. 9? Which shows the more rapid increase of pressure with temperature, a gas or saturated water vapor?

---

[1] This difficulty can be avoided by substituting a large thin-walled glass beaker for the pail, in which case the apparatus should be suspended in the water bath.

[2] Appendix 13, Table H.   [3] Appendix 13, Table E.

4. In what physical state is water when its pressure and temperature correspond to a point below the curve? above the curve?

5. To what extent does the presence of mercury vapor in the bulb vitiate your data?

6. Assume that the laws for ideal gases hold also for vapors up to the very point of saturation. Then, with the aid of the known density of air at $0°$ C, 1 A_s (namely, 1.293×10^{-3} g · cm^{-3}), the density of water vapor in terms of air (namely, 0.624), and the experimental values for the pressures of saturated water vapor, calculate the densities of saturated water vapor at $10°$ C, at $40°$ C, at $70°$ C, and at $100°$ C. Compare these calculated values with the experimentally determined values of the densities given in tables and in this way show how closely the gas laws apply to saturated vapors.

○

EXPERIMENT XIB. PRESSURE OF A SATURATED VAPOR BY THE KINETIC METHOD

The object is to observe, by the kinetic method, the pressures exerted by saturated water vapor at different temperatures. The kinetic method, which is due to REGNAULT, consists in the direct observation of the temperatures of the vapor that rises above a liquid made to boil under various measured pressures. In Fig. 131, B represents a boiler, which may be either a balloon-form boiling flask of about 1-l capacity or an airtight metal boiler. C is a Liebig condenser, through which a slow current of water is passed from a tap. It is only by virtue of the immediate condensation of the vapor as it forms that the pressure within the boiler can be kept constant. (Which of the two tubes in the condenser jacket should be connected to the water inlet, and why?) R is an airtight reservoir of sufficient capacity to smooth out small fluctuations in pressure. The only other essential features of the apparatus are an open-tube manometer M and some kind of vacuum pump, such as a water aspirator.

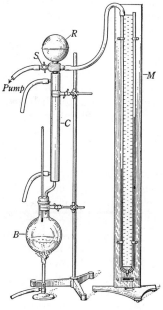

Fig. 131. REGNAULT's form of vapor-pressure apparatus

a. Have the boiler about half-full of water. Test the system for leaks by exhausting to the highest vacuum attainable by means of the pump, and then closing the stopcock *S* and letting the system stand a few minutes.

b. Slowly turn the stopcock *S* so that the system is placed in communication with the atmosphere. Start the circulation in the condenser and heat the water in *B* to boiling. After the conditions have become stationary, read the barometer and its temperature,[1] and observe the temperature of the steam as given by the thermometer in the boiler.

c. Turn the stopcock *S* so as to connect the system with the pump, and exhaust until a difference in height of 5 to 10 cm has been produced in the arms of the manometer. Then close *S* entirely, and after about two minutes of continued boiling observe the temperature of the steam and at once observe the positions of the mercury surfaces in the manometer.

d. Again connect the system with the pump, and repeat the operations described in *c* after the pressure has been reduced to a lower value. Continue in this way until the boiling temperature has fallen to about 75° C, below which temperature the difficulty of boiling with bumping is encountered. (Why?) It is best to make each reduction of pressure smaller than the one that precedes it, so that the pressure steps represent approximately equal changes of temperature — for instance, about 5°.

e. If the instructor so directs, attach a small pressure pump at *S* and investigate the temperatures of the steam for pressures somewhat higher than that of the atmosphere.

f. Correct the observed temperatures of the steam for the errors of the thermometer, exactly as in Exp. XIᴀ, *f.*

g. From the corrected barometer reading and the manometer readings calculate the total pressures above the surface of the liquid. Remember that the open-tube manometer indicates the so-called *gauge pressure*, or difference between the pressure in the system and the pressure of the atmosphere. The pressures should be reduced to 0° C and corrected for capillary depression, as in Exp. XIᴀ, *g* and *h.*

1. State precisely what two quantities you have observed in this experiment and what relation they bear to the pressure and temperature of saturated water vapor.

2. Plot your experimentally determined values of the pressure and temperature of saturated water vapor on the sheet of coordinate paper used in Exp. XIᴀ, question 1.

---

[1] Appendix 6.

3. How may the boiling point of water be determined from your observations?

4. What determines the highest temperature for which this apparatus is applicable? the lowest temperature?

5. Just how does the reservoir R perform its function? What are the causes of the small fluctuations of pressure in the system?

6. A simpler but less exact form of the CLAPEYRON-CLAUSIUS equation (Sec. **133**) can be derived by making the assumptions that the volume of a unit mass of the liquid v_l is negligible in comparison with that of the saturated vapor v_v and that the saturated vapor conforms to the ideal gas laws. Thus, putting $v_l = 0$ and $v_v = R'T/P$, we get $dP/dT = JLP/R'T^2$, or

$$\frac{dP}{P} = \frac{JL\,dT}{R'T^2}.$$

This equation can be integrated if it be assumed either that the heat of vaporization L is a constant or that it is proportional to the first power of the absolute temperature (which is more nearly the case if the temperature range is large). On the first assumption one obtains $\log P = A - (B/T)$, where A and B are constants, and on the second, $\log P = A - (B/T) + C \log T$, where A, B, and C are constants. These equations fit curves like Fig. 125 well over a limited range of temperatures. How well does the equation $\log P = A - (B/T)$ fit your experimental data? Using the pressures which you determined in this and in the preceding experiment for a temperature near zero and a temperature near 100° C, compute values of A and B. Then compute several intermediate values of P and see how closely they fit your curve.

o

OPTIONAL LABORATORY PROBLEM

Acceleration Due to Gravity from Boiling-Temperature Measurements. At two different places, where the accelerations due to gravity are g and g_s respectively, the barometric pressures B and B_s will be different even when the atmospheric pressures at the two places are the same; for if the pressures were the same, then $B\rho g = B_s\rho g_s$, where ρ is the density of mercury. If now the dependence of the boiling temperature of a given substance upon the barometric height B_s is known from formulas or tables for a place where the acceleration due to gravity is g_s, then it is possible from a determination of the boiling temperature at a second place to find the barometric height which would correspond to the atmospheric pressure at that place if the acceleration due to gravity there were g_s. The actual barometric height B at the time and place of the boiling-temperature measurement can be determined in the usual way with a barometer. The unknown acceleration due to gravity g can then be calculated from the equation $g = B_s g_s/B$. This method has yielded valuable results in gravitational surveys.

Use a reliable thermometer accurate to at least 0.01° C to determine the boiling temperature t_b of water. Then find B_s corresponding to this value of t_b from tables (*International Critical Tables*, Vol. III, p. 112) or else calculate it from the interpolation formula $t_b = 100 + 0.0375(B_s - 760)$, where B_s is in millimeters of mercury; this formula is quite accurate within the range $715 \leqq B_s \leqq 775$. Read the barometer, reduce the reading to 0° C, and calculate g.

○

EXPERIMENT XIc. HYGROMETRY

The object is to determine the hygrometric state of the atmosphere in two ways: with the dew-point hygrometer and with the psychrometer.

PART I. The Dew-Point Hygrometer. The form of dew-point hygrometer that offers the greatest possible precision in the determination of the dew-point temperature is that designed originally by ALLUARD (Fig. 132). A metal tube having a nickel-plated and highly polished face is mounted on a heat-insulating support. The tube contains ether, through which a stream of air is bubbled by means of an aspirator bulb until the temperature is lowered to the point at which dew begins to form on the polished surface. To facilitate determination of the moment at which condensation begins, there is mounted close to the polished surface, but not in thermal contact with it, a second polished nickel surface upon which no dew is formed. Thermometers are provided for determining the temperatures of the ether and of the atmosphere. The instrument should be protected from air currents, for otherwise the observed dew points are too low.

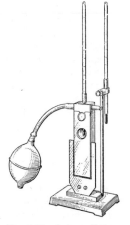

FIG. 132. ALLUARD's form of dew-point hygrometer

1. Why too low?

In order to avoid error due to body heat and moisture, the polished surface and thermometer should be observed through a telescope placed several meters away. At the instant when the face of the tube begins to look cloudier than the adjacent polished surface, the temperature of the ether is noted. Then, with the current of air stopped, the temperature at which the cloudiness disappears is observed. With a little practice the temperatures at which the dew appears and disappears can be made to approach each other to within about 0.1° C. The mean of these two temperatures [1] is taken as the dew point τ.

a. Place the hygrometer in a room in which there is no evaporating water, pour enough ether in the tube to cover the window, and turn the polished face into as favorable a light as possible. Make a quick approximate determination of the dew point. In subsequent observations regulate the evaporation of the ether so that the temperature falls very slowly in the neighborhood of the point sought. In taking observations with rising temperatures an occasional air bubble may be allowed to pass through the ether in order to keep it stirred.

b. Make at least five careful determinations of τ. Observe each time the temperature of the atmosphere near the instrument.

c. Use the average value of τ to find the pressure P of the water vapor in the atmosphere. This is simply the pressure of saturated water vapor at the temperature τ, and is obtained at once from tables[2]; for, although the cooling of the layers of atmosphere which are in contact with the metal surface causes an increase in the density both of the air and of the water vapor of which these layers are composed, yet, since the barometric pressure is in no way affected by the cooling, it is evident that the pressure both of the air and of the water vapor within these layers must remain precisely the same as outside, where no cooling takes place.

d. Find the absolute humidity ρ. A table of *Densities of Saturated Water Vapor* [2] gives the value of this quantity *within the cooled layer*, but in order to obtain its value at a distance from the instrument the density within the cooled layer must be multiplied by T_τ/T_a, where T_τ and T_a are the absolute temperatures of the dew point and of the atmosphere respectively; this follows from the law that, for constant pressure, density is inversely proportional to absolute temperature.

Instead of obtaining ρ in this way from tables, it may be calculated directly from P with the help of Eq. [171], Chap. 10. For it can easily be shown that, since the density of air at 0° C and 1 A$_s$ is 1.293×10^{-3} g · cm^{-3}, and since under like conditions of temperature and pressure the density of water vapor in terms of air is 0.624, Eq. [171] becomes

$$\rho = 0.00029 \, \frac{P}{T_a}, \qquad [184]$$

---

[1] The observed temperatures, if carefully taken, will be slightly below the actual dew point. This is because the initial deposit is in the form of minute droplets, whose vapor pressure is greater than that for a flat surface at the same temperature (Sec. **133**). At ordinary temperatures and for droplets of radii 10^{-4} cm, the error introduced by this effect is approximately 0.02° C. It may therefore be neglected ordinarily.

[2] Appendix 13, Table E.

where P is expressed in millimeters of mercury. This extension to unsaturated vapors of an equation which holds rigorously only for ideal gases must be permissible in practice, since the results of Exp. XIA, question 6, show that at ordinary temperatures Eq. [171] may be applied with very little error even to saturated water vapor.

e. Calculate the relative humidity r.

PART II. The Psychrometer. *a.* Have the wet bulb thoroughly saturated with water. If the psychrometer is of the stationary type, in which the thermometers are mounted on a base, hang the instrument from a cord and allow it to swing as a pendulum for several minutes. If it is of the sling type (Fig. 133), whirl the thermometers rapidly for half a minute. The rate of rotation should be a natural one, for if it is too violent or irregular the instrument may be damaged. Take care not to strike the instrument on any object.

Stop the instrument, and quickly read the wet-bulb and then the dry-bulb thermometer. Immediately set the instrument in motion and then take a second set of readings. Repeat until at least two successive readings of the wet-bulb thermometer are found to agree closely, thereby showing that the wet bulb has reached its lowest temperature. Observe the barometric pressure B.

b. Calculate the pressure P of the water vapor in the atmosphere by means of the empirical formula

$$P = P_w - 0.00066\,B(t_a - t_w), \qquad [185]$$

where t_a and t_w are the readings, in degrees centigrade, of the dry bulb and wet-bulb thermometers respectively, and P_w is the pressure of saturated water vapor as given by tables for the temperature t_w.

Fig. 133. The sling psychrometer

c. With the help of Eq. [184] and tables, calculate ρ, r, and τ.

2. Derive Eq. [184].

3. If the temperature of the air at sunset on a clear day is 10° C, and if the wet-bulb thermometer reads 8° C, at what temperature will dew form? Need there be fear of frost during the night?

4. If, in question 3, the wet-bulb thermometer had read 4.5° C, what would the dew point have been? In this case frost would have been almost certain. Why?

OPTIONAL LABORATORY PROBLEM

Simple Chemical Hygrometer. Fill three 250-ml flasks with small fragments of pumice stone. Saturate two of them with strong sulfuric acid and the third with distilled water. Weigh very carefully the flasks containing the acid. Connect the three flasks to a water aspirator, with the one containing the water in the middle, and pass a gentle stream of air through them for a considerable time. Disconnect and again weigh the flasks containing the acid. The ratio of the gains in weight obviously will be the relative humidity. What are the advantages and the disadvantages of this type of hygrometer? A still better form of chemical hygrometer is described in connection with Prob. 13 at the end of this chapter.

○

QUESTION SUMMARY

1. What is the nature of the departures of actual gases from the equation of state for ideal gases? If the product PV at constant room temperature is plotted against P, what is the nature of the curve obtained (*a*) for an ideal gas? (*b*) for hydrogen and helium? (*c*) for all other gases?

2. How are these departures explained by the kinetic theory? What other evidence is there that this explanation is correct?

3. State the VAN DER WAALS equation of state; make clear the meaning of each constant in the equation and the reasons for its introduction. How closely does this equation fit the facts for real gases?

4. Illustrate the continuity of the liquid and gaseous states by describing ANDREWS'S experiments on carbon dioxide. Define *critical temperature*.

5. What is the distinction between a *gas* and a *liquid*? Contrast the kinetic-theory picture of a liquid with that of a gas.

6. On the basis of kinetic theory, how is evaporation explained? Why is it a cooling process?

7. Define *saturated vapor*. Upon what factors do the density and pressure of a saturated vapor depend, and how do they depend upon them?

8. What is meant by the *boiling temperature*? Why must a thermometer which is to indicate the true boiling temperature be placed not in the boiling liquid itself but in the vapor arising from it?

9. How does the presence of a dissolved solid affect the vapor pressure of a liquid? the boiling temperature? the freezing temperature?

10. Name and define the four quantities involved in hygrometric determinations. How are they determined experimentally? Explain in detail how each may be calculated with the help of a table of densities and pressures of saturated water vapor when one of these is known. Why is it that if the dew point τ is known, the pressure P can be obtained directly from the table, or vice versa, whereas the density ρ cannot?

PROBLEMS

1. *Isobaric curves* are the curves that connect the volume and the temperature when the pressure is kept constant. Indicate the form of the isobars (*a*) for ideal gases; (*b*) for actual gases.

2. In the VAN DER WAALS equation, Eq. [183], when V is the volume of the gas in terms of its volume at 0° C and 1 A_s as the unit, and P is the pressure in atmospheres, the constants a and b for air have the values 0.0026 and 0.0021 respectively. (*a*) Given 1.00 l of air at 0° C under a pressure of 1 A_s, what will be its new pressure if it is compressed to 0.150 l and brought back again to the same temperature? (*b*) How does the pressure at the smaller volume compare with that predicted by BOYLE'S law?

Ans. (*a*) 6.65 A_s; (*b*) 6.67 A_s.

3. By writing the VAN DER WAALS equation in powers of V and making use of the fact that the three roots of this cubic equation are equal at the critical point, show that the critical temperature T_c, the critical pressure P_c, and the critical volume V_c are, in terms of the VAN DER WAALS constants, $T_c = 8\,a\,/\,27\,NRb$; $P_c = a\,/\,27\,b^2$; $V_c = 3\,b$.

4. Explain why, from the standpoint of kinetic theory, a lower temperature can be reached by fanning an open vessel of ether than by fanning an open vessel of water.

5. *Isometric curves* are the curves that connect the pressure and the temperature when the volume is kept constant. Indicate the form of these curves for (*a*) ideal gases; (*b*) actual gases; (*c*) a saturated vapor in contact with the liquid.

6. In a uniform barometer tube in which the mercury stands but 40.0 cm high, the space above the mercury is 40.0 cm long and contains at first only dry air and mercury vapor. A few drops of ether are then introduced into the tube. If the pressures of saturated ether vapor and of saturated mercury vapor at the existing temperature are 30.0 cm and 9.0×10^{-5} cm of mercury respectively, to what point above the mercury in the cistern will the mercury in the tube ultimately fall? Assume that the barometer stands at 76 cm. *Ans.* 21 cm.

7. If 150 cm³ of oxygen are collected over water at a pressure of 740 mm of mercury and a temperature of 20.0° C, what volume would the dry oxygen occupy under standard conditions of pressure and temperature? The maximum pressure of aqueous vapor at 20.0° C is 17.6 mm of mercury.

Ans. 133 cm³.

8. Find the boiling temperature of water on a mountaintop at a time when the barometer reads 22.0 in. of mercury. *Ans.* 91.6° C.

9. How much work is done against the external atmospheric pressure in evaporating 100 g of water at 98° C and 707.3 mm of mercury, the density of steam under these conditions being 0.560 kg · m⁻³?

Ans. 1.68×10^{11} ergs.

The Properties of Vapors 261

10. Is it correct to say that a hot-air furnace dries the air, in the sense of lowering the absolute humidity? Does it lower the relative humidity?

11. Two thousand cubic centimeters of dry air at 15° C and 1 A_s are passed through flasks which contain a known mass of carbon disulfide at 15° C, and the resulting mixture of air and carbon disulfide vapor is allowed to escape into the room at a pressure of 1 A_s. When the flasks are reweighed, the decrease in mass is found to be 3.011 g. What is the vapor pressure of carbon disulfide at 15° C?   *Ans.* 242 mm of mercury.

12. (*a*) How would you proceed in order to raise the humidity in a house that is heated by means of a hot-air furnace? (*b*) How much water would have to be evaporated in order to raise the relative humidity of a room of volume 500 m³ from 20 to 50 percent, the temperature being 20° C?
                     Ans. (*b*) 2.6 kg.

13. With the help of BOYLE'S law and Eq. [184], derive the laboratory equation for use with the chemical hygrometer of Fig. 134, namely

$$\Delta m = 0.00029 \frac{P}{T_a} \frac{B - P_a}{B - P} V,$$

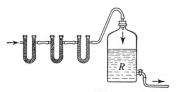

FIG. 134. The chemical hygrometer, first used by C. BRUNNER in 1844. Air is drawn through drying tubes and the increase Δm in the mass of the tubes is measured

where P_a is the pressure of saturated water vapor at the temperature of the atmosphere, and V is the volume of water that has been drawn out of R during the experiment. Show that P is the only unknown quantity in this equation.

14. When the relative humidity is 0.47 at 21° C, what will be the dew point?   *Ans.* 9.3° C.

15. The density of dry air at 18.0° C and 755 mm of mercury is 1.205×10^{-3} g · cm⁻³. Find the density of the atmosphere at this temperature and pressure when the dew point is 10.0° C.
                     Ans. 1.199×10^{-3} g · cm⁻³.

○

WENT TO the Royal Institution last night in hopes of hearing Faraday lecture, but the lecture was given by Mr. Pereira upon crystals, a subject of which he appeared to be master, to judge by his facility and fluency. . . . Met Dr. Buckland and talked to him for an hour, and he introduced me to Mr. Wheatstone, the inventor of the electric telegraph. . . . There is a cheerfulness, an activity, an appearance of satisfaction in the conversation and demeanour of scientific men that conveys a lively notion of the *pleasure* they derive from their pursuits.

Greville's Memoirs, March 17, 1838

CHAPTER TWELVE

QUANTITY OF HEAT AND CALORIMETRY

IT WAS formerly a common supposition, that the quantities of heat required to increase the heat of different bodies by the same number of degrees, were directly in proportion to the quantity of matter in each; and therefore, when the bodies were of equal size, the quantities of heat were in proportion to their density. But very soon after I began to think on this subject, (anno 1760) I perceived that this opinion was a mistake, and that the quantities of heat which different kinds of matter must receive, to reduce them to an equilibrium with one another, or to raise their temperature by an equal number of degrees, are not in proportion to the quantity of matter in each, but in proportions widely different from this, and for which no general principle or reason can yet be assigned.... This opinion was first suggested to me by an experiment described by Dr. Boerhaave (Boerhaave, Elementa Chemiae, exp. 20, cor. 11). After relating the experiment which Fahrenheit made at his desire by mixing hot and cold water, he also tells us that Fahrenheit agitated together quicksilver and water unequally heated. From the Doctor's account, it was quite plain, that quicksilver, though it has more than 13 times the density of water, produced less effect in heating or cooling water to which it was applied, than an equal measure of water would have produced. He says expressly, that the quicksilver, whether it was applied hot to cold water, or cold to hot water, never produced more effect in heating or cooling an equal measure of the water than would have been produced by the water equally hot or cold with the quicksilver, and only two-thirds of its bulk. He adds, that it was necessary to take three measures of quicksilver to two of water, in order to produce the same middle temperature that is produced by mixing equal measures of hot and cold water.

> JOSEPH BLACK, *Lectures on the Elements of Chemistry*, delivered at the University of Edinburgh, Vol. I, p. 79

○

That bodies change in temperature is a fact of direct observation, but the notion that a something called heat passes between bodies of changing temperature is of the nature of a hypothesis. This hypothesis has taken two rival forms (Sec. **54**). One is that heat is an imponderable and indestructible fluid, called by the French chemists DE MORVEAU, LAVOISIER, BERTHOLLET, and DE FOURCROY[1] the *caloric*, the passing of which into or out of a body is the cause of temperature change. The other is that a rise in temperature is an increase, not in the quantity of a contained heat fluid, but simply in the mean kinetic energy of the molecules themselves (Sec. **116**). A knowledge of the caloric theory is now important only because of the

---

[1] *Méthode de Nomenclature Chimique* (Paris, 1787), pp. 30–31; tr. by J. St. John, *Method of Chymical Nomenclature* (London, 1788), pp. 22–23. The term *calorimeter* is due to LAVOISIER (1743–1794); see *Traité Elémentaire de Chimie* (Paris, 1789), Vol. II, pp. 389–390; tr. by R. Kerr, *Elements of Chemistry* (Edinburgh, 1790), p. 345.

Joseph Black as Caricatured by John Ray
and as Portrayed by Henry Raeburn

Good caricatures, like good portraits, are a valuable addition to the history of science. They may depict facts in the life of the person caricatured or may exaggerate personal traits or habits in a way that a true portrait cannot do. Thus, from the friendly caricature reproduced here, one can conclude, quite correctly, that Black was a deliberate and complacent lecturer of methodical habits, quiet manner, and precise speech. These characteristics are embodied in Raeburn's portrait, but in the portrait they are not humorously accentuated by the wig, glasses, and gown, as they are in the caricature. The following word picture given by Black's pupil, colleague, and friend, John Robison, is interesting in this connection:

" His [Black's] personal appearance and manner were those of a gentleman, and peculiarly pleasing. His voice in lecturing was low, but fine; and his articulation so distinct that he was perfectly well heard by an audience consisting of several hundreds. His discourse was so plain and perspicuous, his illustration by experiments so apposite, that his sentiments on any subject never could be mistaken, even by the most illiterate; and his instructions were so clear of all hypothesis or conjecture, that the hearer rested on his conclusions with a confidence scarcely exceeded in matters of his own experience."

A Group of German Physicists
of the 19th Century

Photograph from *Die Naturwissenschaften* **13**, 36 (1925). By permission of the publishers

RUDOLPH CLAUSIUS (lower right) (1822–1888) has been called by MAXWELL the principal founder of the kinetic theory of gases ; he also helped make thermodynamics a science and advanced the dissociation hypothesis to explain electrolysis. AUGUST KUNDT (left) (1839–1894) was HELMHOLTZ's successor at Berlin; the KUNDT tube for comparing the speeds of sound in various mediums is known to every student of physics. GEORG QUINCKE (center) (1834–1924) of Heidelberg is known for his work on surface tension, magnetic permeabilities, and dielectric constants. FRIEDRICH WILHELM KOHLRAUSCH (upper right) (1840–1910) was director of the Reichsanstalt from 1895 to 1905, succeeding HELMHOLTZ ; his texts on practical physics, which have been translated into many languages, did much to stimulate experimental physics.

light that it throws upon the terminology of heat. The theory was altogether abandoned after JOULE'S demonstration of the equivalence of heat and work (Secs. **54, 55**).

138. Unit Quantity of Heat. The heat unit employed by the calorists was the quantity of heat that must enter a unit mass of water in order to raise its temperature one degree. This definition has been retained, but the old concept of the transfer of a heat fluid has been replaced by the concept of a transfer of molecular energy, kinetic or potential or both. In other words, the heat unit is now to be regarded simply as an arbitrarily chosen unit that is especially convenient for measuring energy which is in the form of heat. The heat unit may also be defined as the quantity of heat that must *leave* a unit mass of water in order to lower its temperature one degree; for experiments show that when a given mass of water cools one degree, it gives out an amount of heat equal to that absorbed when it is heated through the same range of temperature.

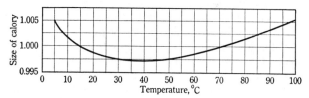

FIG. 135.   Relative values of the calory at different temperatures, taking the 15° calory as the unit

Accurate experiments show that the amount of heat which must enter or leave a unit mass of water in order to produce a change of one degree in the temperature is not the same at all temperatures (Fig. 135); hence the following more precise definitions of the thermal units of energy. The cgs unit of heat is the *calory*. There are two calories in common use. The *15° calory* (cal_{15}) is defined as the quantity of heat required to change the temperature of 1 g of water from 14.5° to 15.5° C. The *mean calory* is defined as one one-hundredth of the quantity of heat required to change the temperature of 1 g of water from 0° to 100° C. In each case the pressure is taken to be 1 A_s. Experiments show that the mean calory is practically equivalent to the 15° calory. As stated in Sec. **55**, the latter is equivalent to 4.185 j. In the fps system the unit of heat is the *British thermal unit* (B.t.u.), which is the heat required to raise 1 lb of water through 1° F.

EXAMPLE. Show that 1 B.t.u. is equivalent to 252 cal.

Thermal Capacity

139. Thermal Capacity. The first investigator to draw a sharp distinction between *heat* and *temperature* was JOSEPH BLACK, who by his discovery of specific and latent heats laid the foundations of the quantitative science of heat.[1] In about the year 1760 BLACK[2] arrived at the important conclusion that "the quantities of heat which different kinds of matter must receive . . . to raise their temperatures by an equal number of degrees, are not in proportion to the quantity of matter in each · · ." In other words, the quantity of heat given up by 1 g of water in falling through one degree would raise very different masses of other substances through one degree — for example, about 30 g of mercury or 9 g of iron. The calorists explained these facts by the assumption that equal masses of different substances possess different capacities for the heat fluid. Thus the *thermal capacity of a body* came to be defined as the number of heat units required to raise the temperature of the body through one degree, and the *thermal capacity of a substance*, or *specific heat* (symbol c), as the number of heat units required to raise the temperature of a unit mass of that substance through one degree.

These definitions are still retained now that heat is regarded as molecular energy; but the fact that different amounts of this energy must be communicated to equal masses of different substances in order to produce the same increase in temperature, or, what is the same thing, the same increase in the average kinetic energy of translation of the molecules (Sec. **116**), is attributed to two factors:

a. The differences in the number of molecules contained in equal masses of different substances.

b. Other differences in the internal work which are incidental to an increase of temperature. The term *internal work* includes not only (1) the work ΔW_t done in raising the temperature of the body but also (2) the work ΔW_a done in augmenting the energy, kinetic or potential, of the atoms and electrons inside the molecules, and (3) the work ΔW_i done in increasing the distances between the molecules, in case the average distance separating them is increased.

---

[1] See *D. McKie and N. H. de V. Heathcote, *The Discovery of Specific and Latent Heats* (Arnold, 1935).

[2] See the quotation at the beginning of this chapter. These *Lectures, published in 1803, after BLACK's death, were written out from his own notes, supplemented by those of some of his students, under the editorship of John Robison. BLACK's heavy duties, ill-health, lack of initiative, and almost morbid horror of hasty generalization prevented him from going further than forming a plan of the work, and during his lifetime most of his great discoveries on heat remained unpublished.

The first of these two factors can easily be investigated, for if this were the only cause of the differences in the specific heats of different substances, these differences would disappear in a comparison of quantities which represent, not equal masses, but equal numbers of molecules. Such a quantity evidently can be obtained by taking, in each case, a mole of the substance. The number of

TABLE VI · *Molar Thermal Capacities of Various Substances.* (*The values are for constant pressure and, in the case of gases, for 15° C and 1 A_s*)

Group	Substance	Formula	Molecular heat, cal · mole^{-1} · deg^{-1}
A	Helium	He	4.96
	Argon	A	4.96
B	Hydrogen	H$_2$	6.86
	Nitrogen	N$_2$	6.93
	Oxygen	O$_2$	7.04
	Carbon monoxide	CO	6.94
	Nitric oxide	NO	7.00
	Hydrochloric acid	HCl	7.15
	Chlorine	Cl$_2$	8.15
C	Carbon dioxide	CO$_2$	8.79
	Nitrous oxide	N$_2$O	8.85
D	Sodium	Na$_2$	13.5
	Potassium	K$_2$	12.9
	Copper	Cu$_2$	12.1
	Mercury	Hg$_2$	13.3
E	Nickel monoxide	NiO	11.9
	Cupric oxide	CuO	11.3
	Mercuric oxide	HgO	11.2
F	Calcium chloride	CaCl$_2$	18.2
	Zinc chloride	ZnCl$_2$	18.6
	Barium chloride	BaCl$_2$	18.6
G	Calcium sulfate	CaSO$_4$	26.7
	Lead sulfate	PbSO$_4$	26.4

units of heat required to raise the temperature of one mole of a substance through one degree is called its *molar thermal capacity* or *molecular heat.* Evidently this quantity is simply the product of the specific heat and the molecular weight. Table VI shows that many of the differences in thermal capacities do in fact disappear upon comparison of equal numbers of molecules. The differences which are still left must be attributed wholly to the second factor. It is true, of course, that when a substance is heated under atmospheric pressure a certain amount of *external work* ΔW_e must also be done in expanding against the atmospheric pressure; but this may be neglected in the case of solids and liquids, and for gases it is practically a constant quantity. (Why constant?)

In 1831 F. E. NEUMANN [1] investigated the molar thermal capacities of chemically similar substances the molecules of which possess the same number of atoms, and found them to be nearly the same in a given state of aggregation — for example, in the gaseous state. This is known as *Neumann's law*. It must of course be inferred from this law that the internal work is about the same for such similar substances. The law is not exact, nor could it be expected to be in view of the differences in the attractions which exist between different sorts of molecules.

Since the molecules of *gases* are not subjected to appreciable mutual attractions, it might be expected that with molecules of equal complexity the molecular work would be less in the gaseous than in the liquid or solid condition (groups B and D, Table VI). Again, it would be natural to conclude that for substances in the same state of aggregation the internal work would, in general, increase with the complexity of the molecule, and this inference is also in accord with the facts presented in the table.

140. Variations of Specific Heats with the Temperature. Experiments show that the specific heat of a given substance is not constant, but that in general it increases steadily with the temperature. The rate of increase, fortunately, is slight for water and for most solids with the exception of carbon, boron, and silicon, so that in ordinary work at moderate temperatures it may be disregarded. But for most liquids it is far from negligible. For example, REGNAULT found that the specific heat of ethyl alcohol is 0.548 cal · g^{-1} · deg^{-1} at 0° C and 0.648 cal · g^{-1} · deg^{-1} at 40° C. The fact of the dependence of the specific heat of water upon temperature, which has already been referred to in Sec. **138** and Fig. 135, was first established by REGNAULT.[2] The first extensive work upon this variation was done by H. A. ROWLAND [3] of Johns Hopkins University, in 1879.

It is apparent that the quantity usually obtained by experiment is not the specific heat at a given temperature but rather the *mean* specific heat between two specified temperatures. Thus, if Q repre-

---

[1] Poggendorff's *Annalen der Physik und Chemie* **23**, 1 (1831).

[2] Poggendorff's *Annalen der Physik und Chemie* **79**, 241 (1850); *Relations des Expériences* (Paris, 1847), Vol. 1, p. 729.

[3] *Proceedings of the American Academy of Arts and Sciences* **15**, 75 (1879–1880); *Physical Papers* (Baltimore, 1902), p. 387.

sents the quantity of heat passing into or out of a mass m while it changes in temperature from t_1 to t_2, then the *mean specific heat c between t_1 and t_2* is

$$c = \frac{Q}{m(t_2 - t_1)}.$$ [186]

For example, the mean specific heat of water between 14.5° and 15.5° C is equal to the 15° calory. If we take infinitely small intervals of temperature we obtain, for the *specific heat at temperature t in the interval dt,*

$$c_t = \frac{Q}{m\,dt}.$$ [187]

It is usually permissible to put c_t, the specific heat at a certain temperature, equal to the mean specific heat c of an adjoining interval of moderate size.

141. Specific Heats at Constant Volume and at Constant Pressure. As has already been pointed out in Sec. 139, a quantity of heat Q communicated to any body may expend itself in one or more of several ways. A portion of it, ΔW_t, may be employed in raising the temperature of the body, and another portion, ΔW_a, in increasing the energy inside the molecules. Since an increase in temperature is in general accompanied by an increase in volume, a part of the heat energy Q may also be expended in two other ways. For if the body be subjected to external forces, external work of expansion ΔW_e will be done against these forces while the volume is changing. So also internal work of expansion ΔW_i will be done against internal forces, such as molecular attractions, while the volume is changing. In general, then,

$$JQ = \Delta W_t + \Delta W_a + \Delta W_i + \Delta W_e,$$ [188]

where the quantities in the right-hand member are expressed in mechanical units, Q is expressed in thermal units, and J is JOULE'S equivalent (Sec. 55). This equation is simply an expression for the particular case at hand of the *first law of thermodynamics* (Sec. 103). A more general mathematical statement of the first law is

$$JQ = \Delta U + \Delta W,$$

an equation which states that a quantity of heat Q absorbed by a system is, in general, used up partly to produce an increase ΔU in the internal energy of the system and partly to cause the system to do external work of amount ΔW.

It is evident from Eq. [188] that the specific heat of a substance is an indefinite quantity unless one specifies the conditions under which the heating is carried out. There are two important cases: when the substance is heated at constant volume, in which case

$\Delta W_i = \Delta W_e = 0$, and when it is heated at constant pressure. We therefore distinguish between the *specific heat at constant volume, c_v,* and the *specific heat at constant pressure, c_p.* (Is c_p always larger than c_v?)

With the volume kept constant, let a mass m of a substance be heated through Δt degrees. The quantity of heat Q which must be communicated to the substance is $c_v m \Delta t$; and, since $\Delta W_e = \Delta W_i = 0$ when the volume is constant, Eq. [188] becomes, upon rearrangement,

$$c_v = \frac{\Delta W_t + \Delta W_a}{Jm \Delta t}.$$ [189]

Suppose, on the other hand, that the substance be heated at constant pressure. In this case $Q = c_p m \Delta t$. By inserting this expression for Q in Eq. [188] and subtracting Eq. [189] from the result, we get for the difference of the two specific heats

$$c_p - c_v = \frac{\Delta W_i + \Delta W_e}{Jm \Delta t}.$$ [190]

In the case of gases, ΔW_i is usually very small (Sec. **128**), whereas ΔW_e is very large because of the large expansivities of gases. In the case of solids the reverse is true: ΔW_i cannot be neglected because of the large cohesive forces between the molecules, whereas ΔW_e is negligibly small because of the small expansivities of solids. The differences between the two specific heats c_p and c_v are found to be much larger in the case of gases than for solids and liquids, however, and this is the only case to which we shall give further consideration.

142. Specific Heats of a Gas. It will now be shown that for any ideal gas, if one of these specific heats is determined by experiment, the other can be deduced. Let us first consider the difference $c_p - c_v$. Imagine a mass m of an ideal gas to have been heated at constant pressure P until the temperature and volume have changed by the amounts ΔT and ΔV respectively. Then, since $\Delta W_i = 0$ for an ideal gas (Sec. **113**) and $\Delta W_e = P \Delta V$ when the pressure is constant, Eq. [190] becomes $c_p - c_v = P \Delta V / Jm \Delta t$. But, by Eq. [131], Chap. 9, $PV = mR'T$, and therefore $P \Delta V = mR' \Delta T$ for constant pressure; hence

$$c_p - c_v = \frac{R'}{J},$$ [191]

where R' is the gas constant for unit mass of the ideal gas in question. Essentially this equation was used by MAYER[1] in 1842 to make the

---

[1] "Bemerkungen über die Kräfte der unbelebten Natur," *Annalen der Chemie und Pharmacie* **42**, 233 (1842). See Sec. **55** and footnote 3, p. 76.

JAMES PRESCOTT JOULE, 1818–1889

From the painting by G. Patten, in the possession of the Manchester Literary and
Philosophical Society

Facsimile of a Portion of a Letter
from J. CLERK MAXWELL to BALFOUR STEWART

Reproduced through the courtesy of Professor H. Lowery

IN THIS LETTER, which is preserved among the historical items in the Department of Pure and Applied Physics in the College of Technology, Manchester, England [see the interesting paper on "The Joule Collection in the College of Technology, Manchester," by H. Lowery in the *Journal of Scientific Instruments* 7, 369 (1930); 8, 1 (1931)], MAXWELL expresses appreciation for JOULE's work as follows:

"There are only a very few men who have stood in a similar position and who have been urged by the love of some truth which they were confident was to be found though its form was as yet undefined to devote themselves to minute observations and patient manual and mental toil in order to bring their thoughts into exact accordance with things as they are."

first computation of J, the other quantities in the equation having been determined by experiment.

Direct determinations of the specific heat at constant volume are difficult to make because of the smallness of the mass of the gas which can be enclosed in any container, as compared with the mass of the container itself. For this reason, values of c_v are usually obtained from those of c_p, either by use of Eq. [191] or from a knowledge of the value of the ratio γ of the two specific heats for the gas in question, namely,

$$\gamma = \frac{c_p}{c_v}. \qquad [192]$$

The value of this ratio can be determined experimentally[1] or computed from a knowledge of the speed of sound in the gas (Chap. 15).

143. Theoretical Values of the Specific Heats of Ideal Gases. Degrees of Freedom and Equipartition of Energy. A simple calculation made with the help of Eq. [191] will show that the *molar* thermal capacity of an ideal gas at constant pressure exceeds that at constant volume by the amount R/J, or 2 cal · mole^{-1} · deg^{-1}, approximately. By applying this result to the experimental data for gases which appear in Table VI, we see that for the *monatomic* gases (group A) the molar thermal capacities at constant volume all have the value 3 cal · mole^{-1} · deg^{-1}, whereas for the *diatomic* gases (group B) they have the value 5 cal · mole^{-1} · deg^{-1}, approximately. Let us see how well the theoretical predictions afforded by kinetic theory agree with these experimental values.

The molar thermal capacity at constant volume is, by definition (Sec. 141), the quantity of heat that must be communicated to one mole of a substance held at constant volume in order to raise its temperature one degree. In the case of a monatomic gas heated at constant volume, $\Delta W_a = \Delta W_i = \Delta W_e = 0$, and therefore all this communicated heat goes to produce an increase in the kinetic energy of translation ΔW_t of the molecules. Now for a mole of gas the total kinetic energy of translation is $N_o E_{av}$, where N_o is the AVOGADRO number and E_{av} is the average kinetic energy of each molecule; and $N_o E_{av} = \frac{3}{2} RT$, by Eq. [160], Chap. 10. Hence the additional kinetic energy which must be imparted to a mole to raise its temperature one degree ($\Delta T = 1$) is $\frac{3}{2} R$. Since the value of R in thermal units is roughly 2 cal · mole^{-1} · deg^{-1}, this gives for the molar thermal capacity at constant volume the theoretical value 3 cal · mole^{-1} · deg^{-1}, in agreement with experiment.

---

[1] See *J. R. Partington and W. G. Shilling, *The Specific Heats of Gases* (Benn, 1924), Chap. 2, for a detailed discussion of the experimental methods for determining c_p, c_v, and γ

Before proceeding to the cases of other gases, it will be found helpful to introduce the concept of *degrees of freedom* of a body. As we well know, a *particle* in space can move in three, and only three, directions such that its motion along any one of these has no component along either of the other two. These three independent directions, obviously, are mutually perpendicular. In view of this fact, a particle is said to have three degrees of freedom, or three possibilities in the way of motion. It cannot rotate, because it has no appreciable dimensions. If the particle is constrained to move in a single plane, it is said to have two degrees of freedom; and if it is still further limited to motion in a straight line, it then possesses only one degree of freedom. A ball thrown through the air has, on the other hand, six degrees of freedom, of which three are translatory and the remaining three are associated with the independent rotations of the body about three mutually perpendicular axes. If the ball were rolling on a table, it would then have only five degrees of freedom, two of translation and three of rotation. A very thin straight rod thrown into the air may be regarded as having five degrees of freedom, three of translation and two of rotation, the rotation of the rod about its geometrical axis being neglected because of the small moment of inertia of the rod about that axis. In the case of a system of *unconnected* bodies, the number of degrees of freedom of the system is the sum of the number of degrees of freedom possessed by the several bodies. Thus, if there are N monatomic molecules in a given mass of gas, the system has $3N$ degrees of freedom, for each monatomic molecule has three degrees of freedom.

The degrees of freedom of a system of monatomic molecules may be divided into three groups, corresponding to the three mutually perpendicular directions of translatory motion, and with each of these groups is associated kinetic energy. According to the theorem of *equipartition of energy*, which has already been mentioned in Sec. 117, when such a gas is heated each degree of freedom in one group receives, on the average, the same amount of kinetic energy as does a degree of freedom in any other group. Thus, since the molar thermal capacity of a monatomic gas kept at constant volume is 3 cal · mole^{-1} · deg^{-1}, it follows, on the assumption of equipartition, that in the heating of a mole of the gas through one degree of temperature each of the three groups corresponding to the three degrees of freedom takes up 1 cal of heat.

Consider now the motions of a molecule of a *diatomic* gas like hydrogen, a simple model of which is shown in Fig. 136. The center of mass of such a molecule has a motion of translation, which involves three degrees of freedom. Besides this, the molecule has appreciable moments of inertia about axes at right angles to the line joining its two atoms, and hence the energies associated with these two rotational degrees of freedom must be taken into account. If equipartition be assumed, each of the five degrees of freedom has associated with it 1 cal · mole⁻¹ · deg⁻¹.

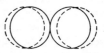

Fig. 136. Simple model of a vibrating hydrogen molecule

The molar thermal capacity of a diatomic gas should therefore be 5 cal · mole⁻¹ · deg⁻¹, which observation shows to be approximately the case. In a *polyatomic* gas there are three rotational degrees of freedom and hence six degrees of freedom in all. The resulting theoretically predicted value of 6 cal · mole⁻¹ · deg⁻¹ is also in reasonable agreement with experiment.

In a similar manner the numerical values of the ratio γ of the two specific heats can be calculated. Thus, by Eqs. [189] and [190],

$$\gamma = \frac{c_p}{c_v} = \frac{\Delta W_t + \Delta W_a + \Delta W_i + \Delta W_e}{\Delta W_t + \Delta W_a}. \qquad [193]$$

For any gas, ΔW_i is very small (Sec. **128**) and

$$\Delta W_e = P \, \Delta V = R \, \Delta T = 2 \text{ cal} \cdot \text{mole}^{-1} \cdot \text{deg}^{-1},$$

approximately. Hence, for a monatomic gas,

$$\gamma = \frac{3 + 0 + 0 \text{ (approx.)} + 2}{3 + 0} = \frac{5}{3} = 1.7;$$

for a diatomic gas,

$$\gamma = \frac{3 + 2 + 0 \text{ (approx.)} + 2}{3 + 2} = \frac{7}{5} = 1.4;$$

and for a polyatomic gas,

$$\gamma = \frac{3 + 3 + 0 \text{ (approx.)} + 2}{3 + 3} = \frac{8}{6} = 1.3.$$

These values are in excellent agreement with the experimental values given in Table VII for the gases of simple molecular structure. For very complex molecules, such as ether, ΔW_a should become large (for reasons given in the next paragraph) and γ should approach unity, which is also verified.

TABLE VII. *Experimental Values of γ for Various Gases at 15° C and 1 A$_s$*

Gas	γ	Gas	γ
Helium	1.666	Oxygen	1.396
Argon	1.666	Air	1.403
Hydrogen	1.408	Carbon dioxide	1.302
Nitrogen	1.405	Ether vapor	1.024

Some diatomic gases, such as chlorine, and also various polyatomic gases of complex molecular structure, possess thermal capacities that are distinctly larger than the theoretically predicted values. It is found that these discrepancies tend to disappear if it be assumed that the molecules are not rigid but that the atoms composing them can vibrate with respect to one another under the action of binding forces which act like springs and which vary in magnitude with the substance and the temperature. On the assumption of equipartition such intramolecular vibrations receive their share of the energy. Now, as in the case of an ordinary pendulum, the energy of these vibrations is at any moment part kinetic and part potential, and it can be proved that if the vibrations are of the simplest type, the average amount of potential energy associated with each vibrational degree of freedom is the same as its average kinetic energy (Chap. 14). Each *vibrational* degree of freedom must therefore have associated with it not 1 but 2 cal · mole^{-1} · deg^{-1}. For example, a diatomic molecule having six degrees of freedom, one of which is vibrational, would have a theoretical thermal capacity of 7 cal · mole^{-1} · deg^{-1}.

It turns out, then, that if the vibrational degrees of freedom are taken into account, the theoretical values of the thermal capacities are no longer lower than the experimental values. As a matter of fact, they are a little too high, although this discrepancy tends to disappear as the temperature increases. It would appear from this that a part of the molecules, but not all, in their collisions with other molecules receive enough energy from the impacts to start them vibrating. This would also explain why many diatomic molecules, such as hydrogen, appear to have no vibrational degrees of freedom at ordinary temperatures; their thermal capacities do not exceed 5 cal · mole^{-1} · deg^{-1} simply because their molecules are too tightly bound to be set into vibration by impacts at ordinary temperatures. It is interesting to note in this

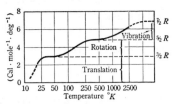

FIG. 137. Variation of the thermal capacity of diatomic hydrogen with temperature. [See F. K. Richtmyer, *Introduction to Modern Physics*, ed. 2 (McGraw-Hill, 1934), p. 297]

connection that when hydrogen is cooled below 60° K, its thermal capacity drops to 3 cal · mole^{-1} · deg^{-1} (Fig. 137), which means that it then behaves like a monatomic gas; at this temperature even its rotational degrees of freedom appear to be suppressed.

This idea that the energy of impact must exceed a certain critical value, which depends upon the kind of molecules, before a certain degree of freedom becomes active, obviously gives a simple explanation of the variation of the thermal capacities of gases with temperature. It presents the very great difficulty, however, that it is actually *contrary to the classical laws of mechanics.* According to the classical theory, every degree of freedom would receive its share of energy at all temperatures, just as a pendulum can be set into vibration, though possibly minutely, by even the smallest blow. The results of experiments in various fields of physics since 1900 have forced the conclusion that the laws of classical physics must be extended if they are to be able to interpret molecular and atomic phenomena satisfactorily, and that a molecule which has a natural frequency of ν vibrations per second will not be put into oscillation until it receives a quantity of energy equal to $h\nu$, where h is a new universal constant known as the *constant of action.* This condition was first enunciated by MAX PLANCK [1] in 1900. The quantity $h\nu$ is called a *quantum of energy,* and the theory the *quantum theory.* According to quantum theory, then, hydrogen molecules, which have relatively small moments of inertia and consequently high natural frequencies of rotation ν, do not receive enough energy $h\nu$ by impact with other molecules below 60° K to set them into rotation. Larger and heavier molecules, like oxygen, have larger moments of inertia and therefore smaller natural frequencies, so that the energy $h\nu$ required for rotation is smaller; hence we should expect their rotational degrees of freedom to remain active at lower temperatures than in the case of hydrogen, a conclusion that is borne out by experiment.

144. Atomic Thermal Capacities.   The Law of Dulong and Petit. NEUMANN'S law for molar thermal capacities, referred to in Sec. **139**, is historically an extension of an important law discovered in 1818 by PIERRE LOUIS DULONG and ALEXIS THÉRÈSE PETIT,[2] in accordance with which the *atomic thermal capacities,* or products of the specific heat and atomic weight, of all the *solid* elements are nearly the same, amounting to about 6.4 cal per gram atomic weight per degree (group D, Table VI). The fact that this law makes the *atomic* heat capacity, rather than the molar, the invariable quantity in the case of solids would seem to indicate that it is the *atoms,* and not the molecules, that are concerned in the kinetic energy of agitation which governs the temperature of solids. Such a conclusion has since received support by evidence from numerous sources, one of the most

[1] The most important of PLANCK'S early papers on radiation are republished in Ostwald's *Klassiker der Exakten Wissenschaften,* No. 206 (Leipzig, 1923). See also PLANCK'S Nobel Prize Address, delivered at Stockholm, June 2, 1920, and entitled "The Origin and Development of the Quantum Theory," tr. by H. T. Clarke and L. Silberstein (Oxford University Press, 1922).

[2] *Annales de Chimie et de Physique* (2) **10**, 395 (1819). An excerpt appears in *A Source Book in Physics* (1935), pp. 179–181.

striking of which is the work of MAX VON LAUE and of the BRAGGS, who showed by an x-ray method that it is the atom rather than the molecule which is the unit in solid crystals.[1] Now, because of the definiteness of the structure of a solid, the atoms must be held in place between their neighbors, and thus the only sort of heat motions which they can undergo are motions of vibration about their positions of equilibrium. Hence the atoms have only three degrees of freedom, all vibrational. As in the vibration of gas molecules, each vibrational degree of freedom has 2 cal associated with it, for equipartition of energy divides the vibrational energy for each direction into equal parts of kinetic and potential energy. The atomic thermal capacity of a solid should therefore be 6 cal per gram atomic weight per degree, which is, to a first approximation, the value given by DULONG and PETIT'S experimental law and is close to the value approached asymptotically by the experimental curves in Fig. 138 as the temperature increases.

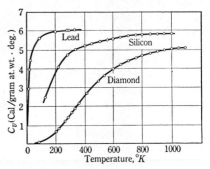

FIG. 138. Variations of atomic thermal capacities of solids with temperature. The curves for nearly all solids lie between the curve for lead and the curve for diamond. It will be observed that the thermal capacity for lead at room temperature is approximately that predicted by DULONG and PETIT'S law, but below 100° K it drops rapidly to zero. The curves for the other substances differ from lead mainly in the temperatures above which they acquire the DULONG and PETIT value and in their slopes below these temperatures

It was known very early that nonmetallic elements of small atomic weights and high melting points, such as carbon, boron, and silicon, were notable exceptions to DULONG and PETIT'S law. Extended research, especially after low-temperature measurements became possible, showed that at least one reason for this was the fact of the variation of thermal capacity with temperature. It was found that at low temperatures the thermal capacities of all solids examined approach zero and that it is only at relatively high temperatures that they approach values in accord with DULONG and PETIT'S law (Fig. 138). As with gases, these variations with temperature have their explanation in the quantum theory. If the binding forces acting in the directions of the three vibrational degrees of freedom are dif-

[1] See W. H. Bragg and W. L. Bragg, *X-Rays and Crystal Structure* (Bell, 1924).

ferent, the atom will have correspondingly different frequencies of vibration ν in those directions. Hence the energy $h\nu$ imparted to the atom by an impact may be sufficient to start the vibration characteristic of one degree of freedom but not of another, and the latter remains inactive. At low temperatures very few of the impacts supply enough energy to start vibration in any of the degrees of freedom, and accordingly the value of the thermal capacity approaches zero at such temperatures. The fact that light elements with high melting points do not have thermal capacities in accord with the DULONG and PETIT value is also explained; the atoms of such elements are bound together by relatively large forces, their natural frequencies are large, and hence a high temperature is required to awaken all degrees of freedom.

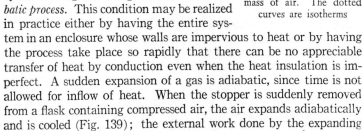

FIG. 139. Curves showing the adiabatic relation between P and V for a given mass of air. The dotted curves are isotherms

145. Adiabatic Processes. Any physical or chemical process carried out in such a way that no heat is allowed to enter or leave the system during the change is called an *adiabatic process.* This condition may be realized in practice either by having the entire system in an enclosure whose walls are impervious to heat or by having the process take place so rapidly that there can be no appreciable transfer of heat by conduction even when the heat insulation is imperfect. A sudden expansion of a gas is adiabatic, since time is not allowed for inflow of heat. When the stopper is suddenly removed from a flask containing compressed air, the air expands adiabatically and is cooled (Fig. 139); the external work done by the expanding gas has been at the expense of its internal energy.

The relation between the pressures and volumes of an ideal gas in an adiabatic change was obtained by LAPLACE[1] and SIMÉON DENIS POISSON[2] before 1823. To derive this relation, consider a given mass m of ideal gas. Let the gas undergo an adiabatic change until its pressure, volume, and temperature have changed from P, V, T to $P + dP$, $V + dV$, $T + dT$ respectively. Then, since $Q = 0$ in an adiabatic process, and since $\Delta W_i = 0$ for an ideal gas, Eq. [188] becomes $dW_t + dW_a + P\,dV = 0$, or

$$d(W_t + W_a) = -P\,dV. \qquad [194]$$

[1] Published in the last volume of his famous *Traité de Mécanique Céleste* (1823), Vol. V, Bk. XII, Chap. III.

[2] *Annales de Chimie et de Physique* (2) **23**, 5 (1823).

This equation tells us that the change in the internal energy is equal to the external work done, the negative sign indicating that if the external work is done *by* the gas, the internal energy is *decreased*. Now, by Eq. [189], $d(W_t + W_a) = c_v Jm \cdot dt$; therefore

$$c_v Jm \cdot dt = - P \, dV. \qquad [195]$$

From Eq. [131], Chap. 9, we obtain

$$P \, dV + V \, dP = mR' \, dT. \qquad [196]$$

Solve this equation for dT and substitute in Eq. [195] for dt; then, remembering that $R' = J(c_p - c_v)$ by Eq. [191] and that $c_p/c_v = \gamma$, we have

$$\frac{dP}{P} + \gamma \frac{dV}{V} = 0 \qquad [197]$$

Integration of this differential equation gives

$$\log P + \gamma \log V = \text{constant}, \qquad [198]$$

or $\qquad\qquad P \cdot V^\gamma = \text{constant}. \qquad [199]$

This is the equation of the adiabatic curves, like those in Fig. 139. As doubtless will have been observed, its derivation involves the assumption that c_v and γ are independent of the temperature, something which is not implied in our definition of an ideal gas. But even if the assumption is not valid, Eq. [199] will hold over a small range of temperatures for which c_v and γ are sensibly constant. This equation was used by CLÉMENT and DESORMES [1] in their experimental determination of γ in 1819.

EXAMPLE. Prove that for small strains taking place adiabatically the volume modulus of elasticity of an ideal gas is

$$k = \gamma P, \qquad [200]$$

where P is the initial pressure of the gas.

Solution. Employ a method similar to that used in the treatment of isothermal elasticity (Sec. 93).

○

Heats of Transformation

146. Energy Changes Associated with Changes in the State of Aggregation. Up to BLACK'S time it was generally supposed that the rise in temperature of a substance in contact with a hot body was continuous; but BLACK pointed out that while ice or snow is changing

---

[1] *Journal de Physique (de la Métherie)* 89, 321, 428 (1819). See also Laplace, *Traité de Mécanique Céleste* (1823), Vol. V, Bk. XII, Chap. III.

into water it maintains, if well stirred, a perfectly constant temperature. To the relatively large amounts of heat which he found had to be added at constant temperature to solids to change them into liquids and to liquids to change them into gases, or, inversely, which are given out in the freezing of liquids and the condensation of gases, BLACK gave the name "latent heats." In order to explain these latent or hidden heats the calorists assumed that the caloric and the solid, say, formed a kind of chemical compound, namely the liquid, thus suppressing the properties of the caloric so that it could not produce a rise in temperature. From the modern point of view, as exemplified by Eq. [188], it is evident that fusion and vaporization simply represent changes for which ΔW_i is zero. While fusion or vaporization is progressing, the temperature of the body remains constant because all of the energy of motion communicated to its molecules by the source of heat is at once transformed into potential energy; that is, the heated molecules immediately break away from the forces which have been holding them in the given state, solid or liquid as the case may be, and thereby lose their increased kinetic energy of translation as quickly as they receive it.

By *heat of vaporization* (symbol L) is meant the energy required to change a unit mass of a substance from a liquid at a certain temperature to a vapor at the same temperature under a specified pressure. Similar definitions exist for *heat of fusion* and *heat of sublimation*.

147. Crystalline and Amorphous Solids. The change from solid to liquid at a definite temperature, which depends only on the nature of the substance and the pressure, is characteristic of solids having a definite *crystal structure*. Solids that do not have such a structure, called *amorphous solids*, of which waxes, glass, and most alloys are examples, pass gradually through all stages of viscosity in melting or solidifying. In such cases the temperature changes continually, there being no definite point at which the substance may be said to melt. Newer methods of analysis, involving the use of x-rays, have shown that many apparently amorphous solids are really made up of extremely small crystals, whereas others appear to represent *undercooled* liquids of very high viscosity. For example, quartz occurs ordinarily in the crystalline form, but when melted and allowed to cool it undercools rather than freezes and becomes sufficiently viscous to assume, to casual inspection, all the properties of a solid.

The existence of a definite melting point in the case of a crystalline solid does not necessarily mean that the process of melting is a sharp transition from the crystalline to the liquid state. The two states undoubtedly are continuous, and the transition from the one to the other probably goes through a sequence of unstable states, characterized by the same hooklike shape of the *PV*-curves as occurs in the VAN DER WAALS isotherms for a fluid below the critical temperature (Fig. 126). There is considerable evidence that near the melting point the solid is no longer entirely crystalline, and that the liquid after melting is still to a great extent crystalline but gradually becomes less so as the temperature is raised.[1]

Besides having definite melting points, many crystalline substances have been found to exhibit other definite transformation temperatures at which they undergo changes in crystal structure. At each of these transformation points, ΔW_t is zero and there exists a heat of transformation (Fig. 140). Thus, if a piece of iron wire heated to bright orange color is allowed to cool, it becomes gradually redder until a temperature of about 900° C is reached, at which point it suddenly brightens, and then becomes redder again until all color has disappeared. This sudden flash of color, which

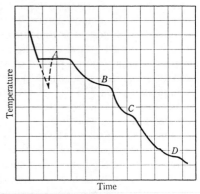

FIG. 140. Typical cooling curve for a crystalline substance that can exist in several crystal forms. At A the substance is freezing, and at each of the points B, C, and D it is undergoing a transition from one crystal form to another. By taking suitable precautions it is often possible to undercool the liquid before freezing, as shown by the dotted portion of the curve

indicates a reheating, occurs because iron exists in at least two crystalline forms, known as α-iron and γ-iron. As it cools down through the temperature 900° C, it changes from the γ to the α form, and this change is accompanied by the evolution of heat, which reheats the wire.

○

Calorimetry

Calorimetry [2] is concerned with the measurement of energy that is in the form of heat. Two general methods have been adopted for making such measurements, and these may be referred to as *thermo-*

---

[1] See *J. Frenkel, "Continuity of the Solid and the Liquid States," *Nature* **136**, 167 (1935).

[2] See the excellent article on "Calorimetry" in *A Dictionary of Applied Physics*, ed. by R. Glazebrook (Macmillan, 1922), Vol. I, pp. 32–75.

metric calorimetry and *change-of-state* calorimetry. In the first method the estimation is reduced to observations of temperatures and the thermometer is the instrument of prime importance; in the second method fixed temperatures are employed and no thermometer is needed.

148. The Method of Mixtures. An example of the thermometric method having considerable historical and practical interest is the *method of mixtures*. It consists in mixing known masses of substances of different temperatures, observing the resulting temperature, and then writing out an equation that contains in one member all the heat quantities lost by the cooling bodies, and in the other all the heat quantities gained by the warming bodies. For example, suppose that it be required to find the heat of vaporization of steam L from an experiment in which a mass m_1 of steam at temperature t_1 is condensed in a mass m_2 of water of temperature t_2 and specific heat c_2. Let t_m be the final temperature attained by the mixture. Then the quantity of heat lost by the vapor in condensing is $L \cdot m_1$; that lost by the condensed vapor in passing from its temperature of condensation t_1 down to t_m is $c_2 m_1(t_1 - t_m)$. The heat gained by the water is $c_2 m_2(t_m - t_2)$. But heat is also gained by the calorimeter vessel which holds the liquid during the experiment, and by the thermometer and stirrer. If the combined thermal capacity of these last bodies be represented by c', the quantity of heat absorbed by them is $c'(t_m - t_2)$. Hence, if it be assumed that no heat is lost to, or gained from, the surroundings during the experiment, the equating of the heat losses and gains would give

$$Lm_1 + c_2 m_1(t_1 - t_m) = c_2 m_2(t_m - t_2) + c'(t_m - t_2). \qquad [201]$$

Unless special precautions are taken, the transfer of energy to or from the surroundings is relatively large. It is due to *convection currents* set up in the air immediately in contact with the calorimeter vessel, to the emission or absorption of *radiant energy* by the vessel, and to the *conduction* of heat through the surrounding medium. In general this transfer can be reduced by enclosing the calorimeter vessel in stagnant air surrounded by thermally insulated constant-temperature walls, by using such large quantities of liquid that the temperature change is small, by making the initial temperature of the liquid about as much below the temperature of the enclosure as the final temperature t_m is above it, and by having the outside surface of the calorimeter vessel highly polished so that it emits and absorbs radiation very weakly.

By observing the foregoing precautions with the form of calorimeter shown at H in Fig. 141, the *cooling correction* can be reduced considerably. The correction is made with the help of an empirical law, announced by

NEWTON [1] in 1701, according to which the time-rate of cooling of a body is proportional to the temperature difference between the body and its surroundings. Since the rate of cooling will also be nearly proportional to the rate of loss of heat if the thermal capacity of the cooling body is constant, it follows *that the time-rate of loss of heat is proportional to the difference in temperatures.* NEWTON'S law of cooling is not even approximately correct,

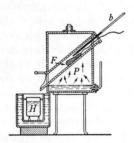

FIG. 141. REGNAULT'S apparatus for measuring the specific heat of a solid or liquid by the method of mixtures [*Annales de Chimie et de Physique* (2) **73**, 5 (1840)]. If the specimen is a solid, it is placed, in finely divided form, in a wire net which is heated in the inclined tube F by means of the temperature bath P, and then is lowered on a thread into the calorimeter. If the specimen is a liquid or powder, or is soluble in water, it is first sealed up in a thin-walled tube of good conducting material

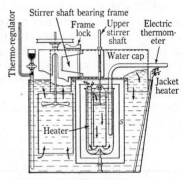

FIG. 142. A highly developed calorimeter, carefully designed to reduce convection and conduction losses and stirring and lag errors. It is closed against evaporation and completely surrounded by a shielding mass of circulating water whose temperature is controlled thermostatically. [Reproduced by permission from W. P. White, *The Modern Calorimeter* (Reinhold Publishing Corporation, N. Y.), a classic treatment on this subject]

but, when the cooling correction is small and the temperature difference is not more than 5° C, its incorrectness will not generally introduce an appreciable error into the result.

In modern calorimeters designed for the greatest possible precision, the correction for radiation is eliminated by the use of a jacket maintained at the temperature of the calorimeter throughout the whole of the experiment. The greatest sources of error which then remain to be guarded against are the loss of water by evaporation and the conduction of heat out of the calorimeter along such metallic paths as the shaft of the stirrer or the thermometer leads. Fig. 142 shows such a modern calorimeter with which a precision as great as 0.1 percent can be obtained.

---

[1] *Abridged Philosophical Transactions* (London, 1749), Vol. IV, Part II, p. 1. This paper is reproduced in **A Source Book in Physics* (1935), pp. 125–128.

The method of mixtures is also used to determine the amount of heat evolved or absorbed in various chemical reactions, and for this purpose special forms of calorimeters have been developed. In the bomb calorimeter, first devised by MARCELLIN PIERRE EUGÈNE BERTHELOT [1] in 1881, *the heat of combustion* of, say, coal is determined by exploding the coal in a closed vessel, or bomb, which contains oxygen under high pressure, the explosion having been started by an electric spark. The explosion method has also been used for measuring the specific heat at constant volume of a gas at very high temperatures, the gas to be investigated being placed with a known amount of explosive mixture in a special type of steel bomb.[2]

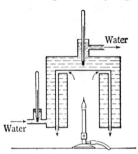

FIG. 143. One form of the JUN-KERS continuous-flow calorimeter, arranged for determining the heat of combustion of illuminating gas

149. The Method of Continuous Flow.

The important method of *continuous flow* was used by H. L. CALLENDAR and H. T. BARNES [3] in their work in 1902 on the variation of the specific heat of water with temperature (Fig. 135). In the form of apparatus shown in Fig. 143 a steady stream of water is allowed to flow past the point where heat is being evolved, in such a manner that all the heat is absorbed by the stream. If the rise in temperature and the rate of flow of the water are observed, the quantity of heat carried away by the water in unit time can be calculated. The great advantage of this method is that the cooling correction (Sec. 148) can be obtained simply by making experiments with different rates of flow but with the rise of temperature kept constant.

The method of continuous flow has also been used to determine JOULE'S equivalent, the specific heats of gases at constant pressure, and the conductivity of heat down a metal bar. In such experiments it is now the general practice to use electrical energy as the source of heat; this has many advantages when experiments of the highest precision have to be made, on account of the facility with which the heat supply can be controlled.

150. The Method of Cooling.

A convenient way to determine specific heats is the *method of cooling*. It consists in comparing the times required for

---

[1] *Annales de Chimie et de Physique* (5), **23**, 160 (1881), et seq. See the excellent article on "Bomb Calorimeters" in *A Dictionary of Applied Physics*, ed. by R. Glazebrook (Macmillan, 1922), Vol. I, pp. 26–31; also T. C. Sutton, *Journal of Scientific Instruments* **10**, 286 (1933), for a recent design capable of high precision.

[2] See *J. R. Partington and W. G. Shilling, *The Specific Heats of Gases* (Benn, 1924), pp. 112–132.

[3] *Philosophical Transactions* **199**, 55, 149 (1902).

a given closed vessel to cool in air, through a given number of degrees, first when filled with water and then when filled with the substance whose specific heat is sought. If t_1 and t_2 are the times of cooling and c_1 and c_2 the corresponding heat capacities of the vessel and contents in the two cases, then it may be shown that

$$\frac{t_1}{t_2} = \frac{c_1}{c_2}.$$

For, since the nature and area of the cooling surface are the same in both cases, at a given temperature and given outside conditions this surface must always lose heat at the same rate, no matter what substance may be enclosed in it. Let, then, ΔQ denote the quantity of heat that passes out of a given surface at temperature T in an infinitely short element of time. This loss in heat will be associated with a small fall in temperature ΔT_1, which will be determined by $\Delta Q = c_1 \Delta T_1$. If now the heat capacity be changed from c_1 to c_2 by a change in the contents of the vessel, the new fall in temperature ΔT_2 in the same infinitely short interval of time and at the same temperature T will be determined by $\Delta Q = c_2 \Delta T_2$. Hence

$$\frac{\Delta T_1}{\Delta T_2} = \frac{c_2}{c_1};$$

that is, at any given surface temperature the two *changes in temperature* during the same small interval of *time* are *inversely* proportional to the heat capacities. This is exactly equivalent to the statement that, at a given temperature, the two *intervals of time* required for the small change in *temperature* are *directly* proportional to the heat capacities. Since, then, the times required to pass through each small element of the scale, say from 60° to 59°, are proportional to the heat capacities, the *total* times required to pass through any interval of temperature made up of these small elements must be proportional to the heat capacities.

It is to be observed that this conclusion involves no assumption whatever regarding the nature of the law of cooling, or regarding the relation between the roles played by convection currents and by true radiation in the cooling process. It rests solely upon the assumption of similarity in the outside temperature conditions and uniformity of temperature in all parts of the cooling body. This last condition is difficult to fulfill when the cooling vessel contains solids; hence the method has not in general proved satisfactory for the determination of the specific heats of such substances.[1] For liquids, however, it has been found both accurate and convenient.[2]

151. Methods Depending upon Change of State. These methods consist essentially either in determining the mass of ice m that a heated body will melt while its temperature is falling to 0° C or in finding the mass of steam m' that a cold body will condense while its temperature is rising to the boiling temperature of water. The heat given up by the body in the first

[1] See, however, W. G. Marley, *Proceedings of the Physical Society* **45**, 491 (1933).

[2] See A. Ferguson and J. T. Miller, *Proceedings of the Physical Society* **45**, 194 (1933), for a recent application of the method.

case is then $79.6 \cdot m$ cal$_{15}$, since the heat of fusion of ice is 79.6 cal $\cdot$ g^{-1}. That taken up in the second case is $L \cdot m'$ (Sec. **146**).

The ice calorimeter is as old as BLACK, but the modern form is due to ROBERT WILHELM BUNSEN [1] (1811–1899). It depends on the fact that a change of volume occurs during fusion. In Fig. 144 the mantle of ice b is formed in the air-free water w by inserting a freezing mixture into the tube D. The point to which the mercury M rises in the graduated capillary tube T is then noted. The heated specimen which is to be tested is next dropped into D, where it melts a certain amount of ice. The movement of the mercury in T to the right because of the contraction of the ice on fusion is proportional to the amount of ice melted. The value in thermal units of one division of T is determined by inserting in D a substance of known thermal capacity or by placing an electrical heater in D and measuring the heat supplied as electrical energy. Although this calorimeter has proved itself valuable for determin-

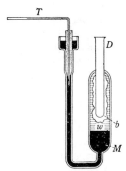

FIG. 144. The BUNSEN ice calorimeter

ing the specific heats of very small specimens, the objection to its use for precise work is that the density of ice appears to vary, probably owing to the presence of traces of dissolved air in the water and perhaps to strains set up in the ice during its formation.

The steam calorimeter (Fig. 145) was used by J. JOLY,[2] in 1886, to make the first accurate determinations of the specific heats of gases at constant volume. Two similar hollow copper bulbs hang in a closed chamber from the two arms of a balance. One of the bulbs is exhausted and the other contains the gas under investigation. When steam is admitted into the chamber, it condenses upon the bulbs and the walls until their temperature reaches that of the steam. The bulbs are provided with pans to catch the water condensed upon them. More water is condensed on the bulb containing the

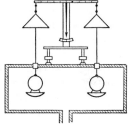

FIG. 145. JOLY's differential steam calorimeter, very much simplified

gas than on the empty one, and the difference is measured by adding weights until balance is restored. If m is the difference in mass of water condensed, Lm is the quantity of heat required to raise the mass of gas contained in the full bulb from its initial temperature to that of the steam. Corrections for the expansion of the bulbs are necessary.

---

[1] Poggendorff's *Annalen* **141**, 1 (1870); **142**, 616 (1871). *Annales de Chimie et de Physique* (4), **23**, 50 (1871); *Philosophical Magazine* (4), **41**, 161 (1871).

[2] *Proceedings of the Royal Society* **41**, 352 (1886); **47**, 218 (1889); *Philosophical Transactions* **182**, 73 (1891).

The steam calorimeter, equipped with only one bulb, has been used to measure the heat of vaporization of water. The temperature of the empty bulb is raised from the ice point to the steam point by surrounding it with steam, a mass m of steam being condensed on the bulb; the bulb is then filled with a mass M of water and the process repeated, a mass m' of steam being condensed this time. The student can easily show that, if the mean calory is taken as the thermal unit, $L = 100\,M/(m' - m)$. This method is of importance because it makes possible a determination of L directly in terms of the mean calory.

○

The Second Law of Thermodynamics

The convertibility of heat and work as expressed by the first law of thermodynamics (Secs. 103, 141) places heat on the same basis with all forms of mechanical energy. According to the law, for example, the heat generated when a rotating flywheel is brought to rest by friction is equal to the loss of kinetic energy of the wheel; or again, the heat generated when a falling object hits the ground is accounted for by the loss in kinetic energy of the object. Now, so far as the first law is concerned, the reverse of these processes would also be possible: a flywheel at rest could suddenly start rotating, thus gaining kinetic energy while the bearings lost an equal amount of heat energy and became cooler; or an object lying on the ground could suddenly jump up into the air while the ground at the same time became cooler. Neither of these processes violates the first law. Yet they never have been observed to happen. Considerations of this kind lead to the conclusion that when heat is one of the forms of energy involved, conservation of energy alone is not a sufficient condition for the occurrence of a process in nature, although it is a necessary one.

The restricting principle which, together with the first law, provides a sufficient criterion of allowable processes grew out of the work of CARNOT (Sec. 103). It was first formulated by CLAUSIUS and KELVIN, and was called by the former the *second law of thermodynamics*.[1] One of its simplest formulations, due to KELVIN, is as follows: *There is no natural process the only result of which is to cool a heat reservoir and do work as exemplified by raising a weight.* The law really is a generalization of the principle that when two bodies at different temperatures are placed in contact, the flow of heat is always from the hotter to the colder. It does not exclude the transfer of heat from a colder to a hotter body, a process which happens, for example, in mechanical refrigerators, but it does say that this

---

[1] See footnote 3, p. 76.

transfer cannot occur unless some outside agent does work or there is some other additional effect that is a compensation for this un- natural result. Obviously if it were not for the principle set forth by the second law, heat from the condenser of an engine could be col- lected, raised to boiler temperature without doing work, and used again as a source of energy for the engine. This would result in a kind of perpetual motion which, because its existence is denied by the second law, is called *perpetual motion of the second kind.*

No attempt will be made here to develop the quantitative expres- sion for the second law; . this is not due to any special mathematical difficulties but because an adequate discussion of the concepts in- volved would require more space and time than we have at our dis- posal.[1] The quantitative formulation has to do with the efficiency of any engine for transforming heat into work. No heat engine can have a better efficiency than $(Q_1 - Q_2)/Q_1$, where Q_1 is the heat drawn from the boiler and Q_2 is the heat rejected at the condenser during a given time; this will be clear if one notes that $(Q_1 - Q_2)/Q_1$ is the fraction of the heat received that is transformed into work. Now the second law states that even if an engine were entirely free from friction its efficiency $(Q_1 - Q_2)/Q_1$ cannot exceed $(T_1 - T_2)/T_1$, where T_1 and T_2 are the absolute temperatures of the source of heat and the condenser, respectively; that is,

$$\frac{Q_1 - Q_2}{Q_1} < \frac{T_1 - T_2}{T_1}. \qquad [202]$$

For example, with the boiler of a steam engine at 160° C and the condenser at 70° C, the efficiency cannot be more than 21 percent, *even if no frictional or other losses are involved*; the actual efficiency probably would be about 15 percent. The point evidently is that only a part of the total heat Q_1 drawn from the boiler can be con- verted into work. The remainder Q_2, although not annihilated, is wasted in the sense that it has descended to the temperature of the surroundings and has become unavailable for doing work. The second law provides a measure of this loss of availability or *degrada- tion* of energy. After all, it expresses quantitatively what all of us feel intuitively: that although energy is conserved whenever heat is evolved, yet there is something that has been lost; the heat in an object is not so available for doing work as is the kinetic energy of

---

[1] For an interesting, elementary account of the history and implications of the second law, see *M. Mott-Smith, *The Story of Energy* (Appleton-Century, 1934). An elementary mathematical treatment will be found, for example, in *A. W. Barton, *A Textbook on Heat* (Longmans, Green, 1933), Chap. 13.

the object when it is in motion as a whole. Heat is unique among the various forms of energy; it can be converted into other forms only partially and temporarily. If all temperature differences in the universe are eventually wiped out, then gross mechanical motion will no longer be possible and we must approach what the philosophers of the nineteenth century called the "Wärmetod," or "heat death," — a sort of thermodynamic Götterdämmerung.

In 1854 CLAUSIUS [1] gave another interpretation of the second law, in the course of which he introduced a new concept. The inequality [202] can be rearranged in the form

$$\frac{Q_1}{T_1} < \frac{Q_2}{T_2}, \qquad [203]$$

where Q_1/T_1 depends only on the source of heat and Q_2/T_2 only on the condenser. The very important quantity Q/T CLAUSIUS [2] called the *entropy* of the heat at the temperature T. The foregoing inequality, then, states that in any actual engine, entropy of amount Q_1/T_1 is taken from the source of heat and a larger amount of entropy, Q_2/T_2, is given to the condenser. Such considerations lead to another way of stating the second law: *In any isolated system every change results in an increase in the entropy of the system.*

CLAUSIUS summed up all of this in his famous statements of the two laws of thermodynamics: (*a*) The energy of the universe is constant; (*b*) the entropy of the universe is always increasing. Today we should be more cautious about extrapolating our laws into regions where they have not been tested; experiments have not been made in all parts of the universe. Moreover, it must be remembered that thermodynamics ignores the mechanisms of a process and makes no assumptions with regard to the structure of matter (Sec. 103). When the behavior of the molecules and atoms of matter is taken into account by means of the methods of statistical mechanics, the state of maximum entropy predicted for any isolated system by the second law comes to be interpreted as the *most probable state*. Thus the second law, with its implication of constantly increasing unavailable energy of the universe, applies to the *average probable* condition to be met over exceedingly long periods of time. From this point of view the "Wärmetod" need not be regarded as inevitable.

---

[1] Poggendorff's *Annalen der Physik und Chemie* **93**, 500 (1854).

[2] Poggendorff's *Annalen der Physik und Chemie* **125**, 390 (1865). A part of this paper appears in *A Source Book in Physics (1935)*, pp. 234–236.

EXPERIMENT XIIA. DETERMINATION OF γ BY THE METHOD OF CLÉMENT AND DESORMES

The method of CLÉMENT and DESORMES is of interest not only because it represents the earliest experimental work on adiabatic processes but because it affords a *direct* determination of the ratio γ. A quantity of air is compressed into a vessel until its pressure has a value which we will designate by P_1. The vessel is then momentarily put into communication with the atmosphere until the pressure inside falls to the atmospheric pressure, here called P_2. During this expansion, which may be assumed to be adiabatic, the temperature of the gas falls. After the vessel is closed again, the gas takes up heat from its surroundings until it reaches its initial temperature, and in so doing attains a final pressure P_3 which will be above that of the atmosphere. Consider a unit mass of the gas, and regard the gas as ideal. During the adiabatic expansion its volume changes from V_1 to V_2 and, according to the adiabatic equation for pressure and volume, Eq. [199],
$$P_1 V_1^\gamma = P_2 V_2^\gamma.$$

Now the volume of the gas in the closed vessel remains constant while the gas is warming, and hence the final volume of unit mass of the gas is also V_2. Thus, since the initial and final temperatures are the same, we have
$$P_1 V_1 = P_3 V_2.$$

By eliminating V_1/V_2 between the two foregoing equations, we obtain finally
$$\gamma = \frac{\log P_1 - \log P_2}{\log P_1 - \log P_3}. \qquad [204]$$

The desired ratio γ may therefore be obtained experimentally by observing the values of the three pressures.

It must be pointed out that the method of CLÉMENT and DESORMES is open to serious objections.[1] The most important of these is that when the vessel is opened to the atmosphere momentarily, oscillations of the air occur and it is impossible to know at what moment to close the vessel so as to have the pressure inside exactly equal to the atmospheric pressure.

Apparatus and Measurements. The apparatus is shown in Fig. 146. The vessel V is a flask or carboy of about 5-l capacity. The mouth of this vessel is plane ground and is covered with a plate of metal or glass C, also plane ground, which is provided with a hose-nipple. Means should be provided for holding the cover firmly against the

---

[1] For a comprehensive discussion of the sources of errors to be avoided in the experiment, the student is referred to *J. H. Poynting and J. J. Thomson, *Heat* (Griffin, 1911), pp. 288–294.

ground rim of the vessel; a small rod R, held in a clamp and with a rubber stopper on the end, will answer the purpose. The hose-nipple is connected through a T-tube to an oil manometer M and to a bicycle pump, or other small compression pump, as shown in Fig. 146.

A drying agent, such as strong sulfuric acid, calcium chloride, or phosphorus pentoxide, is put into the vessel V in order to dry thoroughly the enclosed air.

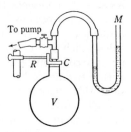

FIG. 146. Modified form of CLÉMENT and DESORMES'S apparatus

a. Compress the air in the vessel until the difference in pressure $P_1 - P_2$ is between 20 cm and 40 cm of oil, and then close the connecting tube to the pump. Allow about 15 min for the compressed air to regain temperature equilibrium with the room, as shown by the pressure's becoming constant, and then read the manometer and the barometer.

b. Momentarily open the vessel to the atmosphere by sliding the cover C sidewise and, after an interval of about a half-second, sliding it back. During this operation it may be found desirable to improve the thermal insulation of the vessel by surrounding it with cotton batting contained in a wooden case.

c. Wait until the temperature of the air in the vessel has risen again to that of the room, as indicated by the pressure's becoming constant, and then read the manometer.

d. Repeat *a*, *b*, and *c* at least twice.

e. From each set of data calculate the value of γ for air by means of Eq. [204], and take the mean of the results. Remember that the readings of the barometer and of the oil manometer must be reduced to the same unit of pressure; if the density of the oil is not known, it will, of course, be necessary to determine it.[1]

    1. Do you see any objection to an initial exhaustion of the gas in place of the compression? Actually this was the method followed by CLÉMENT and DESORMES in 1819, the method of initially compressing the gas having been employed at a later date by GAY-LUSSAC and J. J. WELTER.[2]

    2. When WILHELM KONRAD RÖNTGEN [3] repeated this experiment in 1873 he employed a device similar to an aneroid barometer to measure the changes of pressure. Why would such an instrument be preferable to a liquid manometer?

    [1] The density may be determined with a Mohr-Westphal balance, as in Exp. XIIIA.

    [2] See P. S. Laplace, *Annales de Chimie et de Physique* (2) **20**, 267 (1822); also *Traité de Mécanique Céleste*, Vol. V, Bk. XII, Chap. III.

    [3] Poggendorff's *Annalen der Physik und Chemie* **141**, 552 (1870); **148**, 580 (1873).

3. Construct on a PV diagram the curves that will represent the changes which occurred in your experiment (Fig. 147). This may be done as follows. Let specific volumes, or volumes per gram, be the abscissas. Taking the density of air at 0° C and 1 A_s as 1.2930 g · l⁻¹, calculate with the help of the gas law the specific volumes corresponding to the values of room temperature and P_1, P_2, and P_3 found in your experiment. Construct the corresponding isotherm. Draw the vertical line through the point corresponding to P_3 on

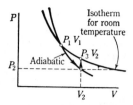

Fig. 147. Graphical representation of the operations performed in the experimental determination of γ

this isotherm. Draw the horizontal line through P_2. The intersection of these two straight lines will be P_2, V_2. Why? Draw the adiabatic curve and interpret the complete diagram.

○

EXPERIMENT XIIB. THE HEATING VALUE OF ILLUMINATING GAS [1]

The *heating value*, or quantity of heat evolved by the combustion of a unit volume of the gas, is usually determined with a JUNKERS [2] continuous-flow gas calorimeter of the type shown in Fig. 148. A measured volume V of gas under an observed pressure P is burned in the calorimeter, and the rise in temperature from $t_1°$ to $t_2°$ of a mass M of water as it flows through the calorimeter is determined. If Q be the heating value of the gas, expressed in thermal units per unit volume, and c be the mean specific heat of water between $t_1°$ and $t_2°$,

$$Q = \frac{cM(t_2 - t_1)}{V}. \qquad [205]$$

Since the calorimeter is so constructed that the heat liberated by the condensed water which is formed by the combustion of the hydrogen and hydrocarbons in the gas is not allowed to escape, the quantity Q in Eq. [205] gives the maximum utilizable heat evolved by the gas; this quantity Q is called the *heat of combustion* or *gross heating value* of the gas. There are many industrial operations where the water formed in combustion escapes as steam, however, and hence it is frequently the practice to deduct the amount of heat thus lost from

[1] See *Circular of the Bureau of Standards*, No. 48 (1914), "Standard Methods of Gas Testing"; No. 65 (1917), "Gas Calorimeter Tables"; also C. W. Wardner and E. F. Mueller, "Industrial Gas Calorimetry," *Technologic Papers of the Bureau of Standards*, No. 36 (1914); also C. G. Hyde and F. E. Mills, *Gas Calorimetry* (Benn, 1932).

[2] *Zeitschrift für Instrumentenkunde* **15**, 408 (1895).

the gross heating value. The resulting lower value, Q', is called the *net heating value*. Thus if the mass of the condensate is m and its temperature as it emerges from the condensate drain (Fig. 148) is t', then evidently the net heating value Q' of the gas is given by

$$Q' = \frac{cM(t_2 - t_1) - mL_t - cm(t - t')}{V}, \qquad [206]$$

where L_t is the heat of vaporization of water at the temperature of condensation t.

1. Explain the physical significance of each term in Eq.
[206].

*Adjustments and Measure-
ments.*[1] Connect a pressure-
regulator and a direct-reading
gas meter in the line supply-
ing gas to the burner under
the calorimeter. Adjust the
pressure of the gas to about
1.5 in. of water by adjusting
the weight of the floating cylinder in the pressure-regulator.
Before lighting the gas, turn
on the water and regulate
its flow until there is a constant small stream passing
out through the overflow E.
Then remove the burner from
the calorimeter, light it, and
replace it in the calorimeter,
taking care to center it properly with the help of a mirror placed below the burner.
Regulate the flow of water
and gas and the exhaust
damper D, until the burned
gas leaves the calorimeter at
approximately the tempera-

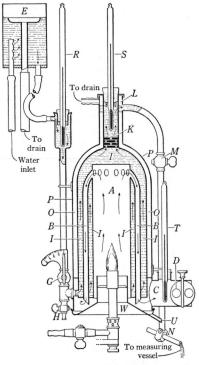

Fig. 148. Improved form of the Junkers gas calorimeter. A, combustion chamber; B, condenser tubes; C, outlet for exhaust; D, damper for exhaust; E, inlet overflow weir; G, inlet water valve; I, water space; L, out-let overflow weir; O, air space; T, exhaust thermometer; U, condensate drain; W, de-tachable exhaust chamber

ture of the entering gas, and until there is a difference of from 20° to

---

[1] If the calorimeter at your disposal differs in its details from the one described here, consult the operating instructions furnished by the manufacturer.

40° F in the inflowing and outflowing water. Never attempt to adjust the gas or to light the burner while the latter is inside the calorimeter.

Wait until the temperatures indicated on the various thermometers have become constant. Then, at an instant when the large hand of the gas meter passes through zero, begin to collect in vessels both the heated overflow water and the condensate water. End the run when 1 ft³ of gas has been used. Read thermometers R and S frequently during the run; if thermometer S fluctuates rapidly, read it as often as possible so as to obtain a fairer average. If the temperature readings have not stayed constant within a few tenths of a degree, discard the results and attempt to better the conditions.

Before making any further adjustments in the apparatus, be sure that you have obtained all of the following data: the gas pressure; the temperatures of the entering gas, of the burned gases, of the condensate, and of the room; the average temperatures of the inflowing and of the outflowing water; the masses of the empty overflow vessels, of the vessels and overflow water, of the empty condensate vessel, and of the vessel and condensate; the barometer reading; and the volume of the gas burned.

Calculate the heat of combustion and the net heating value of the gas in British thermal units per cubic foot; in doing this the volume V in Eqs. [205] and [206] should be reduced to standard conditions of temperature and pressure; in industrial and commercial tests made in this country it is customary to take these as 30 in. of mercury and 60° F.

Make a second complete run, this time with a gas pressure of about 2.0 in. of water.

2. Express the gross heating value of the gas in calories per cubic meter of gas measured at 0° C and 1 A_s.

3. Why is it the practice to reduce the volume of gas V to standard conditions of pressure and temperature?

4. Explain why an attempt is made to adjust the apparatus so that the temperature of the products of combustion as they leave the calorimeter is the same as that of the gas which enters. If these two temperatures were not the same, how would you correct your results for the error thus introduced?

5. The exhaust gases are saturated with water vapor, whereas the air that enters is not. What error will this introduce in your results, and how would you correct it?

6. What factors determine the amount of condensate water?

7. Why is it not necessary to take into account the thermal capacity of a continuous-flow calorimeter?

OPTIONAL LABORATORY PROBLEMS

1. Specific Heat of Mercury by the Method of Mixtures. Enclose the mercury in a small bottle, warm it in a steam heater, and then lower it into the water in the calorimeter vessel. Observe the precautions mentioned in Sec. **148.** Allow the boiler to run throughout the experiment so that the temperature of the surroundings will remain constant. The thermometers in the water jacket and in the calorimeter should be read to tenths and to hundredths of a degree respectively. After the bottle containing the mercury has been placed in the calorimeter, stir the water continuously and take its temperature at half-minute intervals until the temperature begins to fall, and then at longer intervals.

The mean thermal capacity of the bottle can be obtained by making a separate experiment with the empty bottle. That of the calorimeter vessel and stirrer can be calculated from a knowledge of the masses of the vessel and stirrer and the specific heat of the material of which they are made. In the case of the thermometer, the thermal capacity is equal to the volume of the immersed portion multiplied by 0.45 cal · deg^{-1}, the mean thermal capacity of one cubic centimeter of the glass and mercury.

To find the cooling correction (Sec. **148**), employ the equation $Q = k(t - t')D$, in which Q is the quantity of heat lost to or gained from the water jacket during the experiment, t is the *mean* temperature of the water in the calorimeter vessel during the experiment, t' is the mean temperature of the water jacket, D is the time of duration of the experiment, and k is a constant of proportionality. The constant k can be determined by filling the calorimeter to the same level as in the original experiment with a known mass of water which has been heated to the highest temperature t_m reached by the mixture in the original experiment. The time required, with continuous stirring, for this water to cool through, say, a half-degree is then noted, and from this and the temperature of the water jacket the constant k can be determined.

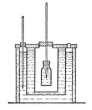

Fig. 149. Cooling calorimeter

2. Specific Heat of Alcohol by the Method of Cooling. Partly fill a nickel-plated cooling bottle (Fig. 149) with a known mass of the alcohol to be tested. Gently heat the bottle over a burner until the temperature of the alcohol is about 37° C, and then suspend it in the air space in the inner calorimeter vessel. Read the thermometer about once a minute until the temperature has fallen perhaps twelve degrees; be sure to observe the exact time of each reading. Also observe the temperature of the water jacket from time to time.

Make a similar set of readings with the bottle filled to the same level as beforehand, but this time with warm water.

Plot the two cooling curves to the same large scale, and read off upon them the two times included between any two selected temperatures. The specific heat of the alcohol can then be calculated from a knowledge of these times, the masses of the alcohol and of the water, the specific heat of the water, and the thermal capacity of the bottle and thermometer (Sec. 150). If the thermal capacity of the bottle is not known, it will have to be determined by a separate experiment. The mean thermal capacity of the thermometer may be taken as 0.45 cal · deg^{-1} per cubic centimeter of the glass and mercury that was immersed in the liquid during the experiment.

o

QUESTION SUMMARY

1. Distinguish clearly between *temperature* and *quantity of heat*. Contrast the two hypotheses which have been proposed to account for the fact that bodies change temperature.

2. Define *calory*; *British thermal unit*. These are units of what?

3. Define *specific heat*; *molar thermal capacity*; *atomic thermal capacity*. How can the fact that different substances have different thermal capacities be explained?

4. Why is the specific heat at constant pressure greater than the specific heat at constant volume? How much greater is it in the case of gases?

5. On the kinetic theory what value would be expected for the molar thermal capacity at constant volume of an ideal gas? Do actual gases have this value?

6. What is meant by *degrees of freedom*? In general, how many degrees of freedom does a particle have? a solid body? What is the law of equipartition of energy?

7. If the law of equipartition of energy holds, what value would be expected for the molar thermal capacity at constant volume of a diatomic gas? of a polyatomic gas? Is this found to be the case?

8. What value would be expected for the ratio of the specific heats at constant pressure and at constant volume for an ideal gas? for a monatomic gas? for a diatomic gas? for a polyatomic gas? Is this found to be the case? How do you account for values of γ less than 1.33?

9. Do the specific heats of all gases vary with the temperature? What types of gases show the largest variations? Account for these variations.

10. Explain why the molar thermal capacity of hydrogen decreases from 5 cal · mole^{-1} · deg^{-1} at ordinary temperature to 3 cal · mole^{-1} · deg^{-1} at about 60° K. Why does not oxygen or nitrogen behave similarly?

11. In the case of solids at high temperatures, is it the molar or the atomic thermal capacity that is the same for all substances? What is the significance of this fact for the kinetic theory?

12. State DULONG and PETIT'S law. What is the nature of the exceptions to it? Account for the fact that the atomic thermal capacities of solids are approximately 6 rather than 3 cal per gram atomic weight per degree.

13. Describe in detail how the atomic thermal capacities of solids vary with the temperature. How is this variation explained? Why do substances with a high melting point and low atomic weight, like silicon, depart from DULONG and PETIT'S law at higher temperatures than substances with a low melting point and a high atomic weight, such as lead?

14. What is an adiabatic process? What is the relation between the pressure and the volume of an ideal gas in an adiabatic change? How is this relation made use of in determining γ?

15. How does the mechanical theory of heat account for the fact that the temperature remains constant while ice is melting or while water is boiling? Define *heat of vaporization*; *heat of fusion*; *heat of sublimation*.

○

PROBLEMS

1. A flywheel of mass 300 kg and diameter 2.00 m has an angular speed of 1.00 rev · sec^{-1}. If the whole mass of the wheel is assumed to be concentrated in the rim, how much heat is produced in stopping it by friction?

Ans. 1.41 × 10^3 cal.

2. In one of the first of JOULE'S experiments (Sec. **55**), the work done by falling weights was expended in stirring water (Fig. 150). After allowances were made for the friction of the pulleys, the speed of the weights at the end of their descent, etc., the data obtained in the first experiments were: total mass of weights, 57.8 lb; height of fall, 5.00 ft; number of times weights were allowed to fall, 21; rise in temperature of water, 0.563° F; thermal capacity of water and containing vessel, 13.9 B.t.u. per deg F. Calculate the value of JOULE'S equivalent from these data.

Ans. 775 ft · lbwt/B.t.u. = 4.17 j · cal^{-1}.

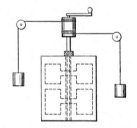

FIG. 150. Diagram of JOULE'S apparatus for determining the mechanical equivalent of heat

3. Will the drop in temperature of 10 g of water from 65° to 35° C heat a given mass of cold water as much as the drop of 5 g from 95° to 35° C?

4. In the manufacture of lead pipes the solid lead is forced through an annular die by applying a pressure of, say, 2.0 × 10^4 lbwt · in.$^{-2}$. Given that the mean specific heat of lead is 3.13 × 10^{-2} cal · g^{-1} · deg^{-1} and that its density is 710 lb · ft^{-3}, how much will the temperature of the lead rise in passing through the die?   *Ans.* 93° C.

5. (*a*) By following the method of MAYER, compute JOULE'S equivalent from the experimental data for oxygen in Tables VI and VII. (*b*) Show that this method involves the assumption that no heat is expended in internal **work of** expansion. (*c*) Was MAYER justified in making this assumption?

Ans. (*a*) 4.16 × 10^7 erg · cal^{-1}.

6. When heat is applied to a quantity of chloroform kept at constant volume, for which γ is 1.15, what fraction of it goes into increasing the kinetic energy of translation of the molecules? How is the remainder expended? *Ans.* 23 percent.

7. By combining Eq. [199], which gives the relation between the pressures and volumes of an ideal gas in an adiabatic change, with Eq. [133], the equation of state of an ideal gas, show that the relation between volumes and temperatures of an ideal gas in an adiabatic change is

$$TV^{\gamma-1} = \text{constant},$$

and that the relation between pressures and temperatures is

$$TP^{\frac{1-\gamma}{\gamma}} = \text{constant}.$$

8. A given mass of air, initially at 27° C and 1 A_s, is suddenly compressed to half its initial volume. Assuming that the compression is adiabatic, find what temperature the gas will attain. What will be the final pressure? *Ans.* 123° C; 2.6 A_s.

9. (*a*) Show that the external work done by an ideal gas at the temperature T during an isothermal expansion from volume V_1 and pressure P_1 to volume V_2 and pressure P_2 is given by

$$\Delta W_e = \int_{V_1}^{V_2} P\,dV = NRT \log_e \frac{V_2}{V_1} = P_1 V_1 \log_e \frac{V_2}{V_1} = P_2 V_2 \log_e \frac{V_2}{V_1}.$$

(*b*) Calculate the work done in compressing 2.0 liters of gas isothermally until its volume is 1.0 liter, the initial pressure being 72 cm of mercury. *Ans.* (*b*) 1.3×10^2 j.

10. Show that the external work done by an ideal gas during an adiabatic expansion from pressure P_1 and volume V_1 to pressure P_2 and volume V_2 is given by

$$\Delta W_e = \int_{V_1}^{V_2} P\,dV = \frac{1}{\gamma - 1}(P_1 V_1 - P_2 V_2).$$

11. A quantity of hydrogen is contained in a cylinder fitted with a movable piston, both the cylinder and the piston being made of some material that is impermeable to heat. The pressure and volume of the gas are 2 A_s and 1 ft³ respectively. If the piston is allowed to move out until the pressure is halved, (*a*) by how much is the volume increased, and (*b*) by how much is the heat content of the gas decreased? *Ans.* (*a*) 0.6 ft³; (*b*) 6×10^2 cal.

12. At 100° C and 1 A_s the heat of vaporization of water is 538.7 $\text{cal}_{15} \cdot \text{g}^{-1}$ and the density of steam is 5.98×10^{-4} g · cm⁻³. (*a*) What part of the heat of vaporization of water is used in doing external work against the atmospheric pressure, and (*b*) what part is used in increasing the internal energy of the water? *Ans.* (*a*) 7.5 percent; (*b*) 92.5 percent.

13. How many British thermal units would be required to convert 1.0 lb of snow at − 10° C into steam at 98° C? The specific heat of ice is 0.50 cal · g⁻¹ · deg⁻¹. *Ans.* 1.3×10^3 B.t.u.

14. When a substance that expands on freezing is subjected to an increased pressure, the melting point is lowered. The melting point of a substance that contracts on solidifying is raised in similar circumstances. (*a*) Is this what you should expect? (*b*) Explain the phenomenon, first observed by FARADAY, that when two pieces of ice are pressed together and then released, they are frozen onto each other. (*c*) It is observed that where lava forms a pool the crust sinks in when it is broken; can you use this fact to account for the interior of the earth's being solid even though it is at very high temperatures?

15. (*a*) Which are the more likely to burst in freezing weather, the hot-water or the cold-water pipes in a house? (*b*) Calculate the greatest average force per unit area which can be exerted by pure water in freezing; take the density of ice as $0.917 \text{ g} \cdot \text{cm}^{-3}$ and neglect the variation of the heat of fusion of ice with pressure. Assume also that the temperature is low enough so that the water will freeze no matter what the pressure.

Ans. (*b*) 3.7×10^{10} dyne $\cdot$ cm^{-2} or 270 ton wt $\cdot$ in.$^{-2}$

○

I CAN never forget the British Association at Oxford in the year 1847, when in one of the sections I heard a paper read by a very unassuming young man who betrayed no consciousness in his manner that he had a great idea to unfold. I was tremendously struck with the paper. I at first thought it could not be true because it was different from Carnot's theory, and immediately after the reading of the paper I had a few words of conversation with the author James Joule, which was the beginning of our forty years' acquaintance and friendship. On the evening of the same day that very valuable Institution of the British Association, its conversazione, gave us opportunity for a good hour's talk and discussion over all that either of us knew of thermodynamics. I gained ideas which had never entered my mind before, and I thought I too suggested something worthy of Joule's consideration when I told him of Carnot's theory. Then and there in the Radcliffe Library, Oxford, we parted, both of us, I am sure, feeling that we had much more to say to one another and much matter for reflection in what we had talked over that evening. But what was my surprise a fortnight later when, walking down the valley of Chamounix, I saw in the distance a young man walking up the road towards me and carrying in his hand something which looked like a stick, but which he was using neither as an Alpenstock nor as a walking stick. It was Joule with a long thermometer in his hand, which he would not trust by itself in the char-à-bancs coming slowly up the hill behind him lest it should get broken. But there comfortably and safely seated on the char-à-bancs was his bride — the sympathetic companion and sharer in his work of after years. He had not told me in Section A or in the Radcliffe Library that he was going to be married in three days, but now in the valley of Chamounix, he introduced me to his young wife. We appointed to meet again a fortnight later at Martigny to make experiments on the heat of a waterfall (Sallanches) with that thermometer: and afterwards we met again and again and again, and from that time indeed remained close friends till the end of Joule's life. I had the great pleasure and satisfaction for many years, beginning just forty years ago, of making experiments along with Joule which led to some important results in respect to the theory of thermodynamics. This is one of the most valuable recollections of my life, and is indeed as valuable a recollection as I can conceive in the possession of any man interested in science.

WILLIAM THOMSON (LORD KELVIN), Address delivered on the occasion of
the unveiling of JOULE's statue in Manchester Town Hall, December 7, 1895

CHAPTER THIRTEEN

THE MECHANICS OF FLUIDS

POSTULATE 1.

Let it be supposed that a fluid is of such a character that, its parts lying evenly and being continuous, that part which is thrust the less is driven along by that which is thrust the more; and that each of its parts is thrust by the fluid which is above it in a perpendicular direction if the fluid be sunk in anything and compressed by anything else.

POSTULATE 2.

Let it be granted that bodies which are forced upwards in a fluid are forced upwards along the perpendicular [to the surface] which passes through their centre of gravity.

ARCHIMEDES, "On Floating Bodies." Translation by T. L. HEATH [1]

○

A fluid is distinguished from a solid by the fact that the fluid cannot permanently sustain a shearing stress (Sec. **94**). This and many other properties of fluids have been discussed in connection with the phenomena of elasticity, thermal expansion, diffusion, viscosity, and change of state. Indeed, some of the mechanical principles of fluids that are treated in the present chapter are already known to the student from his study of gases and from his experiences in the laboratory with the barometer, water aspirator, and other instruments that utilize these principles.

○

Archimedes' Principle and the Statics of Fluids

152. Archimedes' Principle. In the first book of his treatise "On Floating Bodies," in which he established the principles of hydrostatics, ARCHIMEDES stated his famous discovery that a solid body immersed in a fluid at rest under gravity is buoyed up by a vertical force equal in magnitude to the weight of the displaced fluid. The law was restated in 1586 by STEVIN,[2] who advanced for it the following proof. Within a body of fluid, isolate in thought some mass by means of an imaginary bounding surface S, Fig. 151. Since the mass

---

[1] T. L. Heath, *The Works of Archimedes* (1897), pp. 253–300. By permission of The Macmillan Company, publishers. ARCHIMEDES founded his whole theory of hydrostatics upon these two postulates.

[2] *De Beghinselen des Waterwichts* ("Principles of Hydrostatics") (Leiden, 1586). For a French translation, see *Les Œuvres Mathematiques de Simon Stevin de Bruges*, ed. by A. Girard (Leiden, 1634).

SIMON STEVIN of Bruges has been called by SARTON "the most original man of science of the second half of the sixteenth century," and "one of the greatest mathematicians of the sixteenth century and the greatest mechanician of the long period extending from Archimedes to Galileo" [see *Isis* **21**, 241–303 (1934); this is an excellent discussion of STEVIN's life and achievements]. He was best known to his contemporaries for his works on the science of fortification (readers of Sterne's *Tristram Shandy* will remember that STEVIN's book on this subject was constantly quoted by Uncle Toby), and for his invention of a carriage that was propelled by sails and ran on the seashore, carrying twenty-eight passengers faster than a horse could gallop. He was also chiefly responsible for the introduction of a proper system of bookkeeping in the Dutch and French national accounts.

Of more fundamental and lasting importance, however, were STEVIN's contributions to mathematics and mechanics. He wrote the first systematic treatise on decimal fractions, established the essential value of such fractions, and suggested for the first time (1585) the extension of the decimal idea to weights and measures. In addition to giving a very simple and original demonstration of the law of equilibrium on an inclined plane, based on the postulate of the impossibility of perpetual motion (see Plate 8), STEVIN deduced the laws of hydrostatic pressure and applied them to floating bodies, as well as to the computation of the pressure on the sides and bottoms of vessels containing liquids. He discovered the hydrostatic paradox, that the pressure at the bottom of a liquid depends only upon the height and density of the liquid and not at all upon the size and shape of the liquid column producing it, and proved it experimentally by making one pan of a balance the bottom of a vessel which could be filled with various liquids and be given different shapes.

SIMON STEVIN, 1548–1620

Reproduced from *Isis* **21**, 244 (1934), by permission of the Editor

of fluid within this boundary is in equilibrium, its weight must be neutralized by forces exerted by the surrounding fluid. But these latter forces depend only upon the conditions that exist outside of S, and are wholly independent of the nature of the substance within S. Hence any immersed body whatever which has the surface S must be buoyed up by forces the sum of which is equal to the weight of the displaced fluid.

FIG. 151. STEVIN'S proof of ARCHIMEDES' principle

153. Weight of a Body in Vacuum. From ARCHIMEDES' principle it follows, for example, that when a body is balanced upon scales in air, the balancing weights do not accurately represent the *true weight* of the body, that is, its weight in vacuum. For both the body and the weights are buoyed up by the air, and since, in general, the volume of air displaced by the body is not the same as that displaced by the weights, the buoyant effects upon the two sides of the balance must be different. The true weight W can be obtained easily from the apparent weight W_a, the volume of the body V, the density of air ρ_a, and the density of the weights ρ_w. For, since the resultant downward force on the body is $W - V\rho_a$ and that on the weights $W_a - (W_a\rho_a/\rho_w)$, we have, by assuming that the balance arms are equal and applying the principle of moments (Sec. **72**),

$$W = W_a + \left(V - \frac{W_a}{\rho_w}\right)\rho_a. \qquad [207]$$

154. Density Determinations by Hydrostatic Weighing. ARCHIMEDES' principle also furnishes a convenient and accurate method for determining the densities of irregular solids and of liquids. Thus, if any solid body the weight of which in vacuum is W be found to undergo an apparent loss of weight ΔW when immersed in a liquid of density ρ_l, the density ρ of the solid is evidently

$$\rho = \frac{W\rho_l}{\Delta W}. \qquad [208]$$

In order to find ΔW accurately an air correction must, of course, be applied. For if the apparent weight of the body in air be W_a and its apparent weight in the liquid be W_l, then the equation of balance for this case, in which the body hangs in the liquid while the weights hang in air, is evidently $W - V\rho_l = W_l - (W_l\rho_a/\rho_w)$. By substituting the value of W found in Eq. [207], there results at once

$$\Delta W = V\rho_l = W_a - W_l + \left(V - \frac{W_a - W_l}{\rho_w}\right)\rho_a. \qquad [209]$$

EXAMPLE. It is found that a certain solid body undergoes an apparent loss of weight ΔW_1 when immersed in a liquid of known density ρ_1 and an apparent loss of weight ΔW_2 when immersed in another liquid of unknown density ρ_2. Show that the unknown density is given by the formula

$$\rho_2 = \frac{\Delta W_2}{\Delta W_1}\,\rho_1. \qquad [210]$$

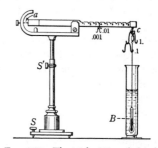

FIG. 152. The Mohr-Westphal balance. It is described in Exp. XIIIA

In using Eq. [210] the weighings in the two liquids may be made with an ordinary balance, of course, but for the sake of rapidity a modified form of balance due to the German pharmacist KARL FRIEDRICH MOHR (1806–1879) is commonly used (Fig. 152). Strictly speaking, the apparent losses of weight ΔW_1 and ΔW_2 are subject to the air corrections given by Eq. [209], but since in practice the densities ρ_1 and ρ_2 usually differ little from each other, the influence of these corrections upon the result is negligible.

155. Two Fundamental Principles of Fluids at Rest. The explanation which STEVIN gave for ARCHIMEDES' principle (Sec. **152**) involves the hypothesis that the fluid exerts a force on each element of a body with which it is in contact and that this force still exists when any portion of the fluid itself is regarded as a body immersed in the rest of the fluid; it is because a simple explanation can be found for ARCHIMEDES' principle on the basis of this hypothesis that the principle is so important theoretically. The force which a fluid exerts can be traced to various causes. When the fluid completely fills a vessel, forces exerted on the fluid surface, as by atmospheric pressure or by a piston, are a source of force in the fluid. Gravity is a second source, since the weight of the upper layers of the fluid is sustained by the lower layers. Molecular attractions also produce force in a fluid, but this will not be considered until later.

Since there can be no shearing stress in a fluid at rest (Sec. **94**), the force exerted by a fluid obviously can have no component parallel to the surface upon which it acts; in other words,

1. *The direction of the force exerted by a fluid at rest upon any element of surface is perpendicular to the surface.*

Now our study of the elastic properties of fluids (Sec. **91**) made it clear that pressure is a more useful and important concept in the mechanics of fluids than is force. What is more, we found that in a fluid at rest this pressure is a *hydrostatic pressure*; in other words,

2. *At any point in a fluid at rest the pressure is the same in all directions.*

This principle can easily be proved by applying the conditions for equilibrium given in Sec. 73 to any small portion of the fluid.

156. Fluids at Rest under Gravity. In order to study the effects of gravity on the pressure of a fluid at rest, consider first the forces that keep a horizontal cylindrical portion of the fluid in equilibrium [(a) in Fig. 153]. Let the area of each of the vertical end faces of the cylinder be A and let P_2 and P_3 be the pressures at these two ends. Then, since the forces on the curved side surfaces

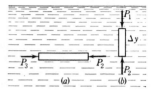

Fig. 153. Pressure in a fluid at rest under gravity

of the cylinder have no components in the direction of the length of the cylinder, the conditions for equilibrium give $P_2 = P_3$; or

3. *The pressure in a fluid at rest under gravity is the same at all points in the same horizontal plane.*

Next consider the equilibrium of a vertical fluid cylinder [(b) in Fig. 153]. Let the height Δy of this cylinder be so small that the density ρ of the fluid may be considered as constant throughout the cylinder. If P_1 and P_2 denote the pressures on the upper and lower faces respectively, the force $(P_2 - P_1)A$ must be equal in magnitude to the weight $\rho g A \, \Delta y$ of the fluid cylinder, or

$$P_2 - P_1 = \rho g \, \Delta y. \qquad [211]$$

(In what units must the pressures P_1 and P_2 be expressed here?) The result of integrating Eq. [211] enables us to conclude that

4. *The difference in pressure between two points at different levels in a mass of fluid at rest under gravity is equal to the weight of a column of the fluid of unit cross-sectional area reaching vertically from one level to the other.*

If we assume what is practically true for the case of most *liquids*, namely, that the density ρ does not vary with the depth, Eq. [211] gives the difference in pressures for any value of Δy without the necessity of integrating. Indeed, the student will recall having thus used the equation in the calculation of fluid pressure as indicated by a manometer.

These laws of equilibrium of fluids were demonstrated in the most simple manner by the French philosopher, mathematician, and man of letters BLAISE PASCAL and amply confirmed by experiments in his *Traitez de l'équilibre des liqueurs . . .*,[1] which was published in 1663 but was probably written before 1653. Except in the matter of simplicity and elegance of expression, however, PASCAL made little advance over STEVIN.

EXAMPLE. On the assumption that the earth's atmosphere is an ideal gas and that the temperature does not vary with altitude, show that the atmospheric pressure B at any altitude y and temperature T is given by

$$B = B_0 e^{-\frac{Mgy}{RT}},$$ [212]

an equation due to LAPLACE,[2] in which B_0 is the atmospheric pressure at the level where $y = 0$, e is the base of natural logarithms, and M is the molecular weight of atmospheric air.

Solution. By writing Eq. [211] and the equation of state of an ideal gas in the forms $dP = -\rho g\, dy$ and $PM = \rho RT$, respectively, and eliminating ρ between them, one obtains the differential equation

$$\frac{dP}{P} = -\frac{Mg}{RT}\, dy.$$

The negative sign signifies that the pressure P decreases as the altitude increases.[3] Indicating the integration of both members of this equation, one has

$$\int_{B_0}^{B} \frac{dP}{P} = -\frac{Mg}{RT} \int_{0}^{y} dy,$$

which yields $\log B - \log B_0 = -Mgy/RT$, and hence Eq. [212]. In carrying out the integration the value of g was treated as constant because its variation with altitude is small and may be neglected. In practice a temperature correction will have to be applied to Eq. [212], since in general the temperature decreases about $6°$ C for each kilometer increase in altitude.[4]

157. Pascal's Principle and the Hydraulic Press. According to Eq. [211] the difference in pressure between two points in a fluid at rest under gravity depends *only* on the difference in the levels of the points and on the density of the fluid, and not at all on the size and shape of the vessel.[5] It follows from this that if the pressure at any

---

[1] Excerpts appear in *A Source Book in Physics* (1935), pp. 75–80.

[2] *Traité de Mécanique Céleste*, Vol. V, Bk. XII.

[3] This was proved conclusively in 1648 by PASCAL and his brother-in-law, FLORIN PÉRIER, in the famous experiments carried out on the Puy-de-Dôme, a high mountain in Auvergne; see *A Source Book in Physics* (1935), pp. 73–75.

[4] See W. J. Humphreys, *Physics of the Air* (McGraw-Hill, 1929), Part I, Chap. III.

[5] This *hydrostatic paradox*, as it is often called, was clearly enunciated by PASCAL in 1653 and applied by him to the invention of the hydrostatic press; but, as SARTON has pointed out [*Isis* 21, 241 (1934)], STEVIN in 1586 had given a good statement of the paradox and had unmistakably indicated the application of it that later led to the hydraulic press.

point in the fluid be increased, there will be an equal increase of pressure at every other point, provided the density does not change appreciably. This statement is commonly referred to as *Pascal's principle of the transmissibility of pressure.*
It is practically true for liquids, since liquids are so nearly incompressible. Although gases are much more compressible, their densities are usually so small that the pressure in a small volume is everywhere nearly the same, and hence the principle holds well for gases also. It is in the hydraulic press [1] (Fig. 154) that

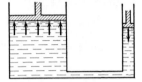

Fɪɢ. 154.   Simple hydraulic press

PASCAL'S principle finds its most beautiful experimental demonstration. The pressure applied at the pump of the hydraulic press usually is so great that the pressure due to gravity is insignificant by comparison, and may be neglected.

○

Capillary Phenomena

The foregoing principles show that if the pressure exerted by the atmosphere upon the free surface of a *liquid* of density ρ is B, then the pressure P at any depth y beneath the surface is given by

$$P = B + \rho g y. \qquad [213]$$

There follows at once the result, in general confirmed by observation, that the surfaces of a liquid at rest in a series of communicating vessels lie in the same horizontal plane. But it was observed as early as 1490 by LEONARDO DA VINCI that if a tube with a very narrow bore, called a *capillary tube* because the bore is as "fine as a hair," is placed with one end in water, the liquid rises some distance above the level in the outer vessel, as in Fig. 157, *b*. Later and more careful investigation has shown the existence of a large number of other phenomena to which the ordinary laws of hydrostatics do not apply. Thus a drop of mercury, instead of spreading out into a thin film, as it would do if gravity alone acted upon its molecules, is held together in an approximately spherical form, and the tendency to sphericity becomes more pronounced as the drop becomes smaller and its weight less important in comparison with the cohesive forces. These are all manifestations of the same intermolecular forces as were taken into account by VAN DER WAALS in arriving at his equation of

---

[1] Patented by JOSEPH BRAMAH in 1795.

state of a fluid (Secs. **128**, **129**), and as were assumed in Sec. **130** in order to reconcile the existence of liquid surfaces with the theory of molecular motion.

158. Additional Evidence as to the Existence and Nature of Intermolecular Forces. The simplest experiments show that, whereas the intermolecular attractive forces have enormous values at short range,[1] they diminish so rapidly with the distance as to become wholly inappreciable at distances that are certainly much less than 10^{-2} mm and that are probably less than 10^{-5} mm. Thus a sheet of glass may be brought extremely close to a surface of water without appearing to be attracted toward it sensibly, but as soon as contact is made the glass clings to the water with remarkable tenacity. The surface of two metal blocks may be pressed together without showing any appreciable attraction, but as soon as they are brought somewhat nearer, as by high pressure or welding, it may require tons of force to pull them apart again. In order to account for this rapid diminution with the distance it has been found necessary to assume that forces between molecules vary inversely with the distance according to a power higher than the fifth and perhaps up to the seventh, so far as present knowledge goes. Such laws of variation should be no cause for surprise. Even if these forces turned out to be gravitational forces, it does not necessarily follow that the ordinary law of inverse squares (Eq. [25], Chap. 2) will hold for them; for the law of inverse squares was arrived at from experiments on the attraction of masses of finite volume, the distances between the nearest points of which are almost infinitely large in comparison with the distances between the molecules of the bodies.

159. The Theory of Molecular Pressure. In Sec. **128** and Fig. 122, which should be reviewed at this time, it was shown that there exists in the layers of a *gas* next to the walls of the containing vessel a resultant inward pressure P' due to the intermolecular forces. Similar considerations evidently hold for those molecules of a *liquid* that are nearer to the surface than the radius r of the sphere of molecular attraction. For while the molecules within the space *cefd* of the sphere (Fig. 155) exactly neutralize the effects on the molecule m of the molecules within the space *acdb*, the downward resultant of the intermolecular forces of all the molecules in *efg* is wholly unbalanced. This force continually urges the molecule m into the interior of the liquid. The result of the action of all these unbalanced

---

[1] Repulsive forces which cause the molecules to occupy space do not come into play until the distances are still smaller.

forces upon all the molecules contained in a surface layer of thickness *r*, called the *active layer*, must be, then, an interior pressure of uncertain, perhaps enormous, magnitude. It has been estimated for water at something like 10^4 atmospheres, but it has never been

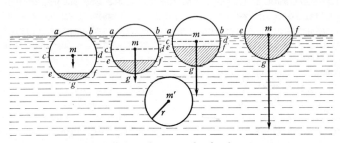

Fig. 155. An explanation of molecular pressure

measured directly and never can be. For, since a liquid is always bounded on all sides by a surface, this molecular pressure usually balances itself and therefore cancels out in hydrostatic measurements. Hence it is that Eq. [213], which leaves the existence of molecular pressure altogether out of account and treats the liquid molecules as though they were so many independent grains of sand, nevertheless gives, in general, correct results. But it is exactly such apparent violations of the ordinary hydrostatic laws as are shown in capillary phenomena that furnish a beautiful proof of the existence of molecular pressures in liquids.

160. Variation of Molecular Pressure with Curvature of Surface. It is easy to show that the molecular pressure must be greater underneath a convex surface and less underneath a concave surface than it is beneath a flat one. Thus let *m*, Fig. 156, represent a molecule that lies in the active layer at a given distance beneath a surface, and let the circle drawn about *m* represent the sphere of influence of intermolecular forces. The surface will first be assumed to be flat (*acb*), then convex (*ecf*), and finally concave (*gch*). In the first case, since *pidjq* neutralizes *pacbq*, the resultant downward force acting upon *m* is due to the attraction of the molecules lying within the

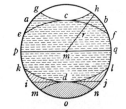

Fig. 156. Effect of surface curvature on the molecular pressure

segment *iojd* of the sphere. In the second case *pkdlq* neutralizes *pecfq*, and the resultant downward force is due to *kold*, a volume that is greater than *iojd*. In the third case the resultant force is due to *mond*, a volume that is less than *iojd*. Hence the resultant downward force upon the molecules in the active layer is greatest beneath the convex surface and least beneath the concave surface. It is evident also from the same kind of reasoning that the greater the convexity the greater the pressure. All of this may be summarized by saying that if the molecular pressure beneath a plane surface be represented by P_p', that beneath a curved surface is $P_p' \pm P_c'$, in which the magnitude of P_c' depends upon the nature of the liquid and the magnitude of the curvature, its sign being plus or minus according as the surface is convex or concave. LAPLACE [1] proved by a mathematical analysis of the forces exerted by segments of the kind shown in Fig. 156 that P_c', expressed in terms of a characteristic constant S of the liquid and the two principal radii of curvature [2] R_1 and R_2 of the surface, is

$$P_c' = S\left(\frac{1}{R_1} + \frac{1}{R_2}\right).$$ [214]

Since the quantity in parenthesis is, by definition, the curvature of the surface, this equation states that P_c' is proportional to the curvature.

161. Capillary Ascension and Depression. By starting with the fact of observation that a liquid in a capillary tube has a curved instead of a flat surface, its rise or fall in the tube, according as its surface is concave or convex, follows as a matter of course from a simple consideration of the pressures involved. Thus, the correct value of the pressure at a distance y below a *plane* surface is not that given by Eq. [213] but rather that given by the expression

$$B + \rho g y + P_p',$$

and the pressure at the same distance y beneath a *concave* surface in a capillary tube (Fig. 157, *a*) is given by

$$B + \rho g y + P_p' - S\left(\frac{1}{R_1} + \frac{1}{R_2}\right).$$

Hence there can be no equilibrium until the larger molecular pressure beneath the flat outside surface has pushed the liquid in the tube to

---

[1] *Traité de Mécanique Céleste*, Supplement to Bk. X (1806); tr. by N. Bowditch (Boston, 1829–1839).

[2] The principal radii of curvature of a surface at any nonsingular point may be found geometrically as follows. Let the intersection of two mutually perpendicular planes lie along the normal to the surface at the point considered. These planes will intersect the surface in curves of radii of curvature R_1 and R_2. Rotate the two perpendicular planes about the normal until R_1 and R_2 are a maximum and a minimum. R_1 and R_2 are then the principal radii of curvature of the surface at the point considered. (What are the principal radii of curvature of a spherical surface? of a cylindrical surface?)

such a height h that the total pressures at any two points in the same horizontal plane of the liquid are equal; that is, until

$$B + \rho g y + P_p' = B + \rho g y + P_p' - S\left(\frac{1}{R_1} + \frac{1}{R_2}\right) + \rho g h,$$

or
$$\rho g h = S\left(\frac{1}{R_1} + \frac{1}{R_2}\right). \qquad [215]$$

The quantities R_1 and R_2 are the principal radii of curvature at the point considered (for example, 1, 2, or 3, Fig. 157, b), and h is the elevation of this point above the outside plane surface.

It thus appears that, correctly speaking, a liquid does not rise in a capillary tube because of a "capillary attraction," any more than it rises in a suction pump because of the attraction of the vacuum created by the lifting of the piston. In both cases the liquid is *pushed*

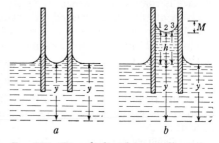

Fɪɢ. 157.   Rise of a liquid in a capillary tube

up by a pressure existing outside. In the case of the pump this is the atmospheric pressure acting on top of the water in the well; in the case of the capillary tube it is the pressure acting in the surface layer of the outside liquid.

162. Measurement of the Capillary Constant S. If it were possible to remove entirely the molecular pressure within the capillary tube, the height of rise would be a measure of P_p', just as the height of rise of the water in a long evacuated tube is a measure of the atmospheric pressure. Since, however, nothing more can be done than to obtain a curved surface within the capillary tube, it is only the *capillary constant S* that can be found from observations of the quantities h, ρ, R_1, and R_2 which appear in Eq. [215]. In the general case it is difficult to measure R_1 and R_2, but if the tube is circular in cross section, it then follows from symmetry that at the middle of the meniscus, $R_1 = R_2 = R$, so that Eq. [215] reduces to

$$\rho g h = \frac{2\,S}{R}. \qquad [216]$$

If, further, the tube is so small that the height of the meniscus (M, Fig. 157, b) is negligible in comparison with h, in which case h is practically constant for all points of the surface, then it follows from Eq. [215] that the curvature $(1/R_1) + (1/R_2)$ also is practically constant. But the only surface of constant curvature that can fulfill the condition imposed by Eq. [216] is a section of a sphere of radius R. Hence, finally, if r is the radius of the tube and α is the *angle of contact*, or angle at which the liquid surface meets the immersed portion of the tube wall (Fig. 158), then $r = R \cos \alpha$ and Eq. [216] becomes

$$\rho g h = \frac{2S}{r} \cos \alpha. \qquad [217]$$

If the liquid can be made to *wet* completely the interior of the tube, so that its angle of contact α with the walls is zero, then the meniscus must be a hemisphere, and the radius R of the surface is simply the radius r of the tube. It is left to the student to show that, for zero angle of contact, if the height of the

Fig. 158. Rise of a liquid in a capillary tube

meniscus is not wholly negligible in comparison with h, the mean value of the latter can be obtained by adding $\frac{1}{3} R$ to the height of the lowest point of the meniscus. Eq. [217] thus modified is found to hold for tubes of as much as 1-mm diameter.[1]

Eq. [217] asserts that, other things remaining the same, the height of rise h is inversely proportional to the radius of the tube, a law discovered experimentally by JAMES JURIN (1684–1750) and known as *Jurin's law*. JURIN also found (and this too follows from the equation) that the value of h depends on the section of the tube at the position of the meniscus and is independent of the form of the remainder of the tube. Eq. [217] makes the measurement of the capillary constant S a very simple matter in the case of liquids that wet the solids of which capillary tubes can be made.

163. Angle of Contact. It remains only to show why a liquid in a capillary tube assumes a curved surface, a task of no difficulty when it is remembered that a liquid surface can be in equilibrium only when it is perpendicular to the resultant force (Sec. **155**). Consider a molecule of the liquid at the point O where the liquid touches the solid wall of the containing vessel (Fig. 159). Let f_c represent the vector sum of all the forces exerted upon this molecule by such portion of the liquid as lies within the sphere of molecular attraction when the liquid surface is assumed horizontal, and let f_a represent the vector sum of the forces exerted upon the molecule at O by the molecules of the wall which lie either above or below the horizontal

---

[1] When the radius of the tube becomes larger than 1 mm, Eq. [217] requires correction because the meniscus is no longer spherical. Tables for making this correction are given in *N. K. Adam, *The Physics and Chemistry of Surfaces* (Oxford University Press, 1930), pp. 295–301, and in the *International Critical Tables*, Vol. IV, p. 435.

line passing through *O*. The attraction f$_c$ between like molecules is called *cohesion*, and the attraction f$_a$ between unlike molecules, *adhesion*. The angle at which the liquid surface meets the immersed portion of the solid is known as the *angle of contact*, and will be denoted by α. Three cases may now be distinguished:

a. If α is to be 90°, the vector sum F of the three forces must be parallel to the wall. From Fig. 159 it will be seen that this is possible only if $f_c = 2 f_a$, or if the attraction of the liquid for itself is twice that for the solid.

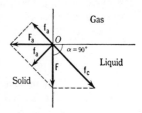

F$_{IG}$. 159. The case where the angle of contact is 90°

b. If $f_c > 2 f_a$, the vector sum F is directed into the liquid, as shown in Fig. 160, and equilibrium cannot exist until the surface near *O* has become convex and the angle of contact α obtuse. This is true for paraffin and water and also for mercury and glass, the angles of contact in these two cases being approximately 105° and 140° respectively. An angle of 180° would indicate no adhesion between the liquid and solid and is never realized. Angles of contact are rarely, if ever, quite definite, but may have any value between two extremes. If the liquid edge is advancing along the solid, the angle is a maximum; if receding, the angle is a minimum. This may be observed

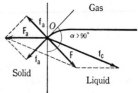

F$_{IG}$. 160. In this case the angle of contact is obtuse

clearly in the case of a raindrop traveling down a dirty windowpane.

c. If $f_c < 2 f_a$, the vector sum F is directed into the solid, as shown in Fig. 161. Hence equilibrium cannot exist until the surface near *O* has become concave and the angle of contact α acute. If the liquid attracts the solid as much or more than it attracts itself (that is, if $f_c \gtrless f_a$), the angle of contact is necessarily zero and a thin film of the liquid lies flat against the face of the solid. Liquids that do this are said to *wet* the solid.

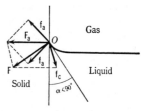

F$_{IG}$. 161. The angle of contact is here acute

That the angle of contact must be zero for this case is evident from the consideration that when a partially immersed body is raised from a liquid, the angle of contact will remain constant at a value greater than zero only if the liquid retreats down the side of the body as rapidly as the body rises; that is, the angle of contact will exceed zero only if the liquid is one that does not wet the solid.

164. Formation of Thin Films on the Surface of a Solid or Liquid. In accordance with PASCAL'S principle (Sec. 157) the molecular pressure P_a existing because of adhesion in the region O where the liquid meets the solid is transmitted undiminished in a direction parallel to the surface of the solid. There exists, therefore, a force that tends to move the molecules in this region along the solid surface (Fig. 162), and the only opposition to this force is the component parallel to the wall of the cohesive force f_c. Hence unless the cohesion exceeds the adhesion a thin film of the liquid must spread out indefinitely over the surface of the solid. This conclusion is not surprising, since it means simply that a body that attracts a liquid strongly will tend to draw every particle of it as near as possible to itself.

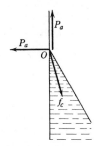

FIG. 162. Formation of a liquid film on the surface of a solid

Thus it is that a drop of water spreads out indefinitely over a perfectly clean glass or mercury surface, that a drop of olive oil spreads over water, or, in general, that any liquid spreads out over any perfectly clean surface which it wets. Clean surfaces are difficult to obtain, of course, and that is true on account of the prevalence of this very phenomenon. Thus, the least drop of oil placed on a mercury or glass surface spreads over it quickly and completely changes the effect of adding a drop of water. However, such familiar facts as the rapid spreading of oil over water and over the surface of a glass container attest the correctness of the foregoing conclusions. Of course, when but a drop of the liquid is present or when the surface is of great extent, a limit to the spreading must be reached when the film is one molecule thick. LORD RAYLEIGH,[1] in 1890, measured films of olive oil on water that had a thickness of but 1.6×10^{-6} mm; this therefore sets an upper limit to the diameter of a molecule of olive oil. More recently, IRVING LANGMUIR [2] and others [3] have shown that films of oil and other insoluble substances on water are only one molecule in thickness and that the molecules often orient themselves very simply in these films. Measurements on such films therefore provide a method of great simplicity for studying the properties of

---

[1] *Proceedings of the Royal Society* **47**, 364 (1890).

[2] *Journal of the American Chemical Society* **39**, 1848 (1917).

[3] See *N. K. Adam, *The Physics and Chemistry of Surfaces* (1930), Chap. II, for an excellent discussion of the experimental methods used in the more recent work and a summary of the interesting conclusions drawn from the study of surface films.

the molecules themselves, such as their size, shape, flexibility, and the forces around them. It is becoming increasingly clear that these forces are chemical in nature and identical with those which cause chemical reaction and solution.

165. Work Done in Extending a Liquid Surface; "Surface Tension." Another important result of intermolecular attraction is that liquid surfaces tend to contract to the smallest possible area and therefore behave *as though* they were in a state of tension. For, since every molecule in the active layer (Sec. **159**) is always being urged into the interior, it follows that as many molecules as can do so will leave this layer and pass within; that is, a liquid will tend to draw together into a shape which makes the surface area, and therefore the potential energy, least for a given volume. Thus it is that all bodies of liquid which are not distorted by gravity or other external forces always assume the spherical form, as, for example, a raindrop, a soap bubble or a globule of oil floating beneath the surface of a liquid of the same density.

This tendency to assume the form of smallest surface should result in a sensible contractility in the case of a liquid film with gas on both sides, as, for example, a soap bubble, because in this case the area of the surface is so large. Experiment amply supports the conclusion. Thus, a soap bubble may be observed to begin to draw back into the bowl of the pipe as soon as the blower removes his mouth. A wet loop of thread laid upon a soap film formed in a wire ring (Fig. 163) is drawn out at once into a circular form as soon as the film within the loop is pierced with a hot wire. A film formed in the wire frame *abdc*, Fig. 164, snaps the movable wire *ab* back toward *cd* as soon as the stretching force *f* is removed.[1]

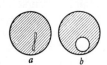

Fig. 163. Loop in a soap film

The fact that the loop of thread in Fig. 163 takes an accurately circular form shows that it is subjected to precisely the same force at all points on its circumference; yet the fact that such a film exhibits varying colors shows that it has a widely varying thickness. One must therefore conclude that *the contractility of liquid films is independent of their thickness.* Thus the force *f* which must be applied

---

[1] For further examples of this fundamental tendency of liquid surfaces to contract to the smallest possible area, see the extended studies of JOSEPH ANTOINE FERDINAND PLATEAU (1801–1883), the blind physicist, which are described in his *Statique expéri-mentale et théorique des Liquides soumis aux seules Forces moléculaires* (1873); for translations, see Taylor's *Scientific Memoirs* (London, 1844 and 1852), Vols. IV and V, and the *Smithsonian Reports* (1863–1865). See also *C. V. Boys, *Soap Bubbles* (Macmillan, 1928).

to the wire *ab* in Fig. 164 to keep the film from contracting must always be the same, whether the film has been stretched little or much; that is, whether it is thick or thin. Now, if the liquid surface really were an elastic membrane, HOOKE'S law (Sec. **90**) should be applicable to it and the foregoing result would be a very strange one. But if it be remembered that when a liquid surface is stretched, the increase in surface is made possible by the transfer of new molecules from the interior to the surface, and that these are acted upon by precisely the same forces as the original surface molecules, the fact that *f* is independent of the extent of the surface is explained.

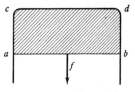

FIG. 164. Stretching a film

It follows from the constancy of *f* that if the wire *ab*, Fig. 164, be pulled down a distance *d*, the work done by *f* is equal to *fd* and is proportional to the increase in surface. Let *σ* represent the work which must be done against the intermolecular forces in order to bring enough molecules into the active layer to form one new unit area of surface; then, since the total increase in surface (considering both sides of the film) is $2\,ab \cdot d$, it follows that $fd = \sigma \cdot 2\,ab \cdot d$, or $\sigma = f/2\,ab$. But $2\,ab$ is simply the length of the line of surface to which the stretching force *f* is applied. Hence, since $\sigma = f$ when $2\,ab = 1$, it follows that *in any liquid surface the work required to bring up into the active layer, against intermolecular attraction, enough molecules to form one new unit area of surface is equal numerically to a tangential contractile force across an imaginary line of unit length in the surface* (Fig. 165).

FIG. 165. The coefficient of surface tension *σ* is equal to a tangential contractile force across an imaginary line of unit length in the surface

The work necessary to extend a liquid surface is of fundamental importance in dealing with liquid surfaces, and many problems can be attacked without knowing more than the magnitude *σ*. In the solution of such problems, however, the mathematical device of substituting for this work a hypothetical contractile force acting in all directions *parallel* to the surface is used to simplify the calculations.[1] This is what is generally known as the *surface tension*. The term should not be taken so literally, however, as to imply an actual mechanism in liquid surfaces such as exists in a stretched membrane. The fundamental mechanism is the perpendicular inward attraction exerted on the surface molecules by those lying immedi-

---

[1] This method was introduced by THOMAS YOUNG in an essay on the "Cohesion of Fluids," *Philosophical Transactions* **95**, 65 (1805).

ately under them; there is no need to speculate how this can be transformed into a surface tension parallel to the surface, for the surface tension does not exist as a physical reality, but is only the mathematical equivalent of the work done against the molecular pressure in enlarging the surface.

166. The Coefficient of Surface Tension. The quantity σ is called the *coefficient of surface tension*. This surface energy[1] can be expressed either in ergs per square centimeter or in dynes per centimeter. As we shall see in the next section, it is identical with the constant S in Eqs. [214] to [217]. The value of σ depends upon the purity of the liquid in the surface, upon the temperature, upon the age of the surface, and upon the gas with which the surface is in contact.

Experiment shows that the surface tension almost invariably decreases with rise in temperature; although this is to be expected, since a rise of temperature corresponds to a pushing apart of the molecules, clear conclusions as to its exact meaning in terms of molecules have not yet been reached.

At the critical temperature t_c, where a liquid and its vapor become identical, the surface tension at the interface should diminish to zero. The following formula for the coefficient of surface tension at temperature $t°$ C is found to hold well:

$$\sigma_t = \sigma_0 \left(1 - \frac{t}{t_c}\right)^n,$$

where σ_0 is the value of σ at $0°$ C and n is a constant which varies slightly from liquid to liquid but has a mean value of about 1.2. The surface tension depends only slightly on the nature of the gas above the surface. A liquid, of course, is always in contact with its own vapor or the gas in which it is immersed, so that all surface tensions are in reality *interfacial tensions*. Such interfacial tensions exist also at the interface of two liquids.

167. Relation between Surface Tension and the Molecular Pressure under a Curved Surface. As an illustration of how "surface-tension" problems may be solved simply from a knowledge of the magnitude σ of the work necessary to form one new unit area of surface, let us derive Eq. [214] from such considerations instead of by the very difficult mathematical method employed by LAPLACE (Sec. **160**). The pressure under a convex liquid surface must be greater than that under one which is plane simply because the displacement of a curved

---

[1] The student should perhaps be warned that the coefficient of surface tension σ is not equal to the total surface energy ϵ per unit area. Since the surface tension decreases with rising temperature, a liquid film must cool on extension; hence in the isothermal extension represented in Fig. 164 a quantity of heat q per unit area was absorbed from the surroundings. The total energy ϵ of formation of unit surface is therefore $\sigma + q$, and hence $\sigma = \epsilon - q$. The quantity $\epsilon - q$ is what is known as the *free energy* per unit area. It can be shown by a thermodynamic argument to equal $\epsilon + T (d\sigma/dT)$, where T is the absolute temperature.

surface parallel to itself and toward the convex side results in an increase in area, and the work involved in this increase must be done by the pressure difference which moves the surface. Let $ABCD$, Fig. 166, be a small rectangular element of the surface and let it be displaced parallel to itself a small distance ds into the position $A'B'C'D'$. Let the normals to the surface at A and B meet at O_1 and the normals to the surface at B and C meet at O_2. Denote the radius of curvature of the arc AB by R_1 and that of BC by R_2. Draw BE parallel to AA'. The increase in length of the arc AB is therefore $EB' = ds \cdot \angle EBB' = ds \cdot \angle AO_1B = ds \cdot AB/R_1$.

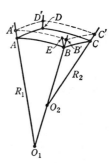

Fig. 166. Work done in displacing a convex liquid surface

Similarly, the increase in length of the arc BC is $ds \cdot BC/R_2$. The increase in area of the element of surface after the displacement is therefore

$$\left(AB + \frac{AB}{R_1} \cdot ds\right)\left(BC + \frac{BC}{R_2} \cdot ds\right) - ABCD,$$

or, if second-order terms be neglected,

$$ABCD \cdot ds\left(\frac{1}{R_1} + \frac{1}{R_2}\right).$$

If σ be the coefficient of surface tension, the work done in bringing about this increase in area is

$$\sigma \cdot ABCD \cdot ds\left(\frac{1}{R_1} + \frac{1}{R_2}\right),$$

and this must equal the work done by the pressure difference P_c' under the curved surface. This work is the pressure difference multiplied by the change in volume, or $P_c' \cdot ds \cdot ABCD$. Hence

$$P_c' = \sigma\left(\frac{1}{R_1} + \frac{1}{R_2}\right). \qquad [218]$$

If this equation is compared with Eq. [214], it will be seen that *the capillary constant S is simply the coefficient of surface tension σ.*

Instead of using the foregoing energy method, Eq. [218] could have been derived also by treating the surface tension as a hypothetical contractile force acting in all directions parallel to the surface and finding the normal component of this force. That this is legitimate was shown in Sec. **165**. Consider an infinitely small rectangular element of a convex surface bounded by the arcs Δs_1 and Δs_2, Fig. 167. Let these arcs correspond to the two principal radii of curvature R_1 and R_2. If the coefficient of surface tension of the liquid composing the surface be σ, then the force that acts on each of the arcs Δs_1 is $\sigma \cdot \Delta s_1$.

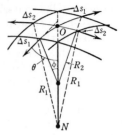

FIG. 167. A convex sur-
face element

These forces are, of course, tangential to the surface and perpendicular to the arcs. Similarly, the force that acts on each of the arcs Δs_2 is $\sigma \cdot \Delta s_2$. Since the surface is curved, all four of these forces have small components in the direction of the normal ON, and the pressure P_c' beneath the element is evidently the sum of these normal components divided by the area $\Delta s_1 \cdot \Delta s_2$ of the element. Now the component of $\sigma \cdot \Delta s_1$ parallel to ON is $\sigma \cdot \Delta s_1 \cos \theta = \sigma \cdot \Delta s_1 \sin \phi$. But in the limit, $\sin \phi = \Delta s_2 / 2\, R_2$. Hence the sum of the normal components of the two forces $\sigma \cdot \Delta s_1$ is $\sigma \cdot \Delta s_1 \cdot \Delta s_2 / R_2$. Similarly, the sum of the normal components of the two forces $\sigma \cdot \Delta s_2$ is $\sigma \cdot \Delta s_2 \cdot \Delta s_1 / R_1$. Hence the pressure P_c' due to the curvature is

$$P_c' = \frac{\dfrac{\sigma \cdot \Delta s_2 \cdot \Delta s_1}{R_1} + \dfrac{\sigma \cdot \Delta s_1 \cdot \Delta s_2}{R_2}}{\Delta s_1 \cdot \Delta s_2} = \sigma \left(\frac{1}{R_1} + \frac{1}{R_2} \right),$$

which is Eq. [218]. This demonstrates the usefulness of the notion of surface tension; although it is only a mathematical fiction, it enables us to deal with surface effects without making explicit reference to the more fundamental concept of molecular attractions upon which the phenomena in reality depend.

EXAMPLE. Show that the pressure beneath a convex spherical film, such as a soap film, is given by

$$P_c' = \frac{4\,\sigma}{R}, \qquad\qquad [219]$$

where R is the radius of curvature of the film.

Eq. [219] gives the amount by which the air pressure within a spherical soap bubble must exceed the outside pressure in order to counteract the force due to the surface tension of the soap film. If the spherical bubble has only one surface, as in the case of a bubble of steam in water, the pressure within needed to counteract surface tension evidently is $2\,\sigma/R$.

168. Summary of Methods of Dealing with Surface-Tension Problems. Surface-tension problems can usually be solved by any one of the three methods which have been discussed, for they have been shown to be mathematically equivalent. These methods are: (*a*) the classical method [1] of LAPLACE, GAUSS, and POISSON (Sec. **160**); this

---

[1] Excellent reviews of the classical theories are given in Maxwell's article on "Capillary Action" in the *Encyclopaedia Britannica*, ed. 9 [reprinted in *The Scientific*

method is highly difficult mathematically and, as it is based on
the assumption of infinite subdivisibility of the liquid, gives no
information regarding the properties of the individual molecules;
(b) the energy method, which considers simply the work per unit
area σ done against intermolecular attractions in forming new surfaces; this is a simple and powerful method and gives much information regarding the nature of molecules and molecular forces;
(c) the surface-tension method, founded by YOUNG, which considers σ
as a hypothetical tension acting in all directions parallel to the surface; this method is exceedingly useful as a mathematical device,
but, taken alone, it can never tell us anything about the mechanisms
that exist in a liquid surface.

○

Fluids in Motion

Because of the complexity of the laws of moving fluids,[1] we shall
make no attempt to deal with any except the simplest cases of steady
flow. The motion of a fluid is said to be *steady* if the velocity of flow at
any point is constant in magnitude and direction. It is to be noticed
that the term *steady flow* does not imply that the velocity of any
particular particle of the fluid is necessarily constant but only that
the velocity at a particular *point* remains constant. At very low
speeds the motion is usually steady, but as the speed is increased
the velocity at any given point begins to vary in an irregular manner.
The motion is then said to be *turbulent*. For any given case there is
some definite *critical speed* at which the change from the one type
of motion to the other takes place. In the case of steady flow, the
particles of the fluid move along definite *stream lines* which remain
fixed in position. Any narrow tubular portion of the fluid that is
bounded by stream lines is called a *tube of flow*. It is evident that
such a tube has the property that any particle of fluid that is in a
given tube remains in it during the whole course of the flow.

**169. Bernoulli's Theorem for an Incompressible Nonviscous Liquid in
Steady Flow.** Let A_1A_2, Fig. 168, be a section of a tube of flow in

---

Papers of James Clerk Maxwell (Cambridge University Press, 1890), Vol. II, pp. 541–
591, and revised by LORD RAYLEIGH in the *Encyclopaedia Britannica*, ed. 11] and in
Kapillarität und Oberflächenspannung, by G. Bakker [Vol. VI of the Wien-Harms
Handbuch der Experimentalphysik (Leipzig, 1928)].

[1] A classical treatise in this field is H. Lamb, *Hydrodynamics*. For a comprehensive
elementary discussion, see *E. Edser, *General Physics for Students* (Macmillan),
Chaps. 11–13. A more advanced treatment will be found in A. G. Webster, *The
Dynamics of Particles and of Rigid, Elastic, and Fluid Bodies* (Teubner, 1904).

an incompressible liquid that is flowing steadily without friction in
a pipe of varying cross section. When a volume V flows in through
A_1, an equal volume flows out through A_2.

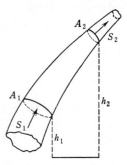

The liquid below A_1 acts like a piston in
forcing the volume V of the liquid across
the boundary A_1, and if the pressure at this
point be P_1, the work done *on* the liquid is
P_1V. In the same time the liquid between
A_1 and A_2 does work P_2V in forcing liquid
out across A_2. The net amount of work
done *on* the liquid in A_1A_2 is $(P_1 - P_2)V$,
and, since the friction is assumed to be negligibly small, all of this work must appear
in the form of increased kinetic and potential energy. Let the heights of A_1 and A_2

FIG. 168. Section of a tube
of flow

above some arbitrarily chosen level be h_1 and
h_2 and let the speeds of the liquid at these points be v_1 and v_2.. The
increase in the potential energy between A_1 and A_2 is then $V\rho g(h_2 - h_1)$
and the increase in the kinetic energy is $\frac{1}{2}V\rho(v_2^2 - v_1^2)$. Hence

$$(P_1 - P_2)V = V\rho g(h_2 - h_1) + \frac{1}{2} V\rho(v_2^2 - v_1^2),$$

or $$P_1 + \rho gh_1 + \frac{1}{2} \rho v_1^2 = P_2 + \rho gh_2 + \frac{1}{2} \rho v_2^2 = \text{constant}. \qquad [220]$$

This is the expression at which DANIEL BERNOULLI arrived in his
Hydrodynamica (1738). Its physical meaning is that, when an incompressible liquid is in steady motion with-
out friction along a tube of flow, the energy
required to force a unit mass of the liquid
into one end of the tube plus the potential
and kinetic energy of the unit mass is a
constant. This *constant* is of course not the
same for different stream lines.

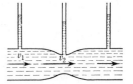

As an application of BERNOULLI's theorem,
consider the steady flow of water through a
horizontal pipe containing a constriction,
such as is shown in Fig. 169. If the pipe

FIG. 169. The pressure in
a horizontal pipe is least
where the speed of the
fluid is greatest

is of small cross section, the differences in level of the liquid at
various points may be neglected, and Eq. [220] becomes

$$P_1 + \frac{1}{2} \rho v_1^2 = P_2 + \frac{1}{2} \rho v_2^2. \qquad [221]$$

If the pipe is entirely filled with the liquid, the speed v_2 at the constriction is greater than the speed v_1 in the larger part of the pipe.
Hence the pressure P_2 inside the constriction is less than that in the
larger part. This is exactly what one would expect if it be remem-

Basel, Switzerland

Photograph courtesy of the Swiss Federal Railroads

BASEL, beautifully located on the Rhine, is famous in the history of science as the birthplace and home of most of the members of the illustrious BERNOULLI family and of LEONHARD EULER. Four of the most famous of the BERNOULLIS — JACQUES, two of the JEANS, and DANIEL — were professors at the University of Basel, which was founded in 1460 and is the oldest university in Switzerland.

Vignette (Redrawn) from the Title-Page
of Daniel Bernoulli's HYDRODYNAMICA (1738)

THE array of hydraulic machines which this vignette depicts shows how extensive
was the practical hydromechanical knowledge of its time.

bered that a particle entering the constriction is accelerated and that this can be accomplished only if the force behind the particle exceeds that in front of it. This principle is utilized in the *aspirator* (Fig. 170) and also in the *Venturi meter* for gauging the flow of water in pipes.

BERNOULLI'S theorem does not apply to gases, because gases are highly compressible; nevertheless, the theorem does predict qualitatively many effects that are observed in a flowing gas. The ordinary atomizer, the forced draft of a locomotive, and the curving of a baseball furnish familiar illustrations.

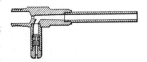

FIG. 170. The aspirator is a form of air pump. Water from a faucet is forced through the constriction while the side tube is connected to the vessel which is to be exhausted

170. Motion of a Viscous Fluid. If the effects of viscosity and of other friction in the pipe are very marked, a straightforward application of BERNOULLI'S theorem would lead to erroneous conclusions. Thus, if a viscous incompressible liquid is flowing steadily in a horizontal pipe of uniform cross section, there is a dissipation of energy throughout the pipe because of the friction and, since the speed in this case must be the same at all points, the decrease in mechanical energy must take place at the expense of the pressure. As a result, there is a fall of pressure along the pipe.

The student should at this time review the basic ideas on viscosity that appear in Sec. 121. In Sec. 122 the kinetic explanation of viscosity in gases is given, and in Exp. XB the theory of the steady flow of a viscous fluid through a capillary tube is developed. Measuring the flow through a capillary tube was the first method used for determining coefficients of viscosity and is still the most generally employed. In the case of a liquid, the coefficient of viscosity may also be found by observing the speed with which a small sphere of known diameter and mass falls in it. The theory of this method was first worked out by GEORGE GABRIEL STOKES,[1] who showed that a small sphere falling in a viscous medium soon attains a constant speed. This is true of raindrops. Finally, the coefficient of viscosity of a liquid can be found by measuring the torque required to rotate one cylinder inside another with constant speed, the space between being filled with the liquid (Fig. 174, p. 323).[2] A knowledge of the viscous properties of various fluids obviously is of importance in such practical problems as the determination of the value of oils for lubricating purposes.

[1] *Transactions of the Cambridge Philosophical Society* **9**, 8 (1850), Part I, Sec. IV; *Mathematical and Physical Papers* (Cambridge University Press, 1922), Vol. III, p. 55.

[2] For a comprehensive consideration of these various methods, see G. Barr, *A Monograph of Viscometry* (Oxford University Press, 1931).

EXPERIMENT XIIIA. DENSITY OF LIQUIDS

The object is to determine the densities of several liquids with a Mohr-Westphal balance and with a constant-weight hydrometer.

PART I. The Mohr-Westphal Balance.[1] The density of a liquid can be determined rapidly and at the same time accurately by means of a Mohr-Westphal balance (Fig. 152). One arm of the beam of this balance is graduated decimally and supports a glass plummet B. The other arm is so counterpoised that the beam will balance when the plummet is in air. If the plummet be immersed in a liquid, it suffers an apparent loss of weight (Sec. 154) and suitable riders must be hung from the various notches on the beam in order to restore the balance.

The absolute weights of the riders need not be known if only their relative weights are 1.0, 0.1, 0.01, and 0.001. For suppose that to balance the beam when the plummet is immersed in water of density ρ_1, riders must be placed as indicated in Fig. 152. The apparent loss of weight ΔW_1 of the plummet, expressed in terms of the weight 1.0, is then evidently 1.1044. Suppose now that when the water is replaced by another liquid *of the same temperature* and of density ρ_2, the beam balances when rider 1.0 is in notch 9, 0.1 in notch 2, 0.01 in notch 4, and 0.001 in notch 9. In this case the apparent loss of weight ΔW_2 is 0.9249. Hence, from Eq. [210],

$$\rho_2 = \frac{0.9249}{1.1044} \rho_1.$$

The density of water at any temperature ρ_1 can be obtained from tables,[2] and hence ρ_2 can be calculated no matter what happens to be the weight of rider 1.0. In practice the plummet is so made that its loss of weight ΔW_1 when immersed in water at 15° C is equal to the weight of the rider 1.0; hence, if no high degree of accuracy is desired, the reading of the balance when the plummet is immersed in any liquid gives at once the density of that liquid.

---

[1] The Mohr-Westphal balance can be used to determine the densities of the manometer oil used in Exp. XIIA and the alcohol used in Exp. XIIIB. It can also be employed to find the density of liquid air, thus enabling the student to calculate the diameter of air molecules and the AVOGADRO number. Observe the following precautions with liquid air: *a.* Handle the fragile Dewar flask with care so that the liquid does not splash; protect your eyes. *b.* Immerse objects in the liquid air very slowly, and do not attempt to make observations until the air has ceased to boil. *c.* Do not close the neck of the flask with a stopper. *d.* Do not allow the liquid air to saturate any organic material.

[2] Appendix 13, Table B.

In using the balance, first turn the base so that the leveling screw S lies in the vertical plane which includes the beam. Clean the plummet B and hang it from the hook c by means of a single strand of fine wire. With the plummet in air, adjust the screw S and, if necessary, the balancing weight at the left end of the beam until the two points at a are accurately together. Immerse the plummet in distilled water, remove with a glass rod any air bubbles that cling to its walls, and then place on the beam such riders as are necessary to bring the points together again. Read the temperature of the liquid. Clean the plummet and take similar readings with it immersed in the liquid of unknown density.

1. What kind of correction must be applied if the temperatures of the two liquids are not the same?

PART II. The Constant-Weight Hydrometer. Take the readings of a direct-reading constant-weight hydrometer (Fig. 171) when it is floated in water and then in each of the other liquids. The theory of this instrument is too simple to require explanation. The directly calibrated scale should be read by looking through the liquid, with the eye placed as little as possible beneath the level of the surface. If the instrument does not read the correct density of distilled water at the observed temperature, a correction amounting to the difference must be applied to its indication of the density of another liquid.

FIG. 171. The constant-weight hydrometer is a convenient instrument for finding the specific gravity of a liquid when precision is unnecessary

2. When a hydrometer having a stem 1 cm in diameter is floated in water, with what force due to the surface tension of the water is it pulled downward? How much deeper does it sink in the water than it would in a liquid of the same density that does not rise on the stem?

3. The *specific gravity* of a substance is defined as the ratio of the mass of a certain volume of the substance to the mass of a like volume of some standard substance. In the case of solids or liquids, the standard usually chosen is water at 4° C. Why is the Mohr-Westphal balance sometimes classed as a specific-gravity balance?

4. How does specific gravity differ from density? What are the specific gravities of the liquids which you tested?

5. If liquid air was one of the substances tested in this experiment, calculate the diameter of the molecules of air and calculate the AVOGADRO number. Make a summary of all the experimental data and calculated molecular constants for air that you have obtained in Exps. VIIIA, XB, and the present experiment.

EXPERIMENT XIIIB. COEFFICIENT OF SURFACE TENSION

The object is to determine by two different methods the coefficients of surface tension for distilled water and for alcohol, at room temperature. Since the presence of the least trace of oil or other contamination upon the surface of a liquid may change completely the value of the surface tension, it is important to clean thoroughly all parts of the apparatus which are to be brought into contact with the liquid and thereafter not to touch these parts with the hands. The beakers and other glassware should be cleaned by washing them first with soap and water, then with an aqueous solution of potassium hydroxide, and finally with distilled water.

PART I. The Capillary-Tube Method. One method of measuring the capillary constant S, and hence the coefficient of surface tension σ, is to observe the rise of the liquid in a glass capillary tube (Sec. 162). In the case of liquids like pure water and alcohol in contact with clean glass, the angle of contact is zero and the cosine term in Eq. [217] becomes unity.

a. Prepare a number of fresh capillary tubes by heating to softness bits of clean glass tubing in a Bunsen flame, and drawing them down to diameters of from 0.1 to 0.5 mm. On account of the difficulty of cleaning tubes of small bore, it is best to employ tubes that have never been soiled.

b. Select two tubes which appear to be of circular cross section and fasten them to a *clean* mirror scale by means of rubber bands, allowing their ends to project beyond the lower end of the scale, as in Fig. 172. Take the reading h_0 of the fixed point O on the scale by placing the eye so that the image of O comes into coincidence with O itself.

c. Immerse the lower end of the scale, with its attached tubes, in distilled water contained in a clean beaker, and raise and lower it several times so as to wet thoroughly the capillaries above the points

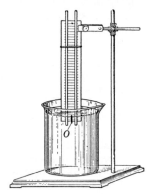

Fig. 172. The capillary-tube method of measuring surface tension

to which the water rises. Clamp the scale in a vertical position and then, preferably by means of a rack-and-pinion adjustment, raise the clamp until the pointer O comes exactly into contact, from below, with the liquid surface. Slip each capillary tube up a trifle (Why?), and then take the scale-reading h_1 of the bottom of the meniscus in each tube. Obviously the height of rise h is given by $h_1 - h_0$.

d. Take the temperature of the liquid.

e. On each tube mark with a bit of wax the point to which the liquid rose. Remove the tube from the scale, scratch it with a sharp file at this point, and break it off as squarely as possible. Attach it upright to the side of a block of wood with soft wax and measure the diameter of the bore at the broken end by means of a micrometer microscope.[1] Repeat this measurement several times, in each case with a different diameter.

f. Repeat operations *b* to *e* with two other tubes and with alcohol as the liquid.

g. Employ Eq. [217] to calculate the coefficients of surface tension σ $(= S)$ for water and for alcohol at room temperature. The density of the water at any temperature can be obtained from tables.[2] If the density of the alcohol is not known, it can be determined by means of a Mohr-Westphal balance (Exp. XIIIA).

1. How do you explain the differences between the results obtained with different tubes?

2. How high would the alcohol tested rise in a tube 0.12 mm in diameter?

3. Would corrections for the heights of the meniscuses improve your values of the surface tensions?

PART II. A Direct Method. The most direct method of determining surface tension obviously is to measure the force necessary to support a film of liquid of known length. For example, if a straight wire of length l be carefully lifted horizontally from the surface of a liquid, it will draw a film of liquid with it. If the force f which is necessary to balance the force exerted by the stretched film be measured with a delicate spring balance, then, in view of Sec. **165**, the coefficient of surface tension is given by

$$\sigma = \frac{f}{2\,l}.$$ [222]

a. Clean the platinum wire frame *a*, Fig. 173, by holding it with a pair of tweezers in a Bunsen flame until it is red-hot. Without touching the frame with the fingers, hang it on the delicate helical spring *s*.

b. Raise the platform *b* carrying the beaker of distilled water until the frame is immersed. Then slowly lower the platform by

---

[1] See Appendix 9, *The Micrometer Microscope.*

[2] Appendix 13, Table B.

means of the rack and pinion *r*. At the same time lower the glass mirror which slides on the scale so that the scratch on the mirror is always left directly back of the flat disk which marks the base of the spiral spring.

Lower the platform until the vertical film of water just breaks, and record the scale-reading at which. the rupture occurs. Be sure to place the eye so that the image of the disk in the mirror is directly back of the disk itself. (Why?)

c. Repeat the foregoing operations until a number of consistent readings have been obtained. If it appears that the liquid surface has become contaminated in spite of all precautions, stir the liquid vigorously after each observation by means of a clean glass rod.

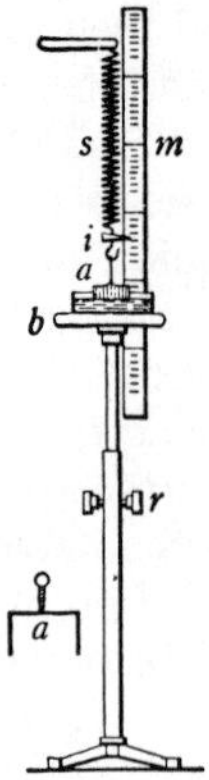

FIG. 173. Apparatus for a direct measurement of surface tension

d. Take the temperature of the liquid.

e. Take the reading of the disk when the frame is out of the liquid and it and the spring are hanging freely. If this reading be subtracted from the readings taken with the film, the difference obviously is a measure of the total force exerted on the wire by the two surfaces of the film.

f. Repeat operations *a* to *e* with alcohol as the liquid.

g. Measure the length *l* of the wire frame.

4. Will an ordinary ruler suffice for the determination of *l*, or should a vernier caliper be employed?

h. Observe the elongation of the spring produced by a known weight of the same order of magnitude as the force exerted by the film.

i. On the assumption that the elongation of a spiral spring is proportional to the stretching force, calculate the total force *f* exerted by each of the liquid films. Then calculate the coefficient of surface tension for each liquid.

5. Do the results of this experiment confirm the conclusion arrived at theoretically (Sec. **167**), that the capillary constant *S* in Eq. [214] is simply the coefficient of surface tension σ?

6. Why is it not necessary to take into account the adhesion between the platinum wire and the liquid?

7. Explain how the direct method of measuring surface tension could be employed to prove experimentally that the coefficient of surface tension is independent of the thickness of the film and also of its length.

EXPERIMENT XIIIc. COEFFICIENT OF VISCOSITY OF A LIQUID

The object is to employ the method of coaxial cylinders to determine the coefficient of viscosity of lubricating oil at room temperature and to study the variation
of viscosity with temperature. The
viscometer, shown in Fig. 174, is similar to the one designed by G. F. C.
SEARLE [1] for the measurement of the
coefficients of viscosity of very viscous
liquids. In the theory that follows it
will be shown that a constant torque
applied to the movable inner cylinder
will cause it to rotate with a speed
that depends only on the viscosity of
the liquid contained in the space between the cylinders.

Suppose that the inner and outer
cylinders (Fig. 175) have radii R_1 and
R_2, respectively, and that the length
of the inner cylinder covered by the
liquid is l. Let the inner cylinder be
rotated with a constant angular speed

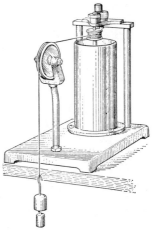

Fig. 174.   Coaxial-cylinder type
of viscometer

ω_1. If it be assumed that the liquid wets the walls of the cylinders,
so that there is no slip between the walls and the liquid, and if it
be assumed that there is no abrupt change of
speed between contiguous particles of the liquid,
then it follows that successive cylindrical layers
of the liquid must be rotating about the common axis of the cylinders with angular speeds
ω which fall off from ω_1 at the inner cylinder
to zero at the outer cylinder. Consider an *imaginary* cylinder of radius r described about the
axis. The speed of the particles on the inner
surface of this cylinder will be greater than that
of the particles on the outer side. The velocity gradient (Sec. 121) normal to this surface,
namely $- du/dr$, evidently will be the same at
every point of it, and hence the dragging torque exerted on the

Fig. 175.   Transverse
section of the two co-
axial cylinders. The
space between them is
filled with the liquid
under test

---

[1] *Proceedings of the Cambridge Philosophical Society* **16**, 600 (1912) ; also G. F. C.
Searle, *Experimental Physics* (Cambridge University Press, 1934), pp. 250–254.

fluid inside this imaginary cylinder by the slower-moving liquid outside of it is

$$L = f \cdot r = \eta \cdot 2\,\pi r^2 l\left(-\frac{du}{dr}\right).$$ [223]

This equation is derived directly from Eq. [165], Chap. 10, by supposing that the cylindrical surface is made up of a large number of plane strips.

Now u, the linear speed of the particles at a distance r from the axis, is related to the angular speed ω by Eq. [88], Chap. 7, and hence

$$\frac{du}{dr} = \frac{d}{dr}(r\omega) = \omega + r\frac{d\omega}{dr},$$ [224]

by the rule for the differentiation of the product of two functions. The first term, ω, in the right-hand member of this equation expresses the difference in the linear speeds of two adjacent particles on the radius r which are moving with the same angular speed ω; since there is no angular displacement of one of these particles relative to the other, no shearing stress is involved in this term. The second term, $r\,d\omega/dr$, does, however, involve a shearing stress, for it expresses a difference in the linear speeds of two particles that are moving with different angular speeds. In view of these considerations, Eq. [223] may be written in the form

$$L = -\eta \cdot 2\,\pi r^3 l\frac{d\omega}{dr}.$$ [225]

Since the fluid is in a steady state of motion, the torque L must also be the torque applied to the inner cylinder in order to keep it in steady rotation.

Upon rearranging the quantities in Eq. [225] and integrating, one has

$$L\int_{R_1}^{R_2}\frac{dr}{r^3} = -\eta \cdot 2\,\pi l\int_{\omega_1}^{0}d\omega.$$

Performance of the indicated integration and solution for η gives

$$\eta = \frac{R_2{}^2 - R_1{}^2}{4\,\pi l R_1{}^2 R_2{}^2} \cdot \frac{L}{\omega_1}.$$ [226]

As indicated in Fig. 174, the torque L is produced by means of a weight attached to a thread wound around a drum on the inner cylinder. If the mass of the weight is m and the radius of the drum is R_3, the torque on the cylinder *when the weight descends with constant speed* is mgR_3. The constant speed v of the weight is given by

$\omega_1 R_3$. By substituting for L and ω_1 in Eq. [226], one obtains finally

$$\eta = \frac{(R_2{}^2 - R_1{}^2)\,gR_3{}^2}{4\,\pi l R_1{}^2 R_2{}^2} \cdot \frac{m}{v}, \qquad [227]$$

in which all the quantities except m, v, and l remain constant throughout the experiment.

The End Correction. From Eq. [227] it appears that the mass m needed to produce a constant speed v varies directly with the length l. Hence, if the experiment be performed with different lengths l of the cylinder covered by the fluid, and if the observed values of the mass m required to produce a given speed be plotted as a function of l, the result will be a straight line. This straight line will not pass through the origin, however, but will intersect the l-axis on the negative side of the origin, which means that force would be required to produce the constant speed v even if the cylinder were of zero length. This is because there is a torque exerted by the viscous fluid on the bottom surface of the cylinder, a fact which was not taken into account in deriving Eq. [227]. This end effect may be expressed as a correction to the length l; that is, if the l-intercept of the plotted curve is l', then $l + l'$ is the corrected length to be used in Eq. [227]. The quantity $l + l'$ evidently gives the length of a cylinder without end area equivalent to the actual cylinder.

PART I. Coefficient of Viscosity at Room Temperature. *Experimental Procedure. a.* Mount the apparatus on the edge of a platform so that it is somewhat higher than the table. Place absorbent paper under the base of the apparatus so as to catch any spilled oil. If the dimensions of the apparatus are not known, carefully remove the inner cylinder and measure its length and also the diameters of the two cylinders and of the drum. Be careful not to injure the bearing pivots. Fill the apparatus to the top of the inner cylinder with the lubricating oil, castor oil, or other liquid to be tested. Be very careful to prevent oil from getting on the outside of the apparatus.

b. Attach a 5-g mass to the end of a silk fishline and wind the line on the drum with as little overlapping as possible. Select a point about 10 cm below the position of the mass m, and as the weight descends with constant speed measure the time required for it to traverse the known distance from this point to the floor. Make at least two trials. Make similar observations with masses of 10, 20, 30, etc., grams used as the driving weights.

c. Remove a little more than half of the liquid from the apparatus. Measure accurately the length l of the inner cylinder that is now covered by the liquid. Then repeat *b*, using as the driving weights masses of 2, 5, 10, 15, etc. grams.

d. Note the temperature of the oil.

Calculations.[1] From the data obtained in *b,* plot a curve with values of *m* as abscissas and of *v* as ordinates. The reciprocal of the slope of this curve gives an average value of m/v. On the same sheet plot a second curve from the data obtained in *c.* Then choose some value of *v* and draw a horizontal line; the intersections of this line with the two curves evidently determine the masses necessary to produce the chosen speed *v* for each length of cylinder.

On a separate sheet of graph paper plot these two masses as ordinates and the corresponding lengths of the cylinder as abscissas. The *l*-intercept of the resulting curve gives the correction *l'* to be added to the actual length *l* of the cylinder.

By making use of this corrected value of *l* and the average value of m/v obtained from the first curve, calculate the coefficient of viscosity.

1. Do the first two curves which you plotted pass through the origin? What information does this give with regard to the friction in the bearings?

2. Does Eq. [227] apply to gases as well as to liquids? Give reasons for your answer.

3. What would have been the percentage error in the final result if you had failed to make the end correction?

4. The rotating cylinder of the viscometer used in this experiment is so designed that its *average* density is approximately the same as that of the oil. What is the advantage?

PART II. Variation of the Coefficient of Viscosity with Temperature. The outer cylinder of the viscometer has built into it an electrical heating coil which operates from a 110-volt source.[2] Fill the apparatus with oil and heat it to about 90° C. Take temperatures immediately before and after each run with a thermometer having a stem thin enough to permit its insertion in the narrow space between the cylinders. Stir the oil and, when its temperature has become steady, remove the thermometer and make the run with a 10-ğ mass as the driving weight.

Repeat at progressively lower temperatures and thus obtain the necessary data for a viscosity-temperature curve.

5. How does the coefficient of viscosity for the liquid tested vary with temperature?

---

[1] It may prove to be more convenient to defer making these calculations until the experimental data required in Part II have been obtained.

[2] If the viscometer is not equipped with a heating coil, heat the oil by playing a Bunsen burner about the lower part of the cylinder.

QUESTION SUMMARY

1. State ARCHIMEDES' principle. Is it an empirical principle or can a rigorous theoretical proof of it be given? What are its practical and its theoretical importance?

2. State four fundamental principles of fluids at rest under gravity.

3. Explain the rise of liquids in a capillary tube. Discuss fully. Calculate the height to which a liquid will rise in a capillary tube of radius r.

4. Describe some other phenomena in liquids that are due to the same cause as the rise in capillary tubes and show how they may be used to determine the capillary constant.

5. Explain why one would not expect the stretching of a liquid film to be described by HOOKE'S law. What sort of law does describe it? Define the *coefficient of surface tension of a liquid*. In what units is it measured? Can it be expressed in any other kind of units?

6. In what three ways is the *capillary constant* defined? Show that the three definitions are equivalent.

7. Distinguish between *steady* and *turbulent* flow in fluids. State and prove BERNOULLI'S theorem. What is its practical importance? Discuss the implications of the various assumptions involved in its proof.

o

PROBLEMS

1. It is said that ARCHIMEDES discovered his principle while seeking to detect a suspected fraud in the construction of a crown made for Hiero of Syracuse. The crown was thought to have been made from an alloy of gold and silver instead of from pure gold. If it weighed 1000 gwt in air and 940 gwt in water, how much gold and how much silver did it contain? Assume that the volume of the alloy was the combined volumes of the components. *Ans.* 811 g of gold; 189 g of silver.

2. The gas bags of a certain large airship have a total volume of 3.5×10^6 ft³. At sea level, under standard conditions, 1.0×10^3 ft³ of air weighs approximately 80 lbwt, and the same volumes of the impure hydrogen and helium used in airships weigh approximately 10 lbwt and 15 lbwt respectively. (*a*) What is the total lift of this airship when it is filled with hydrogen? (*b*) How much larger would the gas bags have to be if filled with helium in order to lift the same total load? *Ans.* (*a*) 2.5×10^5 lbwt; (*b*) 8 percent.

3. A cylinder of cork is floating upright in water. (*a*) If the air above the water be removed, will the cylinder rise or sink in the liquid? (*b*) Derive an expression for the ratio of the lengths of the submerged portions.

4. An uncalibrated constant-weight hydrometer (Fig. 171) is immersed successively in two liquids of densities 1.0 and 1.1 g · cm⁻³, the two points of immersion are marked accurately on the stem, and the intervening stem is then divided into ten equal parts. Assuming that the stem is accurately cylindrical, state whether the instrument so calibrated will give correct readings in liquids of intermediate densities. Explain fully.

5. One arm of an inverted Y-tube (Fig. 176) is placed in a liquid of unknown density, while the other arm is placed in water. A portion of the air is pumped from the tube until the unknown liquid stands at a level of 33.1 cm above that in its open vessel, and the water stands at a level of 30.0 cm above that in the vessel of water. The temperature is 20.0° C. (*a*) Find the unknown density of the liquid. (*b*) Is it possible to find the pressure in the tube above the liquids without knowing the pressure of the outside atmosphere?
Ans. (*a*) 0.905 g · cm⁻³.

Fig. 176. Density of a liquid by the method of balanced columns

6. Explain why a needle or other small body which is much more dense than water may yet float upon water provided the angle of contact α is finite (Fig. 177).

7. In Fig. 178 a drop of water placed in the conical tube is observed to travel rapidly toward the small end, whereas a drop of mercury travels toward the large end; a bubble of air in a conical tube filled otherwise with water moves toward the large end. Explain.

8. (*a*) Deduce Eq. [217] from the consideration that, in a capillary tube in which the liquid is elevated, the total upward force is the vertical component of the surface tension acting upon a line whose length is the circumference of the tube, while the balancing downward force is the weight of the liquid raised. (*b*) How high will water rise in pores that are 0.001 mm in diameter? *Ans.* (*b*) 3×10^3 cm.

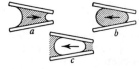
Fig. 177. Heavy body floating on water

9. Given a liquid that wets glass, prove by two different methods that its height of rise between two parallel glass plates is the same as that for a cylindrical tube provided that the distance between the plates equals the radius of the tube.

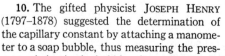
Fig. 178. Motion of drops of liquids and air bubbles in conical tubes. *a*, drop of water; *b*, drop of mercury; *c*, air bubble in water

10. The gifted physicist JOSEPH HENRY (1797–1878) suggested the determination of the capillary constant by attaching a manometer to a soap bubble, thus measuring the pressure existing in the bubble. Assuming the coefficient of surface tension of a soap solution to be 70 dyne · cm⁻¹, find the difference in levels in the arms of a water manometer attached to a bubble 7 cm in diameter. *Ans.* 0.8 mm.

11. What, approximately, would be the boiling point of water if so much of the air had been driven out of the water by heating that the bubbles forming in the water were only 10^{-4} mm in radius? Assume that the coefficient of surface tension is independent of temperature and has a value of 75 dyne $\cdot$ cm^{-1}. Is this a fair assumption? What is the value of the coefficient of surface tension of water at this temperature if the critical temperature is 374° C?   *Ans.* 200° C.

12. How much work must be done against molecular forces to blow a soap bubble 15 cm in diameter?   *Ans.* 9.9 × 10⁴ ergs.

13. Neglecting loss of energy due to viscosity, show that the speed of efflux of a liquid into the atmosphere from an orifice in a large tank (Fig. 179) which is kept filled to a constant level by the continuous addition of fresh liquid is $v = \sqrt{2\,gh}$, where h is the depth of the orifice below the free surface of the liquid. This result was obtained by EVANGELISTA TORRICELLI [1] in 1641.

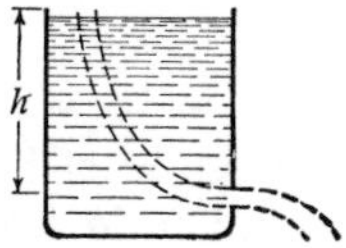

FIG. 179. Liquid flowing from an orifice in a tank

14. A vertical pipe 3.0 in. in diameter contains a constriction 0.50 in. in diameter. If the flow of the water is 0.050 ft³ $\cdot$ sec^{-1} and the pressure at a point 4.0 ft above the constriction is 90 lbwt $\cdot$ in.$^{-2}$, what is the pressure in the constriction? Neglect friction.   *Ans.* 83 lbwt $\cdot$ in.$^{-2}$.

15. In the Venturi water meter, water flows through a horizontal constricted tube, and the pressures at the wide and constricted sections are determined by means of gauges placed at these points (Fig. 169). Neglecting friction, show that V/t, the volume of water flowing in unit time, is given by

$$\frac{V}{t} = A_1 \sqrt{\frac{2\,gh}{\left(\dfrac{A_1}{A_2}\right)^2 - 1}},$$

where A_1 and A_2 are the cross-sectional areas of the wide and constricted sections, respectively, and h is the difference in pressure between these two points expressed in centimeters of water. Because viscosity has been neglected, this result will be in error by as much as 5 percent.

---

[1] See Bibliography.

CHAPTER FOURTEEN

MOTION WITH VARYING ACCELERATION

G IVEN A pendulum composed of any number of weights,[1] if each of the weights be multiplied by the square of its distance from the axis of oscillation and if the sum of these products be divided by that obtained by multiplying the sum of the weights by the distance of the common center of gravity of all the weights from the same axis of oscillation, there results the length of a simple pendulum which is isochronous with the compound pendulum, or in other words the distance between the axis and the center of oscillation of the given compound pendulum.

Translated from HUYGENS's *Horologium Oscillatorium*, Part IV, Proposition V

Ө

In our study of linear acceleration, in Chapter 1, and of angular acceleration, in Chapter 7, we confined our attention for the most part to the very restricted though important cases of *constant* acceleration. In the present chapter we shall first discuss briefly a more general case of linear acceleration in which the velocity of a particle moving in a plane is changing with time in any way whatever, and then shall proceed to the study of several important cases of variable acceleration in which the motion is periodic in nature.

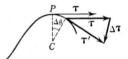

FIG. 180. Motion of a particle in a curved path

171. Motion in Any Curved Path in a Plane. When a particle P moves in any plane curve, its acceleration may be resolved conveniently into two rectangular components (Sec. **39**), one along the tangent and the other along the normal to the path. In Fig. 180 let $\boldsymbol{\tau}$ be a vector of *unit length* which is, at each instant, directed along the tangent in the sense in which the motion is described. Then, if v be the speed of the particle and if $\mathbf{v}$ be its velocity, we evidently may write

$$\mathbf{v} = v\boldsymbol{\tau}. \tag{228}$$

From Sec. **18** we have, as the defining equation for linear acceleration at any instant t,

$$\mathbf{a}_t = \lim_{\Delta t \to 0} \frac{\Delta \mathbf{v}}{\Delta t} = \frac{d\mathbf{v}}{dt}, \tag{229}$$

which now may be written

$$\mathbf{a}_t = \frac{d}{dt}(v\boldsymbol{\tau}). \tag{230}$$

---

[1] HUYGENS did not here distinguish between mass and weight.

This equation becomes, by the ordinary rule for the differentiation of a product,

$$a_t = \frac{dv}{dt}\,\boldsymbol{\tau} + v\,\frac{d\boldsymbol{\tau}}{dt}. \qquad [231]$$

Here, obviously, we have the total acceleration at any time t expressed as the sum of two components. The first of these, which has the magnitude dv/dt, is in the direction of $\boldsymbol{\tau}$ and hence is the component of acceleration *along* the path; it is therefore referred to as the *tangential acceleration*. As for the remaining component, we shall proceed to show that it is along the *normal* to the path and is directed toward the center of curvature of the path; this component is accordingly called the *centripetal*, or *normal*, *acceleration*.

First we must interpret the differential coefficient $d\boldsymbol{\tau}/dt$ in Eq. [231]. Imagine the particle P to be moving along the curve in Fig. 180, and let $\boldsymbol{\tau}$ and $\boldsymbol{\tau}'$ be the values which the *unit vector* $\boldsymbol{\tau}$ has at the beginning and end of the interval of time dt: then the vectors $\boldsymbol{\tau}$, $\boldsymbol{\tau}'$, and $d\boldsymbol{\tau}$ must make up a triangle, as shown. If $d\theta$ denotes the angle in radians through which the unit vector $\boldsymbol{\tau}$ turns in the time dt, it is evident from the figure that $d\boldsymbol{\tau}$ is equal in magnitude to $d\theta$. Moreover, since dt is an infinitesimal time, $d\theta$ is also infinitesimal, and hence the vector $d\boldsymbol{\tau}$ is at right angles to the unit vector $\boldsymbol{\tau}$. Hence, if we denote by n a vector of unit length drawn along the normal to the path toward its center of curvature C, we may write for our differential coefficient

$$\frac{d\boldsymbol{\tau}}{dt} = \frac{d\theta}{dt}\,\text{n}. \qquad [232]$$

Returning to Eq. [231], we have then, for the *total* linear acceleration of a particle moving along any path,

$$a_t = \frac{dv}{dt}\,\boldsymbol{\tau} + v\,\frac{d\theta}{dt}\,\text{n}. \qquad [233]$$

Translated into words, this equation states that the tangential acceleration is equal in magnitude to the time-rate of change of the speed, and that the centripetal acceleration is equal in magnitude to the speed of the particle multiplied by the time-rate at which its linear velocity is changing direction. It is important to note that dv/dt represents only the time-rate of change of the *speed* and

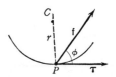

Fig. 181. Force applied at a given instant to a particle P moving in a curved path

hence that it does not give the entire acceleration except in the special case when the direction of the velocity is constant, that is, when the motion of the particle is in a straight line.

THE clock shown in this plate is, in all of its essentials, identical with the pendulum clocks in use today (see any modern article on clocks, such as that in the *Encyclopaedia Britannica*). The last hundred and eighty years have seen only minor changes. Improved workmanship, however, has brought about a steady increase in accuracy, until the best modern Riefler clocks have been found to be in error by no more than 0.02 second at the end of a month's run.

Clocks with hands and trains of wheels driven by weights and regulated by a frictional or an inertial resistance were in existence long before the time of HUYGENS, but they were so unreliable that, for example, GALILEO, in his experiments on falling bodies used in preference the ancient water clock (see Sec. 1, p. 4). Later GALILEO perceived the advantages of using pendulums to measure time, and it was at his suggestion that physicians adopted a simple pendulum of variable length as a "pulse measurer." Evidently what was needed for measuring extended times was a pendulum that would continue to swing and a means of counting the oscillations. GALILEO left behind suggestions for such a device, but it was HUYGENS who was the first to realize it, and by so doing he raised the measurement of time to a new level of refinement.

But in all of HUYGENS's investigations the theoretical aspects of the problem always shared his interest with the practical; in fact "no man of science ever preserved a more harmonious balance between all those sources by which scientific truth can be revealed." Thus his very practical study of pendulums and clock mechanisms — investigations which resulted in a lasting benefit to mankind — arose from his recognition of the clock as an important scientific instrument and led in turn to profound studies of the fundamental laws of dynamics.

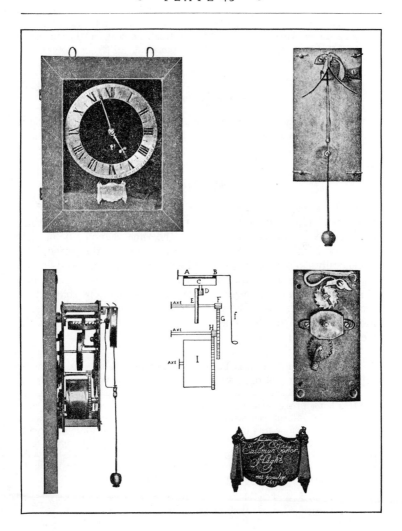

Huygens's Pendulum Clock of 1657

From *Œuvres complètes de Christiaan Huygens*, Vol. XVII, by permission of the Board of Directors
of the Hollandsche Maatschappij der Wetenschappen, Haarlem, Holland

EXAMPLE. Given that f in Fig. 181 is the total accelerating force applied at a given instant to a particle of mass m which is moving in a curved path in a plane, show that the tangential and centripetal components of the force are given by the two expressions

$$f \cos \phi = Km \frac{dv}{dt},$$

$$f \sin \phi = Kmv \frac{d\theta}{dt} = Kmv\omega = Km \frac{v^2}{r}, \qquad [234]$$

where v is the instantaneous speed of the particle and r is the radius of curvature of the element of path over which the particle is passing.

○

Periodic and Vibratory Motions

In many of the physical processes that occur in nature, a definite physical condition constantly recurs after equal intervals of time. Such processes are called *periodic*. The time T required for each repetition is called the *period*, and the reciprocal of the period, $1/T$, is generally called the *frequency*. The planets in their orbits about the sun, a point on a vibrating violin string, the balance wheel of a clock, the piston of an engine — these undergo motions that will be recognized as periodic. We shall also see, in Chapter 15, that wave motions are the result of periodic motions of the particles of the medium which transmits the waves.

172. Motion in a Circular Path with Constant Speed. This familiar type of periodic motion, first investigated by HUYGENS,[1] needs no

---

[1] In a letter to the Secretary of the Royal Society, dated September 4, 1669, HUYGENS said: "I send you herewith appended, some anagrams which I shall be pleased to have you keep in the registers of the Royal Society, which has been so kind as to approve this method of mine for avoiding disputes, and for rendering to each individual that which is rightly his in the invention of new things." Two of the fourteen anagrams which were enclosed give the essential theorems concerning the magnitude of centripetal force. One of them was as follows:

a	b	c	d	e	f	g	h	i	l	m	n	o	p	q	r	s	t	u	x	y	z
3	0	6	0	7	1	0	0	5	1	3	2	3	2	0	6	3	4	4	1		
9	0	1	3	5	1	2	0	6	1	5	2	0	0	2	4	0	6	5	0		

A translation of these anagrams, in Latin, first appeared in the *Horologium Oscillatorium* (1673). Each line of numerals in the anagram represents a line in the Latin theorem, and each numeral denotes the number of times that the letter immediately above it appears in that line. A translation of the theorem reproduced here is as follows: If a body traverses a circumference with the same speed which it would gain in falling from a height equal to one quarter the diameter, the centrifugal force then acting upon the body will be equal to the pull of gravity upon it ; that is, it stretches

extensive discussion, for it is merely a special case of motion in any curved path (Sec. **171**). One may describe it by writing $dv/dt = 0$, and $d\theta/dt = \omega =$ constant. Accordingly, in view of Eq. [233], the total acceleration is of constant magnitude $v\omega$ and is entirely *centripetal*, that is, always directed along the radius and toward the center of the circle. When a particle P revolves in a circle of center O, Fig. 182, the position of P at any moment may be assigned by giving the angle that PO makes with some fixed radius, such as OA. This angle is called the *phase* of P's motion. If, at any moment from which we begin reckoning time, P is at the point A, then after a time t it will have revolved through an angle ωt, where ω is the angular speed of the particle. The angle ωt

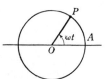

Fig. 182. The phase of a periodic motion is the angle ωt swept out by the line OP since the time when the particle occupied some chosen reference point A

is related to the period T of the motion, which is the time taken by the particle in making one complete revolution, by the equation

$$\omega t = \frac{2\,\pi}{T}\,t.$$

If a particle of mass m moves in a circle of radius r with a constant linear speed v, the *total* force f required to maintain the acceleration obviously is entirely centripetal in direction and has a constant magnitude f which is given by any of the following expressions:

$$f = Kma = Kmv\omega = Km\frac{v^2}{r} = Kmr\omega^2 = Kmr\frac{4\,\pi^2}{T^2}$$
$$= Kmv\frac{2\,\pi}{T} = Kmr \cdot 4\,\pi^2\nu^2, \quad [235]$$

where ω is the angular speed of the particle expressed in radians per unit time and ν is the frequency, or angular speed, expressed in revolutions per unit time. If the quantities which appear in Eqs. [235] are expressed in cgs units and the factor K is thus made unity, the force will be given in dynes. The student should employ Eqs. [22], Chap. 2, [86] and [88], Chap. 7, and [233] in the verification of Eqs. [235].

---

the cord which retains it just as much as if the body were suspended by this cord. Translations of others of HUYGENS's theorems on centripetal force will be found in *A Source Book in Physics* (1935), pp. 28–30. Proofs of these theorems were found among HUYGENS's papers after his death and have been published by the Société Hollandaise des Sciences under the title "De Vi Centrifuga" in *Œuvres complètes de Christiaan Huygens*, Vol. XVI. A German translation is given in No. 138 of Ostwald's *Klassiker der Exakten Wissenschaften* (Leipzig, 1903).

EXAMPLE. Prove that the period of rotation of the conical pendulum shown in Fig. 183 is given by

$$T = 2\pi\sqrt{\frac{h}{g}},$$ [236]

and hence depends only on the depth h of the circular path below the point of suspension and on the acceleration due to gravity.

Solution. The two forces acting on the particle are its weight mg and the force F due to the cord. The acceleration of the particle is entirely centripetal and is of amount ωv, or $r \cdot 4\pi^2/T^2$. Hence the equation of motion for the horizontal direction is $F \sin\phi = mr \cdot 4\pi^2/T^2$, and that for the vertical direction is $F \cos\phi - mg = 0$. Eq. [236] can be obtained from these two equations.

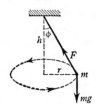

FIG. 183. A conical *pendulum* consists of a material particle of mass m suspended from a flexible, massless cord. As the particle moves with constant speed in a horizontal circular orbit, the cord sweeps out a conical surface

173. Simple Vibratory Motion. When a motion is a to-and-fro motion, it is said to be *vibratory* or *oscillatory*. Among the motions that are both vibratory and periodic the simplest type, and the most important, is that in which the vibrating body is at every instant urged toward some natural position of rest with a force which varies directly as its distance from that position. We shall. call this type of vibration *simple vibratory motion*. It is the type of vibration which is undergone by a weight suspended from an elastic cord, as in Fig. 184, when the weight is pulled down slightly and then released. In fact, all vibrations arising from the elasticity of matter are either simple vibratory motions or are compounded of such motions; for, as was stated in Sec. **90**, the restoring forces called into play by any sort of strains in material bodies are proportional to the displacements, so long as the strains are small and take place under isothermal conditions. Thus the vibrations of tuning forks, of the strings of stringed instruments, of the balance wheel of a watch, are cases of simple vibratory motions. Although there are other kinds of vibratory motion that are not simple vibratory in nature, it was shown by the French mathematician JEAN BAPTISTE JOSEPH FOURIER, in his renowned *Théorie Analytique de la Chaleur* [1] (1822), that any finite periodic motion, however complicated, can be represented as the summation of a number of suitably chosen simple vibratory motions. Simple vibratory motion is usually referred to as *simple harmonic motion*, a

---

[1] *The Analytical Theory of Heat*, tr. by A. Freeman (Cambridge University Press, 1878).

name given to it by LORD KELVIN and P. G. TAIT [1] (1831–1901), probably because the simplest musical sounds are caused by bodies that are executing such vibrations.

Let us now restate the definition of simple vibratory motion in mathematical form. Suppose a particle to be moving back and forth with such a motion over the path $A'OA$, Fig. 184.

Denote by s the displacement of the particle from the center of motion O at any moment, and let values of s be considered as positive when the particle lies between O and A, and as negative when it lies between O and A'. When s is positive, the accelerating force f is toward O and is therefore in the negative direction, and when s is negative, f, being still directed toward O, is in the positive direction. Hence, if the constant of proportionality of the magnitude of f to the magnitude of s is denoted by k, we have, by the definition of simple vibratory motion,

$$f = -ks. \qquad [237]$$

The constant k is called the *force constant* of the system. Its value depends upon the nature of the vibrating system.

In view of NEWTON'S second law of motion, if the force f is expressed in such units that K in Eq. [22], Chap. 2, can be put equal to unity, the defining equation for simple vibratory motion may also be written in the form

$$a = -\frac{k}{m}s, \qquad [238]$$

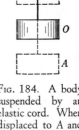

FIG. 184. A body suspended by an elastic cord. When displaced to A and then given its freedom, it undergoes a vertical motion that is simple vibratory in character, since it is acted upon at each instant by a force directed toward O and proportional to the displacement s from that point

where m is the mass of the vibrating particle and a is its linear acceleration. The acceleration is entirely tangential, since the motion of the particle is in a straight line, and hence, by Eq. [233], is equal in magnitude to dv/dt, where v is the speed. This enables us to write Eq. [238] in the alternative form

$$\frac{dv}{dt} = -\frac{k}{m}s. \qquad [239]$$

From either of the last two equations, it is seen that *a motion is simple vibratory if the acceleration is always proportional to the dis-*

---

[1] Kelvin and Tait, *Treatise on Natural Philosophy* (Cambridge University Press, 1912), Part I, pp. 37–38.

placement but opposite in sense to it. The acceleration is greatest when the particle is at the ends A' and A of its path and is momentarily motionless; the acceleration is zero when the particle is passing through O, its natural position of rest. If we integrate Eq. [239], the resulting equation will enable us to deduce additional properties of simple vibratory motion. But before attempting to do this it will be best to familiarize ourselves with the method of solving a still simpler type of differential equation; namely, the defining equation for motion with *constant* linear acceleration.

174. Digression on the Integration of the Defining Equation for Constant Linear Acceleration. The case of motion with constant linear acceleration was discussed in Secs. **19** and **20**, and the expression for the speed v and distance traversed s were deduced in those sections by a method which is rather lengthy, in general, but which avoids the use of the calculus. It is these expressions, namely Eqs. [14] and [16], Chap. 1, that we shall now derive in a very simple manner with the help of the calculus.

The defining equation for motion with constant linear acceleration is

$$\frac{dv}{dt} = a, \quad \text{or} \quad dv = a\,dt, \qquad [240]$$

where a is a constant. Indicating the integration which will give the value of v, we have

$$v = \int a\,dt = a \int dt.$$

By performing the integration, we obtain

$$v = at + C_1, \qquad [241]$$

where C_1 is the constant of integration. If we let $v = v_0$ when $t = 0$, then $C_1 = v_0$ and

$$v = at + v_0, \qquad [242]$$

which is Eq. [14], Chap. 1.

In order to derive the expression for s, the distance traversed by the particle during the time t, we start with Eq. [242], which may be written as follows:

$$\frac{ds}{dt} = at + v_0. \qquad [243]$$

Indicating the integration, we have

$$s = a \int t\,dt + v_0 \int dt.$$

By carrying out the integration, we obtain

$$s = \tfrac{1}{2}at^2 + v_0 t + C_2, \qquad [244]$$

where C_2 is again a constant of integration. If we let $s = 0$ when $t = 0$, we have $C_2 = 0$; hence

$$s = \tfrac{1}{2} at^2 + v_0 t,$$ [245]

which is Eq. [16], Chap. 1.

175. The Equations of Simple Vibratory Motion. We return now to Eq. [239], the defining equation for simple vibratory motion, from which we shall deduce expressions for the speed and displacement of a particle undergoing this kind of motion. The method of solution is similar to that employed in the preceding section.

In order to be able to integrate Eq. [239], multiply both members by $2\,ds$; then, since $ds = v\,dt$, we have

$$2\,v\,dv = -2\,\frac{k}{m}\,s\,ds.$$ [246]

Integration of both members gives

$$v^2 = -\frac{k}{m}\,s^2 + C_1,$$ [247]

where C_1 is the constant of integration. Its value depends upon the initial conditions under which the motion started. Thus, if the particle were pulled out from the origin a distance A (OA in Fig. 184) and let go, then $v = 0$ when $s = A$, and therefore $C_1 = kA^2/m$. By putting this value of C_1 in Eq. [247], we obtain finally

$$v = \sqrt{\frac{k}{m}\,(A^2 - s^2)}.$$ [248]

This gives the speed for any given value of s. The quantity A, which is the maximum value of s and therefore the maximum distance which the vibrating particle moves out away from its position of equilibrium at O, is called the *amplitude* of the motion. Clearly the values of s lie always between A and $-A$.

In order to obtain an expression for s, we substitute ds/dt for v in Eq. [248] and upon rearrangement obtain

$$\frac{ds}{\sqrt{A^2 - s^2}} = \sqrt{\frac{k}{m}}\,dt.$$ [249]

Integration gives

$$\sin^{-1}\frac{s}{A} = \sqrt{\frac{k}{m}}\,t + C_2,$$

or

$$s = A \sin\left(\sqrt{\frac{k}{m}}\,t + C_2\right).$$ [250]

In order to evaluate the constant of integration C_2, *we will agree to reckon time from the instant when the particle reaches the positive end of its path*; that is, when $t = 0$, then $s = A$. When the time is thus reckoned, $C_2 = \sin^{-1} 1 = \pi/2$; and Eq. [250] becomes

$$s = A \sin\left(\sqrt{\frac{k}{m}}\, t + \frac{\pi}{2}\right),$$

or
$$s = A \cos\sqrt{\frac{k}{m}}\, t. \qquad [251]$$

The quantity $\sqrt{k/m}\, t$ is termed the *phase* of the motion, and is evidently a measure of the fraction of a whole vibration completed since the particle occupied the chosen reference point, namely, the positive end of the path.

Eq. [251] shows that a given value of s occurs periodically, since if t is increased by $2\pi/\sqrt{k/m}$, then s becomes

$$A \cos\sqrt{\frac{k}{m}}\left(t + \frac{2\pi}{\sqrt{\frac{k}{m}}}\right) = A \cos\sqrt{\frac{k}{m}}\, t.$$

Thus the period T, or time for one complete cycle of changes in the value of s, is

$$T = \frac{2\pi}{\sqrt{\frac{k}{m}}}; \qquad [252]$$

or, in view of Eqs. [237] and [238],

$$T = 2\pi\sqrt{-\frac{m}{\frac{f}{s}}} = 2\pi\sqrt{-\frac{s}{a}}. \qquad [253]$$

This expression shows that the period of a simple vibratory motion does not depend on the amplitude but only on the ratio s/a. Such vibrations, whose period is independent of the amplitude, are called *isochronous*. Since f and s are always opposite in sign (Why?), the quantity under the radical is always positive.

If the period T is known, either k/m or f/s can at once be determined. Thus, if the vibratory motion is due to the elasticity of a cord, as in Fig. 184, the constant f/s can be determined not only by the static method of observing the force f required to stretch the cord through a given distance s, but also kinetically, by observing the period of vibration T of the system when any mass m is hung from the cord and substituting for m and T in Eq. [253].

By successive differentiation of Eq. [251] and by suitable substitutions, one may obtain the following expressions for the speed and the acceleration of the particle at any given time t:

$$v = \frac{ds}{dt} = -\sqrt{\frac{k}{m}}\, A \sin\left(\sqrt{\frac{k}{m}}\, t\right) = -\frac{2\,\pi}{T}\, A \sin\left(\frac{2\,\pi}{T}\, t\right) \qquad [254]$$

and

$$a = \frac{dv}{dt} = -\frac{k}{m}\, A \cos\left(\sqrt{\frac{k}{m}}\, t\right) = -\frac{4\,\pi^2}{T^2}\, A \cos\left(\frac{2\,\pi}{T}\, t\right) = -\frac{k}{m}\, s. \qquad [255]$$

Their derivation is left as an exercise for the student.

> EXAMPLE. Derive the equations for the displacement, speed, and acceleration of a particle undergoing simple vibratory motion, for the case where time is reckoned from the instant when the particle is passing through its position of equilibrium.

176. The Circle of Reference and Simple Vibratory Motion. It can be shown that when an object, such as the bob of the conical pendulum in Fig. 183, is caused to move with constant speed in a horizontal circle in the beam of light from a projection lantern, the shadow of the object moves back and forth across the screen with a simple vibratory motion. In other words, simple vibratory motion is the component, in a single direction, of motion in a circle with constant speed. To prove this, suppose the particle P' in Fig. 185 to be moving with constant angular speed ω in a circular path of radius A. If P is the projection of P' on any arbitrarily selected diameter $A'OA$, P vibrates once along $A'OA$ in each revolution of P'. Since the motion of P is that part of the motion of P' which is

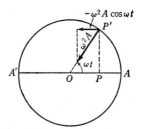

FIG. 185. Illustrating the circle of reference and how it may be employed to deduce the expression for the acceleration of a particle undergoing simple vibratory motion

in the direction of $A'OA$, the acceleration a of P is the component of the acceleration of P' in that direction. The acceleration of P' is $\omega^2 A$ in the direction $P'O$, or $-\omega^2 A$ in the direction OP' (Sec. 172). Hence the acceleration of P is

$$a = -\omega^2 A \cos \omega t, \qquad [256]$$

where ωt is the phase of the motion of P' and hence also of the motion of P. Since $\omega = 2\,\pi/T$, Eq. [256] becomes finally

$$a = -\frac{4\,\pi^2}{T^2}\, A \cos\left(\frac{2\,\pi}{T}\, t\right),$$

which is Eq. [255]. Hence any simple vibratory motion may be regarded as a projection of a uniform motion in a circle. The circle is called the *circle of reference* of the simple vibratory motion.

By projecting the speed and displacement of P' on the diameter $A'OA$, Fig. 185, expressions for the speed and displacement of a particle undergoing simple vibratory motion can also be deduced. The circle of reference therefore affords a means of obtaining all the equations of simple vibratory motion without the use of the calculus.

177. Summary of the Equations of Simple Vibratory Motion. The following equations summarize the results of Secs. **173, 175,** and **176**:

$$\textit{Period:} \qquad T = \frac{2\,\pi}{\sqrt{\dfrac{k}{m}}} = 2\,\pi\,\sqrt{-\dfrac{\dfrac{m}{f}}{s}} = \frac{2\,\pi}{\omega}. \qquad [257]$$

$$\textit{Phase:} \qquad \theta = \sqrt{\frac{k}{m}}\,t = \frac{2\,\pi}{T}\,t = \omega t. \qquad [258]$$

$$\textit{Displacement:} \quad s = A \cos \omega t. \qquad [259]$$

$$\textit{Speed:} \qquad v = \omega\sqrt{A^2 - s^2} = -\,\omega A \sin \omega t. \qquad [260]$$

$$\textit{Acceleration:} \quad a = -\,\omega^2 s = -\,\omega^2 A \cos \omega t. \qquad [261]$$

$$\textit{Force:} \qquad f = -\,ks = -\,m\omega^2 s. \qquad [262]$$

178. Angular Simple Vibratory Motion. A body rotating backward and forward about an axis is said to have a *simple vibratory motion of rotation* when each of its particles moves with simple vibratory motion. The balance wheel of a watch and the torsion pendulum (Fig. 187) afford excellent examples. In order to obtain a defining equation for angular simple vibratory motion, let

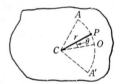

Fig. 186. Illustrating simple vibratory motion of rotation

the body shown in Fig. 186 oscillate about the axis C in such a manner that each particle P follows the equation of simple vibratory motion, namely,

$$a = -\frac{4\,\pi^2}{T^2}\,s.$$

If r be the distance of the particle in question from C, then the linear quantities a and s are related to the corresponding angular quantities α and θ by Eqs. [90] and [87], Chap. 7, and hence the equation $a = -\,4\,\pi^2 s/T^2$ may be at once transformed into

$$\alpha = -\frac{4\,\pi^2}{T^2}\,\theta. \qquad [263]$$

Thus a body may be said to be executing simple vibratory motion of rotation *if its angular acceleration is proportional to its angular displacement and is opposite in direction.*

The combination of Eq. [263] with the rotational analogue of NEWTON's second law of motion, namely Eq. [101], Chap. 7, gives for the accelerating torque L in simple vibratory motion of rotation

$$L = - I \frac{4 \pi^2}{T^2} \theta, \qquad [264]$$

where I is the moment of inertia of the body about the axis of oscillation. It is to be noted that Eqs. [263] and [264] are the rotational analogues of Eqs. [261] and [262].

179. The Torsion Pendulum. The torsion pendulum consists of a vertical wire which is clamped at the upper end and rigidly attached at the lower end to the center of a horizontal disk (Fig. 187). When the disk is twisted around the wire as axis and released, it performs angular vibrations. Now we found in Sec. **96** that the angular displacement θ produced in a given wire by an applied torque L is proportional to L; that is, $L = L_o\theta$, where L_o is the constant of torsion of the wire. But the torque exerted by the twisted wire is equal in magnitude and opposite in direction to the torque required to produce the twist. Hence the accelerating torque exerted *by* the wire *on* the disk is $- L_o\theta$ when the displacement is θ, or

FIG. 187. The torsion pendulum

$$L = - L_o\theta. \qquad [265]$$

Comparison of this equation with Eq. [264] shows that the motion of the disk is one of angular simple vibratory motion of period

$$T = 2 \pi \sqrt{\frac{I}{L_o}}. \qquad [266]$$

It is evident from this equation that the moment of inertia I of any body may be determined experimentally by suspending it from a wire for which the constant of torsion is known and observing the period of vibration.

180. The Physical Pendulum. Any rigid body suspended by a horizontal axis and free to swing about its position of equilibrium under the action of its own weight and the reaction of the axis of support constitutes a *physical,* or *compound, pendulum.* An ordinary clock pendulum is an example. The mathematical theory of the physical pendulum was first worked out by HUYGENS and published in his

Horologium Oscillatorium in 1673. This was the second *great* problem to be solved in kinetics, the first (that of falling bodies) having been solved by GALILEO (Sec. 1). It was the first successful attempt to deal with the kinetics of a rigid body; in fact, it was in this connection that the concept of what we now call the moment of inertia first came to light.[1]

Fig. 188 represents a vertical section through the center of mass, c, and perpendicular to the axis of suspension, S, of such a pendulum.

If M denotes the mass of the entire pendulum, and d the distance Sc, then when the pendulum is pulled aside through an angle θ and set free to vibrate, the only external force having a torque about S will be the weight Mg, applied at c (Sec. 74). This torque is equal in magnitude to $Mgd \sin \theta$ and is opposite in direction to the displacement θ, or

FIG. 188. The physical pendulum

$$L = - Mgd \sin \theta. \qquad [267]$$

The motion therefore is *not* angular simple vibratory motion, for L is not proportional to θ but to $\sin \theta$. However, in the limit, when the vibration is of infinitely small amplitude, $\sin \theta$ is equal to θ expressed in radians. Then Eq. [267] becomes

$$L = - Mgd\theta, \qquad [268]$$

from which it can be concluded that a physical pendulum swinging with a very small amplitude follows the laws of simple vibratory motion of rotation. These laws may be applied to pendulum problems only if the amplitude be kept so small that the error introduced by replacing $\sin \theta$ by θ is less than the necessary observational errors of the experiment. Under these conditions, then, we have for the period of a physical pendulum, from Eqs. [264] and [268],

$$T = 2 \pi \sqrt{\frac{I}{Mgd}}, \qquad [269]$$

where I is the moment of inertia of the body about the axis of suspension.

EXAMPLE. Prove that the period of vibration of a physical pendulum is given by

$$T = 2 \pi \sqrt{\frac{k_c{}^2 + d^2}{gd}},$$

where k_c is the radius of gyration of the pendulum about an axis through its center of mass and parallel to its axis of suspension.

---

[1] See the quotation at the beginning of this chapter.

EXAMPLE. Are the vibrations of a physical pendulum strictly isochronous? the vibrations of a torsion pendulum? Explain in what way the two cases differ.

181. The Ideal Pendulum. By an *ideal pendulum* is meant a material particle suspended by a massless cord. It may be regarded as a special case of a physical pendulum, and the student can easily deduce from Eq. [269] the result that, for small amplitudes, the period is given by

$$T = 2\pi\sqrt{\frac{l}{g}},$$ [270]

where l is the length of the pendulum.

Of course no actual pendulum is ideal, although a small heavy ball suspended from a fixed point by a thread may with little error be treated as such. Indeed, pendulums of this kind were employed prior to the middle of the eighteenth century for the determination of the acceleration due to gravity at a given place. This was accomplished by adjusting the length until the pendulum vibrated a little faster or slower than the pendulum of a standard clock and then making observations by the *method of coincidences*, which is a very accurate method for comparing two periodic motions of nearly the same period.[1] However, despite its simplicity in construction and use, the ideal pendulum does not in actual practice provide an entirely satisfactory method for determining g with a high degree of precision, one great objection being that it is necessary to make a complicated correction for the slackening of the suspension thread which occurs as the pendulum approaches the limits of its swing. Since this defect is absent from the physical pendulum, it is better to use some form of the latter for the determination of g.

EXAMPLE. Derive the expression for the period of an ideal pendulum by two additional methods: (*a*) by equating the accelerating force on the bob to the product of the mass and linear acceleration of the bob and then applying the equations of simple vibratory motion; (*b*) by equating the potential energy of the bob when it is at the extremity of an oscillation to the kinetic energy when it is at the mid-point of an oscillation.

182. The Reversible Physical Pendulum. Given any physical pendulum vibrating about an axis S, as in Fig. 188, our next problem is to find other points along the line Sc about which the pendulum will vibrate with the same period as about S. We will solve this problem by a graphical method. First, with the aid of LAGRANGE'S theorem of parallel axes (Sec. **86**), we rewrite Eq. [269] in the form

$$T = 2\pi\sqrt{\frac{I_c + Md^2}{Mgd}},$$ [271]

---

[1] In Exp. XIVc. g is determined by means of an ideal pendulum and the method of coincidences.

CHRISTIAAN HUYGENS must be placed with ARCHIMEDES, GALILEO, and NEWTON in the first rank of mechanical investigators. The first problems in the dynamics of systems composed of many particles were solved by him. His HOROLOGIUM OSCILLATORIUM (1673) is a dynamical work of importance exceeded only by the *Principia*. In it will be found the solution of the celebrated problem of the "center of oscillation," the invention and construction of the pendulum clock, the determination of the true relation between the length of a pendulum and the time of its oscillation, the determination by pendulum observations of the acceleration due to gravity, the invention of the theory of evolutes and the discovery that the cycloid is its own evolute and is strictly isochronous, the ingenious idea of applying cycloidal cheeks to clocks, and the theorems concerning centrifugal force which formed the necessary prelude to NEWTON's *Principia*. In the proposition, there assumed as an axiom, that the center of gravity of any number of independent bodies can rise to its original height but no higher, is expressed for the first time one of the most fruitful principles of physical science, now known as the principle of the conservation of energy.

For excellent appraisals of HUYGENS's life and work, see P. Lenard's *Great Men of Science* (Macmillan, 1933), pp. 67–83; E. Mach's *The Science of Mechanics* (Open Court, 1893), pp. 155–187; and H. Crew's *The Rise of Modern Physics* (Williams & Wilkins, 1935), pp. 120–133.

o

CHRISTIAAN HUYGENS at the Age of Forty-two

From the portrait, painted in 1671 by KASPAR NETSCHER, which hangs in the Municipal Museum,
The Hague

THIS PORTRAIT of HUYGENS shows him in his prime. It was painted two years before
the publication of his HOROLOGIUM OSCILLATORIUM (1673). HUYGENS was at this
time living in Paris, having been invited by COLBERT on behalf of Louis XIV to
be a member of the newly founded Academy of Sciences (see Plate 47), where he
occupied an influential, if not a dominant, position.

where d is the distance from the center of mass c to the axis of suspension, and I_c is the moment of inertia of the body about an axis through its center of mass and parallel to the axis of suspension. If the period of vibration now be found and plotted for various values of d, curves are obtained similar to those reproduced in Fig. 189. Corresponding to any given axis of suspension S_1 or period T_1 there are, in general, three other parallel axes S_2, S_3, and S_4 along the line Sc about which the pendulum will have the period T_1. The four axes may be grouped in two pairs, S_1 and S_4, S_2 and S_3,

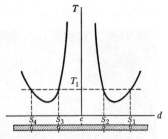

Fig. 189.   Relation between the period and the position of the axis of a physical pendulum. The point c corresponds to the center of mass of the pendulum, and d is measured to the right or to the left of c according as the axis of suspension S is on one side or the other of the center of mass

such that the axes of a pair are equidistant from c. Let $cS_1 = S_4c = d_1$ and $cS_2 = S_3c = d_2$. This enables us to write for the period

$$T_1 = 2\pi \sqrt{\frac{I_c + Md_1{}^2}{Mgd_1}} = 2\pi\sqrt{\frac{I_c + Md_2{}^2}{Mgd_2}},$$

or $$\frac{g}{4\pi^2} T_1{}^2(d_1 - d_2) = d_1{}^2 - d_2{}^2. \qquad [272]$$

If d_1 is not equal to d_2, we may divide both members of Eq. [272] by $d_1 - d_2$; when this is done, there results

$$T_1 = 2\pi \sqrt{\frac{d_1 + d_2}{g}}. \qquad [273]$$

From this it is seen that the period T_1 is the same as that of an ideal pendulum (Sec. 181) of length $d_1 + d_2$, whether the physical pendulum be suspended from an axis through S_1 or S_2, or be inverted and set to vibrate about an axis through S_3 or S_4. Thus, if the pendulum were supported at S_1, its period would be the same as if the whole of its mass were concentrated at the point S_3, Fig. 189. The point S_3 was therefore called by

Fig. 190.   KATER'S reversible pendulum

HUYGENS the *center of oscillation* corresponding to the *center of suspension S_1*. It is obvious that the centers of oscillation and suspension are interchangeable and that the distance between them is equal to the length of the equivalent ideal pendulum. Hence if $d_1 + d_2$ be measured, g can be found from Eq. [273]. This property of the pendulum was pointed out by HUYGENS in 1673, but it was first used by HENRY KATER,[1] an English sea captain, in 1817, in the reversible pendulum (Fig. 190) with which he made his celebrated determination of the value of g.

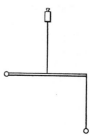

It is clear from what precedes that a physical pendulum does not swing as if its mass were concentrated at the center of mass c. Although it is true that the potential energy of the pendulum, in any position, is the same as if its mass were concentrated at c, the pendulum does not swing as if all its mass were at

FIG. 191. A block with a bifilar suspension. When swinging parallel to the plane of the figure, the block does not rotate and the period is that of an ideal pendulum of length equal to the length of the cords. When swinging perpendicular to the plane of the figure, the block has energy of rotation and the period is that of an ideal pendulum of length SS', where S' is the center of oscillation

FIG. 192. Simplified schematic diagram of the Eötvös balance. A light horizontal rod, suspended at the middle by a delicate fiber, supports at its extremities two weights which are at different vertical heights

that point because its kinetic energy is that of its mass supposed concentrated at c plus its energy of rotation about c (Fig. 191).

In modern practice KATER'S reversible pendulum, or some other form of physical pendulum, is used mainly for making *absolute* determinations of g. In making measurements of small *variations* in g from one point to another, such as are caused by variations in geological structure, the instrument usually employed is the Eötvös torsion balance[2] (Fig. 192), which combines the necessary sensitivity with portability and comparative ease in use. It is used commercially in explorations for oil[3] and other valuable natural deposits.

[1] *Philosophical Transactions* 108, 33 (1818). For more detailed discussions of the reversible pendulum see *J. H. Poynting and J. J. Thomson, *Properties of Matter* (Griffin, 1913), p. 12, or *F. H. Newman and V. H. L. Searle, *The General Properties of Matter* (Macmillan, 1933), p. 40.

[2] R. von Eötvös, Wiedemann's *Annalen* 59, 354 (1896). See also F. H. Newman and V. H. L. Searle, *The General Properties of Matter* (Macmillan, 1933), pp. 52–57, and the "*Eötvös*" *Torsion Balance* (published by L. Oertling, Ltd., London).

[3] See *L. L. Nettleton, "Applied Physics in the Search for Oil," *The American Physics Teacher* 3, 110 (1935).

EXPERIMENT XIVA. CENTRIPETAL FORCE

The object is to study the type of periodic motion represented by the motion of a body with constant speed in a circular path. The experiment consists essentially in revolving a body of known mass M in a circular path and adjusting the angular speed until a spring to which the body is attached is extended a definite amount. The force f necessary to stretch the spring this amount is then measured, as are also the radius r and the frequency of revolution ν. One then has all the data needed to test the expression for the centripetal force first developed by HUY-GENS; namely,

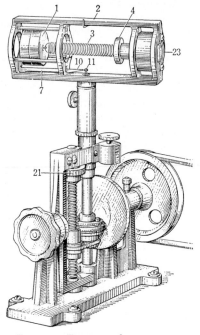

FIG. 193. Centripetal-force apparatus

$$f = K \cdot M \cdot 4\,\pi^2\nu^2r. \quad [274]$$

The apparatus is shown in Fig. 193. The body 1 slides horizontally in the frame 2 and is attached to a spring 3, which in turn is attached to the frame. When the frame is rotated about the vertical axis 11, the body moves out away from the axis until the spring exerts enough centripetal force to keep the body moving in a circle. As the body moves outward it actuates the lever 7, causing the pointed tip 10 to move upward; when the latter has risen to a point opposite the index 11, the apparatus is rotating at the speed for which it was designed. The tension in the spring can be changed by turning the threaded collar 4, thus making it possible to perform experiments at a number of different speeds. The weight 23 is simply a counterweight, to provide balance for the apparatus during rotation.

a. Adjustments. Remove the apparatus from the socket of the rotator and adjust the tension of the spring to its lowest point. By turning the screw 21, Fig. 193, set the friction drive of the rotator at the zero position, so that the spindle does not turn when the motor is running. Reclamp the apparatus on the rotator and start the motor.

Slowly increase the speed until the pointer is even with, but no higher than, the fixed index. To increase the speed beyond this point may harm the apparatus or cause more serious damage. Since the speed of the motor will probably vary slightly, it will be necessary to make such continual slight adjustments of the screw 21 as will keep the tip of the point oscillating freely about the index as an average position.

b. Frequency of Rotation. The manipulations here described should be practiced several times before any observations are recorded. One observer is to pay constant attention to the proper adjustment of the speed. The other observer, holding a stop watch or a watch with a second hand, is to obtain the value of the speed in the following manner: (1) record the initial reading of the revolution-counter; (2) engage the revolution-counter with the gear on the spindle and at the same time start the stop watch; (3) after the lapse of exactly 1 min (3 min if an ordinary watch is used), disengage the counter; to make sure that the gear on the counter does not continue to turn after it is disengaged, apply the finger lightly as a brake; (4) record the final reading of the counter.

Repeat the foregoing set of operations at least five times; use each final reading of the counter as the initial reading for the next trial. If the numbers of revolutions in successive 1-min intervals differ by more than two or three, it is an indication that greater care should be exercised in making the manipulations.

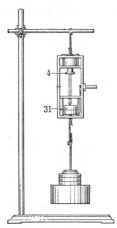

FIG. 194. Method of measuring the force exerted by the spring

c. Force Exerted by the Spring. Remove the apparatus from the rotator and hang it by the hook in the counterweight as shown in Fig. 194. By means of a fine wire attach a weight-hanger to the hook in the body that was revolving. By adding weights to the hanger, stretch the spring until the body, as shown by the indicator, is in the same position as it occupied during the rotation. Record the total force f applied to the spring; this consists of the weight of the hanger and added weights, and the weight of the body that was revolving. If the mass M of the latter is not stamped upon it, remove and weigh it.

d. Radius of Rotation. With the apparatus still in the position shown in Fig. 194, measure with vernier calipers the distance, when the pointer is opposite the index, from the axis of rotation to the mark 31, which designates the center of mass of the body. The average of at least five such measurements is to be taken as the value of the radius of rotation r.

e. Repetitions of the Experiment with the Spring at Different Tensions. Adjust the spring to its maximum tension, replace the apparatus on the rotator, and repeat *b* and *c*.

Adjust the spring to a tension midway between the two values previously used, and repeat *b* and *c*.

f. Calculations. The following calculations are to be made for each of the three sets of observations obtained with different spring tensions.

To calculate the average frequency of revolution ν, divide the revolution-counter readings taken in *b* into two equal groups; for example, if six readings are taken, the first three constitute the first group and the second three constitute the second group. By taking the differences between corresponding readings in the two groups, three values are obtained, the average of which is the number of revolutions during three times the time interval to which the observations apply.

Substitute the experimentally determined values of f, M, ν, and r in Eq. [274] and simplify.[1] Compute the percentage difference between the right-hand and left-hand members of the equation.

   1. What kind of curve would be obtained if the force exerted by the spring were plotted as a function of the square of the frequency of revolution?

   2. For any one of the values of the tension of the spring compute the period of revolution T and angular speed ω of the revolving body, and also the linear speed v of its center of mass.

   3. In calculating the average value of the frequency of revolution ν, why should one not follow the simpler procedure of taking differences between each two consecutive readings of the counter and averaging?

   4. Explain in detail how a centripetal-force apparatus could be used to compare the masses of two bodies. Of which would this be a direct comparison, the inertial masses or the gravitational masses?

   5. How could the apparatus be used to determine the acceleration due to gravity?

---

[1] In addition to the elongation produced in the spring by the revolving body, a slight elongation also takes place because the spring itself has mass. The apparatus is so designed, however, that this possible source of error can be ignored.

EXPERIMENT XIVb. SIMPLE VIBRATORY MOTION

This experiment affords a study of simple vibratory motion. In Part I the object is to study the motion of a loaded spiral spring and to determine the force constant of the spring by two different methods. In Part II a torsion pendulum is employed to determine the torsional properties of a steel rod and the moment of inertia of a body.

PART I. The Loaded Spiral Spring. *a.* When a body is suspended from a vertical spiral spring as in Fig. 195 and is set into vibration along a vertical line, the total force acting on the body at any instant is evidently the vector sum of the force of gravity and the elastic force exerted by the spring. If it can be shown experimentally that this total force is proportional to the displacement of the body from its position of equilibrium, it then follows from Eq. [237] that the vertical oscillations of the body represent a case of simple vibratory motion.

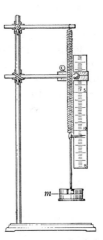

Use a 50-g weight-hanger for the body in question and attach it to the spring, as in Fig. 195. In order to determine whether the force f is proportional to the displacement s, add five 100-g weights to the weight-hanger, one at a time, and observe the resulting positions of the pointer on a graduated mirror placed behind the spring.[1] In taking a reading place the eye in such a position that the tip of the pointer and its image appear to be in line with each other. (Why?)

FIG. 195. A loaded spiral spring

1. Plot a curve of which the abscissas represent the weights on the hanger and the ordinates represent the corresponding positions of the pointer. What does the slope of this curve represent?

b. Calculate the mean displacement of the weight-hanger for 100 gwt added to it. To obtain the mean, use the method of differences described in Exp. VIIIB, *e.*

Calculate the force constant of the spring, $-f/s$, in cgs units.

c. Since it is found that the loaded spring will follow the laws of simple vibratory motion if set to vibrating vertically, it is also

---

[1] The initial load on the spring should be large enough to separate the spirals. If the spring is a heavy one, such as is commonly used to close doors, it may be necessary to employ an initial load of several kilograms and to add weights in increments of several hundred grams.

possible to determine the force constant of the spring by observing the period of vibration T for a given load m, and substituting in Eq. [253].

Hang on the spring a total load of, say, 100 gwt. Note the position of the pointer on the mirror-scale, and use this position for a reference line. Set the mass vibrating vertically through a moderately small amplitude and take with a stop watch the time of 50 vibrations. Count the vibrations by observing the transits in one direction of the pointer across the reference line. Compute the period T in seconds per vibration.

Also determine the periods when the total loads on the spring are 200, 300, and 400 gwt.

Before attempting to calculate the force constant from these data, we must take into account the fact that the spring itself has mass and hence that the period of the system is a little larger than it would be if the spring were massless. In other words, Eq. [253] should for our present purpose be written in the form

$$T = 2\pi \sqrt{-\frac{m+\Delta m}{\dfrac{f}{s}}}, \qquad [275]$$

where m is the suspended mass and Δm is the correction factor for the inertial effect of the spring.

2. With the aid of Eq. [275] prove that if the values of T are plotted against the corresponding values of m, the intercept of the resulting curve on the mass-axis gives the correction factor Δm.

3. Using your observed values of T and m, plot T^2 as a function of m and thus find Δm for the spring used in this experiment.

By substituting corresponding values of T and $m+\Delta m$ in Eq. [275], calculate the values of f/s and average them to obtain the force constant of the spring. Compute the percentage difference between this value and that obtained in *b*.

d. It can be shown theoretically that the correction factor Δm is equal to one third of the mass of the spring. Weigh the spring and compare the value of Δm thus afforded with the value obtained by the graphical method.

4. How much error would have been introduced into the determination of the force constant by failure to correct for the inertial effect of the spring?

PART II. The Torsion Pendulum. The torsion pendulum (Fig. 187) used in this experiment consists of a solid metal disk suspended from one of the steel rods of Exp. VIIIB. In the former experiment both L_o, the constant of torsion of the rod, and n, the shear modulus of the steel in the rod, were determined by a statical method. The present experiment affords a kinetical method of determining not only these two quantities but also the moment of inertia I of the suspended disk. For, in view of Eq. [266], this may be accomplished by observing first the period T of the torsion pendulum alone, and then the period T_1 when it is caused to vibrate after the addition to the disk of a second body of known moment of inertia I_1. The period of the system after the addition of I_1 is given by

$$T_1 = 2\pi \sqrt{\frac{I + I_1}{L_o}}. \qquad [276]$$

The elimination of L_o from Eqs. [266] and [276] gives

$$I = T^2 \frac{I_1}{T_1{}^2 - T^2}, \qquad [277]$$

and the elimination of I gives

$$L_o = 4\pi^2 \frac{I_1}{T_1{}^2 - T^2}. \qquad [278]$$

a. Use the rod of greatest length and smallest diameter and be sure that it is the one which you tested in Exp. VIIIB. With the disk alone attached to the rod, observe with a stop watch the time of, say, 25 vibrations and compute the period T in seconds per vibration.[1] In setting the disk into oscillation do not at the same time set it swinging and do not twist the rod through an angle of more than 10°.

5. Why should the amplitude be made moderately small?

b. Add to the disk a heavy metal ring and observe the new period T_1. Be sure that the rod passes through the center of the ring. (Why?)

Calculate I_1, the moment of inertia of the ring for the axis about which it was rotating. Obtain the mean diameter of the ring by measuring the inside and outside diameters, adding these, and dividing by 2. If the mass is not found stamped upon the ring, determine it by weighing.

---

[1] If greater accuracy in the determination of the period is desired, the *method of coincidences* should be employed (see Exp. XIVc). Also see *D. C. Miller, *Laboratory Physics* (Ginn, 1903), p. 102; *B. Stewart and W. W. H. Gee, *Elementary Practical Physics* (Macmillan, 1885), Vol. I, p. 187; or *J. S. Ames and W. J. A. Bliss, *A Manual of Experiments in Physics* (American Book Co., 1898), p. 168.

6. Beginning with Eq. [103], Chap. 7, the general equation for moment of inertia, prove that the moment of inertia of a hollow cylinder about its geometrical axis is $I_1 = \frac{1}{8} M(D_1{}^2 + D_2{}^2)$, where D_1 and D_2 are the outside and inside diameters of the cylinder. Hence show that $I_1 = \frac{1}{4} M(D^2 + t^2)$, where D is the mean diameter and t is the thickness of the cylinder. Can t^2 be neglected in calculating I_1 for the ring used in this experiment?

c. Calculate with the aid of Eq. [277] the moment of inertia I of the disk about the axis of rotation. Also measure the diameter and mass of the disk and calculate I by means of Eq. [102a], Chap. 7; the mass of the disk will probably be found stamped upon it. Compute the percentage difference in the two values of I.

Calculate by means of Eq. [278] and Eq. [119], Chap. 8, the constant of torsion of the suspension rod and the shear modulus of the steel in the rod. Compare these values with those obtained for the same rod by the statical method.

7. Why was it not necessary to take into account the moment of inertia of the suspension rod and its fittings?

8. Derive a laboratory equation which would enable one to employ the torsion pendulum for the comparison of the unknown moments of inertia of two bodies.

○

OPTIONAL LABORATORY PROBLEM

Determination of Moments of Inertia by Various Methods. Determine by several different experimental methods, and also, if possible, by calculation, the moment of inertia about a given axis of some common object or some object of which it may be important to know the moment of inertia, such as an ordinary hand wrench, an automobile flywheel or piston rod, the wheel of an Atwood's machine, or the wheel of the apparatus used for studying the torsion of a wire by the statical method (Exp. VIIIB). In the case of either the Atwood's machine or the torsion apparatus, do not attempt to remove the wheel from its bearings.

○

EXPERIMENT XIVC. DETERMINATION OF g BY MEANS OF AN IDEAL PENDULUM

In Exp. I an approximate determination of the acceleration due to gravity was made by the direct method of observing the acceleration of a freely falling body. The pendulum furnishes an indirect method of measuring this quantity, which is capable of yielding much more accurate results. The advantage of using an ideal pendulum for the determination is that its length l can be measured directly, whereas the equivalent length of a physical pendulum is not easily obtained. The unknown period T of the ideal pendulum is obtained by comparing it with the known period T_s of a standard pendulum.

Since the period T enters Eq. [270] as a second power whereas the length l enters as a first power, the value of T must be known to a higher degree of precision than the value of l. For this reason we shall determine T by the valuable *method of coincidences*, which is a very accurate method for comparing two periodic motions of nearly equal periods.

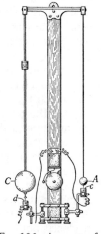

The apparatus for determining the instant of coincidence is shown in Fig. 196. When the ideal pendulum A and standard pendulum C are in coincidence, they pass through the mercury contacts c and d at the same instant, thus completing an electric circuit and causing the electric bell to sound. Since the two pendulums do not have exactly the same period, no further signal is heard until the faster pendulum has gained one half-vibration upon the slower. If between two such coincidences the standard pendulum has made N half-vibrations, then the ideal pendulum

FIG. 196. Apparatus for determining the period of an ideal pendulum by the method of coincidences

has made either $(N + 1)$ or $(N - 1)$ half-vibrations, depending upon whether its period T is shorter or longer than the period T_s of the standard pendulum. Hence the time interval between two successive coincidences is

$$\frac{T}{2}(N \pm 1) = \frac{T_s}{2} N,$$

from which
$$T = \frac{T_s N}{N \pm 1}.$$
[279]

However, on account of the difficulty of observing accurately the exact instant of coincidence, it is preferable to observe the number of half-vibrations N of the standard pendulum that occur in several successive intervals n. In this case one pendulum will have gained n, instead of one, half-vibrations on the other, and Eq. [279] must be modified to

$$T = \frac{T_s N}{N \pm n}.$$
[280]

Adjustment of Apparatus. No adjustment of the standard pendulum should be made except under the supervision of an instructor. Always keep the electric circuit open whenever both pendulums are motionless.

The length of the ideal pendulum should be such that the distance from its knife-edge to the center of its bob differs from the corresponding distance on the standard pendulum by not more than 0.5 cm. The coincidences will then not occur more than once in about 5 min, which is desirable. The contact points on the two pendulums should be clean, and they should rest exactly in the centers of the mercury globules when the pendulums are at rest. There must be enough clean mercury in the cups to insure good contact but not so much that the mercury is visibly drawn to one side by the passage of the contact point through it.

By means of a thread tie back the bob of the ideal pendulum a distance of about 4 cm, taking care not to twist the suspension wire in the least. Complete the electric circuit as shown in Fig. 196. Set the ideal pendulum in motion by burning the thread. If a signal is heard at every passage, set the standard pendulum into vibration with an amplitude somewhat smaller than that of the ideal pendulum. (Why smaller?) It will be noted that there is not just one signal at a coincidence, but that signals occur for several successive swings. This is because one pendulum gains only slightly on the other and the passages of the contact points through the mercury are not instantaneous. The mean of the first and last of these successive signals is to be regarded as the coincidence.

Observations. Start counting the half-vibrations of the standard pendulum at any convenient time, preferably just before a coincidence occurs. Record the ordinal number of each half-vibration that is accompanied by a signal; for example, if signals accompany the counts 14, 15, 16, record these ordinal numbers. Continue this count uninterruptedly for at least 20 min.

Note whether the ideal pendulum gains or loses on the standard pendulum.

Find the length l of the ideal pendulum; this may be taken, without appreciable error, as the distance from its knife-edge to the center of its bob. Do not remove the pendulum from its mounting for this purpose unless necessary. Measure the diameter of the bob with vernier calipers. Obtain the length of the suspension wire by means of the scale and vernier which are mounted back of the pendulum; usually the zero of this scale is exactly at the knife-edge. If the apparatus is not equipped with a scale and vernier, use a meter stick or, for greater accuracy, a cathetometer.

Record the period T_s of the standard pendulum; its value will be found marked on the apparatus.

Calculations. Compute the ordinal numbers of the particular half-periods at which it is estimated that exact coincidences occurred; for example, if the successive signals occurring during some one time of approximate coincidences were numbered 81, 82, 83, 84, then $82\frac{1}{2}$ is to be taken as the ordinal number that corresponds to exact coincidence.

Calculate the mean number of half-vibrations N of the standard pendulum in n successive intervals between coincidences. This is done by dividing the ordinal numbers which correspond to exact coincidence into two groups of n numbers each and then subtracting the first reading of the first group from the first reading of the second group, and so on to the end. The average of these differences is N.

Calculate T and g by means of Eq. [280] and Eq. [270] respectively. Compare the value of g thus determined with the accepted value for your locality.

1. In what ways does the pendulum used in this experiment fall short of the requirements of a simple pendulum?

2. Derive the expression for the period of a physical pendulum that consists of a spherical bob of radius R suspended from a wire of negligible mass and of length $l - R$. Then find by actual calculation whether an appreciable error [1] was introduced into your determination of g by the assumption that the pendulum used was an ideal pendulum of length l.

3. Explain in detail why it was necessary to confine the vibrations of the pendulums to small amplitudes.

4. A more exact expression for the period of an ideal pendulum vibrating with an amplitude θ is [2]

$$T = 2\pi\sqrt{\frac{l}{g}}\left\{1 + \frac{1}{4}\sin^2\frac{\theta}{2} + \frac{9}{64}\sin^4\frac{\theta}{2} + \cdots\right\}.$$

Find by actual calculation whether an amplitude of 4 cm for the ideal pendulum used in this experiment was large enough to produce an appreciable error in your determination of g.

---

[1] For a discussion of the corrections for air resistance, curvature of knife-edges, yielding of support, etc., which must be made in very accurate pendulum determinations, see *F. H. Newman and V. H. L. Searle, *The General Properties of Matter* (Macmillan, 1933), pp. 43–49.

[2] For a derivation of this equation see A. G. Webster, *The Dynamics of Particles and of Rigid, Elastic and Fluid Bodies* (Teubner, 1904), p. 48.

QUESTION SUMMARY

1. What is the vector expression for the total instantaneous acceleration of a particle moving in any plane curve whatever in terms of the speed of the particle and the time-rates of change of speed and direction? State clearly what each component and each term of this expression means physically. How would you calculate the magnitude of the total acceleration? its direction? What is the vector expression for the total instantaneous acceleration in terms of the speed, its time-rate of change, and the radius of curvature?

2. What is meant by a *periodic motion*? Define the *period, frequency,* and *phase* of a periodic motion.

3. If a particle of mass *m* moves in a circle of radius *r* with a constant linear speed *v*, what is the magnitude of the total acceleration? What is its direction? What are the magnitude and direction of the force required to maintain this acceleration?

4. Define *simple vibratory motion* in words and by means of an equation. What is meant by the *displacement* in such motion? by the *amplitude*?

5. Upon what does the period of a simple vibratory motion depend? Give expressions for calculating the displacement, speed, and acceleration at any time *t*, and the speed and acceleration for any displacement *s*; state clearly the meaning of each symbol used.

6. Upon what does the period of a torsion pendulum depend? Does it depend upon the amplitude?

7. Starting from the expression for the period of a simple vibratory motion of rotation, derive expressions for the periods of physical and ideal pendulums; point out clearly any simplifying assumptions made and the limits within which these assumptions are valid; state in words the exact meaning of each symbol used.

8. Define *center of oscillation*. What is its importance?

○

PROBLEMS

1. A particle of mass 5 g is moving in a curved path, and its total acceleration at a given moment is $(3\,\tau + 4\,n)$ cm · sec^{-2}. Find (*a*) the tangential acceleration; (*b*) the centripetal acceleration; (*c*) the magnitude of the total acceleration; (*d*) the angle ϕ which the total acceleration makes with the tangent to the curve; (*e*) the tangential component of the accelerating force; (*f*) the centripetal component of the accelerating force; (*g*) the total accelerating force.

2. A projectile is fired from a gun with a muzzle velocity of magnitude *V* and elevation ϕ above the horizontal. (*a*) Disregarding the resistance of the air, derive expressions for the speed and for the tangential acceleration of the projectile at the end of the time t. (*b*) At what point in the trajectory will the centripetal acceleration be equal to the total acceleration? (*c*) What is the value of the total acceleration?

3. Assume for the sake of simplicity that the bodies of the solar system are particles and that the planets move about the sun in circular orbits. (*a*) Show that it follows at once from KEPLER's second empirical law of planetary motion (Sec. **33**) that the total orbital acceleration of any planet is directed along the line connecting the sun and the planet. (*b*) Prove that it follows from KEPLER's third law and HUYGENS's expression for centripetal acceleration that the accelerations of the planets toward the sun vary inversely as the squares of their distances from the sun. (*c*) Show that the foregoing results, taken together with NEWTON's second law of motion, lead to the conclusion that the gravitational force of attraction of the sun for any planet is a central force which varies inversely as the square of the distance between the two bodies.

4. (*a*) Assuming that the moon revolves around the earth with constant speed in a circular orbit of radius approximately sixty times the radius of the earth, calculate the centripetal force which must act upon each unit mass of the moon in order to hold it in its orbit. Take the period of revolution of the moon as 27 da 8 hr. (*b*) Compare the foregoing result with the force of the earth's attraction upon a unit mass at the distance of the moon as computed from NEWTON's law of gravitation; it was this computation that led NEWTON to assert his law of gravitation (Sec. **34**).

Ans. (*a*) 0.271 dyne · g⁻¹; (*b*) 0.272 dyne · g⁻¹.

5. (*a*) Taking the radius of the earth as 6370 km, calculate the centripetal force required to hold a 1-g mass upon the surface at the equator, and also in latitude 45°. Hence find what would be the values of g at the equator and in latitude 45° if the earth did not rotate. (*b*) What would the period of rotation of the earth have to be in order that bodies at the equator might have no weight? (*c*) The weight of any body in absolute units is, by definition, mg, where m is the mass of the body and g is the acceleration due to gravity. In view of this definition is it correct to say that the weight of a body is equal to the gravitational pull of the earth upon it? Is it correct to say that g at any point is equal to the gravitational attraction of the earth for a unit mass placed at that point? Speaking literally, is g the "acceleration due to gravity"? *Ans.* (*a*) 3.37 dynes, 2.38 dynes, 981.41 cm · sec⁻², 982.31 cm · sec⁻²; (*b*) 1.4 hr.

6. When a train is rounding a curve, it is desirable that the forces exerted on the two rails should be equal and hence that the total force exerted by the train should be perpendicular to the plane of the track. Prove that for a train rounding a curve of radius r with speed v, the angle ϕ, Fig. 197,

FIG. 197. Elevation of the outer rail

which the plane of the tracks should make with the horizontal is $\phi = \tan^{-1}(v^2/rg)$.

HUYGENS settled in Paris in 1666, where for more than fifteen years he occupied an influential position among the group of men who founded and composed the Academy of Sciences. He resided and worked in the philosophic seclusion of the Royal Library, in which was set aside also the room shown in Plate 47 for meetings of the academy.

The plate shown is a reproduction, considerably reduced in size, of a copperplate engraving made by Sébastien Le Clerc and used as the frontispiece of each volume of the sumptuously printed editions of the early work of the Academy of Sciences which were distributed as personal gifts by Louis XIV. It is the earliest known representation of an actual meeting of a learned society. The figures in the center are Louis XIV and COLBERT.

The French Academy of Sciences was the outcome of informal gatherings of a group of philosophers and mathematicians in Paris near the middle of the seventeenth century. The group included such men as DESCARTES, FERMAT, HUYGENS, MERSENNE, and PASCAL. Finally COLBERT proposed to Louis XIV the establishment of a regular academy, and the first meeting was held December 22, 1666. Members received pensions from the king and financial assistance with their researches. These researches were divided into the two groups of *mathematics*, including mechanics and astronomy, and *physics*, which at that time included chemistry, botany, anatomy, and physiology. After COLBERT's death the work of the academy was diverted to the practical details of constructing "the new paradise at Versailles." The suggestion of this task is artfully introduced into the background of Le Clerc's picture, while in the foreground are many symbols of the materials and instruments of the sciences in which the academy was interested. The air pump invented by ROBERT BOYLE in 1660 appears on the table at the left in the picture.

For further details, consult A. Wolf's *A History of Science, Technology, and Philosophy in the 16th and 17th Centuries* (Allen & Unwin, 1935), pp. 63–67; M. Ornstein's *The Role of the Scientific Societies in the Seventeenth Century* (University of Chicago Press, 1928), pp. 165–193.

●

A Meeting of the French Academy of Sciences
on the Occasion of a Visit by Louis XIV

7. A car, after descending a steep incline, runs around the inside of a vertical circle of large diameter D, as in Fig. 198. (*a*) Show that if there were no friction, the car must start from a point 1.25 D above the lowest point of the circle in order that it may keep to the track at the highest point of the circle. This is essentially one of HUYGENS'S theorems on the magnitude of centripetal force (see the footnote on page 332). (*b*) What force does the track exert on the car when the latter first reaches the lowest point of the circle? (*c*) when the car is halfway up the circle?   *Ans.* (*b*) 6 *mg*; (*c*) 3 *mg*.

FIG. 198. Problem 7

8. A particle moves with a simple vibratory motion of period 2 sec in a path 10 cm long. Calculate the displacement, speed, and acceleration of the particle at half-second intervals during a complete vibration, and plot upon a single sheet of graph paper the results of the several sets of calculations.

9. A particle of mass 5 g describes a simple vibratory motion with a period of 0.6 sec and an amplitude of 18 cm. Find the phase, speed, and accelerating force at an instant when the displacement of the particle is -9 cm.   *Ans.* 120° or 240°; 1.6×10^2 cm · sec^{-1}; 5.0×10^3 dynes.

10. The gravitational force of a solid sphere of uniform density upon a particle embedded in it varies directly as the distance of the particle from the center of the sphere (Prob. 20, p. 52). If the earth were such a sphere, and if a hole passed completely through it along a diameter, how long a time would be required for a body dropped through the hole to reach the other side? Take the radius of the earth as 4.0×10^3 mi.   *Ans.* 42 min.

11. A wrought-iron shelf is moved horizontally with a simple vibratory motion. A block of soft steel rests on the shelf, and it is observed that this block just begins to slide when the vibration has a period of 2.5 sec and an amplitude of 1.0 ft. What is the coefficient of static friction for steel on iron?   *Ans.* 0.20.

12. (*a*) By evaluating the expression

$$\frac{1}{T} \int_0^T E \, dt,$$

where E is the instantaneous value of the kinetic energy of a particle which is executing simple vibratory motion, show that the time average of E, averaged over one complete vibration, is $mA^2\omega^2/4$. (*b*) In a similar manner derive an expression for the time average of the potential energy averaged over one complete vibration. (*c*) Is it true that the total energy of the particle is on the average half kinetic and half potential, as was asserted in Secs. **143** and **144**, in the discussion of the theory of thermal capacities?

13. The period of a certain pendulum was 1.002 sec at sea level. It was carried to the top of a mountain and the period found to be 1.003 sec. Find the height of the mountain above sea level.   *Ans.* 4 mi.

14. A thin, uniform hoop of diameter D hangs on a nail. It is displaced through a small angle in its own plane, and then released. Assuming that the hoop does not slip on the nail, prove that its period of vibration is the same as that of an ideal pendulum of length D.

15. A thin homogeneous rod of length l is oscillating about a horizontal axis passing through one end of it. Find the length of the equivalent ideal pendulum and the position of the center of oscillation.   *Ans.* $2 l/3$.

O

IT IS difficult for us, at this late date, to realize how great a stride was made by Huygens when he discovered how, by simply weighing and measuring a given body, say a lath, one could sit down and compute its period of vibration when suspended about any particular line as axis.

The problem of a freely falling body, solved by Galileo, is the first great solution in dynamics; that of the physical pendulum by Huygens is the second and an immensely more difficult one. It was, in fact, the first dynamical system, as distinguished from a particle, to be studied. It was too complex for Galileo; and it was a stumbling block to Descartes; but perplexity is the beginning of knowledge. Some of the most important concepts and principles of dynamics center about the pendulum. . . .

Mechanics is in a peculiar sense an Italian science. Its first and simplest chapter — Statics — was written by Archimedes in Sicily. Its earliest applications to engineering were made by Vitruvius, Frontinus, and Leonardo da Vinci. Its second and most fundamental chapter — Dynamics — was written by Galileo who conceived the idea of acceleration as the criterion of force. The final chapter (in a certain sense; in the true sense, there is no final chapter in science) of classical mechanics was written by Lagrange, a native of Turin. The pendulum is the first great dynamical problem solved outside of Italy.

H. CREW, *The Rise of Modern Physics* (The Williams & Wilkins Company, 1935), pp. 131–132. By permission of the publishers

CHAPTER FIFTEEN

WAVE MOTION AND SOUND

THE VELOCITIES of pulses propagated in an elastic fluid, are in a ratio compounded of the subduplicate ratio of the elastic force directly, and the subduplicate ratio of the density inversely; supposing the elastic force of the fluid to be proportional to its condensation.

Proposition XLVIII, Book II, of NEWTON's *Principia*, as translated by ANDREW MOTTE in 1729

O

If we consider what experience has taught us about the various ways in which energy can be transferred from one point to another, we find that these ways fall into two general classes. Perhaps the most obvious of these is the passage of matter from point to point, as when a projectile possessing kinetic energy passes from a gun to a target, when a fuel possessing stored-up heat of combustion is transported from the mine to the consumer, or when energy is carried by the wind, by running water, or by the charges in an electric current. The second class may be illustrated by the transmission of mechanical energy by means of a moving axle or shaft, by the propagation of sound energy through matter, and by the transmission of energy of motion by ocean waves. This second method, in all cases that come to our notice in ordinary experience, is characterized by the fact that it involves the existence of a continuous, material medium which extends from the source of the energy to the point where the energy is utilized. In some cases the transfer of energy through the continuous medium consists solely of a propagation of a state or condition of some kind from the one point to the other without any motion of the medium as a whole. This important process is called *wave motion*, and it is the one to which we shall give our attention in this chapter.

When a rod is struck on the end with a hammer, a single *wave-pulse* in the form of an endwise compression of the rod travels along its length. If the rod is struck periodically, a succession of similar pulses is sent along it, and this constitutes what is called a *wave-train*. The idea that sound energy from, say, the vibrating string of a musical instrument is propagated through air in this manner was familiar to ARISTOTLE.[1] With modern apparatus it is possible actually to

---

[1] See D. C. Miller, *Anecdotal History of the Science of Sound* (Macmillan, 1935), p. 3.

measure the exceedingly minute variations of pressure[1] that constitute an ordinary sound wave; and, what is more striking, these waves have been photographed (Fig. 199) and the photographs applied to the development of scientific methods of determining the acoustical properties of large buildings and auditoriums.[2]

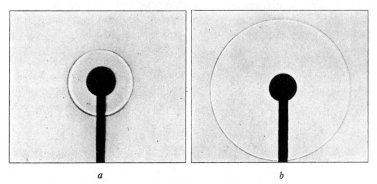

a   *b*

FIG. 199. Photographs of sound pulses in a gas. The pulse is produced by an electric spark which is placed behind a shield so that the light from the spark will not reach the photographic plate. The pulse is photographed by passing light through it from a second spark which is produced automatically a fraction of a second after the first one. The time interval between the two sparks may be varied at will. In *a* the spherical pulse has just emerged from behind the light shield and is spreading out so that after a few thousandths of a second it occupies the position shown in *b*. [A. L. Foley and W. H. Souder in *Physical Review* **35**, 373 (1912)]

A good illustration of how waves originate and upon what properties of a medium their propagation depends is afforded by the familiar water waves that emanate from a point where a stone dropped into the water strikes the surface. Series of photographs taken at

---

[1] A sound whose frequency is 500 vib · sec⁻¹ will be just audible if it produces a variation in the pressure of the air of 10^{-3} dyne · cm⁻² or 10^{-9} A_s. Such a variation requires an amplitude of about 10^{-9} cm [see H. Fletcher, *Speech and Hearing* (Van Nostrand, 1929), pp. 132–166]. The energy of sound waves is thus very small, and the ear must be extremely sensitive in order to detect such minute variations in pressure. "If a million persons were to talk steadily, and the energy of their voices were to be converted into heat, they would have to talk for an hour and a half to produce enough heat to make a cup of tea" (R. L. Jones in *Bell Laboratories Record*).

[2] Various methods of recording and photographing sound waves are described in D. C. Miller, *The Science of Musical Sounds* (Macmillan, 1916), Lecture III. For a description of the method of spark photography see, for example, E. G. Richardson, *Sound* (Longmans, Green, 1927, 1935), p. 17; A. B. Wood, *A Textbook of Sound* (Macmillan, 1930), p. 340; or A. Wood, *Sound Waves and their Uses* (Blackie, 1930), p. 19.

various stages of such disturbances [1] show that the surface of the water under the stone is forced down, leaving a hole, and that the water forced out of the hole first piles up about the edges and then begins to fall back into the hole and also forward upon the outer undisturbed surface; because of the inertia of the water, this action does not cease when the hole has been filled but goes on until a depression is formed where the water previously was heaped up, and the liquid piles up on each side of this new depression. The whole process is then repeated until the viscous resistance brings the oscillating liquid to rest.

It should be evident from these considerations that if energy is to be transmitted through a medium by means of waves, the medium must possess either the following properties or else properties that are physically analogous to them:

a. There must be a tendency for the medium to return to its original condition after being disturbed. In large water waves gravity is the cause of the return, and in ripples it is the surface tension. In sound waves the requisite restoring force is due to the elastic nature of the matter through which sound travels.

b. The medium must be capable of storing up energy; that is, it must possess inertia.

c. The frictional resistance must not be so great that the oscillatory motion cannot take place. The absorption of wave-energy by friction is called *damping*.

o

Speeds of Waves in Elastic Mediums

The study of waves that are propagated through matter by virtue of the fact that matter possesses elasticity very evidently has its foundation in the application of the theory of vibratory motion (Chap. 14) to the fundamental laws of elasticity as revealed by dynamical and statical investigations (Chaps. 5 and 8). Since our present discussion will be confined almost entirely to such types of waves, we shall for the most part be on quite familiar ground, for we have merely to extend our study of elastic vibrations so as to be able to describe their transmission *through* a medium. Studies of this kind were begun early in the history of elastic theory by GALILEO,[2] and since then have been associated with such great names as MARIN

---

[1] Many such photographs are reproduced in A. M. Worthington, *A Study of Splashes* (Longmans, Green, 1908).

[2] * *Two New Sciences*, "First Day." Much of our knowledge of GALILEO's achievements in the field of acoustics is derived from MERSENNE.

Mersenne,[1] Newton,[2] Young,[3] Ernst Chladni,[4] the brothers Weber,[5] Félix Savart,[6] and, more lately, Helmholtz,[7] Karl Rudolph Koenig,[8] and Rayleigh.[9]

183. Speed of a Compressional Pulse in Any Homogeneous, Isotropic Medium. We will first consider the speed of propagation of a compressional pulse in a homogeneous, isotropic, elastic medium. For convenience of analysis imagine a portion of the medium to be contained in a rigid tube of unit cross-sectional area and of infinite length. Let the density of the medium be represented by ρ and the pressure under which it stands before any compression is produced by P. Let the medium be conceived to be divided into unit cubes, 1, 2, 3, etc., as in Fig. 200. Imagine one end of the tube to be closed by a massless, frictionless piston, and let the piston suddenly be started forward by the application to it of a constant pressure slightly greater than P, namely, $P + dP$. There will then be started down the tube a compressional pulse which will travel through the medium with some speed v. It is this speed which it is desired to determine.

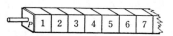

Fig. 200. Model used in deriving the theoretical expression for the speed of a compressional wave in an elastic medium

As soon as the piston starts to move forward, cube 1 begins to be compressed and the pressure inside it rises. When this pressure has reached the value $P + dP$, the cube will cease to be compressed any farther and thenceforth will merely transmit pressure to cube 2. Call $- dV$ the change of volume, measured in fractions of a unit of length, which cube 1 thus experiences. Under the action of the pressure $P + dP$ transmitted from cube 1 to cube 2, the latter also

[1] *Harmonie Universelle* (Paris, 1636). An abridgment in English is given in J. Hawkins, *A General History of the Science and Practice of Music* (London, 1853, 1875), pp. 600–616.

[2] *Principia*, Bk. II, Sec. VIII.

[3] *Philosophical Transactions* **90**, 106 (1800); *Lectures on Natural Philosophy and the Mechanical Arts* (1807); *Miscellaneous Works*, ed. by G. Peacock (London, 1855), Vol. I, p. 64; Vol. II, p. 141.

[4] *Entdeckungen über die Theorie des Klanges* (Leipzig, 1787); *Die Akustik* (Leipzig, 1802); *Neue Beyträge zur Akustik* (Leipzig, 1817).

[5] W. E. and E. H. Weber, *Wellenlehre auf Experimente gegründet* (Leipzig, 1825).

[6] *Annales de Chimie et de Physique* (1819–1840).

[7] *Vorlesungen über die mathematischen Principien der Akustik* (1898); *Die Lehre von den Tonempfindungen* (1862), *On the Sensations of Tone*, tr. by A. J. Ellis (Longmans, Green, ed. 1, 1875; ed. 2, 1885; ed. 3, 1895; ed. 4, 1912).

[8] *Quelques Expériences d'Acoustique* (Paris, 1882).

[9] *The Theory of Sound* (Macmillan, 1877, 1894).

will be compressed an amount $-dV$ and thereafter will merely transmit the pressure $P + dP$ to cube 3. Similar reasoning may be applied to the remaining cubes in their numerical order. Since the tube has unit cross section, a reduction $-dV$ in the volume of any particular cube, such as 6, is accompanied by a forward motion of the piston through a distance numerically equal to $-dV$, and each of the cubes previously compressed (namely, 1, 2, · · ·, 5) will move forward a distance numerically equal to $-dV$. Thus, as the pulse moves down the tube, the piston, and with it all the compressed cubes, will move forward uniformly. Since v represents the speed of the pulse, and since the cubes have unit length, it follows that in unit time each of v cubes will experience the compression $-dV$. During the same time the piston will therefore have moved forward a distance $v(-dV)$. The average speed of the *piston* is therefore $v(-dV)$. Evidently this is also the expression for the speed with which all the cubes which have been compressed are moving forward at the end of unit time.

Now the speed v with which the *pulse* moves forward may be found by an application of NEWTON'S principle of work (Sec. 46) as follows. In the foregoing operation the acting force is numerically equal to the pressure applied to the piston, namely, $P + dP$. The work done by this force in unit time is $(P + dP) \cdot v(-dV)$. This work accomplishes two things: (*a*) all of the substance contained in v cubes acquires potential energy by being compressed from a condition in which it exerts a pressure P to one in which it exerts a pressure $P + dP$; and (*b*) all of the mass contained in v cubes acquires an average speed of amount $v(-dV)$. The mean force overcome in the first operation is half the sum of the initial and final forces, namely, $P + \frac{1}{2} dP$, and the work done against this force in unit time is $(P + \frac{1}{2} dP) \cdot v(-dV)$. In the second operation, since the mass of each cube is ρ, and since in unit time v cubes are set in motion with an average speed $v(-dV)$, the average kinetic energy imparted in unit time is $\frac{1}{2} \rho v \cdot v^2 (-dV)^2$. Now, by the work principle, the work done by the acting force is equal to the sum of the potential and kinetic energies imparted to the substance; hence we have

$$(P + dP) \cdot v(-dV) = \left(P + \frac{dP}{2}\right) \cdot v(-dV) + \frac{1}{2} \rho v^3 (-dV)^2,$$

or
$$v = \sqrt{-\frac{1}{\rho}\frac{dP}{dV}}. \qquad [281]$$

Since we are dealing with unit cubes, the quantity dP/dV is the force applied per unit area divided by the change in volume per unit

COLLADON and STURM'S
Determination of the Speed of Sound in Water
Lake of Geneva, Switzerland, 1827

From a restoration in the Deutsches Museum, Munich

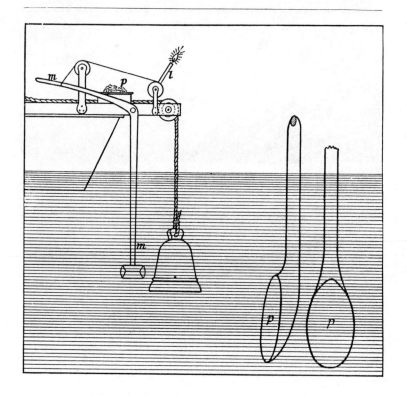

Diagram of COLLADON and STURM's Apparatus
for Determining the Speed of Sound in Water

TWO BOATS were stationed 13,487 m apart (see Plate 48). A bell suspended from
one of the boats was struck under water by means of a lever m which at the same
moment caused the candle l to ignite the powder p and set off a visible flash of light.
An observer in the second boat with a listening tube measured the time which
elapsed between the flash of the light and the sound of the bell. The mean of a
number of such observations gave a value of 9.4 sec. The temperature of the water
was 8° C. The speed of sound in water at 8° C is therefore 13,487/9.4, or
1435 m · sec^{-1}. The results were reported in detail in the *Annales de Chimie et de
Physique* (2) **36**, 236 (1827) and in Poggendorff's *Annalen der Physik und Chemie* **12**,
171 (1828).

volume, and this, by Eq. [110], Chap. 8, is equal to — k, where k is the volume modulus of elasticity of the medium. Hence, finally, we have for the speed of a compressional wave in any homogeneous, isotropic medium

$$v = \sqrt{\frac{k}{\rho}}.$$ [282]

This result was obtained by NEWTON by a very difficult type of reasoning but one that clearly shows his great genius; it appears in the *Principia*.[1] (In what units must k be expressed in Eq. [282]?)

Since the foregoing expression involves only the constants k and ρ of the medium, it is evident that a pulse once started will travel on and on down the tube at a rate which has nothing whatever to do either with the size or shape of the tube or with whether the piston continues to move forward or not. In other words, we have deduced a general expression for the speed of a compressional pulse in the kind of medium under consideration. The expression must hold for the speed of propagation of a sound pulse that originates at a point within the medium and spreads radially from the center of disturbance; for in this case, as in the case just discussed, the pulse is one of pure compression, since the particles are free to move only in one direction, namely, along radii emanating from the point of disturbance.

It will be interesting to see how well the results obtained by the use of this formula, which has been deduced from purely theoretical considerations, agree with the results of direct experiment. In 1893 AMAGAT[2] published results of his experiments on the compressibility of water which showed that at 10° C a change in pressure from 1 to 50 atmospheres produced a change in volume from 1.0000 to 0.99757 cm³. From these data we get as a theoretical value for the speed of a compressional pulse in water at 10° C

$$v = \sqrt{\frac{k}{\rho}} = \sqrt{\frac{49 \times 76 \times 13.6 \times 981}{1 \times 0.00243}} = 143,000 \text{ cm} \cdot \text{sec}^{-1} = 1430 \text{ m} \cdot \text{sec}^{-1}.$$

J. D. COLLADON and J. K. F. STURM made a direct measurement of the speed of sound in Lake Geneva in 1827 and obtained a value of 1435 m · sec⁻¹ at 8° C (see Plates 48 and 49). The difference between this value and the one calculated from Eq. [282] is well within the limits of observational error.

Water, obviously, is the only liquid in which large-scale measurements of the speed of sound are practicable. In recent years the methods of submarine signaling have been applied to experiments in sea water. Charges

---

[1] See the quotation at the beginning of this chapter.
[2] *Comptes Rendus* **116**, 41 (1893).

are exploded under water, and simultaneously radio signals, instead of visible-light signals, are sent out. The sound pulses are received by microphones.[1]

Since both k and ρ in Eq. [282] in general vary with temperature, the speed of a compressional wave also depends on the temperature.[2] In water above 4° C, k increases with temperature whereas ρ decreases; hence v increases with temperature for water above 4° C.

184. Wave-trains and Wavelengths. If in the case of the tube and piston of Fig. 200 the applied pressure had been $P - dP$ instead of $P + dP$, the piston would have started back instead of forward, and cube 1 would have expanded until its pressure reached the value $P - dP$, which expansion would have been followed by a similar expansion of cube 2, etc. Thus a pulse of *rarefaction* instead of one of condensation would have traveled down the tube, and reasoning in every respect identical with that which precedes shows that the speed of this pulse also would have been $\sqrt{k/\rho}$.

It is important to observe that in a pulse of *condensation* the particles of the medium are always moving in the same direction as the pulse, whereas in a pulse of *rarefaction* the direction of motion of the particles is always opposite to the direction of propagation of the pulse. If the piston is moved alternately forward and backward at regular intervals, a *wave-train*, or succession of compressions and rarefactions, will follow one another down the tube. In this case it is evident that the motions of all the particles of the medium follow, *in succession*, exactly the motions of the piston; that is, each particle moves forward for an interval of time that is just equal to that required for the piston to move forward, and backward for an interval just equal to that required for the piston to move backward. If then the piston is replaced by the vibrating prong of a tuning fork or by any other body that vibrates under the influences of its own elasticity, the backward motion will begin at exactly the instant at which the forward motion ends, and hence at the end of one complete vibration of the prong — that is, at the end of the time required for the prong to go from A to C and back again to A (Fig. 201) — the whole of the medium between the prong A and some point a to which a pulse travels during the period of one vibration may be divided into two

---

[1] E. B. Stephenson, *Physical Review* (2) **21**, 181 (1923); E. A. Eckhardt, *Physical Review* (2) **24**, 452 (1924); A. B. Wood and H. E. Browne, *Proceedings of the Physical Society of London* **35**, 183 (1923); *Proceedings of the Royal Society* **103**, 284 (1923).

[2] For data on the speed of sound in various substances and at various temperatures, see *International Critical Tables*, Vol. VI, pp. 461–467.

equal parts, *ac* and *cA*, such that all the layers between *c* and *a* are moving forward and are in a state of compression, while all the layers between *c* and *A* are moving backward and are in a state of rarefaction. The relative velocities of these layers are represented in the figure by the arrows. As the fork continues to vibrate, the whole region about it becomes filled with a series of such waves, each wave consisting, as in Fig. 201, of a condensation and a rarefaction. The distance between the beginnings of two successive condensations or two successive rarefactions, or, in general, the distance between any two successive particles that are in the same condition or *phase* of vibration, is called a *wavelength*. It is obvious that if *v* represents the

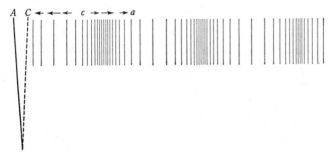

Fig. 201. Wave-train maintained by the vibrating prong of a tuning fork

speed of the wave-train, ν the frequency of the fork in vibrations per unit time, and λ the wavelength, then the following relation holds:

$$v = \nu\lambda. \tag{283}$$

A distinction often made between a musical note and a mere noise is that the former consists of a train of waves, whereas the latter consists either of a single pulse or of irregularly timed pulses. This distinction is, however, not sufficient. Many sounds due to regularly timed pulses are interpreted as noises because they are too complex in structure to be analyzed and understood by the ear.

185. Speed of Sound in an Ideal Gas. In applying Eq. [282] to the calculation of the speed of sound in a gas, NEWTON assumed that the small volume strains that make up a sound wave were proportional to the stress. This would be the case, according to BOYLE'S law, if sound waves produce no temperature changes in passing through a gas. On this assumption the bulk modulus of an ideal gas, as we have already seen (Sec. 93), is equal simply to the pressure *P* under which the gas stands. Hence

$$v = \sqrt{\frac{P}{\rho}}. \tag{284}$$

If the gas is air under ordinary circumstances, then P is simply the barometric pressure expressed in absolute cgs or fps units. By using this formula and the data at his command, NEWTON[1] calculated a value for the speed of sound in air which was nearly 16 percent smaller than the best experimental values. He ascribed this discrepancy to the fact that the molecules of the actual atmosphere occupy space, and tried to explain it by assuming that 16 percent of the linear distance traversed was thus occupied, and that the sound, passing instantly through the molecules, required time only to traverse the interspaces. There is no justification for this assumption, however. In 1738 a commission appointed by the French Academy of Sciences made the first really accurate experiments on the speed of sound in air.[2] Cannons were fired at two stations about 18 mi apart, and observers at these stations recorded the time interval between flashes and reports. By combining the various observations, the effect of the wind was eliminated. The results were still so much larger than NEWTON'S theoretical value that the discrepancy could not be explained by departures from the ideal gas laws.

It was more than one hundred and twenty years after NEWTON published his calculations when LAPLACE[3] pointed out that the passage of sound through a gas is an adiabatic process and not an isothermal one, as NEWTON had assumed it to be.[4] Now it was stated in Sec. 145 that for small strains taking place adiabatically the volume

---

[1] *Principia*, Bk. II, scholium to Sec. VIII. For a discussion of the experimental values which NEWTON had at his disposal, see *A. Wolf, *A History of Science, Technology, and Philosophy in the 16th and 17th Centuries* (Allen & Unwin, 1935), p. 286; also *D. C. Miller, *Anecdotal History of the Science of Sound* (Macmillan, 1935), pp. 27–29.

[2] *Mémoires de l'Académie des Sciences* (1738).

[3] *Annales de Chimie et de Physique* (2) **3**, 238 (1816); **20**, 266 (1822); *Traité de Mécanique Céleste* (1823), Vol. V, Bk. XII. See also S. D. Poisson in *Journal de l'École Polytechnique* (1807); *Annales de Chimie et de Physique* (2) **23**, 5 (1823).

[4] It is sometimes said that the process is adiabatic because the sound vibrations are so rapid that the temperature inequalities produced by the compressions do not have time enough for equalization. But this statement implies that NEWTON'S expression, Eq. [284], would give correct values provided the frequency were made low enough, whereas experiment shows that just the opposite is the case. RAYLEIGH [*The Theory of Sound* (Macmillan, ed. 2, 1896), Vol. II, p. 28] has shown that the heat conduction actually increases with the frequency, and an analysis based on kinetic-theory considerations [E. U. Condon, *The American Physics Teacher* **1**, 18 (1933)] shows that heat conduction should cause appreciable departures from the adiabatic behavior only for waves of such high frequency that the wavelength is comparable with the mean free path of the gas molecules, that is, for a wavelength of 10^{-5} cm approximately. Ordinarily sound is of too low, rather than too high, a frequency for the process to be isothermal.

modulus of elasticity of an ideal gas is given by $k = \gamma P$, where γ is the ratio of the two specific heats of the gas and P is the pressure as before. Hence on this view the theoretical expression for the speed of sound in an ideal gas is

$$v = \sqrt{\frac{\gamma P}{\rho}}, \qquad [285]$$

which gives results in conformity with experimental values.[1] Conversely, this formula makes it possible to obtain the value of γ with great accuracy from experiments on the speed of sound.

In view of the equation of state of an ideal gas (Sec. 105) and the definition of density, Eq. [285] may be rewritten in the form

$$v = \sqrt{\frac{\gamma RT}{M}}, \qquad [286]$$

in which M is the molecular weight of the gas. From this equation we may conclude that the speed of sound in ideal gases depends only on the kind of gas and the temperature, and is wholly independent of changes in pressure. Thus its speed on a mountain top is the same as that at the foot of the mountain if the temperature is the same at both places. If we denote by v_t the speed of sound in a given gas at $T°$ K and by v_0 the speed in the same gas at $T_0°$ K, and apply Eq. [286], there results

$$v_t = v_0 \sqrt{\frac{T}{T_0}}. \qquad [287]$$

By substituting in this equation the best determination of the speed of sound in dry air at $0°$ C, namely [2] 331.45 m · sec^{-1}, one finds that the speed of sound in air increases about 0.6 m for each degree centigrade of rise in temperature.

186. Speed of Compressional Waves in Thin Solid Rods. The analysis in Sec. 183 shows that if a medium is confined in a rigid tube so that there is no possibility of its expanding laterally, the speed of a compressional wave depends only upon the volume modulus of elasticity and the density of the medium. This condition is realized also when a disturbance originates in the interior of an elastic medium of great extent in all directions. But when the wave travels along a thin rod of solid material, slight lateral expansions and contractions occur in that portion of the rod which is

[1] *For discussions of the experimental determinations of the speed of sound, see *A Dictionary of Applied Physics*, ed. by R. Glazebrook (Macmillan, 1923), Vol. IV, pp. 687–691; A. B. Wood, *A Textbook of Sound* (Macmillan, 1930), pp. 229–271; E. G. Richardson, *Sound* (Longmans, Green, 1935), Chap. 1; J. H. Poynting and J. J. Thomson, *Sound* (Griffin, 1899), pp. 22–31.

[2] *International Critical Tables*, Vol. VI, p. 462.

undergoing the volume strain. Hence if we imagine a rod of unit cross section divided into unit cubes after the fashion of Sec. **183**, and apply a small pressure dP by means of a piston at one end (Fig. 200), then while each cube is undergoing a volume compression of amount $- dV$ the piston will move forward not now $- dV$, but some distance ds numerically a trifle larger than $- dV$. From reasoning identical with that given in Sec. **183** an equation results which differs from Eq. [281], that is, from

$$v = \sqrt{ - \frac{1}{\rho} \frac{dP}{dV} },$$

in no respect save that ds replaces $- dV$. We obtain, then,

$$v = \sqrt{ \frac{1}{\rho} \frac{dP}{ds} }. \qquad [288]$$

But dP is the force applied per unit area, and ds is the change in the length of the rod per unit length. Hence, by Eq. [107], Chap. 8, dP/ds is YOUNG'S modulus Y. Thus in thin rods compressional waves move with a speed given by the expression

$$v = \sqrt{ \frac{Y}{\rho} }. \qquad [289]$$

187. Speed of a Transverse Wave along a Stretched Wire. It will be noted in what precedes that we have considered merely the longitudinal motion along the axis of the rod and, safely enough if the rod is thin, have not taken into account the fact that a solid has shear elasticity as well as volume elasticity (Sec. **94**). In substances that possess shear elasticity an additional type of elastic wave motion is possible — namely, *transverse* wave motion, which is characterized by the fact that the particles of the medium vibrate in paths which are perpendicular to the direction of propagation of the wave. The waves that travel along a rope when one end is caused to vibrate by the hand are of this sort. Transverse waves obviously may be regarded as having two degrees of freedom (Sec. **143**), for the particles can vibrate in two independent directions at right angles to the direction of propagation. Compressional, or longitudinal, waves, on the other hand, have only one degree of freedom, since the particles move in the line of propagation of the wave itself. In substances that do not possess shear elasticity, longitudinal waves are the only kind of elastic waves possible.

Let us now consider the case of a transverse wave-pulse sent down a stretched string, such as the string of a violin. It will be assumed that the string is perfectly flexible and uniform and that it is stretched

so tightly by a force f, as in Fig. 202, that the effect of the weight of the wire can be neglected in comparison with the stretching force. We will limit the problem to small displacements which do not change appreciably the tension of the string; this simplifies matters and is, after all, the only case of practical importance. Let the string be plucked at one end, so that a transverse pulse is sent down it. The expression for the speed of this pulse can be deduced most satisfactorily with the aid of the calculus,[1] but can also be obtained by the following method, which is due to P. G. TAIT.[2] Let the curve *mno* in Fig. 203 represent a portion of the stretched string over which the de-

FIG. 202. Transverse wave-pulse in a string. The string may be placed under a constant stretching force of known magnitude by passing one end over a pulley and hanging a weight from it

formation is being propagated. Let Δs be an element of the string so small that it may be considered as the arc of a circle of radius r. Since the string is assumed to be flexible, the only force of appreciable magnitude that is urging the element toward the center C arises from the force f in the string, and this may be regarded as a pull acting tangentially at each end of the arc Δs. The sum of the centripetal components of these two forces is $2f \sin \theta = 2f \cdot \frac{1}{2} \Delta s/r = f\,\Delta s/r$, since θ is small. But since the deformation *mno* is propagating itself unchanged in character along the string, each element of the string must assume in succession the positions occupied at any instant by all the other elements. In other words, at the instant which we have been considering the configuration associated with the element Δs is not moving at all in the direction of C, but is instead moving into the position of the adjacent element on *mno*; that is, it is moving with a speed v along the circumference of the circle which has r for its radius. Hence its acceleration is entirely centripetal, and the force in absolute units needed to

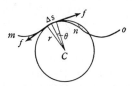

FIG. 203. Speed of a transverse wave in a stretched string

produce this acceleration is, by Eq. [234], Chap. 14, $\sigma \cdot \Delta s \cdot v^2/r$, where σ is the mass of unit length of the string. But since this force is also equal to $f\,\Delta s/r$, we have finally $\sigma \cdot \Delta s \cdot v^2/r = f\,\Delta s/r$, or

$$v = \sqrt{\frac{f}{\sigma}}. \qquad [290]$$

---

[1] See Fig. 207.   [2] *Dynamics* (A. & C. Black, 1895), p. 283.

Thus the speed of a transverse wave in a string is a function only of the stretching force and the linear density.

Transverse waves may also be sent along heavy rods, the wave traveling as a result of the shear elasticity of the substance of which the rod is composed. A consideration of the speed of propagation of these waves, however, is beyond the scope of this book.

188. Waves in Solid Bodies. Because a disturbance will, in general, be transmitted through a solid body by transverse as well as

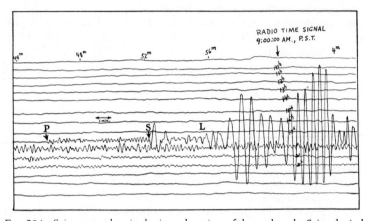

FIG. 204. Seismogram showing horizontal motions of the earth at the Seismological Laboratory of the California Institute of Technology and the Carnegie Institution of Washington, Pasadena, California, on July 17, 1934. The source of the disturbance was 4900 km away. At the time marked P the disturbance began with small longitudinal vibrations which were followed 6 min 26 sec later by stronger transverse vibrations S. Both of these wave motions came through the body of the earth. The surface waves L arrived 3 min 30 sec later. This shock was destructive in Panama

longitudinal waves, the general problem of wave motion in a solid is exceedingly complicated, even for the case where the solid is isotropic as regards elasticity. The problem [1] first attracted the attention of AUGUSTIN LOUIS CAUCHY and SIMÉON DENIS POISSON, two mathematicians of the highest order, because of certain important developments in the theory of light which had just been advanced by THOMAS YOUNG and by AUGUSTIN JEAN FRESNEL; this occurred in 1821, in the same year that NAVIER published his formulation of

---

[1] A. L. Cauchy, *Exercices de Mathématique* (1830); S. D. Poisson in *Mémoires de l'Académie des Sciences* **8** (1829); Rayleigh, *The Theory of Sound*, Chap. XXII. The problem was treated exhaustively by G. G. Stokes, "On the Dynamical Theory of Diffraction," *Transactions of the Cambridge Philosophical Society* **9**, 1 (1849), also *Mathematical and Physical Papers* (Cambridge University Press, 1883), Vol. II, p. 243.

the differential equations of elasticity (Chap. 8). We will briefly set down some of the chief results, without proof.

For a large mass of matter of volume modulus of elasticity k, shear modulus n, and density ρ, the speed of the *longitudinal* wave turns out to be

$$v = \sqrt{\frac{k + \frac{4}{3}n}{\rho}}. \qquad [291]$$

On the other hand, the purely *transverse* waves, such as those that would be set up by the torsion of a cylinder, travel with the speed

$$v = \sqrt{\frac{n}{\rho}}, \qquad [292]$$

which is always considerably less than that of the longitudinal waves. (What do these general expressions reduce to in the case of a fluid?)

Seismic waves are both transverse and longitudinal. Seismograph records (Fig. 204) taken at an observing station some distance from the source of an earthquake indicate three distinct sets of waves: (*a*) longitudinal waves which have come directly through the body of the earth with a speed of about 8 km · sec^{-1}; (*b*) transverse waves, also direct, with a speed of about 4.5 km · sec^{-1}; and (*c*) large-amplitude surface waves, known as "RAYLEIGH waves," which are analogous to water waves and penetrate to a small depth only. Since the speeds of the various types of waves are known, the time between their arrivals gives the distance of the disturbance from the station.

o

The Equation of a Wave

Before we proceed with further analysis, it is important to get clearly in mind the general characteristics of wave motion. Fig. 205 should prove helpful for this purpose, and the student should also review Secs. 173 to 177, which deal with the kinetics of simple vibratory motion.

A summary of these important characteristics, in so far as they apply to waves in an elastic medium, reveals that

a. The particles of a medium through which a wave is passing are in vibration, and for a *simple wave* in an elastic medium these motions are simple vibratory in type (Sec. 173).

b. There is a progressive change of phase of the motions of the particles as one goes along the wave, but at regular intervals there are particles in the same phase; the distance between two such successive particles is called the wavelength λ.

c. Although the particles themselves merely vibrate about their positions of equilibrium, they form at any instant a definite configuration, called the *wave-form*. This wave-form travels through the medium with a speed v that is determined, in the case of an elastic

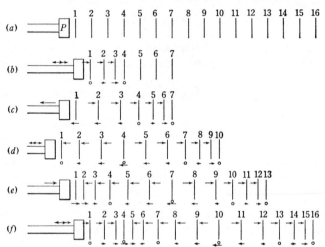

FIG. 205. Motion of the particles of a medium due to the oscillatory motion of a piston. The unit cubes of Fig. 200 are here replaced by vertical lines. (*a*) represents the state of the medium before the piston has begun to move. (*b*) represents its state when the piston has undergone its greatest displacement to the right and is ready to return, a condition which is represented by the double arrow. In the upper line of arrows each arrow represents the direction of *displacement* of the layer toward which it points. The small arrows below the vertical lines show the directions of the *motions* of the layers. A zero below any line indicates either that the medium is at its equilibrium position or that it is just changing the direction of its motion at the end of its path. The succeeding diagrams show the progression of the initial condensation and the subsequent rarefaction for the indicated positions and directions of motion of the piston. Obviously a condensation exists, for example, at 4 in (*b*), since the layers are there crowded together, and a rarefaction exists at 4 in (*d*), since the layers are there separated

wave, by the elasticity and density of the medium. The wave travels a distance of one wavelength during the time that a single particle makes one vibration.

d. The period T of a wave is the same as the period of vibration of any one of the particles, and the frequency v of the wave is equal to the number of such vibrations in unit time.

e. The *amplitude* of the wave, or maximum displacement of the particles, depends upon the intensity of the disturbance which the particles are propagating.

189. Change of Phase along a Wave. Equations of a Simple Elastic Wave. We will now consider these characteristics more analytically in connection with the transmission of transverse waves. Thus, if all the particles in the line XX', Fig. 206 (a), are connected elastically,

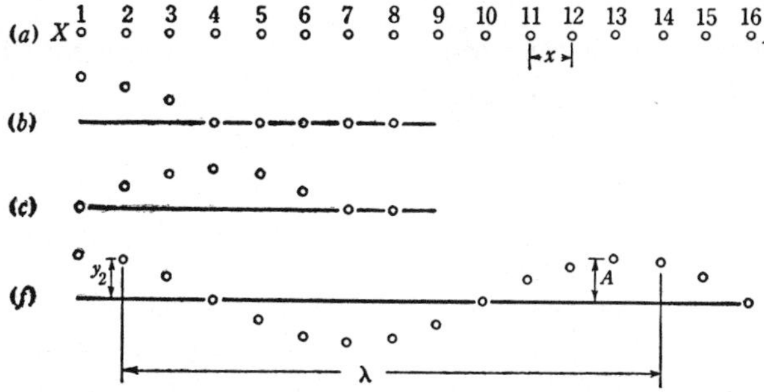

FIG. 206. Displacements of the particles in a transverse wave

and if particle 1 is given a displacement in a direction perpendicular to XX', then this displacement will be communicated successively to particles 2, 3, 4, etc. Further, if particle 1 is caused to vibrate with simple vibratory motion across XX', then all the particles 2, 3, 4, etc. will in succession take up this simple vibratory motion across XX'; in other words, a transverse wave will travel along XX'. Now, it was shown in Sec. 175 that the displacement s of a particle which is executing simple vibratory motion is given by the equation

$$s = A \sin\left(\sqrt{\frac{k}{m}}\, t + C_2\right) = A \sin\left(\frac{2\,\pi}{T}\, t + C_2\right),$$

in which A is the amplitude of the vibration, $(2\,\pi t/T) + C_2$ is the phase, and the constant C_2 is the so-called *phase constant*, or phase at the instant $t = 0$. For our present purposes we shall find it most convenient to agree to reckon times t from some instant when the particle is at the *center* of its path, so that $s = 0$ when $t = 0$. This makes the phase constant C_2 equal to zero, and the foregoing equation becomes

$$s = A \sin \frac{2\,\pi}{T}\, t. \qquad [293]$$

Thus the vertical displacement of any particle such as 2 in Fig. 206 (f) at any time t after it has left its position of equilibrium is

$$y_2 = A \sin \frac{2\,\pi}{T}\, t.$$

The displacement at this instant of another particle such as 3 in Fig. 206 (f) is, of course, different. Thus, suppose that this particle leaves its equilibrium position t' units of time after particle 2 has left its equilibrium position. The displacement of 3 at the instant considered is then obviously

$$y_3 = A \sin 2\pi \frac{t - t'}{T}.$$

And similarly the displacement of any particle is represented by an expression of the form $y = A \sin 2\pi(t - t')/T$, or

$$y = A \sin 2\pi \left(\frac{t}{T} - \frac{t'}{T} \right),$$ [294]

where t'/T is called the *phase difference* between the particle under consideration and the reference particle.

It is evident that the particle for which t' is equal to T will be in a state of motion similar to that of the reference particle 2 and therefore will be the one whose distance from the reference point is a wavelength λ. While the disturbance is traveling forward this distance λ [for example, in Fig. 206 (*f*) from 2 to 14], the particle 2 makes one complete vibration; that is, the wave travels a distance λ in time T, or has a speed v which is given by

$$v = \frac{\lambda}{T}.$$ [295]

Finally, let us denote by x the distance of any particle under consideration (for example, 3 in Fig. 206) from the reference particle. The phase difference t'/T in Eq. [294] may then be expressed in terms of x and the wavelength λ by means of the evident relation

$$\frac{t'}{T} = \frac{x}{\lambda}.$$ [296]

Substitution of this expression for the phase difference in Eq. [294] gives

$$y = A \sin 2\pi \left(\frac{t}{T} - \frac{x}{\lambda} \right).$$ [297]

This is the equation of a simple elastic wave of amplitude A, period T, and wavelength λ, which is traveling in the positive direction along the X-axis. An instantaneous "picture" of the wave at any given time t_0 can be obtained by putting t equal to t_0 in this equation and plotting y as a function of x; the result obviously will be a sine curve that repeats itself in intervals of time T and in intervals of space λ [Fig. 206 (*f*)]. If, on the other hand, we select some particular value of x_0 and watch the motion at this point as time progresses, substitution of x_0 for x in Eq. [297] tells us that the motion will be simple vibratory, with initial phase $C_2 = -2\pi x_0/\lambda$.

EXAMPLE. Show that the equation for a simple wave may be expressed in the alternative form

$$y = A \sin \frac{2\,\pi}{T}\left(t - \frac{x}{v}\right),$$ [298]

where v is the speed of the wave. As convenience suggests, either this expression or Eq. [297] may be used to express the displacement.

Although these equations have been derived explicitly for a transverse wave, they hold equally well for a longitudinal wave, except that in the latter case the displacement is parallel to the X-axis. Hence the confusion resulting in the diagram for compressional waves (Fig. 205) from the fact that the displacements are parallel to the direction of wave motion may be obviated by plotting those displacements vertically. With that convention in mind, Fig. 206 (b), (c), (f) may be taken to represent the progression either of a longitudinal or of a transverse wave.

190. The General Equation of a Wave. Eqs. [297] and [298] are usually referred to as *integrated forms of the wave equation* because, as we will now show, they are merely one solution of a much more general and very powerful differential equation, called the *wave equation*. This equation can be deduced by differentiating both members of Eq. [298] twice with respect to the time t and also twice with respect to the distance x, and then combining the two resulting equations. In performing these differentiations it must be kept in mind that y in Eq. [298] is a function of the two independent variables t and x, and hence that a differentiation of y with respect to t, say, amounts to allowing t alone to vary while x is temporarily held constant. In such a case the derivative is called the *partial derivative of y with respect to t*, and is denoted by the symbol [1] $\partial y/\partial t$.

Thus, taking the partial derivative of y with respect to t in Eq. [298], we have

$$\frac{\partial y}{\partial t} = A\,\frac{2\,\pi}{T}\cos \frac{2\,\pi}{T}\left(t - \frac{x}{v}\right).$$ [299]

Differentiating a second time with respect to t gives

$$\frac{\partial^2 y}{\partial t^2} = -\,A\,\frac{4\,\pi^2}{T^2}\sin \frac{2\,\pi}{T}\left(t - \frac{x}{v}\right),$$

or, in view of Eq. [298], $$\frac{\partial^2 y}{\partial t^2} = -\,\frac{4\,\pi^2}{T^2}\,y.$$ [300]

---

[1] Given $y = F(x, t)$, then by definition

$$\frac{\partial y}{\partial t} = \lim_{\Delta t \to 0} \frac{F(x,\, t + \Delta t) - F(x,\, t)}{\Delta t};$$

similarly, $$\frac{\partial y}{\partial x} = \lim_{\Delta x \to 0} \frac{F(x + \Delta x,\, t) - F(x,\, t)}{\Delta x}.$$

In a similar manner, partial differentiation of Eq. [298] twice with respect to x gives

$$\frac{\partial^2 y}{\partial x^2} = -\frac{1}{v^2}\frac{4\pi^2}{T^2}y. \qquad [301]$$

By eliminating the quantity $(-4\pi^2 y/T^2)$ from Eqs. [300] and [301], we have finally

$$\frac{\partial^2 y}{\partial t^2} = v^2 \frac{\partial^2 y}{\partial x^2}, \qquad [302]$$

in which v is the speed of propagation of the wave. This is the differential equation of a transverse wave which is traveling along the X-axis and in which the particles are vibrating in the y-direction. From either Fig. 207 or Eq. [290] we know that $v = \sqrt{f/\sigma}$.

It will be noted that the wave equation in differential form does not contain the sine function which appears in Eq. [298], the equation from which it was deduced. In other words, the equation in differential form applies not only to sinusoidal waves but also to waves in which the vibrations of the particles are other than simple vibratory in nature. Moreover, although we have deduced the equation for the case of a certain transverse wave, the same line of reasoning might be applied to a longitudinal wave by replacing the transverse displacement of the string, here designated by y, by either the longitudinal displacement or the pressure variation. Thus we might write

$$\frac{\partial^2 P}{\partial t^2} = v^2 \frac{\partial^2 P}{\partial x^2}, \qquad [303]$$

where P represents the variation of the pressure from the normal pressure. From these considerations it is easy to see why the differential form of the wave equation is so powerful and why it is one of the most frequently invoked equations in all of theoretical physics.

EXAMPLE. Compute the total kinetic and potential energy per unit length of a stretched string along which a transverse wave is traveling.

Solution. The transverse speed of any particle of the string is given by Eq. [299], and hence the kinetic energy per unit length is

$$E = \frac{1}{2}\sigma\left(\frac{\partial y}{\partial t}\right)^2 = \frac{1}{2}\sigma A^2 \frac{4\pi^2}{T^2}\cos^2\frac{2\pi}{T}\left(t - \frac{x}{v}\right). \qquad [304]$$

The restoring force per unit length is, in view of Fig. 207, $\sigma \cdot \partial^2 y/\partial t^2$, and this may be written (Eq. [300]) as $-\sigma \cdot 4\pi^2 y/T^2$; hence the potential energy per unit length is

$$V = -\int_0^y \text{force} \cdot \text{distance} = \int_0^y \sigma\frac{4\pi^2}{T^2}y \cdot dy = \sigma\frac{4\pi^2}{T^2}\frac{y^2}{2},$$

or

$$V = \frac{1}{2}\sigma\frac{4\pi^2}{T^2}A^2\sin^2\frac{2\pi}{T}\left(t - \frac{x}{v}\right). \qquad [305]$$

The total energy per unit length is therefore

$$E + V = \frac{1}{2}\sigma A^2 \frac{4\pi^2}{T^2}\left[\cos^2\frac{2\pi}{T}\left(t - \frac{x}{v}\right) + \sin^2\frac{2\pi}{T}\left(t - \frac{x}{v}\right)\right]$$

$$= \frac{1}{2}\sigma A^2 \frac{4\pi^2}{T^2}. \qquad [306]$$

The total energy of a string of given density and period, and having a given amplitude, is therefore constant. The energy is seen to vary with the square of the amplitude, the latter being determined by the vigor with which the string is plucked. Note that the *average* kinetic and potential energies are the same.[1]

191. Pitch and Loudness of Sound. In the preceding sections we have seen that any wave motion possesses, besides its speed of propagation, the characteristics of frequency, amplitude, and wave-form. Instead of the amplitude, it is often more useful to speak of the *intensity* of the wave, this being measured at any point by *the amount of energy that passes in unit time through unit area taken normally to the direction of propagation at the point.* In the case of sound, intensities ordinarily

Fig. 207. Method of deducing the speed of a transverse wave in a stretched string with the aid of the calculus. An element Δx of the string has a vertical acceleration $\partial^2 y/\partial t^2$ because there is a difference between the y-components of the two stretching forces f acting tangentially at the two ends of the elements; this difference is $f \sin \alpha_2 - f \sin \alpha_1 = f(\sin \alpha_2 - \sin \alpha_1) = f[\partial (\sin \alpha)/\partial x] \Delta x$. For small displacements, $\sin \alpha = \tan \alpha = \partial y/\partial x$ (see Appendix 3), and therefore the difference may be written $f[\partial^2 y/\partial x^2] \Delta x$. By Newton's second law this force is equal to the product of the mass and acceleration of the element, namely, $\sigma \cdot \Delta x \cdot \partial^2 y/\partial t^2$, where σ is the linear density. Hence, finally, $\partial^2 y/\partial t^2 = [f/\sigma] \cdot [\partial^2 y/\partial x^2]$. By comparing this equation with Eq. [302], we have $v = \sqrt{f/\sigma}$

are measured in microwatts per square centimeter of the sound wave striking an ear or other acoustical receiver. One can show with little difficulty (see Sec. 190) that the intensity of any wave is proportional both to the square of its amplitude A and to the square of the frequency ν.

In dealing with sound phenomena one must distinguish carefully between the foregoing characteristics of the wave motion, which are purely physical and objective in nature, and the psychological characteristics of the resulting sound as they are perceived through the human ear. These psychological characteristics are pitch, loudness, and quality (Sec. 206). The *pitch* of a note depends chiefly on the frequency of the wave striking the ear,[2] although recent studies [3] have shown that our judgments of pitch are also affected somewhat

---

[1] Compare with Prob. 12, Chap. 14.

[2] Obviously, it is incorrect to say that it depends chiefly on the wavelength; as can be seen, for example, from the fact that a tuning fork vibrating with a given frequency produces a note of the same pitch regardless of whether it is in a cold room or a warm one.

[3] For a brief summary of this work, see H. Fletcher, *Bell Laboratories Record* **13**, 130 (1935).

by the intensity and form of the incoming wave. For example, the note from a tuning fork of frequency 262 vib · sec^{-1}, which corresponds to middle C on the international musical scale, will decrease markedly in pitch when the loudness of the note is doubled, even though the frequency of the fork has remained unchanged.

The *loudness* of a sound is closely related to the intensity of the wave entering the ear; but experiment shows that it increases not as the intensity increases, but roughly as the logarithm of the intensity: on the average, any given intensity must be increased by about 26 percent of itself before the ear will record a difference in loudness. Moreover, just as pitch is not dependent on frequency alone, so is loudness dependent not only on intensity but also on the pitch and wave-form. For example, thousands of times as much energy is required to produce an audible sound of 30 vib · sec^{-1} as would be required at 2000 vib · sec^{-1}. In the range 800–1800 vib · sec^{-1} the effect of the frequency is not large, however, and here the loudness and intensity are related in a relatively simple way. There is at the present time no theory of hearing that is adequate to explain these and all the other complicated auditory phenomena which have been revealed by modern experimentation. Suffice it to say, the simple picture of the action of the human ear given in many books on elementary science is far from being adequate.

o

Interference Phenomena

192. Superposition of Two or More Simple Waves. Interference. Our treatment so far has been concerned with simple waves in which each particle of the transmitting medium is acted upon by an elastic restoring force which varies directly with the displacement. This is the kind of wave that is set up in a medium by a vibrating tuning fork, or by any other source which vibrates with a single frequency. But we shall see eventually that most bodies can vibrate in many different ways at the same time. For example, a trained ear can detect several tones of different frequencies in the note from a large bell or from a violin string. Such vibrating bodies consequently set up in the surrounding medium a *complex wave* consisting of simple waves of different frequencies and amplitudes which traverse the same portion of the medium simultaneously.

As a result, each particle of the medium is subjected to as many forces as there are component simple waves, and since these forces are wholly independent of one another, each force will produce its own simple vibratory motion. The problem of describing a complex wave therefore reduces to one of learning how to add the several component displacements that make up the complex displacement.

The simplest case is that of two simple sinusoidal waves of the same period and phase but unequal amplitudes A' and A'', which are traveling together along the same line. The displacements y' and y'' in the two component simple waves are, by Eq. [297],

$$y' = A' \sin 2\,\pi \left(\frac{t}{T} - \frac{x}{\lambda}\right),$$

and

$$y'' = A'' \sin 2\,\pi\left(\frac{t}{T} - \frac{x}{\lambda}\right);$$

hence the resultant displacement y in the complex wave is

$$y = y' + y'' = (A' + A'') \sin 2\,\pi \left(\frac{t}{T} - \frac{x}{\lambda}\right). \qquad [307]$$

In other words, the resultant wave is a sinusoidal wave of the same period and phase, but its amplitude is the sum of the amplitudes of the component waves (Fig. 208).

Any modification of amplitude due to the superposition of waves is called *interference*.

If the component waves differ in phase or in period, the addition is most easily accomplished graphically; this is done simply by

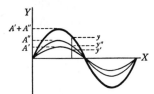

Fig. 208. Illustrating the addition of two simple waves which have the same periods and phases but different amplitudes and which are traveling in the same direction

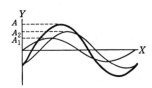

Fig. 209. Addition of two simple waves which have the same period but different phases and amplitudes and which are traveling in the same direction

adding the ordinates of the sine curves which represent the component simple waves.

If the components differ in phase and in amplitude but not in period, as shown in Fig. 209, the resultant wave is a sinusoidal wave with the same period as the components but an intermediate phase.

If the periods of the component waves are different, a wave of the type shown in Fig. 210 is obtained; the resultant wave is not sinusoidal in this case.

This whole problem of the addition of simple waves to form complex ones was generalized by FOURIER [1] in 1822, in a very important mathematical principle known as *Fourier's theorem*. FOURIER found that any periodic disturbance or wave-form of permanent type can

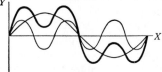

FIG. 210. The resultant of two sinusoidal waves having different periods is not a sinusoidal wave

be represented as a summation of simple vibratory terms of the type

$$y = A_1 \sin (\theta + C_1) + A_2 \sin (2\,\theta + C_2) + A_3 \sin (3\,\theta + C_3) + \cdots, \quad [308]$$

where the $A's$ and the $C's$ are the amplitudes and phase differences, respectively, of the component waves. By means of this theorem a complex wave may be analyzed into its components, and ingenious methods have been devised for doing this mechanically. Such FOURIER analyses have proved to be of great practical value [2] in the design of musical instruments, radios, and alternating-current dynamos.

● **193. Beats.** Fig. 211 represents a case of interference that has particular interest in the case of sound waves. Since the two component wave-trains here do not have quite the same period, their phase difference, and hence the amplitude of their sum, undergoes periodic variations at any given point along the wave. Thus, suppose that the waves are compressional waves coming from two tuning forks of slightly different frequencies ν_1 and ν_2 and suppose that the ear is placed at some

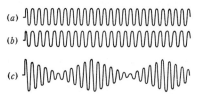

FIG. 211. (*a*) and (*b*) represent two simple waves of slightly different period, and (*c*) is the resultant complex wave formed when they traverse the same portion of a medium simultaneously

point along the wave. Obviously, if a compression due to one source reaches the ear at the same time as a rarefaction from the other

[1] See Sec. **173**.

[2] See *D. C. Miller, *The Science of Musical Sounds* (Macmillan, 1916); also *H. Fletcher, *Speech and Hearing* (Van Nostrand, 1929), which describes the extensive work done in this field in the Bell Telephone Laboratories. W. E. Byerly, *An Elementary Treatise on Fourier's Series* (Ginn, 1893), pp. 63–64, is also recommended.

source, there will be *destructive* interference. But since one source is vibrating slightly more rapidly than the other, an instant later two compressions or two rarefactions will be in coincidence at the ear, and there results then a reinforcement or *constructive* interference. These alternations in the intensity of the sound when two tones interfere in this way are called *beats*. The interval between two successive beats is simply the time needed, at a given point, for one of the wave-trains to gain 2π radians in phase over the other train. Hence the number of beats occurring in unit time will be equal to $\nu_1 - \nu_2$, the difference in frequency of the two sources.

As was pointed out by HELMHOLTZ,[1] the phenomenon of beats is the physical basis of dissonance. So long as the number of beats produced by sounding two musical notes together is not more than five or six per second, the effect is not particularly unpleasant. From this point on, however, the beats begin to become indistinguishable as separate beats and pass over into a discord. HELMHOLTZ showed that the unpleasantness becomes worst at a difference of frequency of about 30 vib · sec⁻¹. When the difference is as much as 60 vib · sec⁻¹, the effect is again harmonious. The fact that the number of beats heard per unit time equals the difference of the frequencies of the component vibrations provides a ready method of measuring small differences of frequency, and is the principal means employed in tuning such instruments as the piano, organ, and violin. The superheterodyne radio receiving set makes use of beats produced by electric oscillations of different frequencies.

194. Stationary Waves. The phenomenon known as stationary waves is the result of the action on the particles of a medium of two waves of the same period and amplitude that are traveling in *opposite* directions. Fig. 212 (*a*) shows two such oppositely moving trains *A* and *B* at a particular instant when the crests of *A* are opposite to the troughs of *B*, and vice versa. The heavy horizontal line shows the resultant displacement of the particles at this instant. Obviously each one of the particles transmitting the motion is under the action of two disturbances that tend to produce equal and opposite displacements, and as a result the particles suffer no displacement at all. In Fig. 212 (*b*) is shown the situation at a later moment, when each of the waves has progressed an eighth of a wavelength; that is, when the waves have become displaced a quarter of a wavelength with respect to each other. Similarly, when one wave has moved a half-wavelength past the other, the resultant is again zero at every point, as in Fig. 212 (*a*).

---

[1] *Die Lehre von den Tonempfindungen* (1862), tr. by A. J. Ellis under the title *On the Sensations of Tone* (Longmans, Green, ed. 4, 1912), Chap. VIII.

At the points marked N, N', N'', distant from one another by a half-wavelength, the resultant displacement is *always* zero. All such points where the interference is such that there is never any motion of the particles are called *nodes*. At points midway between the nodes the two component waves are in exactly the same phase; these

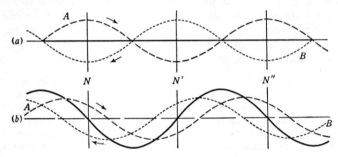

Fig. 212. Showing the resultant of two oppositely directed similar waves at two different moments during their passage

points are called *antinodes* or *loops*, and here there is a maximum of motion. At all points between the nodes the particles are in constant vibration, but all pass through their positions of equilibrium at the same time. There is therefore *no phase difference between successive particles*, and hence the wave is entirely devoid of progressive character. It is for this reason that the wave resulting from the superposition of two oppositely directed similar wave-trains is called a *stationary*, or *standing*, wave. It is important to see the distinction

between a wave that is progressing and one that is stationary. When, for example, a *progressive* transverse wave passes along a stretched string, no portion of the string remains stationary; instead, all the particles move in precisely the same way and to the same extent, but not simultaneously, each succeeding particle being a little later in its movements. If, on the other hand, two simi-

Fig. 213. Stationary transverse wave in a stretched string. The heavy line shows the position of the string at a given instant. When the segment NN' is moving up, that between N' and N'' is moving down, and vice versa

lar but oppositely directed transverse waves are passing along the string, thus forming a stationary wave, the string never moves at the nodes and the amplitude of motion is always a maximum at the antinodes; between any pair of adjacent nodes the particles move up or down together, and on opposite sides of a node the displacements at any instant are in opposite directions (Fig. 213).

Similar considerations hold for a standing wave of the longitudinal type. As will be evident from Fig. 214, the greatest variations of pressure occur in such a wave at the nodes, where there is the least motion, whereas the smallest pressure variations occur at the anti-nodes, where there is the greatest motion. In other words, *our definitions of nodes and anti-nodes refer to displacements.* If we were inter-ested in the distribution of pressure along a standing longitudinal wave, we might make use of the idea of "pressure nodes" and "pressure antinodes." A pressure node would then be defined as a point where the pressure does not change; that is, it is a displacement antinode.

These ideas about standing waves which we have obtained from a study of the diagrams may also be developed from a consideration of Eq. [297]. Let

$$y' = A \sin 2\pi\left(\frac{t}{T} - \frac{x}{\lambda}\right)$$

Fig. 214. Graphical representation of the displacements of the particles in a station-ary longitudinal wave. The particles have just begun to return from their positions of max-imum displacement, from the heavy to the light line. At the node N a condensation is forming, and at N' a rarefaction. After half a period conditions will be reversed. In the neighborhood of any given antinode, however, the particles are moving in the same direction with approx-imately the same speed, so that their relative po-sitions are only slightly changed

represent the equation for the displacement given to the successive particles by the one component wave-train. The oppositely directed component wave-train must be one which at some given instant of time, say $t = 0$, imparts to the same particles displacements that are equal in magnitude but opposite in direction to those imparted by the first component [Fig. 212 (a)]. The equation

$$y'' = A \sin 2\pi\left(\frac{t}{T} + \frac{x}{\lambda}\right) \tag{309}$$

satisfies this condition and therefore represents the equation of the second component wave. The resultant displacement y is therefore

$$y = y' + y'' = A \sin 2\pi\left(\frac{t}{T} - \frac{x}{\lambda}\right) + A \sin 2\pi\left(\frac{t}{T} + \frac{x}{\lambda}\right).$$

By expansion [1] and addition this becomes

$$y = \left(2A \cos 2\pi \frac{x}{\lambda}\right) \sin 2\pi \frac{t}{T}. \tag{310}$$

---

[1] Employing $\sin(\theta \pm \phi) = \sin\theta \cos\phi \pm \cos\theta \sin\phi$.

In order to be able to interpret this equation, consider first the equation for a simple wave, namely

$$y = A \sin 2\pi \left(\frac{t}{T} - \frac{x}{\lambda}\right),$$

and notice that it consists of two parts: the first part is the amplitude A and the second is a sine function of the time t and a phase difference x/λ. Now in the expression for y just obtained (Eq. [310]), $\sin 2\pi t/T$ is the sine function of the time. And evidently, since the phase difference does not appear in this expression, the particles must all be in the same phase of vibration. Similarly, $2A \cos (2\pi x/\lambda)$ represents the amplitude. But since x represents the distance of the particles under consideration from the reference particle, it is evident that the amplitude varies for successive particles. Also, since $\cos (2\pi x/\lambda)$ is zero when x is an odd multiple of $\lambda/4$, it follows that there are nodes, or points of zero amplitude, at points differing successively by half-wavelengths. Furthermore, the algebraic sign of $\cos (2\pi x/\lambda)$ changes at these same points. All these conclusions are in agreement with those previously obtained with the aid of the diagrams.

o

Huygens's Principle and the Phenomena of Reflection, Refraction, and Diffraction of Waves

195. Some Preliminary Ideas about Reflection and Refraction. When a wave arrives at the surface of separation between two different mediums, part of the wave is reflected into the medium in which it originally was traveling (Fig. 215) and part is transmitted into the new medium. The amount of reflection depends on the relative speeds of the wave in the two mediums, none occurring if these speeds are equal.

Thus when sound passes from air to water, less than a thousandth part of the energy is transmitted, nearly all of it being reflected, and the same is true of sounds produced under water when they reach the surface. On the other hand, when sound passes from dry air to air saturated with water vapor, or the reverse, practically all of the energy is transmitted. This is also the case when the difference between the two mediums is one of temperature merely. In an auditorium most of the sound heard is reflected sound, and for this reason reflection is an important factor in architectural acoustics. The phenomenon of reflection is made use of in deep-sea soundings and in so-called echo-prospecting, which includes the various methods of determining the depth of mineral deposits below the surface of the earth by reflection of sound waves from the boundary between the less dense earth and the denser mineral layers.

THIS ILLUSTRATION, which was taken from ATHANASIUS KIRCHER'S *Musurgia Universalis sive Ars Magna Consoni et Dissoni* (Rome, 1650), through the courtesy of The Huntington Library, San Marino, California, accompanies the earliest known account of the celebrated "bell in a vacuum" experiment. A glass bulb, shown at the top of the woodcut, containing a small bell and iron clapper, was cemented to the top of a long lead tube closed at the bottom by a stopcock. A partial vacuum was produced in the glass bulb by filling the whole apparatus with water, then opening the stopcock and draining off some of the water. The iron clapper was manipulated from outside the glass bulb by means of a loadstone or natural magnet. As Professor D. C. Miller remarks in his interesting *Anecdotal History of the Science of Sound*, "Perhaps this use of a magnet to control a device in a vacuum is of more interest than the bell experiment, especially as the latter was then interpreted as showing that the air is not necessary to the transmission of sound."

The *Musurgia Universalis*, a book of nearly 1200 pages, copiously illustrated with fine copperplates, woodcuts, and music, is "an encyclopaedia of the knowledge of sound of the 17th century." A digest, in English, of this elaborate and entertaining work will be found in J. Hawkins's *A General History of the Science of Sound and Practice of Music* (London, 1853, 1875), pp. 635–644. See also D. C. Miller's *Anecdotal History of the Science of Sound* (Macmillan, 1935), pp. 14–18.

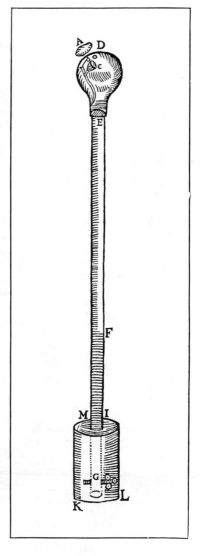

The First "Bell in a Vacuum"
Experiment

One of the Many Fine Copperplates
in A. KIRCHER's *Musurgia Universalis* (Rome, 1650).
It Illustrates Multiple Echoes

When any wave passes from one medium into another in which its speed is different, its line of propagation changes. Such a change in the direction of the transmitted portion of a wave is called *refraction*.

Thus if different layers of the atmosphere are at different temperatures, a sound wave passing through them will tend to travel in a curved rather than a straight path. When sound waves in one gas are made to pass through a lens-shaped bag containing another gas of different density, the results are similar to those observed when light is passed through a lens.

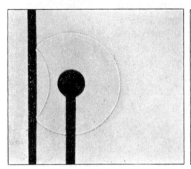 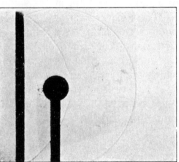

Fig. 215. Showing two stages of a reflected sound pulse, as revealed by spark photography. (Kindness of Professor A. L. Foley)

Refraction experiments of this kind have also been made with acoustic lenses made of pitch, with rubber vessels containing water, and, in one case, with a biconvex lens which consisted of a large balloon containing carbon dioxide.

196. Diffraction. Besides refraction, there is only one other case where waves do not travel in straight lines. This is when they pass by an obstacle. Waves tend to curl around obstacles; that is, they do not form sharp shadows. Thus sound waves can be heard around corners and water waves entering a harbor can be seen to spread into the region behind a breakwater. Any change of this kind in the direction of propagation of waves, not caused by a variation in the properties of the medium but due to the bending of the waves about obstacles, is called *diffraction*. The amount of diffraction in any case depends on the dimensions of the obstacle compared with the wavelength of the waves. A large obstacle, such as a hill, casts a fairly sharp sound shadow, whereas a small object, such as a tree, will not cast an observable sound shadow unless the frequency of the sound is very high.

197. Rays and Wave-Fronts. In our treatment of wave motion up to this point we have found it most convenient to confine our atten-

tion to the lines along which disturbances are traveling. Such a line
which marks the direction of propagation of a wave-train may be
called a *ray* of wave-energy.

Consider, now, S in Fig. 216 to be the point source of a wave
motion in a homogeneous and isotropic medium, that is, a medium
in which the disturbance is propagated with equal speed in all directions. When a disturbance originating at S has just reached a, it
has also then just reached all other points, such
as b and c, which are at the same distance from
S. The spherical surface passing through these
points is called the wave-front of the disturbance.
In general, the *wave-front* may be defined as
the surface passing through adjacent particles
which are in the same phase of vibration; that
is, in a wave-front the particles reach their maximum positive or negative displacements at the
same time.

Fig. 216. A spherical
wave-front

The form of the wave-front under the conditions just mentioned
is *spherical*, but it will be shown later that conditions may arise in
which it has not this form. In Fig. 199 the photographed sound pulse
is spherical; this is because it originated at a point source (the spark)
in an isotropic medium. In Fig. 215 the sound pulse reflected from
the plane surface also is seen to be spherical. Moreover, like the
wave incident on the surface, it is *diverging*, that is, convex toward
the direction in which it is traveling. Later we shall see that under
proper conditions a spherical wave may be *converging*, that is, concave toward the direction in which it is traveling. If the source of a
disturbance is a great distance away, any small portion of the spherical
wave will be sensibly plane. A wave having a plane wave-front is
called a *plane* wave.

In any isotropic medium the rays, or directions of progression of
the wave, are always at right angles to the wave-front. Thus in
Figs. 199 and 216 the rays emanate radially from the source and
are normal to the spherical wave-front. In the case of a plane
wave, the rays are evidently parallel. It will be found that a diagram sometimes conveys clearer ideas when it represents rays, and
at other times it is clearer when it represents wave-fronts. Which
method we shall use in any given case is merely a matter of clearness
and convenience.

198. Huygens's Principle. A means of locating the wave-front at
any time, if its position at some previous time and its speed of propagation are known, is afforded by an important theorem that was
enunciated by HUYGENS in connection with certain problems in the

theory of light. In his *Traité de la Lumière* [1] (1690), HUYGENS points
out that "In considering the propagation of waves, we must remember that each particle of the medium through which the wave
spreads does not communicate its motion only to that neighbor which
lies in the same straight line drawn from the source of the disturbance
but shares it also with all the particles which
touch it and resist its motion. Each particle is
thus to be considered as the center of a wave."
In other words, any particle in the wave-front
of a disturbance may be considered as a point
source from which is spreading out a spherical
wave. Thus let *WW*, Fig. 217, be the instantaneous position of a wave-front which has come
from a disturbance at *S*. In order to find where
this wave-front will be after an interval of time
Δt, draw spheres of radii $v\Delta t$, where v is the
speed of the wave, about each point of the wave-
front *WW*. The envelope of these spherical
secondary waves (namely, $W'W'$) is the required new wave-front. [2]

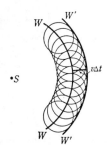

FIG. 217. HUYGENS'S
principle

It would appear from HUYGENS'S construction that a disturbance
should be propagated back to its origin as well as forward. In 1826
FRESNEL [3] pointed out that this did not happen because of interference effects, and later KIRCHHOFF [4] showed by a mathematical
analysis that the secondary waves from the individual sources really
do destroy one another by mutual interference except at the surface
$W'W'$. The wave is therefore propagated only in the direction away
from the origin.

Fig. 218 shows how HUYGENS'S construction explains the phenomenon of diffraction, or the spreading of waves into the region behind an
obstacle. Diffraction is thus seen to be a universal property of waves
which must be exhibited by waves of any type in any medium.

HUYGENS'S principle may also be employed to predict the direction and curvature of a wave reflected from a plane surface, such as

---

[1] There is a translation by S. P. Thompson (Macmillan, 1912). See also *The Wave
Theory of Light*, ed. by H. Crew (American Book Co., 1900) and **A Source Book in
Physics* (1935), pp. 283–294.

[2] A more comprehensive and advanced treatment of HUYGENS'S principle will be
found in P. Drude, *The Theory of Optics*, tr. by C. R. Mann and R. A. Millikan
(Longmans, Green, 1902), Part II, Chap. 3.

[3] *Mémoire Couronné* [*Mémoires de l'Académie des Sciences* (1826), Vol. 5]. A portion of this memoir appears in **A Source Book in Physics* (1935), pp. 323–324.

[4] Wiedemann's *Annalen der Physik* **18**, 663 (1883); *Vorlesungen über mathe-
matische Optik* (Leipzig, 1891).

the reflection photographed in Fig. 215. In Fig. 219, *WWW* is the position the advancing wave from *S* would have reached at a certain instant if it had not encountered the plane surface. Actually, however, when it reached *O*, which is the nearest point on the reflecting

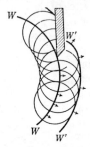

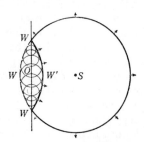

FIG. 218. How HUYGENS's construction explains diffraction. The wavefront curls in behind the obstacle

FIG. 219. Method of applying HUYGENS's principle to trace the path of a wave which strikes a smooth plane surface

surface to *S*, *O* became, according to HUYGENS's principle, a center of a secondary wave. At immediately succeeding intervals the successive points on the reflecting surface were also reached by the incident wave-front and in turn became sources of secondary waves. These secondary waves emitted by each successive point on the reflecting surface have for their envelope the spherical surface *W W' W*. It will be noticed that the reflected wave appears to come from a point behind the reflecting surface. Moreover, in the present case of a plane reflector, this point is as far behind the reflector as the source is in front of it.

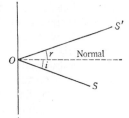

If, in Fig. 219, we center our thoughts on the rays rather than the wave-fronts, it will be evident from the construction that at a given point on the reflecting surface the incident ray and the reflected ray make the same angle with the normal to the reflector

FIG. 220. At any point on any reflector the angle of incidence *i* is equal to the angle of reflection *r*

at that point (Fig. 220); in other words, at each point of a reflecting surface the *angle of incidence* is equal to the *angle of reflection*.

This law of reflection — namely, that the angle of incidence is equal to the angle of reflection — may also be derived from an important principle first enunciated by PIERRE de FERMAT [1] and called *Fermat's principle of*

---

[1] *Œuvres* (Gauthier-Villars, 1891–1912). The paper on the *refraction* of light appears in *A Source Book in Physics* (1935), pp. 278–280.

least time. As applied to reflection, FERMAT'S principle is as follows. Consider that the ray of wave-energy is traveling from a source S to a point S' over the path SOS', Fig. 220. Evidently SO/v is the time of travel before reflection and OS'/v is the time of travel after reflection. Then, according to FERMAT, the path of the wave is such that *the total time* $(SO/v + OS'/v)$ *is a minimum.* A simple geometrical proof will show that this will be true if the angles of incidence and reflection are equal.

> EXAMPLE. Given that a source of disturbance O is at an infinite distance from a smooth plane reflector and in a direction which is not normal to the reflector, prove by drawing a diagram and applying HUYGENS'S principle that the reflected wave will be a plane wave, and that for any ray the angle of incidence is equal to the angle of reflection. What if the surface is not smooth?

199. Change of Phase on Reflection. In order to get a more intimate physical picture of just what happens at a boundary where reflection is occurring, let us consider the particular case of a compressional wave reflected from the end of the rigid tube described in Sec. **183.** If a compressional pulse is sent down the tube, it progresses in a manner which is strictly analogous to that of the direct impact of perfectly elastic balls of equal mass (Sec. **65**). So long as the pulse travels in a medium of uniform density, each layer of particles gives up all its motion to the next layer, which is precisely like it, just as a moving elastic ball striking a stationary elastic ball of the same mass gives up all its motion to the stationary ball and itself comes to rest. But if the pulse at some point O, Fig. 221, strikes a *denser*

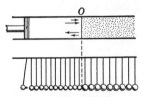

medium than the one in which it is already traveling, the case becomes analogous to the impact of one ball upon another that is more massive; the less dense medium adjoining O on the left, instead of coming to rest in the impact, reverses its motion and starts back. This backward motion is communicated from layer to layer so that a *reflected* compressional pulse travels from O back toward the

FIG. 221. Reflection of a longitudinal pulse from a denser medium, and its analogue

source of the disturbance. Let the long arrows in Fig. 221 represent the directions of propagation of the pulse before and after reflection and the short ones the respective directions of motions of the particles in the pulse. It is evident that the direction of motion of the particles with respect to the direction of propagation of the pulse is the same in the reflected as in the incident pulse. In other words, a pulse of condensation is reflected from a *denser* medium as a pulse of condensation. Another way of stating this is to say that *the dis-*

placements of the particles undergo a change in phase of π radians upon reflection from a denser medium.

But suppose the medium to the right of the interface 0 is *less dense* than the one in which the pulse is initially traveling (Fig. 222); then the case is analogous to the impact of a ball of large mass with one of small mass. Until the pulse reaches 0 each layer in the tube gives up its motion to the next and itself comes to rest. But the layer at 0, instead of coming to rest after the impact, continues to move forward and thus produces a rarefaction; that is, a diminution of pressure at 0. The excess of pressure in the layer to the left of 0 then drives particles toward the right. Thus

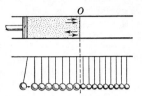

Fig. 222. Reflection of a longitudinal pulse from a less dense medium, and its analogue

a pulse of rarefaction moves back from the boundary 0. In other words, a pulse of condensation is reflected from a *rarer* medium as a pulse of rarefaction. In the pulse approaching 0 the particles move in the direction of propagation of the pulse, whereas in the reflected pulse they move in a direction opposite to that of the propagation of the pulse (Fig. 222). *The displacements of the particles therefore do not undergo any change of phase upon reflection from a rarer medium.*

The same process of reasoning shows that a rarefaction is reflected from a denser medium as a rarefaction, but from a less dense medium as a condensation.

Similar considerations also apply to transverse waves. When a transverse pulse is sent along a string that is either fastened at one end (Fig. 223) or attached to a heavier string, a crest is reflected from the boundary as a crest; just as in the case of a compressional wave reflected from a denser medium, the displacements of the

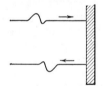

Fig. 223. Reflection of a transverse pulse at the fixed end of a string

particles are reversed in phase by the reflection. If the end of the string is free or is attached to a lighter string, a crest sent along the string is reflected as a trough without change of phase.

○

Sound Phenomena Due to Reflection

200. Resonance of Vibrating Air Columns. Partial Tones. When an air wave traveling in a pipe reaches the open end, it experiences the same sort of reflection as though it passed from a denser to a rarer medium. This statement can easily be proved experimentally. It is also evident from the theoretical consideration that as soon as the wave reaches a point at which *lateral expansion* is possible, the forward movement of the particles is greater than inside the pipe, where lateral expansion is not possible. This increased forward movement at the open end of the pipe means a rarefaction starting back in the pipe.

Consider now a train of waves of wavelength λ approaching the open end of a pipe the other end of which is closed (Fig. 224). In the condensations of the advancing wave-train the motions of the particles are all in the direction of propagation of the wave, whereas

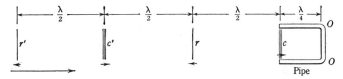

Fig. 224. Wave-train entering a pipe. The first compression has just reached the mouth of the pipe

in the rarefactions they are in the opposite direction. Suppose the pipe to have a length that is exactly one fourth the wavelength of the incident waves. Then the condensation marked c will move into the pipe and be reflected at the closed end as a condensation, that is, as a motion of the particles now from right to left. It obviously will return to the open end at the instant at which the rarefaction marked r, which also consists of a motion of the particles from right to left, reaches the open end of the pipe. Since the reflected condensation which has returned from the closed end now undergoes reflection at the open end as a rarefaction, with a consequent motion of the particles from right to left, it unites as it starts back along the pipe with a rarefaction r which is just entering the pipe, and a wave of rarefaction of increased amplitude is the result (Sec. 192). This wave is reflected at the closed end as a rarefaction (motion from left to right) and again at the open end as a condensation (motion from left to right), exactly in time to unite with the condensation marked c' as it enters the pipe. Thus by this process of continuous union of direct and reflected waves the amplitude of the vibration in the pipe becomes larger and larger until it may be hundreds of times as large as in the original wave. In fact, there would be no limit to the amplitude of the waves traveling up and down the pipe if at each end energy were not partially transmitted to the outside air. The pipe thus becomes, in a way, the source of sound. Any phenomenon of

this kind in which a body is set into vigorous vibration by an agitation having a period that corresponds to one of the natural periods
of the body is called *resonance.*

If the pipe had been only a trifle longer or shorter than λ/4, the
error in the coincidence of the first reflected condensation c with
the incoming rarefaction r would have been slight. Since, however, the
reflected waves now must travel, each time they
traverse the length of the pipe, a distance a little
too great or too small, they soon get completely
out of step with the advancing waves. In this
condition the direct and reflected waves tend to
destroy rather than reinforce one another, and no
resonance is possible. This explains why, when
the length of the pipe is even just a trifle more or
less than the right amount, very little resonance
occurs. If, however, the pipe is continually lengthened, other resonant lengths are obtained. The
length $L = λ/4$ is one that permits c to return to
the mouth of the pipe exactly in time to unite
with r. This length is manifestly the shortest possible resonant length. It is clear that the next
possible resonant length is one which permits c
to return exactly in time to unite with r'. Since r'
is a distance λ behind r, the second resonant pipe
length must be one-half wavelength greater than
the first; that is, $L = 3(λ/4)$. Similarly, it is possible to obtain resonance for $L = 5(λ/4)$, $L = 7(λ/4)$,
and so on; that is, when the length L of the pipe
is an odd multiple of one fourth of the wavelength
λ of the train of waves entering the pipe.[1] Such a
pipe of variable length may be employed, for example, to determine experimentally the speed of
compressional waves in a gas (Fig. 225).

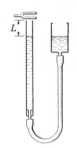

Fig. 225. Method
of determining the
speed v of sound
in air by means of
a closed pipe. The
wavelength of the
incoming wave is
determined by ad-
justing the level of
the water until the
note of the tuning
fork is strongly re-
inforced. Then, by
Eq. [283], $v = νλ$,
where $ν$ is the
known frequency of
the fork. (See Exp.
XVA, Part III)

If, on the other hand, the pipe length is kept constant and the
wavelength of the incoming waves is varied, it follows at once that
a succession of wavelengths $λ_1$, $λ_2$, $λ_3$, and so on, will be found to
which the pipe will respond, and that $λ_1 = 4 L$, $λ_2 = 4 L/3$, etc.

---

[1] Experiment shows that the theoretical value for the shortest resonant length,
namely λ/4, is slightly too large. The discrepancy is explained by the fact that a
condensation returning to the open end does not reach full freedom of lateral expansion until it has passed a short distance beyond the mouth of the tube. RAYLEIGH
estimated that for a tube of radius R this distance is not far from $0.82 R$ [*The Theory
of Sound*, ed. 2 (Macmillan, 1894), Vol. II, pp. 183, 487].

These wavelengths bear the ratios $1 : \frac{1}{3} : \frac{1}{5}$, etc. Hence the frequencies of incoming waves which are able to produce resonance in a closed pipe must bear the ratios $1 : 3 : 5 : 7$, and so on. The tone of lowest frequency to which a given pipe can respond is called the *fundamental* or *first partial* tone of the pipe; the tones of higher frequency are called its *overtones* or *upper partials*. When the frequencies of the overtones, as in this case, are integral multiples of the fundamental frequency, the overtones are also called *harmonics* of the fundamental tone.

> EXAMPLE. By employing an analysis similar to that used for the closed pipe, show that the phenomenon of resonance can also be obtained with a pipe which is open at both ends and that the shortest resonant length must be one half the wavelength of the incoming train. Also show that such an open pipe should produce resonance when its length is any multiple whatever of $\lambda/2$.

The fact that a length of an open pipe can indeed always be found which will respond just as loudly to a given note as any closed pipe, and that this length is twice as great as that of the shortest resonant closed pipe, may be taken as a complete experimental demonstration of the statement made at the beginning of this section as to the nature of the reflection occurring when a wave reaches the open end of a pipe. It is left as an exercise for the student to show that the tones which will produce resonance in a given open pipe of fixed length must bear the frequency ratios $1 : 2 : 3 : 4$, etc. We may say, then, in summary, that in closed pipes only the odd overtones are possible; in open pipes all the overtones, even and odd, are possible. In either case, the overtones are harmonics of the fundamental tone.

201. Natural Periods of Pipes and the Production of Tones by Air Jets. Not only will a given pipe, open or closed, intensify trains of waves of certain definite wavelengths which present themselves at its mouth, but a single pulse entering such a pipe must be returned, by virtue of successive reflections at the ends, as a succession of pulses following one another at equal intervals. In other words, a single pulse must be given back by the pipe as a musical note, of very rapidly diminishing intensity, it is true, but of perfectly definite wavelength. Furthermore, this wavelength must be the wavelength of the train that is capable of producing the fundamental resonance of the pipe. For example, if the pipe is a closed one, then the first time the pulse returns to the mouth after reflection at the closed end it will produce an outward motion of the particles near the mouth, the next time an inward motion, and so on; that is, the pulse must travel four times the length of the pipe in the interval between the appearance of two successive condensations at the mouth. The length of the pipe is thus one fourth of the wavelength of the note given off by it, and this is the relation existing in the case of a train of waves that produces the fundamental resonance. The pipe is therefore said to have a *natural period*; it is capable of producing a note of wavelength four times as great as its own length.

If the pipe is open instead of closed at the farther end, a single condensation entering the mouth *a*, Fig. 226, will emerge at *b* first as a motion of the particles from left to right. The reflected portion will then travel back through the tube as a motion of the particles from left to right (a rarefaction), which will in turn be reflected at *a*, still as a motion from left to right; and thus, after traveling the length of the tube twice, the pulse will again emerge at *b* in its original direc-

Fig. 226. An open pipe

tion (a condensation). Thus in this case the wavelength of the train of waves into which the pipe has transformed the single pulse is twice the length of the pipe; that is, the note given off by the pipe has, as before, the same wavelength as that which will produce the fundamental resonance in the pipe. This tone is, of course, an octave higher than the tone given off by a closed pipe of the same length. It is this ability of a pipe, open or closed, to pick up irregular pulses and transmute them by successive reflections into tones of definite frequency that explains the continuous humming in definite pitch heard when a tube, a seashell, or any similar sort of cavity is held close to the ear.

In order that a pipe may be made to give forth its fundamental note distinctly, however, it is necessary to do more than to start a single pulse in at one end; for the energy of this pulse is dissipated so rapidly in the successive reflections and transmissions that only when the pipe is placed very close to the ear can anything resembling a musical note be recognized at all. If, however, a gentle current of air is directed continuously against one edge of the pipe, as in Fig. 227, the fundamental note can be made, with suitable blowing, to come out strongly. In order to understand this action, consider first a pipe closed at the lower end, and suppose that the original current of air is so directed as to strike the point *a* just inside the edge (Fig. 227). A condensation starts down the pipe and is reflected, when it reaches the bottom, as a condensation or an upward motion of the particles. When this condensation reaches the mouth, it pushes the current of air outside of the edge. This starts a rarefaction down the pipe which, upon its return to the mouth as a rarefaction, draws the current of air inside the edge again. Thus the current is

made to move back and forth over the edge, the period of its vibration being controlled entirely by the natural period of the pipe; for
between two instants of emergence of the jet from the pipe a rarefaction must travel twice the length of the pipe and then a condensation must do the same; in other words, a sound pulse must travel
four times the length of the pipe. Hence the wavelength of the
emitted note is the same as that which corresponds to the natural
period; that is, it is four times the length of the pipe. The source of
the musical note is to be found, then, in the vibra-
tion of the air jet into and out of the end of the
pipe. The pipe itself may be looked upon as merely
a device for enabling the jet to send pulses to the
ear with perfect regularity.

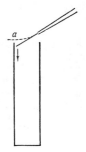

The theory of the open pipe differs only slightly
from that of the closed. If the jet is directed just
inside the edge, a condensation starts down the
pipe, and at the same time, as is indeed also the
case with the closed pipe, the pressure within
the upper end of the pipe begins to rise because of
the influx of air. If the blowing is of just the right
intensity, this pressure may force the jet outside the
edge at the instant when the original condensation

Fig. 227. Produc-
tion of a tone by
an air jet

reaches the lower end and starts back — in this case as a rarefaction.
When this returning rarefaction reaches the mouth, it draws the jet
inside again. At this instant the rarefaction which started down the
pipe when the jet first swung outside has just reached the lower end.
Upon its return to the mouth as a condensation it drives the jet outside again, and thus the jet is alternately forced back and forth over
the edge, its period being controlled entirely by the natural period of
the pipe, for it will be seen that between two successive emergences
of the jet from the mouth of the tube a sound pulse travels down
the tube and back. If the blowing is not of just the right intensity,
so that the pressure reaction near the mouth throws the jet out for
the first time at the instant when the first condensation reaches the
lower end, then the pulses reflected from the lower end do not reach
the mouth at the right instants to set up regular vibration of the jet
over the edge, and consequently no note is produced.

If, in the case of the open pipe, the violence of the blowing is increased to just the right amount, the pressure within the top of the
pipe may be increased so rapidly that the jet is thrown out in one
half its former period. In this case the reflected pulses will get back
to the mouth in just the right time to keep the vibration going, but
the note given forth will be the first overtone of the open pipe, namely

the octave of the fundamental (Sec. **200**). Similarly, still harder blowing of just the right intensity will cause the jet to swing out in one third its former period, and the returning pulses will then get back to the mouth just in time to keep the jet vibrating in the period of the second overtone, the frequency of which is three times that of the fundamental, etc. Blowing of intermediate intensities will produce no notes at all, since the times of return of the reflected pulses are then such as to destroy the vibration which is starting, instead of to keep it going.

The production of overtones in closed pipes is precisely similar, save that in order to produce the first overtone the blowing must be so hard as to cause the jet to swing out of the pipe in *one third* of the time required for the first condensation to travel to the bottom and back, for the first overtone of a closed pipe has a frequency three times that of the fundamental; the second, five times; etc. (Sec. **200**). By blowing with varying degrees of violence across either open or closed tubes, it is generally easy to produce three or four notes of different pitch which are found to have precisely the frequencies demanded by the foregoing theory. If the pipe is long and narrow, it may be quite impossible to produce the fundamental for the reason that the jet is forced out by the increased pressure long before the first pulse returns from the remote end.

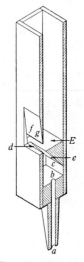

Fig. 228. An organ pipe

202. Types of Wind Instruments. The foregoing theory explains the action of nearly all wind instruments. In organ pipes (Fig. 228) the current of air is forced through the tube *ab* into the air chest *c*, thence through the narrow slit *de* into the *embouchure*, or mouth, of the pipe *E*, where it passes as a narrow jet toward the thin edge or lip *fg*. As a result of small differences in pressure inside and outside of the embouchure, the air jet is caused to deviate to one side or the other of the lip; it moves back and forth across the lip precisely as the jet vibrated across the edge of the pipe in the discussion of Sec. **201**. Flutes and whistles of all sorts are precisely similar to organ pipes in their action. In any of them the air chamber may be either open or closed. In flutes it is open; in whistles it is usually closed; in organ pipes it is sometimes open and sometimes closed. In organs there is a different pipe for every fundamental note, but in flutes a single tube is made to produce a whole series of fundamental notes either by blowing overtones or by opening holes in the side — an

Hermann von Helmholtz
as a Young Man

This picture was taken the year following the publication of his famous paper
on the conservation of energy.

Lord Rayleigh and Lord Kelvin
in the Former's Laboratory at Terling,
July, 1900

Reproduced from the biography, *John William Strutt, Third Baron Rayleigh, O.M., F.R.S.*; by his son
(Edward Arnold & Co., 1924), by permission of the publishers

operation equivalent to cutting off the tube at the hole, since a reflected wave starts back as soon as a point is reached at which there is a greater freedom of expansion than has been met with before.

In the case of some instruments, like the clarinet, the mouthpiece against which the performer blows is almost closed by a reed *l* (Fig. 229) which is clamped at the base and free to swing, under the influence of an outside pressure, so as to close the opening entirely. When the performer blows upon this mouthpiece, a pulse of condensation enters the tube and at the same time the reed closes the opening. This pulse, after reflection from the open end of the clarinet as a rarefaction, and a subsequent reflection at the mouthpiece (closed by the reed), also as a rarefaction, is again reflected at the open end, but now as a condensation; and therefore, after traveling the tube four times, the original condensation returns and forces the reed open, admitting a new pulse. The over-

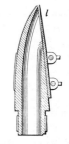

Fig. 229. Mouthpiece of a clarinet

tones which may be produced in such an instrument are evidently those of a closed pipe. It is evident that the vibration frequency is independent of the reed and depends only upon the effective length of the clarinet.

In the trumpet and other brass wind instruments the current of air enters a mouthpiece similar to that shown in Fig. 230. The lips of the performer act as a double reed. A pulse of condensation enters, the lips closing when the reaction of its pressure equals that of the air in the mouth of the performer. This pulse, reflected as a rarefaction from the open end of the trumpet to the lips, reduces the pressure at that point and a new pulse enters. The fundamental depends, then, only upon the length of the instrument. The overtones are produced exactly as in an organ pipe, by blowing more suddenly, and to some extent by increasing the tension of the lips. The possible overtones are those of an open pipe.[1]

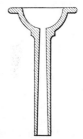

Fig. 230. Mouthpiece of a brass wind instrument

203. Natural Periods of Free Rods. A rod surrounded by air is in every respect analogous to an open pipe, for the reflections at the ends are such as occur when a wave passes from a denser to a rarer medium. Thus a rod will respond to a train

---

[1] *E. G. Richardson, *The Acoustics of Orchestral Instruments and of the Organ* (Arnold, 1929), is a good popular and modern account of this subject.

of waves if it is of such length L that a condensation c incident on the rod at A, Fig. 231, will, after reflection at O, return to A and be again reflected as a condensation at the precise instant when the condensation c' reaches A. The length $2L$ is then the distance that a pulse c travels in the rod before the succeeding pulse c' enters the rod. This is, by definition, one wavelength of the note in the rod. If only one single pulse strikes the rod, the successive reflections of this pulse at A and O will cause a train of waves to be given off at each end. Thus the rod will

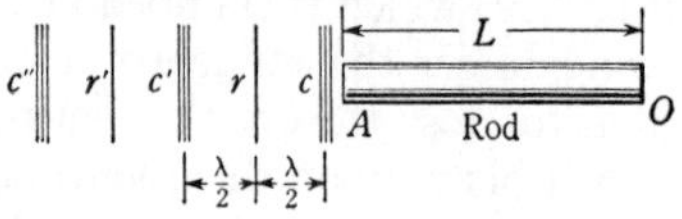

Fig. 231. Wave-train incident on the end of a rod

emit a musical tone the wavelength of which in the rod is twice the length of the rod. The wavelength of this tone in air obviously bears the same relation to its wavelength in the rod as the speed of the wave in air bears to its speed in the rod. If the rod be clamped in the middle, it will respond to and give off precisely the same tone as though it were free, for the compression produced by the clamp at the middle produces at that point the same sort of reflection as occurs at the boundary of a denser medium; hence the clamped rod is equivalent to two closed pipes, each of which gives off the same tone as would an open pipe (that is, a free rod) of double the length. In order to set a rod into *longitudinal* vibrations of this sort, it is customary, instead of striking one end, to clamp it in the middle and stroke it with a rosined cloth if it is of metal, or with a wet cloth if it is of glass.

The foregoing theory suggests an extremely simple and satisfactory means of comparing the speeds of sound in two solids. In order to find the relative speeds of sound in steel and brass, we have only to find the frequencies of two tones produced by stroking steel and brass rods of the same length. For with rods of equal length the number of pulses communicated to the air in unit time by the traveling of pulses up and down the rods is obviously proportional to the speeds of sound in the two rods. Thus, if v_1 and v_2 represent these speeds in steel and brass respectively, and ν_1 and ν_2 the corresponding frequencies produced by the rods of equal length, we have

$$\frac{v_1}{v_2} = \frac{\nu_1}{\nu_2}. \qquad [311]$$

The frequencies ν_1 and ν_2 can be determined by comparing the tones emitted by the rods with those emitted by other vibrating bodies of known, variable frequencies.

204. Nodes and Antinodes in Pipes and Rods. A careful consideration of the resonance of pipes which are giving off the first or higher overtones reveals effects that have thus far been overlooked. For example, it was shown that when a closed pipe has its second resonant length, a condensation c, Fig. 232, must return to the mouth of the pipe at the instant when the rarefaction r' reaches the mouth; but, in this return after reflection, c must somewhere in the pipe collide with the advancing condensation c'. Since c' is one wavelength behind c at the instant of the reflection of c, it is evident that this collision must take place just one half-wavelength from the end of the pipe, namely, at n. Such a collision of two oppositely moving

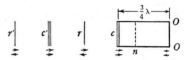

Fig. 232. Formation of nodes and antinodes in a closed pipe

condensations is entirely analogous to the collision of two oppositely moving perfectly elastic balls. These are shown simply to exchange motions (Chap. 5), the effect being the same as though each ball passed through the other without experiencing any effect whatever from it. Thus the waves may be thought of as passing through one another, and their mutual effects may be ignored. As a matter of fact, of course, it is c' that returns to the left after the collision and unites with r' at the mouth, while c is forced back again toward the closed end of the pipe.

One-half period after the collision at n, Fig. 232, of the condensations c' and c ($\rightarrow$ $\leftarrow$) there will occur at n a collision of the rarefactions r' and r ($\leftarrow$ $\rightarrow$). Thus the particles near n are first pushed together by opposing forces, then pulled apart by opposing forces. The result is that they do not move at all. The matter close to n suffers alternate compression and expansion, but the particles at n can never move either to left or to right, because they are always being urged in opposite directions by the oppositely moving waves. In other words, the point n is what we have called

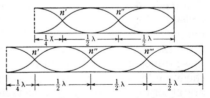

Fig. 233. Stationary waves in closed pipes

a node, and the points between the nodes where the disturbances are greatest are the antinodes (Sec. **194**). If the length of the pipe is $5(\lambda/4)$, $7(\lambda/4)$, etc., it is evident that there will also be nodes at n', n'', etc., Fig. 233. In other words, in any resonant closed pipe (and it is to be remembered that such a pipe is resonant when, and only when, its length is an odd number of fourth-wavelengths) the first node is distant $\lambda/4$ from the open end, and the other nodes follow at intervals of $\lambda/2$. The phenomenon of resonance in a pipe is thus seen to be merely one of the formation in the pipe of stationary waves (Sec. **194**) due to the superposition of direct and reflected waves of the same period and amplitude.

In the case of an open pipe or a free rod the first resonant length is one half-wavelength; and, since a condensation c is reflected as a rarefaction, it is evident that c will collide with r in the middle of the pipe. Hence an open pipe or rod responding to its fundamental has a node in the middle. When it is responding to its first overtone, the nodes are at n' and n'', Fig. 234, each distant $\lambda/4$ from an end and distant $\lambda/2$ apart; similarly for the higher overtones.

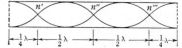

Fig. 234. Stationary waves in an open pipe

205. Melde's Experiment and the Production of Stationary Transverse Waves in Strings. A beautiful illustration of stationary waves in strings is furnished by an experiment devised by F. MELDE[1] in 1859. One end of a light cord is attached to one of the prongs S of a tuning fork (Fig. 235), while the other end O carries a weight of magnitude f. The waves which start down the cord from the vibrating prong are reflected at O, so that two trains of waves moving in opposite directions become superimposed upon the cord. This condition tends to give rise to stationary waves, with nodes at distances from O corresponding to exact multiples of a half-wavelength of the train sent down the cord from the fork. Since, however, the upward-moving train is again reflected at S, the condition for stationary waves in which the nodes are at distances from S corresponding to exact multiples of $\lambda/2$ is also established. It is obvious that both of these conditions can be met, and permanent stationary waves set up in the string, only if the length L of the string is an exact multiple[2] of $\lambda/2$.

Fig. 235. MELDE's experiment

---

[1] Poggendorff's *Annalen der Physik und Chemie* **109**, 193 (1860); **111**, 513 (1860); Wiedemann's *Annalen der Physik* **24**, 497 (1885).

[2] This statement is only approximately correct, since the end of the fork is not exactly at a node, but rather just as near to a node as a point near some other node which has the same amplitude of vibration as the fork.

Instead of varying the length of the string so as to fulfill this condition, it is customary to vary the wavelength by varying the load f. For the wavelength λ is connected with the vibration rate ν of the fork and the speed of propagation v of the train of waves sent down the string by the relation $v = \nu\lambda$, and the speed v is connected with the stretching force f in the string and its mass per unit length σ by Eq. [290], so that by varying the force f it should be possible to find a whole series of values of λ which will give rise to permanent stationary waves in the cord. Thus, when $L = \lambda/2$, the string should vibrate in one segment; when $L = 2(\lambda/2)$, it should vibrate in two segments; when $L = 3(\lambda/2)$, in three segments; etc. (Fig. 236).

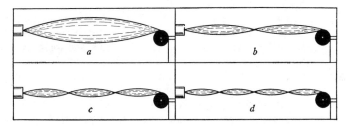

FIG. 236. Showing four modes of vibration of a stretched string

Since, for an appropriate value of f, sharply defined nodes appear in the vibrating string, and since these occur at intervals of $\lambda/2$, it is possible to make a direct measurement of the length λ of the component waves traveling along the string. The speed v of either component wave may then be calculated from the relation $v = \nu\lambda$. MELDE'S experiment therefore makes it possible to determine v by a method that is entirely independent of the method of Eq. [290]. The frequency ν of the fork can best be determined by counting the number of beats per unit time when a standard fork of nearly the same pitch is sounded simultaneously.[1]

206. Partial Tones in Strings. If a stretched string is plucked in the middle, the deformation travels in opposite directions to the two ends and is there reflected; and, since the two reflected portions returning to the middle unite in like phases at this point, the net result of the propagation of the disturbance back and forth over the string is a vibration of the string as a whole in the manner indicated in Fig. 236, a. A string vibrating in this way imparts successive condensations and rarefactions to the air in which it moves, and these, being transmitted to the ear, give rise to a tone of a definite pitch, which is

---

[1] For further details see Exp. XVB, Part II.

called the fundamental tone of the string.[1] Since the time elapsing
between the instant when the string is in the position AcB and the instant when it assumes the position AdB, Fig. 237, is the time required
for the deformation to travel over the paths cBd and cAd, it will be
seen that during the time of one half-
vibration of the string the disturbance
travels on the string a distance exactly
equal to the length of the string. Hence
during the period of one complete vibration of the string the disturbance
travels twice the length of the string.
Thus we arrive, from a wholly differ-

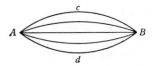

FIG. 237. Stretched string vibrating as a whole

ent point of view, at the conclusion of the preceding section; namely,
that when a string is vibrating as a whole, its length is one half the
wavelength of the waves which are traveling back and forth over it.

If the string is clamped in the middle as well as at the ends and
plucked one fourth of its length from one end, each half vibrates
precisely as the whole string vibrated in the preceding case; but
since the speed of propagation is the same as before, while the distance between reflections is one half as great, the period of vibration
of each half of the string must be one half as great as the preceding
period. Hence the tone communicated to the air is the octave of the
original tone, and the wavelength of the tone is the length of the
string. The tone thus produced by the string is called its first overtone. If the string is not clamped in the middle, but is plucked one
fourth of its length from one end, it still tends to vibrate in two segments, but this vibration is superposed upon the vibration of the
string as a whole, so that the fundamental and the first overtone
can be heard simultaneously (Fig. 238).
Similarly, if the string is plucked one
sixth of its length from one end, it
tends to vibrate in three segments and
the second overtone will be heard with
the fundamental. Thus the string is
capable, under suitable conditions, of
vibrating in any number of segments

FIG. 238. Appearance of a string
which is vibrating so as to produce
its fundamental and first overtone
simultaneously

and of giving out a series of tones whose frequencies bear to the fundamental frequency the ratios $2:3:4:5$, etc. In general, in the

---

[1] Practically, of course, the sound thus derived is of small intensity, and in most
musical instruments the greater magnitude of sound is due to synchronous vibrations
which the string impresses upon its supports and through them upon sounding boards
and resonant volumes of air.

case of the strings of musical instruments, several of these overtones are produced simultaneously with the fundamental. Which
ones are present depends chiefly upon where the string is struck or
bowed.[1] It is to differences in the number and relative prominence
of the overtones that differences in the
qualities or tone colors of different notes
of the same pitch are assigned. In other
words, the quality of the tone from a musical instrument is determined chiefly by
the *form* of the complex wave set up by the
instrument. Overtone structure is not the
only factor, however; although no quantitative measurements have been made as
yet on tone quality, it is known to depend
somewhat on the intensity and probably
also on the pitch. To summarize what we

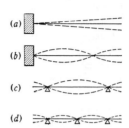

Fig. 239. Transverse waves
in rods

have said here and in Sec. **191**, the psychological characteristics of
pitch, loudness, and quality, although dependent chiefly on frequency,
intensity, and overtone structure respectively, actually depend upon
all three of these physical quantities.

207. Transverse Waves in Rods. In the case of rods the wave travels
as the result of the shear elasticity of the substance of which the rod is
composed (Sec. **187**). The forms assumed by vibrating bars may be seen
from Fig. 239. A bar clamped at the end gives off its fundamental tone when vibrating in the form shown in Fig. 239 (*a*).
If struck more sharply and nearer the free end, it may be
made to give off its first overtone; in this case it vibrates
in the form shown in Fig. 239 (*b*); the relations between the
frequencies of the fundamental and its various overtones are
not simple numbers, however, as is the case with pipes or
strings. If the rod is supported at two points, as in Fig. 239 (*c*),
it will vibrate in the form shown in that figure when yielding
its fundamental tone. The form assumed by the rod when
yielding its first overtone is shown in Fig. 239 (*d*). In this case
also the relation of the frequencies is not simple. If the rod in
Fig. 239 (*c*) is bent, it is found that the nodes are brought
closer together. If it has the form of Fig. 240, the nodes will
occur at the points marked *NN*. The higher overtones are

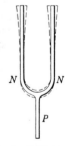

Fig. 240. A
tuning fork

then produced with difficulty and are very much less intense than the
fundamental. A bar bent into this form and supported at *P* is what we have
called a tuning fork. Because of the purity of its tone — that is, the absence

---

[1] For a further discussion of the laws of vibrating musical strings and their experimental verification, see Exp. XVc.

ALTHOUGH private laboratories, owned by individual investigators or their patrons, have existed since the time of ROGER BACON (1214–1294), teaching laboratories and laboratory courses of instruction did not come into existence until the second half of the nineteenth century. The first laboratory for the systematic individual use of students apparently was established at Rensselaer School (now Rensselaer Polytechnic Institute), Troy, New York, by Amos Eaton, in 1824, although similar laboratories developed quite independently soon afterward at the Massachusetts Institute of Technology in this country and at King's College and Oxford University in England.

In Europe university laboratories developed on the whole much more gradually. A few professors, with unusual interest in students, such as HEINRICH GUSTAV MAGNUS (1802–1870) at Berlin, PHILLIP GUSTAV VON JOLLY (1810–1884) at Heidelberg, and WILLIAM THOMSON (later LORD KELVIN) (1824–1907) at Glasgow, permitted and encouraged promising students to work in their own private laboratories and to assist in their own researches. As the number of students increased, the universities began giving financial aid, and university laboratories resulted.

In his own private residence in Berlin [Plate 54], GUSTAV MAGNUS founded in 1842, and conducted until 1870, the first physical institute established in Germany, if not in the world. It gradually evolved into the physical laboratory of the University of Berlin, which was opened in 1863. HELMHOLTZ worked in this laboratory in 1847.

The first laboratory in England that was built and designed specially for the study of experimental physics was the Clarendon Laboratory at Oxford, 1868–1872. The famous Cavendish Laboratory at Cambridge, planned by J. CLERK MAXWELL, was completed in 1874. In the early part of the nineteenth century France was the great center for experimental research, but the work was all carried on in private laboratories. A physical laboratory was built in the old Sorbonne in 1868 and reconstructed in 1894. Excellent laboratories such as that at Zürich, Switzerland, followed. For further details see F. Cajori's "The Evolution of Physical Laboratories," in A History of Physics (Macmillan, 1929), pp. 387–406.

The First German Physical Laboratory

Photograph kindness of Dr. E. Brüche of the A. E. G. Research Institute, Berlin

Physics Laboratory of the
Swiss Federal Institute of Technology, Zürich

Photograph courtesy of the Swiss Federal Railroads

of overtones — it has been adopted for use as a convenient standard of frequency. Any given fork must, of course, be rated first by some absolute method, and then it may be used for comparison with other sources of sound.

○

EXPERIMENT XVA. SPEEDS OF COMPRESSIONAL WAVES IN SOLIDS AND GASES

In this experiment KUNDT'S dust-tube method [1] is employed to determine the speeds of compressional waves in steel and in carbon dioxide at the temperature of the room. The speed v_g of compressional waves in a gas can be determined by enclosing the gas in a closed horizontal tube of adjustable length, sending waves from a tuning fork or vibrating diaphragm of known frequency ν through the gas, and adjusting the length of the gas column until resonance is obtained. The resulting nodes and antinodes formed in the gas column (Sec. 204) may be detected by sprinkling a little cork dust or lycopodium powder along the bottom of the tube. No disturbance of the dust takes place at the nodes, whereas at the antinodes it gathers into ridges. The wavelength λ_g of the waves in the gas is twice the distance between adjacent nodes, and hence the speed of the waves in the gas is given by $v_g = \nu\lambda_g$, where both ν and λ_g are known.

A simple extension of this method makes possible also the measurement of the speed of compressional waves in a solid. This is accomplished simply by replacing the tuning fork or diaphragm by a rod mn of the solid material to be investigated, as in Fig. 241, and setting this rod into longitudinal vibrations, so that it sends compressional waves down the tube. The wavelength λ_r of the wave in the rod is

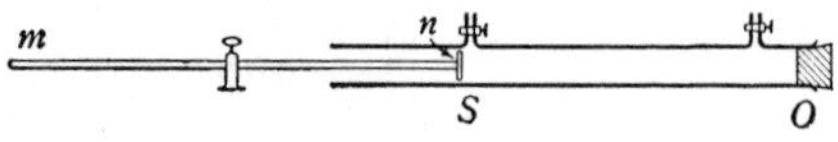

FIG. 241. KUNDT's apparatus

twice the length of the rod (Sec. 203) and hence can be determined. Therefore, if v_r be the speed of the wave in the rod and ν its frequency, we have $v_r = \nu\lambda_r$. Also, if the tube contains air, and v_a is the speed of the waves in air and λ_a their wavelength, then $v_a = \nu\lambda_a$. By combining the two equations so as to eliminate the unknown frequency ν, we obtain as the expression for the speed of compressional waves in the material of the rod

$$v_r = \frac{\lambda_r}{\lambda_a} v_a.$$ [312]

---

[1] Poggendorff's *Annalen der Physik und Chemie* **127**, 497 (1866); **135**, 337, 527 (1868).

Let the length of the rod be L and let the observed distance between adjacent nodes in the air in the tube be s_a; then $\lambda_r = 2L$ and $\lambda_a = 2 s_a$, and, finally,

$$v_r = \frac{L}{s_a} v_a. \tag{313}$$

The value of v_a, the speed of sound in air at the prevailing temperature, can be obtained either from tables or by making the resonance experiment described in Part III of the present experiment.

By similar reasoning, if some other gas is substituted for the air in the tube and v_g and λ_g represent the speed and wavelength of the same tone in that gas, then $v_g = \nu \lambda_g$, or

$$v_g = \frac{\lambda_g}{\lambda_a} v_a, \tag{314}$$

and

$$v_g = \frac{s_g}{s_a} v_a, \tag{315}$$

where s_g is the distance between nodes in the gas in question.

The best way to set a steel rod into longitudinal vibrations is to clamp it in the middle and stroke it with a piece of leather or cloth coated with rosin. It would seem at first thought as though the slipping of the leather along the rod were so irregular that no periodic disturbance could be produced. As a matter of fact, however, the slipping is controlled by the natural period of the rod in much the same way as the vibrations of the air at the mouth of an organ pipe are controlled by the natural period of the pipe (Sec. 201). Thus the first slip starts a pulse down the rod and this pulse, because of the reflections at the ends, returns to the starting point at stated intervals. Of course the tendency to slip is greatest at the instant of the return of the first pulse, so that succeeding slips take place at the instants of return of succeeding pulses. Thus the rod gives off loudly the note corresponding to its natural period.

PART I. Determination of the Speed in Steel. *a.* First pass a gentle stream of air through the glass tube for several minutes in order to eliminate traces of carbon dioxide left by previous experimenters. Then close the stopcocks and scatter fine cork dust evenly throughout the tube. Insert a steel rod in the tube, clamp it at its nodal point, and adjust its position so that the light disk S on the end of the rod (Fig. 241) does not touch the walls of the glass tube. The other end of the tube is closed by a stopper or, in some types of apparatus, by a sliding piston.

b. Move the glass tube back and forth [1] until a maximum of agi-

---

[1] If the end of the tube is closed by a sliding piston, instead of by the stopper O, make this adjustment of the length of the air column with this piston rather than by moving the tube.

tation of the cork dust at the antinodes is produced when the steel rod *mn* is stroked with rosined leather. Exert only a slight pressure on the rod while stroking it.

Measure the distance between O, which is a node, and each of the other nodes. Tabulate these distances in a vertical column, with the corresponding number of half-wavelengths opposite each one. The sum of these distances divided by the sum of the numbers of half-wavelengths gives the best value of a half-wavelength in air of the note given forth by the steel rod.

c. Shake up the cork dust, produce a new set of nodes and antinodes, and obtain another set of readings.

d. Measure the length L of the rod.

e. Calculate v_r from Eq. [313]. Compare this value with that deduced from YOUNG's modulus and the density by use of Eq. [289].

1. Does the temperature have to be taken into consideration in making this comparison?

PART II. Determination of the Speed in Carbon Dioxide. *a.* Pass a gentle current of carbon dioxide through the tube for two or three minutes and repeat the foregoing adjustments and measurements. A one-hole stopper should be slipped over the rod to seal the tube during this part of the experiment.

b. Read the barometer.

c. Calculate the speed of compressional waves in carbon dioxide by means of Eq. [315] and compare the result with the value obtained from the barometric pressure, the density of carbon dioxide, and the value of γ $(= 1.30)$, by use of Eq. [285].

2. Is there an antinode or a node in the vibrating air column at the point S? Discuss with the aid of a diagram.

3. Calculate the frequency of the note given out by the steel rod.

4. What would happen if the rod were clamped at a point other than its nodal point?

PART III. Determination of the Speed in Air by the Resonating-Tube Method. If tables are not available, or if the instructor so directs, determine v_a, the speed of sound in air, with the aid of the apparatus shown in Fig. 225.

a. Set a tuning fork of known frequency into vibration by striking it with a rubber mallet, and hold it three or four millimeters above the mouth of the tube. It is very important always to hold the fork in as nearly as possible the same position. Change the level of the water in the tube until the note of the fork is strongly reinforced. By causing the water to rise and fall rapidly several times in the vicinity of the position of reinforcement, the position for maximum resonance can be determined fairly accurately. Place a rubber band around the tube to mark the level of the water and measure to it from the top of the tube.

b. In a similar manner determine two or three successive reinforcement points.

c. If L_1, L_2, and L_3 represent the first, second, and third measured resonant lengths, and if ΔL represents the correction that must be applied because the reflection at the open end does not take place exactly in the plane of the mouth of the tube (Sec. **200**), then, in view of the theory in Sec. **200**,

$$L_1 + \Delta L = \frac{\lambda_a}{4},$$
$$L_2 + \Delta L = 3\left(\frac{\lambda_a}{4}\right),$$
$$L_3 + \Delta L = 5\left(\frac{\lambda_a}{4}\right).$$

Elimination of ΔL from the first and second of these equations, and also from the first and last, gives

$$L_2 - L_1 = \frac{\lambda_a}{2},$$
$$L_3 - L_1 = \lambda_a.$$

From the mean of these two values of λ_a, the frequency ν of the fork, which should be found marked upon it, and the equation $v_a = \nu \lambda_a$, calculate the speed of sound in air at the prevailing temperature.

5. If you have performed Part III, answer the following questions: (*a*) How does your experimentally determined value of v_a compare with the value deduced from the barometer reading, the density of air for the existing temperature and pressure, and Eq. [285]? (*b*) What is the objection to the resonating-tube method as a means of making an absolute determination of the speed of sound in air?

○

OPTIONAL LABORATORY PROBLEMS

1. Speeds of Compressional Waves in Solids by Kundt's Method. Determine the speed of compressional waves in, say, glass by the method of Exp. XVA. In order to set a glass rod into longitudinal vibration, stroke it with a wet cloth.

2. Relative Speeds of Compressional Waves in Two Solids by a Comparison of Their Natural Frequencies. Select two materials (say, brass and steel) that are available in the form of rods. The rods should be 3 to 4 m long and must be of the same length. Set them successively into longitudinal vibration by clamping them in the middle and stroking with rosined leather. Determine the frequencies ν_1 and ν_2 of the rods by varying the length of a small wire of a sonometer (Exp. XVC) until, when plucked transversely, it produces a tone in tune first with one of the rods, then with the other. The length of the sonometer wire can be varied by means of a sliding bridge. If L_1 and L_2 are the respective lengths of the wires, and v_1 and v_2 the respective speeds in the two rods, then

$$\frac{v_1}{v_2} = \frac{\nu_1}{\nu_2} = \frac{L_2}{L_1},$$

by Eq. [311] and Exp. XVC. Compare your result with the relative value obtained by substituting tabular values of YOUNG'S modulus and the appropriate densities in Eq. [289].

3. Determination of the End Correction for a Resonating Tube. By using the technique of Exp. XVA, Part III, determine the first resonance length L, but in this case hold the fork at least as far away from the end as the radius of the tube. By subtracting this length L from the true values of $\lambda_a/4$, as determined by the method of Exp. XVA, Part III, find the correction ΔL which must be applied to the open end of a pipe to make the first resonance length equal to $\lambda_a/4$. Express this correction as a fractional portion of the radius of the tube. (See page 394, footnote.)

○

EXPERIMENT XVB. WAVES IN STRINGS

In this experiment MELDE'S method (Sec. 205) is employed to determine the speeds of transverse waves in a given string when the latter is subjected to various stretching forces.

PART I. **Speeds of Transverse Waves for Various Stretching Forces.** *a.* Set up an electrically driven tuning fork having a frequency of about 80 to 300 vib · sec⁻¹ and connect it to a single storage cell and a rheostat as shown in Fig. 235. To one prong of the fork attach a light string (for example, a piece of oiled fishline or linen thread) about 120 cm in length. To the other end of this string attach a weight-hanger.

b. Set the fork to vibrating and increase the stretching force by adding weights to the weight-hanger until the string breaks up into some number of vibrating segments. Determine to the nearest gram the stretching force f that produces the sharpest nodes.[1] It is desirable to use values of f for which there will result not more than seven segments.

---

[1] If the force necessary to produce this result is sufficient to stop the vibrations of the fork, first try adjusting the rheostat. If this fails, either shorten the string or increase the number of storage cells; in the latter case it may be necessary to shunt a condenser across the interrupter in order to prevent excessive sparking.

Record the stretching force, the corresponding number of segments existing in the string, and the average length of one of these segments.

c. Vary the stretching force and, in the same manner as in b, determine the weights f corresponding to three or four other wavelengths, say for the string vibrating in 2, 3, 4, etc. segments. It will probably be found necessary to make continual adjustments of the interrupter, since the proper adjustment depends upon the load on tne fork. When through with the apparatus be sure to open the electric circuit.

d. Obtain data for computing the linear density σ of the string by removing the string, weighing it, and measuring its length L.

e. Record the frequency ν of the fork. If this is not marked on the fork, or if the instructor so directs, determine it experimentally by the method described in Part II.

f. For the case of each of the stretching forces employed in b and c calculate the speed v of the transverse wave in the string by two different formulas: by Eq. [290] and by the equation $v = \nu\lambda$. Calculate the mean of each pair of values so obtained. Also compute the percentage of deviation of each member of the pair from the mean.

1. Prove from theoretical considerations that for any particular fork and string the frequency of the fork is given by

$$\nu = \frac{n}{2L}\sqrt{\frac{f}{\sigma}},$$

where n is the number of segments in which the string is vibrating and f is the corresponding stretching force.

2. Show theoretically that the product of the stretching force f and the square of the wavelength λ should be a constant, and that this constant represents the stretching force when the string is vibrating in one segment. What kind of curve should be obtained by plotting corresponding values of λ^2 and f?

3. How well do your experimental results agree with the conclusions arrived at theoretically in question 2?

4. In a stationary wave what is it that is stationary? What moves? Of what is v the speed?

▪ PART II. Frequency of the Tuning Fork by the Method of Beats. Select by ear a standard tuning fork of about the same pitch as that of the fork of unknown frequency. Standard forks are expensive and should be handled with care. Always set such a fork in vibration by striking it with a rubber mallet or with a felt-covered piano hammer.

With the two forks vibrating simultaneously, count with a stop watch the number of beats for several seconds. Then reduce the frequency of the *unknown* fork by attaching to one of its prongs a small piece of soft wax. If this decreases the number of beats, obviously the unknown fork has been brought nearer the standard by weighting it; that is, its natural frequency is larger than that of the standard by the number of beats per second first observed (Sec. 193). If the number of beats is increased when the wax is added, then the unknown fork has the smaller natural frequency; this frequency is that of the standard fork diminished by the number of beats per second first observed.

○

OPTIONAL LABORATORY PROBLEM

Calibration of Tuning Forks by the Method of Beats. Employ the method of Exp. XVB, Part II, to calibrate various tuning forks in terms of standard forks.

○

EXPERIMENT XVc. THE LAWS OF VIBRATING STRINGS

In 1713 BROOK TAYLOR [1] showed that the frequency ν of the *fundamental* tone emitted by a stretched string is given by the formula

$$\nu = \frac{1}{2L} \sqrt{\frac{f}{\sigma}}, \qquad [316]$$

where L is the length of the string, σ is its mass per unit length, and f is the stretching force. This equation can easily be derived on the basis of the discussion of stationary transverse waves in strings which appears in Sec. 205. Most of the laws embodied in TAYLOR'S equation had been discovered experimentally long before, in 1636, by the Jesuit MARIN MERSENNE,[2] and it is probable that even the ancients, particularly PYTHAGORAS (6th century B.C.) and ARISTOTLE, had some knowledge of them.[3]

---

[1] *Abridged Philosophical Transactions* (London, 1749), Vol. IV, p. 391. See also *Methodus Incrementorum Directa & Inversa* (London, 1715).

[2] *Harmonie Universelle* (Paris, 1636). An excerpt in English is given in * *A Source Book in Physics* (1935), pp. 115–116.

[3] See *H. Helmholtz, *On the Sensations of Tone*, tr. by A. J. Ellis (Longmans, Green, 1895), p. 1; also *Lord Rayleigh, *The Theory of Sound*, ed. 2 (Macmillan, 1894), Vol. I, pp. 181–184.

The experiments that follow provide tests of the laws contained in TAYLOR'S equation. These tests can be made conveniently with the aid of a *sonometer*; this is merely a resonance box upon which are stretched several strings whose lengths and stretching forces may be varied at will. The sonometer which we shall employ is equipped with three strings (Fig. 242). Two of these are of steel,

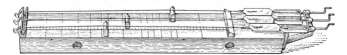

FIG. 242. A sonometer

one with twice the diameter of the other; the third is of brass, of the same diameter as the larger steel wire. We will denote the large and small steel wires by S and s, respectively, and the brass wire by B.

Since the success of the experiment depends largely on developing the facility to tune two strings to the same pitch, the student should begin by practicing this adjustment. First adjust the tensions of the two large wires, S and B, until the pitch of each, when plucked, is nearly the same. Then listen for beats and, when they are heard, practice adjusting the tension in one of the wires until these beats diminish in frequency and finally disappear altogether.

In testing each of the following four laws, compute the percentage of difference between your experimental results and those predicted by the law.

1. *If two strings of the same linear density are to have the same fundamental frequency, their lengths must be proportional to the square roots of the stretching forces.* Adjust the stretching force on s to a little less than one fourth the maximum spring-balance reading. Place a bridge near one end. For purposes of comparison, adjust the force on B until it has the same pitch as s. Measure the length of s. Then decrease this length to exactly half its previous value and re-adjust the force until, by comparison with B, the original pitch is restored.

2. *If two strings of the same length are to have the same fundamental frequency, the stretching forces must be proportional to their linear densities.* Restore the original length and stretching force. Make S and B the same length as s by placing bridges under them. Then increase the force on S until it has the same pitch as s. Finally adjust the force on B until it is in tune with S.

3. *If two wires stretched by forces of equal magnitude are to have the same fundamental frequency, their lengths must be inversely propor-*

tional to the square roots of their linear densities. Decrease the stretching force on S until it is the same as that on s, and then change its length until its pitch is the same as that of s. Similarly, adjust the force on B to that of the other two wires and then change its length until its pitch is the same as that of S.

4. *For a given wire under a constant force the fundamental frequency is inversely proportional to the length.* With B at its full length, adjust s until its pitch is the same as that of B. Then reduce the length of B until its pitch is exactly one octave above the original pitch; until it is exactly two octaves above the original pitch.

1. Derive Eq. [316].

2. Show that each of the foregoing four laws is embodied in Eq. [316].

3. Given a sonometer and one standard tuning fork of known frequency, explain how you would go about determining the unknown fundamental frequency of any other sounding body.

4. Explain why the sonometer strings are mounted on a resonance box.

○

OPTIONAL LABORATORY PROBLEM

Density of a Solid by an Acoustical Method. Determine the density of the steel or brass in a wire by stretching the wire on a sonometer and determining its frequency with the help of a tuning fork of known frequency. Compare the resulting value of the density with that obtained by some other, independent method.

○

QUESTION SUMMARY

1. What is meant by *wave motion*? Give a number of examples of wave motions. In what other general way may energy be transferred from one point to another? What properties must a medium have in order to transmit energy by means of waves?

2. What is the general expression for the speed of a compressional wave in any homogeneous isotropic elastic medium? What form does this expression take in the case of isothermal compressional waves in a gas? in the case of compressional waves in thin solid rods?

3. What is sound? What is the general expression for the speed of a sound wave in a gas? Why does this expression differ from that for an isothermal compressional wave in a gas? How does the speed of sound in air depend upon the temperature, the pressure, the frequency, and the humidity?

4. Define *wave-train*; *wavelength*. What is the quantitative relation connecting wavelength, speed, and frequency in any wave motion?

5. What is the general expression for the speed of transverse waves in stretched wires or cords? for the speed of longitudinal waves in solids? of transverse waves in solids?

6. What is meant by a *simple wave motion*? What are the most important characteristics of simple waves? Give expressions for the displacement of any particle in a simple wave motion at any time t; define carefully each term used. Interpret these equations physically. Show how they may be used to give (*a*) an instantaneous "picture" of the wave and (*b*) an instantaneous "picture" of the motion of any particle participating in the wave motion.

7. Write the general wave equation and explain what each term means physically.

8. How are simple waves combined to form the more complex waves usually encountered in nature? What is the result of compounding (*a*) two simple waves of the same period and phase but different amplitudes? (*b*) two simple waves of different phases and amplitudes but the same period? (*c*) two simple waves of different periods?

9. What is meant by *interference*? Make use of this concept to explain the phenomenon of beats in sound. What relation exists between the number of beats produced per second and the frequencies of the interfering waves?

10. What is meant by *stationary waves*? How are they produced? Describe their characteristics and give an equation for the displacement of any particle at any time t in a stationary wave. Define *node*; *antinode*.

11. Define a *wave-front*. What is meant by a *plane wave*? a *spherical wave*? a *diverging wave*? a *converging wave*? State HUYGENS'S principle.

12. What is meant by the *reflection of waves*? Under what conditions does it take place? By means of HUYGENS'S principle explain what happens when a spherical wave is reflected at a plane surface.

13. Define a *ray*. By means of rays derive the fundamental laws of reflection.

14. What is meant by *diffraction*? Use HUYGENS'S principle to explain it.

15. Discuss the change of phase that takes place on reflection. Make use of this in explaining the resonance of vibrating air columns. With sound of a given wavelength λ, what must be the lengths of a closed pipe to obtain resonance? of an open pipe?

16. What is meant by a *fundamental tone*? an *overtone*? To what overtones does an open pipe respond? a closed pipe? Explain the production of tones in pipes by means of air jets.

17. Discuss the formation of nodes and antinodes in tubes, rods, and strings, and make use of these considerations to explain overtones.

18. Upon what physical quantity does the loudness of a sound depend chiefly? the pitch? the quality? What are some of the distinctions between a noise and a musical tone?

PROBLEMS

1. AMAGAT found that 1 cm³ of alcohol at 14° C and 1 A_s was decreased by 0.000101 cm³ for each atmosphere increase in the pressure to which it was subjected. If the density of alcohol at 14° C is 0.795 g · cm⁻³, what is the speed of sound in this medium?   *Ans.* 1.12×10^3 m · sec⁻¹.

2. (*a*) What physical characteristics must a medium have in order to transmit compressional waves very rapidly? (*b*) Name several substances that would appear to satisfy these requirements and then determine whether you are correct by consulting tables. (*c*) Sound travels about 20 percent faster in "hard" than in "soft" gold. Can you explain this?

3. When a certain tuning fork is sounded in air at 0° C, the wavelength of the compressional waves produced in the air is 130 cm. What is the period of vibration of the fork?   *Ans.* 0.00393 sec.

4. If the speed of a compressional wave in a gas is 320 m · sec⁻¹ at 20° C, what will be the speed at 50° C and twice the pressure?
          Ans. 336 m · sec⁻¹.

5. The E string of a violin is 33 cm long and has a mass of 0.125 g. What force does it exert when tuned to 640 vib · sec⁻¹?   *Ans.* 6.9 kgwt.

6. Find the percentage change in pitch observed by a person standing at a railway station from which a locomotive with whistle blowing is receding at the constant rate of 30 mi · hr⁻¹. This modification in the pitch of a sound due to the relative motion of the source of the sound and the observer is called the *Doppler effect.*[1]   *Ans.* 3.9 percent.

7. The whistle on a certain locomotive has a frequency of 500 vib · sec⁻¹. The speed of the train is 40 mi · hr⁻¹ and the temperature of the air is 20° C. What is the frequency of the sound heard by an observer (*a*) on the train? (*b*) on the track behind the train? (*c*) on the track ahead of the train?
          Ans. (*a*) 500 vib · sec⁻¹; (*b*) 475 vib · sec⁻¹; (*c*) 528 vib · sec⁻¹.

8. The amplitude, frequency, and speed of a certain simple wave are 60 cm, 0.5 vib · sec⁻¹, and 1.5 m · sec⁻¹ respectively. When the displacement y of one of the particles is a maximum in the negative direction, what is the displacement at a point 1.2 m forward in the direction of travel of the wave?   *Ans.* 48 cm.

---

[1] After CHRISTIAN JOHANN DOPPLER, who applied it to the change in color of stars as they approach or recede in the line of sight [*Ueber das farbige Licht der Doppelsterne und einiger anderer Gestirne des Himmels* (Prag, 1842); Poggendorff's *Annalen* **68**, 1 (1846)]. The first acoustical investigation was that of C. H. D. BUIJS BALLOT [Poggendorff's *Annalen* **66**, 321 (1845)]. It is a familiar phenomenon in these days of high-speed traffic, and the student will be able to supply numerous examples. He may be interested also to consider what would happen if the source or observer should move with speeds greater than that of sound; RAYLEIGH states [*The Theory of Sound* (Macmillan, 1896), Vol. II, p. 154] that if an observer moves with a speed twice that of sound a musical piece will be heard in correct time and tune, but backwards.

9. Two musicians stationed some distance apart are playing slightly out of tune, so that 4 beats per second are noticeable. How fast must a person travel from one toward the other in order that no beats shall be noticeable? Calculate for notes of 256 and 384 vib $\cdot$ sec^{-1} and at a temperature of 0° C. *Ans.* 8.5 ft $\cdot$ sec^{-1}; 5.7 ft $\cdot$ sec^{-1}.

10. How much must an organ pipe whose fundamental tone at 0° C is 273 vib $\cdot$ sec^{-1} be heated in order that the frequency of the tone may be changed by 8 vib $\cdot$ sec^{-1}? *Ans.* 16° C.

11. A whistle blown normally with air is blown with hydrogen of density 0.0692 as compared with air at the same temperature and pressure. What change is thus produced? *Ans.* Frequency increased 3.8 times.

12. A brass rod 200 cm long stroked longitudinally is in tune with a 25-cm length of a given sonometer wire. A steel rod 300 cm long stroked longitudinally is in tune with a 26-cm length of the same sonometer wire. Find the relative speeds of sound in steel and brass. *Ans.* $v_s = 1.4 v_b$.

13. Notes of 225 and 336 vib $\cdot$ sec^{-1} are sounded simultaneously. If the even overtones only are present in the first note, and both odd and even overtones in the second note, how many beats per second will occur, and to what overtones will they be due? *Ans.* 6.

14. Given two simple waves of the same frequency and amplitude, traveling in opposite directions in a stretched wire, (*a*) derive an expression for the kinetic and potential energies per unit length of the wire upon which the resulting standing wave is formed, and (*b*) show that the mean total energy is the sum of the energies of the component waves taken separately.

15. When a body is vibrating in a fluid, as in the case of a pendulum swinging in air, there always occurs a dissipation of part of the energy into heat. In so far as this is due to the viscous resistance of the fluid, the only important effect is to cause the amplitude of successive vibrations to decrease. In many cases of such *damped* vibrations, it is found that this decrease in amplitude takes place in such a way that the ratios of successive amplitudes are equal, or $A_0/A_1 = A_1/A_2 = \cdots = A_{n-1}/A_n = c$, where A_0 is the original amplitude and A_n is the amplitude of the nth vibration. (*a*) Show that $A_n = A_0 e^{-nC}$, where e is the base of natural logarithms ($e = 2.718$) and $C = \log_e c$. (*b*) Make a rough sketch of the curve which would be obtained by plotting the displacements of the vibrating body as a function of the time.

O

W. K. CLIFFORD, who was a contemporary of HELMHOLTZ, said of him: "In the first place he began by studying physiology, dissecting the eye and the ear, and finding out how they acted, and what was their precise constitution; but he found that it was impossible to study the proper action of the eye and ear without also studying the nature of light and sound, which led him to the study of physics. He had already become one of the most accomplished physiologists of this century when he commenced the study of physics, and he is now one of the greatest physicists of this century. He then found it was impossible to study physics without knowing mathematics; and accordingly he took to studying mathematics and he is now one of the most accomplished mathematicians of this century."

PROBLEMS FOR REVIEW

In learning the sciences examples are of more use than precepts.

Isaac Newton, *Arithmetica Universalis* (1707)

1. A bullet acquires a speed of 185 m · sec^{-1} while traversing a revolver barrel 20.5 cm long. Find the average acceleration.

Ans. 8.34 × 10^4 m · sec^{-2}.

2. A juggler keeps 5 balls continually in the air, throwing each to a height of 10 ft. What is the interval between each throw, and at what height are the other balls at the moment when one reaches his hand?

Ans. 0.32 sec; 6.4 ft, 9.6 ft, 9.6 ft, 6.4 ft.

3. A stone is dropped from a height of 30 m at the same instant that another is projected upward from the ground. If they meet halfway up, what was the initial speed of the second stone? *Ans.* 17 m · sec^{-1}.

4. An object moves from a point A to a point B with constant acceleration. Show that at a point midway in time between A and B the instantaneous and average speeds of the particle are equal, whereas at a point midway in distance the instantaneous speed is the greater.

5. A man traveling with a velocity of 4 mi · hr^{-1} east finds that the wind seems to blow directly from the north; if he doubles his speed, it appears to come from the northeast. Find the velocity of the wind relative to the ground. *Ans.* 4$\sqrt{2}$ mi · hr^{-1} southeast.

6. A projectile of mass 900 kg struck an embankment with a speed of 400 m · sec^{-1}. It penetrated 4.0 m. Find the average resistance which the embankment offered to its motion. *Ans.* 1.8 × 10^6 kgwt.

7. Two 100-g masses are connected by a string passing over a light frictionless pulley. What mass must be taken from one and added to the other so that the system may move 200 ft in 5.0 sec? *Ans.* 50 g.

8. Two 3-lb masses are connected by a light string hanging over a smooth peg. If a third mass of 3 lb be added to one of them and the system be released, by how much is the force on the peg increased? *Ans.* 2 lbwt.

9. Theory and experiment show that the mass of an electron increases with its speed v according to the Lorentz formula $m = m_0(1 - v^2/c^2)^{-\frac{1}{2}}$, where m_0 is the mass of the electron when at rest and $c = 2.99796$ × 10^{10} cm · sec^{-1}, the speed of light in vacuum. What is the mass of the electron when it is moving with a speed that is nine tenths the speed of light? with a speed of 25,000 km · sec^{-1}? *Ans.* 2.3 m_0; 1.0034 m_0.

10. What is the gravitational force of attraction in poundals between two 1-lb spheres whose centers are 1 ft apart? Is this G?

Ans. 1 × 10^{-9} poundal.

11. In the pulley system shown in Fig. 243, m_1 is 100 g, m_2 is 200 g, and m_3 is 300 g. If friction and the masses of the pulleys and cords may be neglected, what is the acceleration of m_3?

Ans. g/17, down.

12. A train is traveling along a horizontal track at 60 mi · hr^{-1}. Rain, driven by a wind which is in the same direction as the motion of the train, falls with a velocity of 44 ft · sec^{-1} at 30° from the vertical. Find the apparent direction of the rain to a person on the train.   *Ans. 60° from the vertical.*

13. A projectile is fired in a vacuum. Derive an expression for the range in terms of the muzzle speed v_0, the angle of elevation ϕ, and the acceleration due to gravity g. For a given muzzle speed, what angle of elevation gives (*a*) the maximum range? (*b*) the maximum time of flight?   *Ans.* (*a*) 45°; (*b*) 90°.

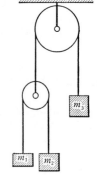

Fig. 243. Problem 11

14. A shell is fired with a velocity of 9100 m · min^{-1} at an elevation of 47°. If the gun is on a cliff 150 m above the sea, calculate (*a*) the time when the shell will strike the water; (*b*) the horizontal distance to the point of impact; (*c*) the greatest altitude reached.

Ans. (*a*) 24 sec; (*b*) 2.5×10^3 m; (*c*) 6.3×10^2 m.

15. A pump delivers M lb of liquid per second through an opening in a pipe of area A ft^2 at a height of h ft above the intake. If the density of the liquid is ρ lb · ft^{-3}, show that the effective horsepower of the pump is $(Mh/550) + (M^3/1100\ g\rho^2 A^2)$.

16. (*a*) By taking the mean distance of the earth from the sun as 9.3×10^7 mi, and the average density and radius of the earth as 5.5 g · cm^{-3} and 4.0×10^3 mi respectively, find the kinetic energy possessed by the earth because of its *orbital* motion. (*b*) If this energy were transformed into heat, how much water would it raise from 0° C to 100° C?

Ans. (*a*) 2.6×10^{33} kg · m^2 · sec^{-2}; (*b*) 6.2×10^{27} kg.

17. (*a*) A 5.0-kg weight is projected up a 30° incline with an initial speed of 3.0 m · sec^{-1}. The coefficient of sliding friction is 0.30. How far up the incline will the weight slide before coming to rest? (*b*) Will it remain at rest or start to slide down? Why?   *Ans.* (*a*) 60 cm.

18. What would be the values of the units of length, mass, and time in a system in which the foot-pound-weight was the unit of work, the horsepower the unit of power, and the acceleration due to gravity the unit of acceleration?   *Ans.* 32/550^2 ft, 550^2/32 lb, 1/550 sec.

19. If a 200-lb man climbs stairs 60 ft high in 2.0 min, what is the average rate in watts at which he works against gravity?   *Ans.* 1.4×10^2 watts.

20. (*a*) Show that an automobile traveling at 65 mi · hr^{-1} goes nearly 100 ft in the second while the driver is reaching for the brake. (*b*) If the coefficient of sliding friction of the tires on the road is 0.75, how far will the

car travel before stopping? (*c*) Obtain some actual data on the starting
and stopping of an automobile, and compute the average acceleration in
each case. Compute the time lost in
making a boulevard stop.

Ans. (*b*) 2.8 × 10² ft.

21. In Fig. 244, find the acceleration
and the force in the cord. Assume that
the coefficient of sliding friction is 0.15
and that the pulley friction is negligible.
Ans. 4.2 × 10² cm · sec⁻² ; 1.1 × 10⁵ dynes.

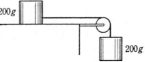

Fig. 244. Problem 20

22. It is found that a certain body slides with constant speed down a
certain inclined plane when the inclination of the plane is ϕ'. Show that
for an inclination ϕ, greater than ϕ', the acceleration of the block has the
magnitude $g \sin (\phi - \phi')/\cos \phi'$.

23. For a given value of the horizontal distance s in Fig. 245, what must
be the value of ϕ for an object to slide down the incline in the shortest time
(*a*) without friction? (*b*) if the coefficient of sliding
friction is 0.20?   *Ans.* (*a*) 45°; (*b*) 51°.

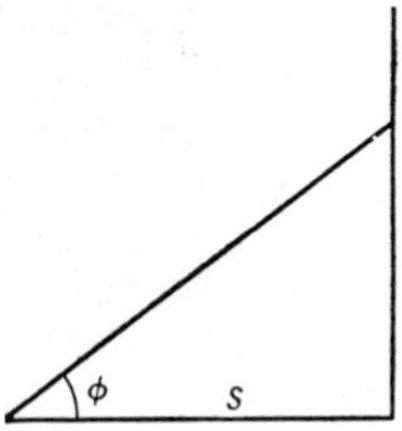

24. An electron is expelled with a speed of
2.90 × 10¹⁰ cm · sec⁻¹ from a radioactive atom which
has a mass of 3.66 × 10⁻²² g. Given that the rest
mass (Prob. 9, p. 419) of an electron is 8.994 × 10⁻²⁸ g,
find the speed of recoil of the atom.

Ans. 2.80 × 10⁵ cm · sec⁻¹.

Fig. 245. Problem 23

25. A bullet of 20.0 g, moving at a speed of
300 m · sec⁻¹, struck and embedded itself in a bird
of 5.00 kg which was flying in the same direction as the bullet with a
speed of 150 km · hr⁻¹. Find the speed of the bird the instant after it
was shot.   *Ans.* 154 km · hr⁻¹.

26. A 20-g rifle bullet is fired into a 4.0-kg block of wood which is suspended by a cord of length 1.0 m. If the block is moved through an angle
of 20°, what is the speed of the bullet?   *Ans.* 2.2 × 10² m · sec⁻¹.

27. A 500-g bird sat on a pole 30 m high. A boy standing 20 m from the
base of the pole shot the bird with a 10-g bullet which had a speed of
150 m · sec⁻¹ when it struck the bird. Assuming that the bullet lodged in
the bird, find (*a*) the distance that the bird rose above the pole and (*b*) how
far from the base of the pole it struck the ground.

Ans. (*a*) 30 cm; (*b*) 4.5 m.

28. Two balls, of mass 10 g and 5.0 g respectively, collide inelastically.
Their velocity after impact is 4.0 m · sec⁻¹, at an angle of 30° with the initial
direction of the 10-g mass. The initial speed of the 10-g mass was
10 m · sec⁻¹. Find the initial velocity of the 5.0-g mass.

Ans. 11 m · sec⁻¹ at 148°.

29. A ball strikes a smooth plane obliquely and rebounds. The velocity before impact makes an angle ϕ_1 with the normal to the plane, and after impact an angle ϕ_2. Obtain the expression for the coefficient of restitution.

Ans. $\tan \phi_1 / \tan \phi_2$.

30. A ball is dropped on the floor from a height h. The coefficient of restitution is e. Find the time and the distance traversed before the ball comes to rest.

$$Ans. \quad t = \sqrt{\frac{2h}{g}} \cdot \frac{1+e}{1-e}; \quad s = h \frac{1+e^2}{1-e^2}.$$

31. Assuming perfectly elastic impact between the water and the blades, compute the efficiency-velocity curve of a Pelton water-wheel.

32. Diagonals are drawn on a square sheet of cardboard of area l^2, and one of the resulting triangles is removed. Find the center of mass of the resulting piece of cardboard. *Ans.* $l/9$ from center.

33. Two solid metal cylinders, each of length 30 cm and having diameters of 10 and 5.0 cm respectively, are joined so that their axes are in the same straight line. Locate the center of gravity and center of mass of the pair if the larger cylinder is made of iron and the other of lead. *Ans.* 6.9 cm from junction.

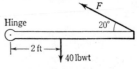

34. The trapdoor in Fig. 246 is 6.0 ft wide and weighs 70 lbwt. A load of 40 lbwt is concentrated 2.0 ft from the hinged edge. Find the force F which, acting at an angle of 20° with the horizontal, is needed to raise the door.

Fig. 246. Problem 34

Ans. 1.4×10^2 lbwt.

35. A door 3.0 ft wide and 7.0 ft high weighs 150 lbwt. If the hinges are 10 in. from the ends and the weight is carried entirely by the upper hinge, what is the total force on each of the hinges?

Ans. 1.6×10^2 lbwt; 42 lbwt.

Fig. 247. Problem 36

36. The axle of a wheel carries a load of 500 kgwt (Fig. 247). What horizontal force must be applied to the axle to raise the wheel over an obstacle 12 cm high, the radius of the wheel being 50 cm? *Ans.* 4.3×10^2 kgwt.

37. The carpenter's square shown in Fig. 248 is supported on a nail at N. Calculate the angle that the larger blade makes with the vertical. *Ans.* 16°.

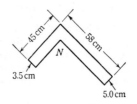

Fig. 248. Problem 37

38. Two parallel forces, each of magnitude 35 poundals, act in opposite directions at the ends of a bar 2.0 ft long. The bar makes an angle of 20° with the direction of the forces. What is the torque of the couple about an axis making an angle of 60° with the normal to the plane of the couple? *Ans.* 12 ft · poundal.

39. Two uniform planks of the same size and weight are hinged at one end (Fig. 249) and stand with their free ends on a smooth floor. They are prevented from slipping by a rope which is tied to each plank at the same distance h above the floor. Find the force in the rope and the force at the hinge.

Ans. $F_r = \frac{1}{2} Wl \sin (\phi/2)[l \cos (\phi/2) - h]^{-1}$.

40. Two masses, m_1 and m_2, are connected rigidly by a rod of length d and negligible mass. Show that the application of the couple of magnitude fd shown in Fig. 250 will cause the system to rotate about its center of mass.

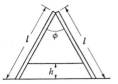

Fig. 249. Problem 39

41. Taking the mass of the earth as 67×10^{20} tons, find the approximate value of the kinetic energy of rotation of the earth on its axis. State all the simplifying assumptions employed in getting the result.   *Ans.* 2.9×10^{29} j.

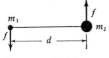

Fig. 250. Problem 40

42. Show that the linear acceleration of a hoop rolling down a plane of inclination ϕ is given by $g \sin \phi/2$.

43. A test made on the flywheel of a certain engine with the help of a Prony dynamometer of the type shown in Fig. 251 yielded the following data: W, 250 kgwt; reading of spring balance, 250 kgwt when the flywheel was at rest and 150 kgwt when its speed was 110 rev · min⁻¹; diameter of flywheel, 1.50 m. Calculate the brake horsepower of the engine.

Ans. 11.3 hp.

44. The shaft of a 30-kw motor carries a gear wheel with 42 teeth that meshes into another gear wheel having 504 teeth. This second wheel is mounted on a shaft 30 cm in diameter and a weight is lifted by a rope wound around this shaft. If the over-all efficiency is 80 percent, how great a load can be lifted when the speed of the motor is 1200 rev · min⁻¹?   *Ans.* 1.6 metric tons.

45. (*a*) In 1851 FOUCAULT hung a long pendulum with a massive bob from a rigid support which imposed no tendency on the pendulum to vibrate more easily in one direction than in the other. By setting this pendulum swinging in a carefully marked direction he was able to demon-

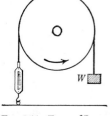

Fig. 251. Form of Prony brake devised by LORD KELVIN. The maximum power which a prime mover can deliver continuously, as determined with such a device, is called the brake horsepower

strate the rotation of the earth. Why? (*b*) HAGEN in 1910 was able to demonstrate quantitatively that the earth is in rotation. A beam was hung horizontally from a long wire. Two masses could be moved at the same time from the middle to the ends of the beam, or the reverse, symmetrically with respect to the axis of rotation. When the masses were displaced, the beam suddenly turned quite perceptibly with respect to its surroundings. Explain.

46. A closed cubical box of volume V is made from thin sheet metal of constant surface density σ. Find the moment of inertia of the box about an axis passing through its center of mass and perpendicular to two opposite faces.   *Ans.* $5\,\sigma\,V^{\frac{4}{3}}/3$.

47. A 50-g clay ball moving with a speed of 30 m · sec^{-1} in a horizontal direction makes inelastic impact with the end of a vertical rectangular bar 1.0 m long and 5.0 cm square which is pivoted at its center of mass. If the mass of the bar is 5.0 kg, what angular speed is communicated to it?   *Ans.* 1.8 rad · sec^{-1}.

48. A homogeneous solid cylinder has wound around it a perfectly flexible cord. One end of the cord is fastened to a fixed point and the cylinder is allowed to fall. Develop the expression for the acceleration. *Ans.* $\alpha = 2\,g/3\,r$.

49. A hoop and a solid disk start down a hill together. Assuming in each case rolling without slipping and neglecting air resistance, find the ratio of their speeds at the bottom.   *Ans.* $\sqrt{3}/2$.

50. (*a*) How could you distinguish a gold sphere from a silver sphere if they had the same radius, the same weight, and were painted the same color? (*b*) If a hard-boiled egg were in the same basket with some uncooked eggs, how would you go about picking it out? (See S. P. Thompson, *Life of William Thomson, Baron Kelvin of Largs*, Vol. II, p. 740.)

51. Explain how a spool may be pulled along a level table top by the end of the thread which is wound on it (Fig. 81) so that the spool (*a*) spins forward; (*b*) spins backward; (*c*) drags without spinning.

52. If the thickness of a solid cylindrical disk of mass M and radius R varies directly as the distance from the axis, what is the expression for the moment of inertia about the axis of the cylinder?   *Ans.* $3\,MR^2/5$.

53. A cylindrical vacuum pump has a bore of 10 cm and a stroke of 15 cm. The volume left at the end of the stroke is 50 cm³. The intake valve has an area of 5.0 cm² and requires a force of 25 gwt to open it. The exit valve has an area of 1.0 cm² and requires a force of 15 gwt to open it. What is the minimum pressure that can be reached by the pump? Assume no leakage. The pump exhausts against an atmospheric pressure of 75 cm of mercury.   *Ans.* 3.5 cm.

54. When a wire of diameter d is stretched, the lateral strain $\Delta d/d$ bears, for a given material, a fixed ratio σ to the longitudinal strain $\Delta l/l$, provided the latter is small. This ratio is called *Poisson's ratio*. It can be shown that

$$\sigma = \frac{3\,k - 2\,n}{2(3\,k + n)}.$$

Show that the value of Poisson's ratio must lie between -1 and $\frac{1}{2}$, and that the three elastic constants Y, n, and σ are connected by the equation $Y = 2\,n(1 + \sigma)$.

55. A tube 100 cm long is half filled with mercury. It is then inverted in a cistern of mercury. If the barometer reads 760 mm, how high above the cistern does the mercury stand in the tube?   *Ans.* 25.2 cm.

56. A simple torsion balance (Fig. 252) consists of a 40-cm crossbar having at each end a spherical lead ball of diameter 20 mm, the bar being suspended by a silver wire 100 cm long and 0.50 mm in diameter. Two spherical lead balls, M and M', each 30 cm in diameter, are brought as close as possible to the small balls but on opposite sides, so that their gravitational attractions tend to turn the bar in the same direction. How much is the silver wire twisted?

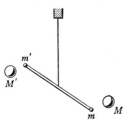

Ans. 0.003°.

57. The carbon filament in an old-style incandescent lamp quivers continuously when cold, but when the lamp is lighted it can be distorted easily by placing the pole of a magnet near the bulb. What is the effect of temperature on the elastic constants of carbon?

FIG. 252. Diagram of the torsion balance used by CAVENDISH for measuring the constant of gravitation G

58. Show that the potential energy per unit volume of an elastic body under hydrostatic pressure is equal to one half of the product of the volume modulus and the square of the volume strain. Does a similar expression hold for the case of a longitudinal strain?

59. The elasticity of a certain rubber cord of length l is such that a weight W attached to its end produces a longitudinal strain of unit amount. Two weights, each of magnitude F, are fastened to the cord, one at an end and one in the middle. The cord with the weights is then lifted from the ground by its free end. Show that the least amount of work which will lift both weights from the ground is $0.5\ Fl[(7\ F/2\ W) + 1]$.

60. At what temperature will the centigrade and Fahrenheit thermometers give readings that are equal numerically but opposite in sign?

Ans. 11.43° F.

61. What would be the value of the expansivity of an ideal gas if the Fahrenheit degree were used instead of the centigrade degree?

Ans. 1/491.72 deg^{-1}.

62. Frame a suitable definition for the *surface* expansivity of a solid body and show that this quantity is approximately twice the linear expansivity.

63. A steel bridge is 200 ft long at 60° F. If tests show the mean linear expansivity of the steel between $-39°$ and 100° C to be 1.03×10^{-5} per deg C, how much allowance should be made for the change in length of the bridge when the temperature ranges from $-20°$ to 100° F? *Ans.* 1.7 in.

64. H. L. FIZEAU (1819–1896) found that when the temperature is defined by a gas thermometer, the linear expansion of copper is given by the empirical formula $l_t = l_0(1 + a't + b't^2)$, where $a' = 1.596 \times 10^{-5}$, $b' = 1.02 \times 10^{-8}$, and l_0 and l_t are the lengths at 0° and $t°$ C respectively. If we were to define temperatures by means of a copper thermometer in which equal increases in length of a copper rod corresponded to equal intervals of temperature, and if we still defined 0° and 100° C in the usual way, what temperature would the copper thermometer indicate when the gas thermometer indicated 50° C?

Ans. 48.5°.

65. A brass wire of length 80 cm and cross-sectional area 3.0 mm² is tightly stretched between two cast-iron clamps rigidly fastened to a cast-iron bar. How much would the force in the wire change if the whole apparatus were cooled from 30° to 10° C? *Ans.* 4.2 kgwt.

66. What would be the value of the expansivity of an ideal gas if the original volume were measured at 100° C instead of at 0° C?

Ans. 0.00268 deg⁻¹.

67. A cubical block of steel 30 cm on an edge is floating on mercury. Find how far the block will sink when the temperature rises from 15° to 75° C.

Ans. 2 mm.

68. A certain mercurial barometer, equipped with a brass scale known to be correct at 15° C, read 735.75 mm at a time when the temperature was 30° C. Reduce this reading to zero; that is, calculate what the reading would be if the temperature were 0° C. *Ans.* 731.59 mm.

69. It is found that a column of mercury at the steam point and 67.5 cm high balances a column of mercury at the ice point and 66.5 cm high (Fig. 99). (*a*) Find the mean expansivity of mercury between 0° and 100° C. (*b*) Given that the density of mercury at 0° C is 13.6 g · cm⁻³, find its density at 100° C. *Ans.* (*a*) 1.5×10^{-4} deg⁻¹; (*b*) 13.4 g · cm⁻³.

70. The variation in the density of water with temperature is given by the equation $\rho_t = \rho_0(1 + at + bt^2 + ct^3)$, where $a = 5.293 \times 10^{-5}$, $b = -6.5322 \times 10^{-6}$, $c = 1.445 \times 10^{-8}$. Find the temperature at which the density of water is a maximum. *Ans.* 4.12° C.

71. Explain atmospheric pressure from the point of view of kinetic theory. Why is the pressure greatest at points near the earth's surface?

72. Show that the pressure exerted by the wind upon a surface placed at right angles to its direction of motion is ρv^2 dyne · cm⁻², provided the wind loses all of its speed v upon impact. What is the difference between wind pressure and ordinary gas pressure?

73. The value of the coefficient of viscosity of air at 19.2° C is 1.828×10^{-4} g · cm⁻¹ · sec⁻¹; calculate the equivalent value expressed in fps units.

Ans. 1.228×10^{-5} lb · ft⁻¹ · sec⁻¹.

74. One gram of nitrous oxide was introduced into an evacuated spherical bulb of radius 10 cm. Find the pressure exerted by the gas on the walls of the bulb when the temperature was 30° C. *Ans.* 10 cm of mercury.

75. Find the increase in the root-mean-square speed of the molecules of hydrogen when the temperature of the gas is raised from 0° to 1° C.

Ans. 330 cm · sec⁻¹.

76. Is the existence of atoms and molecules a hypothesis or a fact? How about the existence of the sun? the stars? Define each of the following terms: *fact; hypothesis; postulate; axiom; theory.*

77. Show that if the pressure exerted by a gas were the result of repulsions between its molecules, the repulsive force would have to vary inversely with the distance in order for BOYLE'S law to hold.

78. Show that if the force between gas molecules were inversely proportional to the distance, the effect of the distant parts of the medium on the molecule would be greater than that of the neighboring parts, and consequently the pressure would depend not only on the density of the gas but on the form and dimensions of the containing vessel.

79. Neglecting air resistance and the presence of any other astronomical bodies, find with what speed a shot would have to be fired vertically upward from the earth's surface so that it never would return.
Ans. $v = \sqrt{2\,Rg}$, where R is the earth's radius.

80. Show that the expressions that were found in Prob. 3, Chap. 11, for the critical constants can also be obtained by finding dP/dV and d^2P/dV^2 from the VAN DER WAALS equation, and setting both equal to zero. Why do these operations yield the expressions for the critical constants?

81. Take as a new unit of temperature the value of T_c, as a new unit of pressure the value of P_c, and as a new unit of volume the *critical volume* V_c. (*a*) Transform the VAN DER WAALS equation to this new system of units. (*b*) What is the remarkable characteristic of this so-called *reduced equation of state* of VAN DER WAALS, and how do you interpret it?
$$Ans. \quad (a)\left(P_1 + \frac{3}{V_1{}^2}\right)(3\,V_1 - 1) = 8\,T_1.$$

82. What is the hygrometric state of the atmosphere on a day when the temperature is 18° C and the dew point is 12° C? *Ans.* $r = 67.7$ percent.

83. The volume of unit mass of a substance is termed its *specific volume*. By the *standard specific volume* of a substance is meant its specific volume if it were an ideal gas under standard conditions of temperature and pressure. What is the standard specific volume of a substance that has a molecular weight of 40? *Ans.* 5.6×10^2 cm$^3 \cdot$ g^{-1}.

84. JAEGER and STEINWEHR found that the specific heat of water in joules per gram per degree at temperature $t°$ C is given by $c_t = 4.2048 - 0.001768\,t + 0.00002645\,t^2$. What error would result from assuming that the specific heat of water at 25° C is the same as its mean specific heat over the range 0° to 50° C? *Ans.* 0.13 percent.

85. Given that the mean atomic thermal capacity of copper in the range 0° to 40° C is 24.5 joules per gram atomic weight per degree, how much energy is required to raise the temperature of 250 g of copper from 10° to 20° C? *Ans.* 226 cal.

86. The mean specific heat of a substance between the temperatures t_1 and t_2 is sometimes defined as the ratio of the quantity of heat required to raise a given mass of the substance from t_1 to t_2 to that required to produce a similar rise in temperature of an equal mass of water. (*a*) Is this definition equivalent to that of Eq. [186], Chap. 12? (*b*) Under what conditions will the two definitions yield quantities that have the same numerical value?

87. What would be the apparent specific heat of a gas that is expanding isothermally owing to the application of heat?

88. (*a*) Find a general expression for the slopes of the curves $PV^\gamma = $ constant. (*b*) Prove that for an ideal gas the ratio of the slope of an adiabatic curve on the PV-diagram to that of an isotherm is equal to the ratio γ of the specific heats. (*c*) Find the ratio of the adiabatic volume modulus of elasticity to the isothermal modulus.

89. (*a*) Does the heat of vaporization of a liquid increase or decrease with increasing temperature? (*b*) What is its value at the critical temperature?

90. Laboratory tests made with a bomb calorimeter show that the *heat of combustion* of a certain grade of coal is 13,000 B.t.u. per pound. If electrical energy can be purchased for 4 ct/kw · hr, what should be the price of the coal in order for the cost of heating a house by the coal and by electricity to be the same? In the case of the coal make an allowance of 50 percent for the heat that goes up the chimney and for other losses.   *Ans.* $152 per ton.

91. A 50-kg block of ice fell 30 m. Assuming that the transformation of mechanical energy into heat is complete and given that the heat of fusion of ice is 79.6 $\text{cal}_{15} \cdot g^{-1}$, find how much ice was melted by the heat generated in the impact with the ground.   *Ans.* 44 g.

92. Calculate the total energy of linear motion of all the molecules in 1 mole of ideal gas at 30° C.   *Ans.* 9×10^2 cal.

93. A given quantity of air is found to have a pressure P_1, volume V, and temperature T_1. After the oxygen has been removed, the remaining nitrogen is found to have a pressure P_2, volume V, and temperature T_2. Express the ratio of the number of moles of nitrogen to the number of moles of oxygen in this sample in terms of the given pressures and temperatures.

Ans. $P_2T_1/(P_1T_2 - P_2T_1)$.

94. Two bulbs containing air are joined by a small tube. The volume of one bulb is three times that of the other, and the initial temperature of each is 273.2° K. Neglecting the expansion of the bulbs, find to what temperature the larger bulb must be raised in order to double the pressure of the air.   *Ans.* 819.6° K.

95. VIOLLE found that the quantity of heat Q required to raise 1 g of platinum from 0° to t° C is given by the formula $Q = 0.0317\,t + 0.000006\,t^2$. Find the temperature of a Bunsen flame if a platinum ball of mass 20 g dropped from the flame into 377 g of water at 0° C raised the temperature of the water 2°.   *Ans.* 1×10^3° C.

96. A piece of ice of mass 50 g is dropped into a brass calorimeter of mass 150 g which contains 800 g of water at 27° C. Calculate the final temperature.   *Ans.* 21° C.

97. Aluminum in finely divided form and of mass 53.8 g is heated to 98.3° C and dropped into 76.2 g of water at 18.6° C contained in a copper calorimeter vessel. The final temperature of the mixture is 27.4° C. The mass of the calorimeter vessel is 123 g and the combined thermal capacity of the thermometer and the metal stirrer is 6.5 cal · deg⁻¹. The cooling correction is found to be negligibly small. Calculate the mean specific heat of aluminum for the temperature range employed. *Ans.* 0.217 cal · g⁻¹ · deg⁻¹.

98. FOURIER and many others have shown that for a thin slab of any material with parallel faces, one of which is at a higher temperature than the other, the rate of conduction of heat through the wall is directly proportional to the area of the surface and to the temperature gradient. For ordinary window glass, the proportionality constant, called the *thermal conductivity*, is found to be 0.0020 cgs unit. How much coal (Prob. 90) must be burned per day to compensate for the loss of heat by conduction through a window-pane of area 1.5 m² and thickness 3 mm when the outer surface is at 2° C and the inner surface at 23° C? *Ans.* 110 lb, if heater is 50 percent efficient.

99. The introduction of 5.0 g of a certain substance at 100° C into a BUNSEN ice calorimeter causes the end of the column of mercury to move 18.5 mm. Given that the mean diameter of the capillary tube is 1.1 mm and that the increase in volume of 1 g of water in freezing is 0.0905 cm³, what is the specific heat of the substance? *Ans.* 0.031 cal · g⁻¹ · deg⁻¹.

100. In a steam calorimeter which is equipped with only one bulb, the bulb is made of copper, weighs 50 gwt, and has a volume of 1 l. Nitrogen at a pressure of 2 A$_s$ is placed in the bulb. If the initial temperature of the chamber is 10° C, how much steam will be condensed on the bulb when steam at 100° C and 1 A$_s$ is admitted to the chamber? *Ans.* 0.87 g.

101. If the density of ice is 0.917 g · cm⁻³ and that of sea water is 1.03 g · cm⁻³, what fraction of the total volume of an iceberg is above water? *Ans.* 0.11.

102. Show that if the density ρ of a body is known roughly, the formula for the weight of the body in vacuum takes the approximately correct form

$$W = W_a\left[1 + \rho_a\left(\frac{1}{\rho} - \frac{1}{\rho_w}\right)\right].$$

103. Find the correction for reducing to weight in vacuum the weight of a quantity of water weighed with brass weights. In the calibration of glassware by means of water, is it imperative to make the reduction to vacuum? *Ans.* 1 mg per g.

104. The *specific gravity* of a substance is defined as the ratio of the mass of a certain volume of the substance to the mass of a like volume of some standard substance. In the case of solids and liquids, the standard usually employed is water at 4° C. (*a*) The density of mercury at 20° C is 13.5462 g · cm⁻³; what is its specific gravity at this temperature? (*b*) Calculate the density of mercury in fps units.
Ans. (*a*) 13.5463; (*b*) 845.64 lb · ft⁻³.

105. A 10.00-g weight placed upon a block of wood weighing 30.00 gwt sinks the block to a certain point in water. In a given salt solution, 15.00 gwt must be placed on the block to sink it to the same point. Find the specific gravity of the solution. *Ans.* 1.125.

106. (*a*) When a film of water exists between two glass plates, as in Fig. 253, it is difficult to draw the plates apart by applying forces normal to their surfaces. Explain. Show that the plates will be pressed together by a force equal to $(2 A\sigma \cos \alpha/d) + C\sigma \sin \alpha$, where α is the angle of contact, A is the area of the film between the plates, d is the thickness of the film, and C is its circumfer-

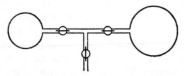

Fɪɢ. 253. Film of water between glass plates

ence. (*b*) A drop of water weighing 0.1 gwt is introduced between two clean pieces of plate glass. With what force will the plates be pressed together when they are 0.0001 cm apart?   *Ans.* 2 tons of force.

107. (*a*) What will be the pressure in an air bubble which is at the bottom of a vessel 15 in. deep, if the diameter of the bubble is 0.040 in. and the barometric pressure is 28 in. of mercury? (*b*) What will the volume of the air in the bubble become when it rises to the surface and mixes with the outside air?   *Ans.* (*a*) 29 in. of mercury; (*b*) 3.5×10^{-5} in.^{3}

108. Two soap bubbles of unequal size are blown separately on the ends of a branched tube which is provided with stopcocks (Fig. 254). Connection with the external atmosphere is then cut off and the two bubbles are put into communication. Explain what happens.

Fɪɢ. 254. Problem 108

109. A solid dam is so built that its cross section is a right triangle having a 3-m base and a 4-m altitude (Fig. 255). If the water comes to the top of the dam, on the vertical side, find the least density which the material composing the dam may have if the dam is to be a stable structure.

Ans. 0.89 g · cm^{-3}.

110. A drop of mercury is placed between two plane slabs of glass which are pressed together until the mercury forms a circular disk 9.5 cm in diameter and 1.0×10^{-3} cm in thickness. Given that the angle of contact between clean mercury and glass is 145° and that the coefficient of surface tension of mercury at 18° C is 520 dyne · cm^{-1}, calculate the force exerted by the mercury on the upper slab of glass.   *Ans.* 6.0×10^7 dynes.

Fɪɢ. 255. Cross section of a dam

111. How much work would one have to do in order to sink a hollow cubical block 1.0 m^3 in volume and weighing 50 kgwt to the bottom of a pond of water 2.0 m deep?   *Ans.* 1.4×10^3 kgwt · m.

112. A battery jar 10 cm in diameter and 12 cm in height is half full of mercury and half full of water. Find the total force against the side walls.   *Ans.* 9.4×10^3 gwt.

113. A horizontal water main 36 in. in diameter contains 27 in. of water. Find the force on the valve that closes the main.   *Ans.* 357 lbwt.

114. A brass weight of density $8.4 \text{ g} \cdot \text{cm}^{-3}$ is dropped into a tank of water 8.0 m deep. Neglecting friction, find how long it takes the weight to reach the bottom.   *Ans.* 1.4 sec.

115. In Prob. 13, Chap. 13, suppose that a horizontal tube containing a constriction is attached to the orifice and that the cross-sectional area of the constriction is equal to one fourth that of the remainder of the tube. What must be the height of the water in the tank in order that the water in the constriction may be on the point of boiling when the temperature is 20° C and the atmospheric pressure is 1 A_s?   *Ans.* 67.3 cm.

116. The coefficient of viscosity of water at 15° C is $0.011 \text{ g} \cdot \text{cm}^{-1} \cdot \text{sec}^{-1}$. Water at this temperature is escaping from the bottom of a tank containing 40 cm of water through a horizontal capillary tube 15 cm long and 0.4 mm in diameter. Calculate the speed of flow with the help of POISEUILLE'S formula.   *Ans.* $12 \text{ cm} \cdot \text{sec}^{-1}$.

117. (*a*) Show that if the resistance of the air, f', be taken into account, the equation of motion of a falling body may be written in the form $m \, dv/dt = mg - f'$. (*b*) Experiment shows that for low speeds the resistance f' which the air offers to a falling body is proportional to the speed v. Thus, letting $f' = kmv$, where k is a constant, show that $v_t = (g/k) + [v_0 - (g/k)]e^{-kt}$, where v_0 is the speed at the time $t = 0$ and e is the base of natural logarithms. (*c*) Show that as time increases the speed approaches asymptotically the limiting value g/k. (*d*) A body falls 110 m from rest in 5 sec. Assuming the air resistance to be proportional to the speed, find the limiting speed which the body approaches asymptotically.

118. (*a*) Show by dimensional reasoning that if the resistance f' experienced by a sphere falling through a fluid is proportional to the first power of the speed v, then $f' = k\eta r v$, where k is a dimensionless constant, η is the coefficient of viscosity of the fluid, and r is the radius of the sphere. An exact analysis first given by STOKES (*Mathematical and Physical Papers*, Vol. III, p. 59) shows that $f' = 6 \pi \eta r v$. What is the expression for f' when it is proportional to v^2? (*b*) By taking into account the weight of the sphere, the buoyant force of the fluid, and the fluid resistance, show that the "terminal speed" of a sphere of radius r and density ρ falling through a fluid of density ρ' and coefficient of viscosity η is 2 $gr^2(\rho - \rho')/9 \eta$. (*c*) Show that the "terminal speed" of a rain drop 0.001 cm in radius is $1.28 \text{ cm} \cdot \text{sec}^{-1}$.

119. Suggest a method for determining the acceleration due to gravity with a good analytical balance. Estimate the accuracy you might expect to attain.

120. A particle of mass m is executing simple vibratory motion. (*a*) Show that the kinetic energy of the particle at any time t is $mA^2\omega^2 \sin^2 \omega t/2$. (*b*) Show that the potential energy of the particle at any time t is $mA^2\omega^2 \cos^2 \omega t/2$. (*c*) Show that the total energy of the particle is constant, and proportional to the squares of both the amplitude and the frequency. (*d*) What is the expression for the kinetic energy for the moment when the particle is at its position of equilibrium?

121. The period and angular amplitude of the balance wheel of a certain watch are 0.40 sec and 30° respectively. (*a*) Find the angular acceleration of the wheel at the instant when its angular displacement is zero. (*b*) If the moment of inertia of the wheel is 1 cgs unit, what is the accelerating torque at the instant when the wheel is at rest?   *Ans.* (*b*) 130 dyne · cm.

122. Show that the number of vibrations lost in a day by an ideal pendulum when its length changes from l to a slightly greater length, $l + \Delta l$, is $\nu \Delta l / 2\,l$, where ν is the number of vibrations completed in 24 hours with length l.

123. A certain rigid pendulum oscillating about a horizontal axis is observed to make 100 complete vibrations in 75 sec. Find its period when the axis is inclined 45° to the horizontal (Fig. 256).
    Ans. 0.89 sec.

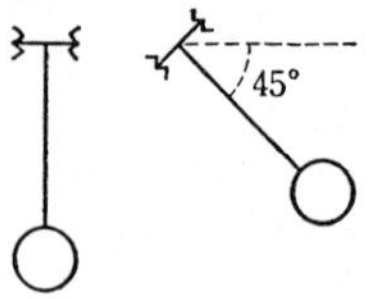

Fig. 256. Problem 123

124. An ideal pendulum is constructed by supporting a small metal sphere by means of a very thin steel wire. At a certain temperature the pendulum beats seconds; that is, the period is 2.0 sec. How many vibrations would be lost in 24 hours if the temperature were to rise through 10° C? Assume the mean linear expansivity of the steel to be 1.0×10^{-5} deg^{-1}.   *Ans.* 2.2 vib.

125. (*a*) Prove, as did NEWTON, that for an ideal pendulum vibrating with a small amplitude, the period varies directly as the square root of the product of the length of the pendulum and mass of the bob, and inversely as the square root of the weight of the bob. (*b*) By means of hollow pendulums filled with various materials, NEWTON demonstrated that pendulums of the same length at the same place had periods which were equal within the narrow limits of error of the measurements. Show that this confirmed, to much greater accuracy, the result which might have been inferred from GALILEO'S experiment with falling bodies, that in a given locality the masses of bodies are proportional to their weights, and that the ratio of mass to weight is independent of the chemical composition of bodies.

126. A skater describes a circle of radius 10 m with a speed of 5 m · sec^{-1}. What must be his angle of inclination from the vertical in order that he may be in equilibrium?   *Ans.* 14°.

127. It is found that a force of 10 gwt elongates a certain elastic cord 50 mm. A 15-g mass is suspended from one end of the cord and is set to vibrating in a vertical line by pulling down on it and releasing it. How far should the suspended mass be pulled down in order that on reaching the highest point of its vibration it may not exert any force upon the cord?
    Ans. 7.5 cm.

128. Explain why it is that the pans of a balance affect the motion of the beam as if their masses were concentrated on their supporting knife-edges.

129. A particle describes a simple vibratory motion of period T and amplitude A. How long does it take the particle to move a distance $A/2$ from its position of equilibrium?   *Ans.* $T/12$.

130. A horizontal wire 1 m in length is clamped at both ends and set into vibration in a vertical plane with an amplitude at the mid-point of 4 mm. What is the shortest period it can have if a rider placed at the mid-point is at no instant to lose contact with the wire?   *Ans.* 0.1 sec.

131. Prove, as did NEWTON (*Principia*, Bk. II, Prop. XLIV), that if water in a U-tube is displaced it will oscillate with a period which is the same as that of a simple pendulum of length equal to one half the length of the water in the U-tube.

132. A cubical block 20 cm on a side is suspended by two cords each of length 15 cm, as shown in Fig. 191. Find the period of vibration (*a*) when the motion is parallel to the plane of the figure and (*b*) when it is perpendicular to the plane of the figure.   *Ans.* (*a*) 0.78 sec; (*b*) 1.1 sec.

133. (*a*) By differentiating Eq. [271], Chap. 14, with respect to *d* and equating the result to zero, prove that the minimum period of vibration of a pendulum is obtained when the distance *d* of the axis of suspension from the center of mass is $\sqrt{I_c/M}$. (*b*) For what position of the axis of suspension does the period have its maximum value, and what information does this give you concerning the nature of the curves in Fig. 189?

134. A circular disk is free to swing about a chord which is fixed in a horizontal position. Find the length of the equivalent simple pendulum and prove that this is least and equal to the radius of the disk when the chord is one side of an inscribed equilateral triangle.

135. A thin wire is bent in the form of a half-circle of radius *R*. It is set oscillating in its own plane about an axis perpendicular to its plane, and passing through the mid-point of the wire. Find (*a*) the length of the equivalent simple pendulum and (*b*) the radius of gyration about the axis from which it is swung.   *Ans.* (*a*) 2 *R*; (*b*) 0.85 *R*

BIBLIOGRAPHY[1]

THERE is no grander nor more intellectually elevating spectacle than that of the utterances of the fundamental investigators in their gigantic power. Possessed as yet of no methods — for these were first created by their labors and are only rendered comprehensible to us by their performances — they grapple with and subjugate the object of their inquiry and imprint upon it the forms of conceptual thought. Those who know the entire course of the development of science will . . . judge more freely and more correctly the significance of any present scientific movement than those who, limited in their views to the age in which their own lives have been spent, contemplate merely the trend of intellectual events at the present moment.

ERNST MACH

O

History and Biography

HISTORIES

BUCKLEY, H. *A Short History of Physics* (Methuen, 1927). A history of achievements rather than of personalities.

CAJORI, F. *A History of Physics* (Macmillan, 1929). A good reference for names and dates.

*CREW, H. *The Rise of Modern Physics* (Williams & Wilkins, 1928, 1935). Modern and readable. The best book of its sort so far published in English.

*DAMPIER WHETHAM, W. C. D. *A History of Science* (Cambridge University Press, 1929, 1930, 1932). Concerned chiefly with the relations of science to philosophy and religion.

DAMPIER WHETHAM, W. C. D. and M. D. *Cambridge Readings in the Literature of Science* (Cambridge University Press, 1924). A book of extracts from the writings of men of science chosen to illustrate the development of definite subjects in the thought of succeeding ages.

GERLAND, E., and TRAUMÜLLER, F. *Geschichte der Physikalischen Experimentierkunst* (Engelmann, 1899). A history of the development of experimental technique.

GUNTHER, R. T. *Early Science in Oxford*, 11 vols. (privately printed, Oxford, 1920–1937). See especially Vol. I, Part III, "Physics"; Vols. VI, VII, VIII, and X, "The Life and Work of Robert Hooke."

*HART, I. B. *Makers of Science : Mathematics, Physics, Astronomy* (Oxford University Press, 1923). Brief, interesting accounts of the lives and works of Aristotle, Archimedes, Roger Bacon, Copernicus, Kepler, Galileo, Descartes, Newton, Boyle, Davy, Faraday, Kelvin, etc. and their relations to each other and to scientific progress.

HEIBERG, J. L. *Mathematics and Physical Science in Classical Antiquity*, tr. by D. C. Macgregor (Oxford University Press, 1922).

HELLER, A. *Geschichte der Physik*, 2 vols. (Emke, 1882–1884). One of the standard histories of physics.

LIBBY, W. *An Introduction to the History of Science* (Houghton Mifflin, 1917). A psychological rather than a logical introduction to the history of science.

---

[1] Throughout this Bibliography a star (*) indicates references that should be especially useful and interesting to the beginner.

435

*MARVIN, F. S. *The Living Past* (Oxford University Press, 1920). The student will find this a valuable aid in fitting science into its proper place in general history.

McKIE, D., and HEATHCOTE, N. H. DE V. *The Discovery of Specific and Latent Heats* (Arnold, 1935). A scholarly study of the foundations of the modern science of heat.

*MILLER, D. C. *Anecdotal History of the Science of Sound* (Macmillan, 1935). The informal character of this book permits the introduction of many interesting comments and reminiscences. A valuable bibliography of original sources is included.

ORNSTEIN, M. *The Rôle of the Scientific Societies in the Seventeenth Century* (University of Chicago Press, 1928). An excellent discussion of the important part played by the scientific societies in the early development of science.

POGGENDORFF, J. C. *Geschichte der Physik* (Barth, 1879). One of the good reference books on the history of physics.

ROSENBERGER, F. *Die Geschichte der Physik*, 3 vols. (Vieweg, 1882–1890). A history to which reference is often made.

SARTON, G. *Introduction to the History of Science* (Williams & Wilkins, 1927–   ). The most scholarly, accurate, and complete work in English. The first three volumes cover the period from Homer to Roger Bacon (13th century).

SEDGWICK, W. T., and TYLER, H. W. *A Short History of Science* (Macmillan, 1917); also later editions. Provides a broad general perspective of the evolution of science. It is not limited to physics.

WHEWELL, W. *History of the Inductive Sciences*, 3 vols. (Parker, 1837, 1847, 1857). An old standard work often quoted.

WOLF, A. *A History of Science, Technology, and Philosophy in the 16th and 17th Centuries* (Allen & Unwin, 1935). This interesting volume treats in detail and with profuse illustrations one of the most fruitful periods of scientific development.

o

BIOGRAPHIES

THOMAS ANDREWS (1813–1885)

P. G. TAIT and A. CRUM BROWN, "Memoir," in *The Scientific Papers of Thomas Andrews* (Macmillan, 1889).

ARCHIMEDES (c. 287–212 B.C.)

T. L. HEATH, *Archimedes* (Macmillan, 1920); also *The Works of Archimedes* (Cambridge University Press, 1897). Much of our information regarding the life of Archimedes comes from Plutarch's Life of Marcellus.

ARISTOTLE (384–322 B.C.)

W. D. ROSS, *Aristotle* (Methuen, 1923, 1930). Other lives in English have been written by J. W. BLAKESLEY (London, 1839); G. H. LEWES (London, 1864); G. GROTE (London, 1872); A. E. TAYLOR (Edinburgh, 1919); J. BURNET (London, 1924).

AMEDEO AVOGADRO (1776–1856)

*B. JAFFE, *Crucibles* (Simon & Schuster, 1930), pp. 157–174.

I. GUARESCHI, *Amedeo Avogadro e la Teoria moleculare* (1901); German translation by O. Merchens, *Amedeo Avogadro und die Moleculartheorie* (Barth, 1903); also in Kahlbaum's Monographien, No. 7, pp. 125–194.

DANIEL BERNOULLI (1700–1782)

M. J. A. N. C. DE CONDORCET, "Éloge de Daniel Bernoulli," *Éloges des Académiciens de l'Académie Royale des Sciences*, ed. by Madame de Condorcet (Paris, 1799).
Philosophical and Mathematical Dictionary (London, 1796), Vol. I, p. 205.

JOSEPH BLACK (1728–1799)

J. ROBISON, "The Editor's Preface" to Black's *Lectures on the Elements of Chemistry* (Edinburgh, 1803).
*W. RAMSEY, *The Life and Letters of Joseph Black* (Constable, 1918).

LUDWIG BOLTZMANN (1844–1906)

*P. LENARD, *Great Men of Science* (Macmillan, 1933), p. 350.
C. H. BRYAN, *Nature* **74**, 569 (1906).

ROBERT BOYLE (1627–1691)

T. BIRCH, *The Works of the Honorable Robert Boyle* (London, 1744, 1772).
*F. MASON, *Robert Boyle, a Biography* (Constable, 1914).

TYCHO BRAHE (1546–1601)

J. L. E. DREYER, *Tycho Brahe* (Black, 1890). A highly interesting account of the state of astronomy at the time.

NICOLAS LÉONARD SADI CARNOT (1796–1832)

M. H. CARNOT, "Life of Sadi Carnot," *Reflections on the Motive Power of Heat*, ed. by R. H. Thurston (Wiley, 1897).
Nature, **130**, 266 (1932).
*E. H. JOHNSON, *Scientific Monthly* **36**, 131 (1933).

HENRY CAVENDISH (1731–1810)

*G. WILSON, *The Life of the Hon. Henry Cavendish* (Cavendish Society, 1851). The standard life of Cavendish.
Introductions by J. CLERK MAXWELL and E. THORPE to Vols. I and II of *The Scientific Papers of the Honorable Henry Cavendish, F.R.S.* (Cambridge University Press, 1921).

RUDOLF JULIUS EMMANUEL CLAUSIUS (1822–1888)

J. W. GIBBS, "Rudolf Julius Emmanuel Clausius," *Proceedings of the American Academy*, New Series **16**, 458 (1889); also *The Collected Works of J. Willard Gibbs* (Longmans, Green, 1828), Vol. II, pp. 261–267.
Proceedings of the Royal Society **48**, i (1890).

NICOLAUS COPERNICUS (1473–1543)

L. PROWE, *Nicolaus Coppernicus* (Weidmann, 1883–1884).

CHARLES AUGUSTIN COULOMB (1736–1806)

T. YOUNG, "Life of Coulomb," *Miscellaneous Works of Thomas Young* (Murray, 1855), Vol. II, pp. 527–541.
*P. LENARD, *Great Men of Science* (Macmillan, 1933), pp. 149–158.

JOHN DALTON (1766–1844)

W. C. HENRY, *Memoirs of the Life and Scientific Researches of John Dalton* (Cavendish Society, 1854). The standard work on Dalton.

H. ROSCOE, *John Dalton and the Rise of Modern Chemistry* (Macmillan, 1895). An especially interesting biography.

R. A. SMITH, *Memoir of John Dalton, and History of the Atomic Theory* (Bailliere, 1856). Emphasizes Dalton's contributions to atomic theory.

RENÉ DESCARTES (1596–1650)

E. S. HALDANE, *Descartes, His Life and Times* (Murray, 1905).

JAMES DEWAR (1842–1923)

H. E. ARMSTRONG, *James Dewar (1842–1923)* (Benn, 1924); *Proceedings of the Royal Society* 111, xiii (1926); *Nature* 111, 472 (1923).

J. CRICHTON-BROWNE, *Science Progress* 18, 126 (1923); *Smithsonian Institution Reports* (1923), p. 547.

LEONHARD EULER (1707–1783)

R. E. LANGER, *Scripta Mathematica* 3, 61, 131 (1935).

M. J. A. N. C. DE CONDORCET, "Éloge d'Euler," *Lettres de L. Euler à une princesse d'Allemagne* (Hachette, 1842).

MICHAEL FARADAY (1791–1867)

*R. APPLEYARD, *A Tribute to Michael Faraday* (Constable, 1931). A eulogy and appreciation of Faraday as a man as well as a scientist.

*J. H. GLADSTONE, *Michael Faraday* (Harper, 1872).   Interesting personal reminiscences.

*B. JONES, *The Life and Letters of Faraday*, 2 vols. (Longmans, Green, 1870). The standard biography of Faraday by one of his intimate friends.

*S. P. THOMPSON, *Michael Faraday, His Life and Work* (Cassell, 1901). The most satisfactory biography of Faraday as a scientist.

*J. TYNDALL, *Faraday as a Discoverer* (Appleton, 1880).   An excellent summary of Faraday's work by a great scientist and expositor.

JOSEPH FOURIER (1768–1830)

F. ARAGO, *Biographies of Distinguished Scientific Men*, tr. by W. H. Smyth, B. Powell, and R. Grant (London, 1857). The memoir on Fourier is reproduced in the *Smithsonian Institution Reports* (1871).

Nature 125, 710 (1930).

GALILEO GALILEI (1564–1642)

W. W. BRYANT, *Galileo* (Sheldon Press, 1925).

*J. J. FAHIE, *Galileo, His Life and Work* (Murray, 1903). The best biography of Galileo in English.

E. NAMER, *Galileo, Searcher of the Heavens*, tr. by S. Harris (McBride, 1931). Treats Galileo's sociological life rather than his scientific achievements.

E. WOHLWILL, *Galilei und sein Kampf für die Kopernikanische Lehre* (1926). This work is based upon recent research and is the best description of Galileo's scientific life written to date.

LOUIS JOSEPH GAY-LUSSAC (1778–1850)

F. ARAGO, *Éloge de Gay-Lussac*.
Proceedings of the Royal Society **5**, 1013 (1843–1850).

JOSIAH WILLARD GIBBS (1839–1903)

H. A. BUMSTEAD, *American Journal of Science* (4) **16**, (1903); also *The Collected Works of J. Willard Gibbs* (Longmans, Green, 1928), Vol. I, pp. xiii–xxvii.
J. JOHNSTON, *Journal of Chemical Education* **5**, 507 (1928).
E. E. SLOSSON, *Leading American Men of Science*, ed. by D. S. Jordan (Holt, 1910), pp. 341–362.
*E. B. WILSON, *Scientific Monthly* **32**, 211 (1931).

HERMANN VON HELMHOLTZ (1821–1894)

*L. KOENIGSBERGER, *Hermann von Helmholtz* (Vieweg, 1902); tr. by F. A. Welby (Oxford University Press, 1906).
Journal of the Optical Society of America **6**, 312, 327, 336 (1922).
A. W. RUCKER, *Fortnightly Review* (1894); reprinted in the *Smithsonian Institution Report* (1894).
Proceedings of the Royal Society **59**, xvii (1895–1896).

ROBERT HOOKE (1635–1703)

*R. T. GUNTHER, *Early Science in Oxford* (privately printed, Oxford, 1930), Vol. VI. This is the life written by R. Waller as an introduction to *The Posthumous Works of Robert Hooke* (1705). To it extracts have been added from J. Ward's *Lives of the Gresham Professors* (1740), and J. Aubrey's *Short Lives*.
The Diary of Robert Hooke, ed. by H. W. Robinson and W. Adams (Taylor & Francis, 1935).

CHRISTIAAN HUYGENS (1629–1695)

F. CAJORI, *Scientific Monthly* **28**, 221 (1929).
*H. CREW, *The Rise of Modern Physics* (Williams & Wilkins, 1935), pp. 120–133.
*P. LENARD, *Great Men of Science* (Macmillan, 1933), pp. 67–83.
Christiaan Huygens, 1629–14 April–1929 (Amsterdam, 1929). In Dutch.

JAMES PRESCOTT JOULE (1818–1889)

*O. REYNOLDS, *Memoirs and Proceedings of the Manchester Literary and Philosophical Society* (4) **6** (1892). "The best biography of a scientist in the English language."
J. DEWAR, *Proceedings of the Royal Institution* **13**, 1 (1890).
J. T. BOTTOMLEY, *Nature* **26**, 617 (1882).

LORD KELVIN (WILLIAM THOMSON) (1824–1907)

*A. GRAY, *Lord Kelvin, an Account of His Scientific Life and Work* (Dent, 1908).
*S. P. THOMPSON, *The Life of William Thomson, Baron Kelvin of Largs*, 2 vols. (Macmillan, 1910). The standard biography of Kelvin, with numerous quotations from original documents and letters which speak for themselves.
*E. KING, *Lord Kelvin's Early Home* (Macmillan, 1909). An intimate account by his sister.
*A. G. KING, *Kelvin the Man* (Hodder & Stoughton, 1925). An intimate picture of Kelvin's personality by his niece.

JOHANN KEPLER (1571–1630)

Johann Kepler, 1571–1630. *A Tercentenary Commemoration of His Life and Work* (Williams & Wilkins, 1931).

J. L. F. BERTRAND, "Kepler: His Life and Work," tr. by C. A. Alexander, *Smithsonian Institution Report* (1869).

W. W. BRYANT, *Kepler* (Macmillan, 1920).

J. E. DRINKWATER, *Life of Kepler* (London, 1833).

KARL RUDOLPH KOENIG (1832–1901)

*D. C. MILLER, *Anecdotal History of the Science of Sound* (Macmillan, 1935), pp. 85–92.

S. P. THOMPSON, *Nature* 64, 630 (1901).

PIERRE SIMON DE LAPLACE (1749–1827)

F. ARAGO, *Biographies of Distinguished Scientific Men*, tr. by W. H. Smyth, B. Powell, and R. Grant (London, 1857). The memoir on Laplace is reproduced in the *Smithsonian Institution Report* (1874).

W. W. ROUSE BALL, *A Short Account of the History of Mathematics* (Macmillan, 1927), pp. 412–421.

LUCRETIUS (TITUS LUCRETIUS CARUS) (*c.* 98–55 B.C.)

*G. D. HADZSITS, *Lucretius and His Influence* (Longmans, Green, 1935). The twelfth volume of the series "Our Debt to Greece and Rome."

EDME MARIOTTE (1620–1684)

L. DARMSTAEDTER, *Journal of Chemical Education* 4, 320 (1927).

*P. LENARD, *Great Men of Science* (Macmillan, 1933), pp. 64–66.

M. J. A. N. C. DE CONDORCET, "Éloge de Mariotte," *Éloges des Académiciens de l'Académie Royale des Sciences* (Paris, 1773).

JAMES CLERK MAXWELL (1831–1879)

*L. CAMPBELL and W. GARNETT, *The Life of James Clerk Maxwell* (Macmillan, 1882). The standard biography of Maxwell. Lewis Campbell was a schoolfellow and lifelong friend of Maxwell; Garnett was his demonstrator at Cambridge.

R. T. GLAZEBROOK, *James Clerk Maxwell and Modern Physics* (Macmillan, 1896). Briefer than the preceding biography.

James Clerk Maxwell, A Commemoration Volume, 1831–1931 (Macmillan, 1931). Brief essays by J. J. Thomson, Planck, Einstein, Larmor, Jeans, Lodge, Glazebrook, and others on various phases of Maxwell's life and accomplishments.

JULIUS ROBERT MAYER (1814–1878)

*P. LENARD, *Great Men of Science* (Macmillan, 1933), pp. 271–286.

J. J. WEYRAUCH, *Robert Mayer, der Entdecker des Princips von der Erhaltung der Energie* (Stuttgart, 1890).

J. J. WEYRAUCH, *Die Mechanik der Wärme von Robert Mayer*, ed. 3, 2 vols. (Stuttgart, 1893).

J. J. WEYRAUCH, *Robert Mayer zur Jahrhundertfeier seiner Geburt* (Stuttgart, 1915).

G. SARTON, *Isis* 13, 18 (1929).

Isaac Newton (1642–1727)

D. Brewster, *Memoirs of the Life, Writings and Discoveries of Sir Isaac Newton* (Edmonston & Douglas, 1860). The standard life of Newton, to which frequent reference is made.

*S. Brodetsky, *Sir Isaac Newton* (Methuen, 1927). Brief and well written; intended for the reader who possesses only a moderate grounding in the elements of science.

W. J. Greenstreet, *Isaac Newton, 1642–1727* (Bell, 1927). A collection of rather technical essays on little-known parts of Newton's work.

*L. T. More, *Isaac Newton, a Biography* (Scribner, 1934). Easily the best biography of Newton so far written, and the only really adequate one.

Sir Isaac Newton, 1727–1927. A Bicentenary Evaluation of His Work (Williams & Wilkins, 1928). A series of papers evaluating Newton's contributions, prepared under the auspices of the History of Science Society.

Heike Kamerlingh Onnes (1853–1926)

F. A. Freeth, *Nature* 117, 350 (1926); *Smithsonian Institution Report* (1926), p. 533.

Blaise Pascal (1623–1662)

L. J. G. Chevalier, *Blaise Pascal* (Longmans, Green, 1930).

W. W. Rouse Ball, *A Short Account of the History of Mathematics* (Macmillan, 1927), pp. 281–288.

*P. Lenard, *Great Men of Science* (Macmillan, 1933), pp. 49–50.

Lord Rayleigh (John William Strutt) (1842–1919)

*R. J. Strutt, *John William Strutt, Third Baron Rayleigh* (Arnold, 1924). An interesting biography, by his son.

A. Schuster, *Proceedings of the Royal Society* 98, i (1920–1921).

Count Rumford (Benjamin Thompson) (1753–1814)

*G. E. Ellis, *Memoir of Sir Benjamin Thompson, Count Rumford* (Macmillan, 1876). A readable account of the life of an exceedingly versatile man.

E. E. Slosson, *Leading American Men of Science*, ed. by D. S. Jordan (Holt, 1910), pp. 9–50.

*M. S. Powell, *The American Physics Teacher* 3, 161 (1935).

J. A. Thompson, *Count Rumford of Massachusetts* (Farrar & Rinehart, 1935). Concerned chiefly with Rumford's nonscientific activities.

Simon Stevin (Stevinus) (1548–1620)

G. Sarton, *Isis* 21, 241 (1934).

P. Lenard, *Great Men of Science* (Macmillan, 1933), pp. 20–24.

*H. Crew, *The Rise of Modern Physics* (Williams & Wilkins, 1935), pp. 86–92.

John William Strutt (1842–1919)
See Lord Rayleigh.

Peter Guthrie Tait (1831–1901)

C. G. Knott, *Life and Scientific Work of Peter Guthrie Tait* (Cambridge University Press, 1911). Tait was a personal friend of Andrews, Kelvin, Joule, Maxwell, Helmholtz, and many other scientists of his period, and this biography throws much light on the personalities of these men as well as upon that of Tait himself. See especially the accounts of Tait's work on the physics of golf and on knots.

BENJAMIN THOMPSON (1753–1814)

See COUNT RUMFORD.

WILLIAM THOMSON (1824–1907)

See LORD KELVIN.

LEONARDO DA VINCI (1452–1519)

*I. B. HART, *The Mechanical Investigations of Leonardo da Vinci* (Open Court, 1925).
E. McCURDY, *The Mind of Leonardo da Vinci* (Dodd, Mead, 1928). An excellent treatment of Leonardo's life, art, and personality.
D. S. MEREJKOWSKI, *The Romance of Leonardo da Vinci*, tr. from the Russian by B. G. Guerney (Random House, New York, 1931); other translations have been published by Constable (1902) and Putnam (1902, 1925). This is an interesting study of Leonardo's personality.

JOHANNES DIDERIK VAN DER WAALS (1837–1923)

H. KAMERLINGH ONNES, *Nature* 111, 609 (1923).

JOHN WALLIS (1616–1703)

W. W. ROUSE BALL, *A Short Account of the History of Mathematics* (Macmillan, 1927), pp. 288–293; *History of the Study of Mathematics at Cambridge* (Cambridge University Press, 1889), pp. 41–46.
Dictionary of National Biography (Macmillan, 1908–1909).
Biographia Britannica (Innys, London, 1747–1766).

JAMES WATT (1736–1819)

H. W. DICKINSON and RHYS JENKINS, *James Watt and the Steam Engine* (Oxford University Press, 1927). A memorial volume prepared for the Committee of the Watt Centenary Commemoration at Birmingham in 1919.
H. W. DICKINSON, *James Watt: Craftsman and Engineer* (Cambridge University Press, 1936).
F. BRAMWELL, *James Watt* (London, 1899); also in the *Dictionary of National Biography*, Vol. 60, pp. 51-62. "The best short biography extant, by an engineer of ripe experience."
T. H. MARSHALL, *James Watt* (London, Boston, 1925).

CHRISTOPHER WREN (1632–1723)

L. WEAVER, *Sir Christopher Wren, Scientist, Scholar and Architect* (Scribner, 1923).
L. MILMAN, *Sir Christopher Wren* (Scribner, 1908).

THOMAS YOUNG (1773–1829)

G. PEACOCK, *Life of Thomas Young* (Murray, 1855).
*F. OLDHAM, *Thomas Young, F.R.S.* (Arnold, 1933).
H. GURNEY, *Memoir of Thomas Young* (1831).
F. ARAGO, *Biographies of Distinguished Scientific Men*, tr. by W. H. Smyth, B. Powell, and R. Grant (London, 1857). The memoir on Young is reproduced in the *Smithsonian Institution Reports* (1869).
H. B. WILLIAMS, *Journal of the Optical Society of America* 20, 35 (1930).

BIOGRAPHICAL COLLECTIONS

ARAGO, F. *Biographies of Distinguished Scientific Men*, tr. by W. H. Smyth, B. Powell, and R. Grant (London, 1857). Biographies of Laplace, Fourier, Young, Watt, and others.

BALL, W. W. ROUSE, *A Short Account of the History of Mathematics* (Macmillan, 1888, 1927). Good short accounts of the lives and works of Stevin, Pascal, Wallis, Huygens, Laplace, Lagrange, Fermat, Bernoulli, Euler, and others.

BELL, E. T. *Men of Mathematics* (Simon & Schuster, 1937). Interesting lives of Archimedes, Descartes, Pascal, Fermat, Newton, Euler, Lagrange, Laplace, Leibniz, and others, with special emphasis on their contributions to the development of modern mathematics.

CAJORI, F. *A History of Mathematics* (Macmillan, 1919, 1931). Brief accounts of the work of Euler, Lagrange, Laplace, Wallis, Fermat, etc.

*CROWTHER, J. G. *British Scientists of the Nineteenth Century* (Kegan Paul, Trench, Trubner, 1935). An attempt to show the effect of social conditions on the development of science through a critical study of the lives of four physicists — Faraday, Joule, Kelvin, and Maxwell — and a physical chemist, Davy.

*HART, I. B. *Makers of Science* (Oxford University Press, 1923). Brief but interesting accounts of the lives and works of Aristotle, Archimedes, Roger Bacon, Copernicus, Kepler, Galileo, Descartes, Newton, Boyle, Davy, Faraday, Kelvin, and others.

Leading American Men of Science, ed. by D. S. Jordan (Holt, 1910). Biographies of Gibbs, Rumford, Rowland, and others.

*LENARD, P. *Great Men of Science* (Macmillan, 1933), tr. by H. S. Hatfield. Excellent short biographies of more than sixty of the important investigators from earliest times until the World War.

*LODGE, O. *Pioneers of Science* (Macmillan, 1910). Dramatically vivid biographies of Copernicus, Tycho Brahe, Kepler, Galileo, Newton, Descartes, Lagrange, Laplace, and others by a master of exposition.

MACFARLANE, A. *Lectures on Ten British Physicists of the Nineteenth Century* (Wiley, 1919). Lectures given at Lehigh University on the lives of Maxwell, Rankine, Tait, Kelvin, and others by one who knew them personally.

○

BIOGRAPHICAL DICTIONARIES

English

Biographia Britannica (Innys, 1747–1766). Excellent biographies of English scientists, such as Boyle, Hooke, Newton, Wallis, and Wren, who lived before 1750.

Dictionary of National Biography (Macmillan, 1908–1909); also several supplements. An excellent reference to Englishmen.

Encyclopaedia Britannica, especially ed. 9 and ed. 11.

Dictionary of Greek and Roman Biography, by W. Smith, 3 vols. (London, 1862–1864).

French

Michaud's *Biographie Universelle* (1854–1865).

Hoefer's *Nouvelle Biographie Générale* (1857–1866).

La Grande Encyclopédie.

Larousse's *Grand Dictionnaire Universel du XIXe siècle français.*

German

Allgemeine Deutsche Biographie.
Brockhaus's *Conversations-Lexikon.*
Meyer's *Grosses Konversations-Lexikon.*
J. C. Poggendorff's *Biographisch-Literarisches Handwörterbuch.*

○

Original Papers, Treatises, and Memoirs

The student should have some acquaintance with most of these works, if only enough so that he may be able to judge whether or not he wants to read them. Although in general they will not be easy to read, and some certainly are too difficult for a beginner, they will repay any effort expended upon them.

ACCADEMIA DEL CIMENTO (1657–1667)

Saggi di Naturali Esperienze fatte nell' Accademia del Cimento (Florence, 1667); tr. by R. Waller, *Essayes of Natural Experiments Made in the Academie del Cimento* (Alsop, London, 1684). This "laboratory manual" of the eighteenth century exerted an enormous influence in spreading the experimental method all over Europe. In it are described the world-famous Florentine thermometers, an improved barometer, a hydroscope, and improved timing devices; as well as classical experiments on air pressure, experiments on the speed of sound, radiant "heat," phosphorescence, the compressibility of water and its expansion on freezing, and the discovery of the rotation of the plane of oscillation of a pendulum (which was later used by Foucault to prove the earth's rotation).

JEAN LE ROND D'ALEMBERT (1717–1783)

Traité de Dynamique (Paris, 1743, 1758); recently republished by Gauthier-Villars in the series *Les Maîtres de la Pensée Scientifique*. In this work d'Alembert enunciates the principle known by his name, namely, that the forces which resist acceleration must be equal in magnitude and opposite in direction to the forces which produce acceleration. The application of this principle enables us to obtain the differential equations of motion of any rigid system.

THOMAS ANDREWS (1813–1885)

"On the Continuity of the Gaseous and Liquid States of Matter," *Philosophical Transactions* **159**, 575 (1869).
Scientific Papers (Macmillan, 1889).

ARCHIMEDES (*c.* 287–212 B.C.)

The Works of Archimedes, tr. and ed. by T. L. Heath (Cambridge University Press, 1897). See especially: "On the Equilibrium of Planes," pp. 189–220; "On Floating Bodies," pp. 253–300.

ARISTOTLE (384–322 B.C.)

The Works of Aristotle, tr. under the editorship of W. D. Ross (Oxford University Press, 1908–1930). See especially: "Physica" (largely metaphysics); "De Caelo" and "De Generatione et Corruptione" (Vol. II); "Meteorologica"

(Vol. III). The prestige of Aristotle's physical works for nearly two thousand years was so great and they have been misquoted so often that the student should read for himself what actually was written; see, in this connection, L. Cooper, *Aristotle, Galileo, and the Tower of Pisa* (Cornell University Press, 1935).

Mechanics, tr. by E. S. Foster (Oxford University Press, 1913). This treatise probably is spurious, but it illustrates the kind of mechanics in vogue at the time of Aristotle.

AMEDEO AVOGADRO (1776–1856)

"Essay on a Manner of Determining the Relative Masses of the Elementary Molecules of Bodies, and the Proportions in which they Enter into these Compounds," *Journal de Physique* **73**, 58 (1811). A translation appears in *Foundations of the Molecular Theory*, Alembic Club Reprint No. 4 (Edinburgh, 1923), pp. 28–51.

DANIEL BERNOULLI (1700–1782)

Hydrodynamica (Argentorati, 1738). The "Bernoulli theorem" is here established, and in the tenth chapter the fundamental ideas of the kinetic theory are set forth. A translation of a portion of the latter will be found in *W. F. Magie, *A Source Book in Physics* (McGraw-Hill, 1935), pp. 247–251.

JOSEPH BLACK (1728–1799)

Lectures on the Elements of Chemistry, delivered in the University of Edinburgh and published from his manuscript by J. Robison (Edinburgh, 1803). Account of the discovery of specific and latent heats.

LUDWIG BOLTZMANN (1844–1906)

Vorlesungen über Gastheorie, 2 vols. (Barth, 1895–1910); French translation by A. GALLOTTI, *Leçons sur la Théorie des Gaz* (Gauthier-Villars, 1902).
Wissenschaftliche Abhandlungen, 3 vols., ed. by F. Hasenöhrl (Barth, 1909).

ROBERT BOYLE (1627–1691)

New Experiments Physico-Mechanicall, touching the Spring of the Air (Hall, Oxford, 1660). A wealth of experiments performed with the air pump built for him by Robert Hooke are here recorded. To the second edition (1662) was added *A Defence of the Doctrine touching the Spring and Weight of the Air*, which gives experimental proof of the relation now known as Boyle's law.

New Experiments and Observations touching Cold, or an Experimental History of Cold, Begun (Crook, London, 1665). A most complete history of everything known about "cold" up to the date of publication. It describes many experiments with freezing mixtures.

Works, ed. by T. Birch, ed. 1, 5 vols. (London, 1744); ed. 2, 6 vols. (London, 1772). Much of this makes entertaining reading.

TYCHO BRAHE (1546–1601)

Astronomiae Instauratae Mechanica (1598). Contains descriptions of the instruments used by Tycho Brahe, which for the first time made possible astronomical measurements that could be called accurate. The great impetus thus given to the study of practical astronomy provided Kepler with his observational material and paved the way for Newton's *Principia*.
De Mundi Aetherei recentioribus Phaenomenis (1588).
Opera Omnia, ed. by J. L. E. Dreyer (Hauniae, 1913–1929).

NICOLAS LÉONARD SADI CARNOT (1796–1832)

Reflections on the Motive Power of Heat, ed. by R. H. Thurston (Wiley, 1897). Gives the solution of one of the most fundamental problems in the entire range of physical science. The modern science of thermodynamics is based upon this paper.

HENRY CAVENDISH (1731–1810)

*"Experiments to determine the Density of the Earth," *Philosophical Transactions* **88**, 469 (1798).
Scientific Papers, 2 vols. (Cambridge University Press, 1921). See especially: "Experiments to determine the Density of the Earth," Vol. II, pp. 249–286; papers on heat, Vol. II, pp. 326 ff.

ERNST F. F. CHLADNI (1756–1827)

Entdeckungen über die Theorie des Klanges (Leipzig, 1787). The acoustic phenomenon known as "Chladni's figures" is here first described.
Die Akustik (Leipzig, 1802). This treatise created experimental acoustics and earned for its author the title of "Father of Acoustics." It contains the discovery of the longitudinal vibrations of strings and rods, and made "Chladni's figures" more generally known.
Neue Beyträge zur Akustik (Leipzig, 1817).

RUDOLF JULIUS EMMANUEL CLAUSIUS (1822–1888)

Papers on the mechanical theory of heat:
"Ueber die bewegende Kraft der Wärme," Poggendorff's *Annalen der Physik und Chemie* **79**, 368, 500 (1850); tr. in *The Second Law of Thermodynamics*, ed. by W. F. Magie (Harper, 1899), pp. 65–107.
Die Mechanische Wärmetheorie, 2 vols., Vieweg (1867–1879); *The Mechanical Theory of Heat*, tr. by W. R. Browne (Macmillan, 1879). Collects Clausius's papers on the subject.
Papers on the kinetic theory of gases:
Poggendorff's *Annalen der Physik und Chemie* **100**, 353 (1857); **105**, 239 (1858); **115**, 1 (1862); tr. in *Philosophical Magazine* (4) **14**, 108 (1857); **17**, 81 (1859); **23**, 417, 512 (1862).
Die Kinetische Theorie der Gase (Vieweg, 1889–1891). Summary of Clausius's work in this field.

NICOLAUS COPERNICUS (1473–1543)

De Revolutionibus Orbium Coelestium (Thorn, 1873). This is the authoritative edition. There is a German translation by C. L. Menzzer (Thorn, 1879) and a French translation by A. Koyré (Paris, 1934). The first edition (1543) differs greatly from the original manuscript. An English translation of a portion of this work will be found in *H. Shapley and H. E. Howarth, *A Source Book in Astronomy* (McGraw-Hill, 1929), pp. 1–12.

CHARLES AUGUSTIN COULOMB (1736–1806)

"Theory of Simple Machines, comprehending the effects of friction and of the stiffness of ropes," *Mémoires des Savants Étrangers* **10**, 161 (1779). Contains Coulomb's researches on friction and on torsion. A translation of a portion of this memoir is given in *W. F. Magie, *A Source Book in Physics* (McGraw-Hill, 1935), pp. 98–105.

JOHN DALTON (1766–1844)

"Experimental Essays on the constitution of mixed gases; on the force of steam or vapour from water and other liquids, in different temperatures, both in a Torricellian vacuum, and in air; on evaporation; and on the expansion of gases by heat," *Memoirs of the Literary and Philosophical Society of Manchester* (1802), Vol. 5, Part 2. A portion of this paper is reproduced in *The Expansion of Gases by Heat*, ed. by W. W. Randall (American Book Co., 1902), pp. 19–22. In this paper Dalton announced two of the most important laws concerning gases: (1) that they expand equally for a given rise of temperature and (2) that at constant volume each gas in a mixture exerts the same pressure as if the other gases were absent.

A New System of Chemical Philosophy, Manchester, Vol. 1, Part 1 (1808); Vol. 1, Part 2 (1810); Vol. 2, Part 1 (1827). The first publication by Dalton of his atomic theory.

JAMES DEWAR (1842–1923)

Collected Papers, 2 vols. (Cambridge University Press, 1927.) Many interesting papers on the liquefaction of gases and low-temperature work, and on soap bubbles.

LEONHARD EULER (1707–1783)

Lettres à une Princesse d'Allemagne sur quelques sujets de physique et de philosophie (St. Petersburg, 1768–1772); (Hachette, Paris, 1842); *Letters to a German Princess*, tr. by H. Hunter (London, 1795, 1802). An elementary exposition of physics, which "for half a century remained a standard treatise on the subject."

Theoria Motus Corporum Solidorum seu Rigidorum (Rostock, 1765). The term "moment of inertia" appears here for the first time.

MICHAEL FARADAY (1791–1867)

*Diary, published by the Royal Institution of Great Britain (Bell, 1932–). A remarkable record of the workings of the mind and the general attitude towards research of "the greatest experimental philosopher the world has ever seen." "Here we have everything that went through Faraday's mind, and we can see the way his thoughts developed and took form in experiment. We see the failures as well as the successes, the trivialities as well as the great achievements. It exhibits his personal qualities, his intellectual honesty, his indefatigable persistence, his modesty and candor better than his papers do and more convincingly than can be done in any biography."

Experimental Researches in Chemistry and Physics (Taylor & Francis, 1859). See especially the papers on the liquefaction of gases.

JOSEPH FOURIER (1768–1830)

Théorie analytique de la Chaleur (Didot, Paris, 1822); *The Analytical Theory of Heat*, tr. by A. Freeman (Cambridge University Press, 1878). The author's most important work; it contains the celebrated theorem named after him.

GALILEO GALILEI (1564–1642)

*Two New Sciences, tr. by H. Crew and A. de Salvio (Macmillan, 1914). Simple and interesting. Every student should have some acquaintance with this pioneer work.

Opere. national edition, prepared under the direction of A. Favaro, 20 vols. (1890–1909). The most complete edition of Galileo ever issued. It contains his scientific and literary work, and voluminous and valuable correspondence.

Louis Joseph Gay-Lussac (1778–1850)

"Researches upon the Rate of Expansion of Gases and Vapors," *Annales de Chimie et de Physique* (1) **43**, 137 (1802). A translation appears in *The Expansion of Gases by Heat*, ed. by W. W. Randall (American Book Co., 1902), pp. 27–48.

"First Attempt to Determine the Changes in Temperature which Gases Experience owing to Changes of Density and Considerations on their Capacity for Heat," *Mémoires de Physique et de Chimie, de la Société d'Arcueil* **1**, 180 (1807). A translation appears in *The Free Expansion of Gases*, ed. by J. S. Ames (Harper, 1898), pp. 3–13.

"Memoir on the Combination of Gaseous Substances with Each Other," *Mémoires de la Société d'Arcueil* **2**, 207 (1809). For a translation see *Foundations of the Molecular Theory*, Alembic Club Reprint No. 4 (Edinburgh, 1923).

Josiah Willard Gibbs (1839–1903)

Collected Works, 2 vols. (Longmans, Green, 1928).

Otto von Guericke (1602–1686)

Experimenta Nova (ut vocantur) Magdeburgica de Vacuo Spatio (Amsterdam, 1672). This work contains descriptions of the first air pump, many interesting experiments on air pressure, the barometer, the Magdeburg hemispheres, the manometer, the lever, thermoscope, and some pioneer experiments on electricity.   There is a German translation of Book III, "De propriis experimentis," in Ostwald's *Klassiker der Exakten Wissenschaften*, No. 59 (Engelmann, 1894). A short excerpt in English will be found in *W. F. Magie, *A Source Book in Physics* (McGraw-Hill, 1935), pp. 80–84.

Hermann von Helmholtz (1821–1894)

"Ueber die Erhaltung der Kraft" (1847). A translation by J. Tyndall appeared in *Scientific Memoirs*, ed. by J. Tyndall and W. Francis (Taylor & Francis, 1853), p. 114.

Die Lehre von den Tonempfindungen als physiologische Grundlage für die Theorie der Musik (1862).   *On the Sensations of Tone as a Physiological Basis for the Theory of Music*, tr. by A. J. Ellis (Longmans, Green, 1875, 1885, 1895, 1912). This work has been called "the principia of physiological acoustics."

Vorlesungen über theoretische Physik, 6 vols. (Barth, 1898–1922).   See especially Vol. 3, *Vorlesungen über die mathematischen Principien der Akustik* (1898).

Wissenschaftliche Abhandlungen (Barth, 1882–1895).

Popular Lectures, tr. by E. Atkinson (Longmans, Green, 1st series, 1873, 1889; 2d series, 1881, 1884).

Hero of Alexandria (c. 1st Century b.c.)

The Pneumatics of Hero of Alexandria, tr. by J. G. Greenwood, ed. by B. Woodcroft (Taylor Walton & Maberly, 1851).   Describes some seventy-eight ingenious pneumatic devices, including the siphon, the force pump, "Hero's fountain," a fire engine, a water organ, and arrangements employing the force of steam.

Robert Hooke (1635–1703)

*Micrographia (Martyn & Allestry, London, 1665).   Hooke was the greatest experimental scientist of his generation. This, his first book, is full of ingenious ideas and remarkable anticipations. Extracts are given in *Old Ashmolean Reprints* VI, which may be obtained from Dr. R. T. Gunther, Old Ashmolean Building, Oxford, England, and in Alembic Club Reprint No. 5 (Edinburgh, 1912).

Micrographia Restaurata, by H. Baker (London, 1745). The text of the *Micrographia* shortened and rewritten.

A Description of Helioscopes, and some other Instruments (London, 1676). Contains the first announcement of Hooke's law of elasticity.

De Potentia Restitutiva (London, 1678). Hooke here elaborates the theory of elasticity and advances a kinetic theory of gases.

Posthumous Works, ed. by R. Waller (London, 1705). Contains Hooke's Cutler Lectures and other discourses.

Philosophical Experiments and Observations, published by W. Derham (London, 1726).

The Life and Work of Robert Hooke, by R. T. Gunther, Vols. VI, VII, VIII, and X of *Early Science in Oxford*. In Vol. VIII will be found facsimile reproductions of the Cutler Lectures which include both the *Description of Helioscopes* and *De Potentia Restitutiva*. These volumes may be obtained from Dr. R. T. Gunther, Old Ashmolean Building, Oxford, England.

CHRISTIAAN HUYGENS (1629–1695)

Œuvres complètes, 18 vols., published by the Société Hollandaise des Sciences (Nijhoff, The Hague, 1888–   ). See especially: "De Motu Corporum ex Percussione," and "De Vi Centrifuga," Vol. 16; "Horologium Oscillatorium," Vol. 18.

Traité de la Lumière (1690); tr. by S. P. Thompson (Macmillan, 1912). The French edition has been republished by Gauthier-Villars in *Les Maîtres de la Pensée Scientifique* (1920).

JAMES PRESCOTT JOULE (1818–1889)

Scientific Papers, 2 vols. (Taylor & Francis, 1884, 1887). See especially the papers on the determination of the mechanical equivalent of heat and the free expansion of gases.

LORD KELVIN (WILLIAM THOMSON) (1824–1907)

Treatise on Natural Philosophy (with P. G. Tait), 2 vols. (Cambridge University Press, 1879); many later editions. "T and T'," as this treatise came to be called, ranks with the classical works of Lagrange and Laplace. The nonmathematical portions, which can and should be read by every student, will be found in *Kelvin and Tait's Elements of Natural Philosophy* ("Little T and T'") (Cambridge University Press, 1875–1912).

Mathematical and Physical Papers, 6 vols. (Cambridge University Press, 1882–1911).

Popular Lectures and Addresses, 3 vols. (Macmillan, 1891–1894).

JOHANN KEPLER (1571–1630)

Astronomia Nova ΑΙΤΙΟΛΟΓΗΤΟΣ, *seu Physica Coelestis, tradita commentariis de Motibus Stellae Martis* (Prague, 1609); German translation by M. Caspar (Munich, 1929). In Chap. 59 will be found Kepler's statement of his first two laws, together with an account of the painful process by which they were deduced.

Harmonices Mundi Libri V (1619). This curious treatise mixes acute observation with whimsical ideas and incoherent mystical hypotheses. A translation of a portion of it will be found in *H. Shapley and H. E. Howarth, A Source Book in Astronomy*

(McGraw-Hill, 1929), pp. 30–40. In Book V, Part 3, Kepler states his third law. Book III, which deals with musical harmony, is abstracted in J. Hawkins, *A General History of the Science and Practice of Music* (London, 1853, 1875), pp. 616–620.

KARL RUDOLPH KOENIG (1832–1901)

Quelques Expériences d'Acoustique (Paris, 1882). Reprints of sixteen articles which had appeared originally in various scientific journals.

JOSEPH LOUIS LAGRANGE (1736–1813)

Mécanique analytique (1788; ed. 2, 2 vols., 1811–1815; ed. 3, 1853–1855; ed. 4, Gauthiers-Villars, 1888–1889), Vols. XI and XII of *Œuvres de Lagrange.* Perhaps the most finished treatise on mechanics in existence. It reduces the whole of dynamics to certain general formulas from which the solution of each separate problem follows by processes so elegant, lucid, and harmonious as to constitute, in the words of Sir William Hamilton, "a kind of scientific poem."

PIERRE SIMON DE LAPLACE (1749–1827)

Traité de Mécanique Céleste (1798–1825), tr. by N. Bowditch (Boston, 1829–1839). "As a monument of mathematical genius applied to the celestial revolutions, the *Mécanique Céleste* ranks second only to the *Principia* of Newton." It is difficult even for the advanced student, however.

Exposition du Système du Monde (1796). Translations, *The System of the World*, by J. Pond (London, 1809), and by H. H. Harte (Dublin, 1830). A portion of the latter is reproduced in *H. Shapley and H. E. Howarth, *A Source Book in Astronomy* (McGraw-Hill, 1929), pp. 155–156. "The *Mécanique Céleste* disembarrassed of its analytical paraphernalia."

Œuvres, 14 vols. (Gauthier-Villars, 1878–1912).

LUCRETIUS (TITUS LUCRETIUS CARUS) (c. 98–55 B.C.)

De Rerum Natura ("On the Nature of Things"), prose translations by H. A. J. Munro (1905–1910), C. Bailey (1910, 1921), and many others; metrical translations by J. Evelyn (1656), W. E. Leonard (1922), and others. Lucretius, a friend and contemporary of Cicero and Julius Caesar, is the most perfect exponent in his time of the natural philosophy of the Greeks who preceded him.

EDME MARIOTTE (1620–1684)

Discours de la Nature de l'Air (Paris, 1670) ; recently republished by Gauthier-Villars in the series *Les Maîtres de la Pensée Scientifique.* Contains the law of compressibility of gases known on the continent of Europe as "Mariotte's law." A translation of a portion of this essay is given in *W. F. Magie, *A Source Book in Physics* (McGraw-Hill, 1935), pp. 88–92.

Traité de la Percussion ou Choc des Corps (Paris, 1677). An interesting discussion of the laws of impact ; referred to by Newton in the *Principia.*

Traité du Mouvement des Eaux et des autres Corps fluides (Paris, 1686) ; *The Motion of Water and other Fluids*, tr. by J. T. Desaguliers (London, 1718). In this work Hooke's law is first applied to "Galileo's problem" of the resistance of beams to rupture.

Œuvres (Leiden, 1717 ; The Hague, 1740).

JAMES CLERK MAXWELL (1831–1879)

*Matter and Motion (Gorham, 1912). Elementary and brief, but a classic on the subject.

*Theory of Heat, ed. 10 (Longmans, Green, 1891). The first four chapters and the tenth are especially valuable for the beginner.

Scientific Papers, 2 vols. (Cambridge University Press, 1890). See especially: "Illustrations of the Dynamical Theory of Gases," Vol. I, p. 377; "On the Viscosity or Internal Friction of Air and Other Gases," Vol. II, p. 1; "Molecules," Vol. II, p. 361; "On the Dynamical Evidence of the Molecular Constitution of Bodies," Vol. II, p. 418; "Atom," Vol. II, p. 445; "Capillary Action," Vol. II, p. 541.

JULIUS ROBERT MAYER (1814–1878)

"Bemerkungen über die Kräfte der unbelebten Natur," Annalen der Chemie und Pharmacie 42, 233 (1842). A translation by G. C. Foster, entitled "Remarks on the Forces of Inorganic Nature," appeared in the Philosophical Magazine (4) 24, 371 (1862).

Die Mechanik der Wärme, 2 vols., ed. 3, ed. by J. J. Weyrauch (Stuttgart, 1893). The best edition of Mayer's collected works. Translations of a number of Mayer's papers appear in the Philosophical Magazine (4) 25, 241, 387, 417, 493 (1863); 28, 25 (1864); also in The Correlation and Conservation of Forces, ed. by E. L. Youmans (Appleton, 1865).

MARIN MERSENNE (1588–1648)

Harmonie Universelle (Paris, 1636). The first extended treatise on sound and music. It treats of "the nature and properties of sound, the movements of bodies, vibrating strings, consonances, dissonances, music, musical instruments, the voice, singing, composition, and other related subjects."

Harmonicorum Libri XII (Paris, 1648). A Latin version of Harmonie Universelle with some changes. An English abridgment will be found in J. Hawkins, A General History of the Science and Practice of Music (London, 1853, 1875), pp. 600–616.

ISAAC NEWTON (1642–1727)

*Mathematical Principles of Natural Philosophy ("Principia"), tr. by Andrew Motte; revised and supplied with a historical and explanatory appendix by F. Cajori (University of California Press, 1934). Other revisions, notably that of W. Davis, 3 vols. (London, 1819), are still available. While the Principia is not easy reading, it will well repay thorough study.

BLAISE PASCAL (1623–1662)

Récit de la grande expérience de l'équilibre des liqueurs (Paris, 1648).

Traitez de l'équilibre des liqueurs et de la pesanteur de la masse de l'air (Paris, 1663). These papers contain Pascal's laws of fluid pressure as well as an account of the famous Puy-de-Dôme experiments which conclusively proved the pressure of air. A translation of a portion of these papers is given in *W. F. Magie, A Source Book in Physics (McGraw-Hill, 1935), pp. 73–80.

Œuvres complètes, 5 vols. (The Hague, 1779); ed. 2 (Paris, 1819).

JOSEPH ANTOINE FERDINAND PLATEAU (1801–1883)

"Statique expérimentale et théorique des Liquides soumis aux seules Forces moléculaires" (1873); "Experimental and Theoretical Researches on the Figures of

Equilibrium of a Liquid Mass Withdrawn from the Action of Gravity," in Taylor's *Scientific Memoirs*, Vols. IV and V (London, 1846, 1852), and in *Smithsonian Institution Reports* (1863, 1864, 1865).

LOUIS POINSOT (1777–1859)

Éléments de Statique (Paris, 1803) ; tr. by T. Sutton, *Elements of Statics* (London, 1848). In this work the idea of couples is introduced.

Théorie Nouvelle de la Rotation des Corps (Paris, 1851) ; *Outlines of a New Theory of Rotary Motion*, tr. by C. Whitley (Cambridge, 1834). Shows that the most general motion of a rigid body can be represented at any instant by a rotation about an axis combined with a translation parallel to the same axis. The original paper was published in Liouville's *Journal de Mathématiques* in 1834.

JEAN LOUIS M. POISEUILLE (1799–1869)

Comptes Rendus 11, 961, 1041, etc. (1840) ; 12, 112 (1841) ; 15, 1167 (1842). *Annales de Chimie et de Physique* (3) 7, 50 (1843). Poggendorff's *Annalen der Physik und Chemie* 58, 424 (1843). *Receuil des savants étrangers* 9, 433 (1846). These papers constitute one of the classics of experimental science. They are frequently quoted as a model of careful analysis of sources of error and painstaking investigation of the effects of separate variables.

LORD RAYLEIGH (JOHN WILLIAM STRUTT) (1842–1919)

The Theory of Sound, 2 vols. (Macmillan, 1877, 1894, 1926). The best treatise on mathematical acoustics in any language.

Scientific Papers, 6 vols. (Cambridge University Press, 1899–1920). Most of these papers are highly technical, but some of the popular scientific lectures will be of considerable interest to the student ; for example, the lecture on "Foam" in Vol. III, p. 351, which gives the earliest satisfactory explanation of the effect of oil in stilling waves.

COUNT RUMFORD (BENJAMIN THOMPSON) (1753–1814)

"An Inquiry Concerning the Source of Heat Which is Excited by Friction," *Philosophical Transactions* 88, 80 (1798). The first serious protest against the caloric theory of heat.

Complete Works, 5 vols. (Macmillan, 1876). Mostly easy and interesting reading.

SIMON STEVIN (STEVINUS) (1548–1620)

De Beghinselen der Weeghconst (Leiden, 1586) ; *De Beghinselen des Waterwichts* (Leiden, 1586). These two papers revived the science of statics and hydrostatics. In the first Stevin enunciates the principle of the triangle of forces and in the second lays the foundations of hydrostatics. A translation of a portion of the first paper will be found in *W. F. Magie, A Source Book in Physics* (McGraw-Hill, 1935), pp. 23–27.

Wisconstige Gedachtenissen (Leiden, 1605–1608) ; Latin translation by W. Snell, *Hypomnemata Mathematica* (Leiden, 1605–1608).

Les Œuvres Mathématiques de Simon Stevin de Bruges, ed. by A. Girard (Leiden, 1634). French translations of the foregoing papers.

JOHN WILLIAM STRUTT (1842–1919)

See LORD RAYLEIGH.

PETER GUTHRIE TAIT (1831–1901)

Properties of Matter (Black, 1885–1907). An excellent elementary treatment of gravitation, elasticity, and the physical properties of fluids.
See also LORD KELVIN.

BENJAMIN THOMPSON (1753–1814)

See COUNT RUMFORD.

WILLIAM THOMSON (1824–1907)

See LORD KELVIN.

EVANGELISTA TORRICELLI (1608–1647)

Trattato del Moto dei Gravi (1641). This paper was republished in his *Opera Geometrica* (Florence, 1644) under the title "De motu gravium naturaliter descendentium et proiectorum." See especially the section "De motu aquarum" in which he records his important results on the discharge of liquids through orifices in the bottom of vessels, which created the science of hydrodynamics. It will be found in Vol. 2, pp. 185 ff. of his collected works (Florence, 1919). A translation of a portion of it is given in *W. F. Magie, *A Source Book in Physics* (McGraw-Hill, 1935), pp. 111–113.
Opere (Florence, 1919). See especially "De motu aquarum," Vol. 2, pp. 185; also Vol. 3, p. 186, which reproduces Torricelli's two famous letters to Ricci describing his experiments on the pressure of the air and the invention of the barometer. A translation of one of these letters will be found in *W. F. Magie, *A Source Book in Physics* (McGraw-Hill, 1935), pp. 70–73.

JOHN TYNDALL (1820–1893)

Heat Considered as a Mode of Motion (Appleton, 1863). Popular experimental lectures by a master expositor.
Sound (Appleton, 1915). Experimental lectures.
Fragments of Science (Appleton, 1871, 1879).
New Fragments (Appleton, 1897).

JOHANNES DIDERIK VAN DER WAALS (1837–1923)

On the Continuity of the Liquid and Gaseous States (in Dutch) (Leiden, 1873) ; German translation by F. Roth (Barth, 1881, 1899) ; tr. by R. Threlfall and J. F. Adair in *Physical Memoirs* (Physical Society of London, 1890).

PIERRE VARIGNON (1654–1722)

Projet d'une Nouvelle Mécanique (Paris, 1687).
Nouvelle Mécanique ou Statique (Paris, 1725). Varignon independently developed the principle of the parallelogram of forces and applied it to all sorts of statical problems.

LEONARDO DA VINCI (1452–1519)

The Mechanical Investigations of Leonardo da Vinci, by I. B. Hart (Open Court, 1925). A readable account of Leonardo's little-known work in mechanics.

JOHN WALLIS (1616–1703)

Philosophical Transactions **3**, 864 (1668); tr. in *Abridged Philosophical Transactions* (London, 1749), Vol. I, Chap. V, p. 457. The laws of inelastic impact.

Mechanica: sive, de Motu, Tractatus Geometricus (London, 1669–1671). An excellent summary of the mechanics of the time. It contains an exhaustive treatment of the theory of inelastic impact.

Philosophical Transactions **12**, 839 (1677); also *Abridged Philosophical Transactions* (London, 1749), Vol. I, Chap. X, pp. 606–620. Experiments on vibrating strings and other musical instruments.

WILHELM EDUARD WEBER (1804–1891)

Wellenlehre auf Experimente gegründet (with E. H. Weber) (Leipzig, 1825). First experimental treatment of stationary waves and interference of sound.

THOMAS YOUNG (1773–1829)

A Course of Lectures on Natural Philosophy and the Mechanical Arts (London, 1807).

Miscellaneous Works, 3 vols., ed. by G. Peacock (Murray, 1855). Most of Young's writings are very difficult, sometimes obscure.

o

COLLECTIONS OF ORIGINAL MEMOIRS

Alembic Club Reprints (Edinburgh): No. 2, *Foundations of the Atomic Theory* (1923), papers by Dalton, Wollaston, and T. Thomson; No. 4, *Foundations of the Molecular Theory* (1923), papers by Dalton, Gay-Lussac, and Avogadro; No. 5, *Extracts from Micrographia by Robert Hooke* (1912); No. 12, *The Liquefaction of Gases* (1912), papers by Faraday.

Classics of Scientific Method (Bell): *Joule and the Study of Energy*, by A. Wood (1925); *The Composition of Water*, by J. R. Partington (1928) [the researches of Cavendish and Lavoisier]; *The Discovery of the Nature of Air*, by C. M. Taylor (1934) [the work of Boyle, Mayow, Black, Priestley, and Lavoisier on this subject].

Harper's Scientific Memoirs:

The Laws of Gravitation, ed. by A. S. Mackenzie (American Book Co., 1900). Annotated extracts from the papers of Newton, Cavendish, and others; excellent historical notes and a comprehensive bibliography.

The Laws of Gases, ed. by Carl Barus (Harper, 1889). Portions of the classic papers of Boyle and Amagat, with brief biographical sketches of these men.

The Expansion of Gases by Heat, ed. by W. W. Randall (American Book Co., 1902). The classic papers of Dalton, Gay-Lussac, Regnault, and Chappuis, with short biographical sketches.

The Free Expansion of Gases, ed. by J. S. Ames (Harper, 1898). The classic papers of Gay-Lussac, Joule, and Kelvin and Joule, with brief biographies.

The Second Law of Thermodynamics, ed. by W. F. Magie (Harper, 1899). The papers of Carnot, Clausius, and Kelvin in which the second law of thermodynamics is developed.

The Wave Theory of Light, ed. by H. Crew (American Book Co., 1900). The memoirs of Huygens, Young, and Fresnel.

Physical Memoirs (Physical Society of London, 1888–1890). Translations of a number of important memoirs such as that of van der Waals.

Scientific Memoirs, 5 vols., ed. by R. Taylor (London, 1837–1852). Translations of a number of important memoirs such as those of Clapeyron, Plateau, and Regnault.

Scientific Memoirs, ed. by J. Tyndall and W. Francis (London, 1853). A continuation of the preceding work. Contains translations of Helmholtz's paper on the conservation of energy and of papers by Clausius and others.

A Source Book in Physics, by W. F. Magie (McGraw-Hill, 1935). Well-selected quotations from the most significant portions of the original papers of more than ninety physicists together with short biographical sketches.

Ostwald's *Klassiker der Exakten Wissenschaften* (Engelmann, Leipzig). German translations of the classic papers of Helmholtz (No. 1), Dalton (No. 3), Avogadro (No. 8), Galileo (Nos. 11, 24, 25), Carnot (No. 37), Lavoisier and Laplace (No. 40), Gay-Lussac (No. 42), Gay-Lussac, Dalton, Dulong and Petit, Regnault (No. 44), Fahrenheit, Réaumur, Celsius (No. 57), von Guericke (No. 59), Clausius (No. 99), d'Alembert (No. 106), de Saussure (Nos. 115, 119), Andrews (No. 132), Huygens (Nos. 20, 138, 192), Doppler (No. 161), Mayer (Nos. 180, 223), Loschmidt (No. 190), Kelvin (No. 193), Smoluchowski (No. 207), Clapeyron (No. 216), and Poiseuille (No. 237).

○

Critical and Historical Expositions

Cox, J. *Mechanics* (Cambridge University Press, 1919). An elementary text written from the historical point of view.

*Mach, E. *The Science of Mechanics*, tr. by T. J. McCormack (Open Court, 1893), ed. 3 (1907), ed. 4 (1919), Supplement (1915). A fascinating critical presentation of the historical development of the concepts and laws of mechanics. It should be read by every serious student.

Mach, E. *History and Root of the Principle of the Conservation of Energy*, tr. by P. E. B. Jourdain (Open Court, 1911). First sketch of a way of regarding science that has become of great importance.

Mach, E. *Die Principien der Wärmelehre* (Barth, 1896, 1899, 1919). An excellent critical and historical presentation.

Todhunter, I., and Pearson, K. *A History of the Theory of Elasticity and of the Strength of Materials*, 2 vols. (Cambridge University Press, 1886–1893). A monumental work.

○

Methodology and Philosophy of Physical Science

A little reading in the methods and philosophy of science is recommended. Good books are:

Bacon, Francis. *Of the Advancement of Learning*, Everyman's Library (Dutton, 1915). *Novum Organum* (1620), ed. by T. Fowler (1878, 1889). These two essays will also be found in *The Works of Francis Bacon*, ed. by J. Spedding, R. L. Ellis, and D. D. Heath (London, 1857–1874; Boston, 1860–1864). Since the student will hear much about the Baconian method, it is reasonable that he find out at first hand what it is.

*Bridgman, P. W. *The Logic of Modern Physics* (Macmillan, 1927). An illuminating discussion of the operational method.

Bridgman, P. W. *The Nature of Physical Theory* (Princeton University Press, 1936).

*BURTT, E. A. *The Metaphysical Foundations of Modern Physical Science* (Harcourt, Brace, 1927).

CAMPBELL, N. R. *Physics: The Elements* (Cambridge University Press, 1920).

DESCARTES, RENÉ. *Discourse on the Method of Rightly Conducting the Reason, and Seeking Truth in the Sciences*, Everyman's Library (Dutton); also The Harvard Classics.

HOBSON, E. W. *The Domain of Natural Science* (Macmillan, 1923).

LENZEN, V. F. *The Nature of Physical Theory* (Wiley, 1931).

LINDSAY, R. B. and MARGENAU, H. *Foundations of Physics* (Wiley, 1936).

*PEARSON, K. *Grammar of Science*, ed. 3, Part I (Black, 1911).

POINCARÉ, J. H. *Science and Hypothesis* (Scott, 1905).

POINCARÉ, J. H. *Science and Method*, tr. by Bertrand Russell (Nelson, 1914).

*POINCARÉ, J. H. *The Foundations of Science* (Science Press, 1935). Contains translations by G. B. Halsted of "Science and Hypothesis," "The Value of Science," and "Science and Method."

SCHILLER, F. C. S. *Formal Logic* (Macmillan, 1912). "Logic is a necessary part of the methodological equipment of the man of science."

○

Handbooks

GLAZEBROOK, R. *A Dictionary of Applied Physics*, 5 vols. (Macmillan, 1922–1923). The best reference book of its kind in English.

CHWOLSON, O. D. *Traité de Physique*, 5 vols. (Hermann, 1912–1914).

GEIGER, H., and SCHEEL, K. *Handbuch der Physik*, 24 vols. (Springer, 1926–1927).

WIEN, W., and HARMS, F. *Handbuch der Experimentalphysik*, 25 vols. (Leipzig, 1926–1930).

WINKELMANN, A. *Handbuch der Physik*, 5 vols., ed. 2 (Barth, 1903–1909).

WÜLLNER, A. *Lehrbuch der Experimentalphysik*, 4 vols. (Teubner, 1895–1899).

○

Tables

Handbook of Chemistry and Physics (Chemical Rubber Publishing Co.).

International Critical Tables.

Kaye and Laby, *Physical and Chemical Constants.*

Landolt-Börnstein, *Physikalisch-Chemische Tabellen.*

Smithsonian Physical Tables (Fowle).

Tables Annuelles Internationales de Constantes et Données Numériques.

APPENDIXES

1. *Significant Figures and Notations by Powers of Ten*

The idea of significant figures provides a useful method of indicating the accuracy of a numerical result and at the same time of minimizing the labor of computations. A *significant figure* is a digit that is believed to be nearer the actual value than any other. Zeros are significant if any other digits precede them in the number, otherwise not. Thus there are three significant figures in 204, 3.40, 100, 4.00, and 0.00540. If, in measuring the length of an object with a measuring rod capable of being read to a tenth of a millimeter, the result is two meters, thirty-four centimeters, and no millimeters, this is to be recorded either as 2.3400 m, 234.00 cm, or 2340.0 mm, but not as 2.34 m, 234 cm, or 2340 mm.

If it is desired to indicate that a value lies between 2400 and 2600, and therefore has two significant figures, one obviously cannot write this as 2500, for 2500 has four significant figures and indicates that the value lies between 2499.5 and 2500.5. To avoid this ambiguity, 2500 should be written as 2.5×10^3 when it has only two significant figures, or as 2.50×10^3 if it has three.

Much unnecessary work of calculation can be avoided by employing the following rules for dropping meaningless or nonsignificant figures:

a. In casting off nonsignificant figures, if the value of the rejected figures is greater than a half unit in the last place retained, increase the last digit retained by 1; if it is less than half, leave this digit unchanged; if it equals half a unit, increase the digit by 1 half the time only — for example, when the last retained digit is odd.

b. In sums and differences drop every digit that falls under a nonsignificant digit in any of the quantities to be added or subtracted. Thus, the sum of 216.526, 16.5, and 2.054 is 235.1, not 235.080.

c. In products or quotients retain the number of significant figures that appear in the least accurately known quantity involved. Thus the product of 314.428 and 11.0 is 3.46×10^3, not 3458.7080.

d. In computing with logarithms, when any of the quantities which are to be multiplied or divided can be trusted no closer than 0.01 percent, use a five-place table; if no closer than 0.1 percent, use a four-place table; if no closer than 1 percent, use a slide rule.

e. Where angles are involved, distances expressed to 2, 3, 4, or 5 significant figures call for angles expressed to the nearest 30 min, 5 min, 1 min, or 0.1 min, respectively, and vice versa.

In writing a number as a power of 10, the number is written as the product of two factors: the first factor contains as many digits as there are significant figures, the decimal point always being made to appear at the right of the first digit; the second factor is a power of 10. Thus the statement that the speed of light in vacuum is 2.99796×10^{10} cm · sec^{-1} implies that this speed has been determined to six significant figures.

○

2. Precision of Measurement

A physical measurement is of little value unless its degree of accuracy is known. Errors are always present, and in accurate work means must be devised for reducing them as much as is necessary for the purpose at hand and for estimating the probable amount of the uncorrected error. A knowledge of the errors involved in various measurements also often saves an experimenter considerable time. Thus, when a result is to be calculated from several measured quantities, it often happens that some of the measured quantities need to be measured with great precision, whereas the determination of the other measured quantities beyond a certain easily attained accuracy would be wasted effort. For example, if a measured quantity has to be raised to the nth power in a given formula, the percentage error in the measured quantity introduces an error n times as great in the calculated result; consequently, this particular quantity would have to be measured with relatively great precision, whereas other quantities appearing in the formula to powers lower than n would not need to be measured so accurately. Again, it can be shown that if the desired accuracy of a measurement requires that more than about ten trials or repeated observations be made, it is better to improve the apparatus itself than to perform this additional labor. This follows from the fact that the accuracy of a result increases only with the square root of the number of observations, so that the labor involved grows much faster than the accuracy obtained.

Inaccuracies in measurements may be due to *determinate errors*, which can be discovered and eliminated, and to *indeterminate errors*, which usually can be estimated. Determinate errors are constant in amount and may be due to imperfections in measuring instruments, to faulty methods, or to bias in the observer. Indeterminate errors are of two kinds: (*a*) errors whose existence is unsuspected and therefore obviously not subject to estimation; (*b*) errors which are known as *accidental errors*, whose magnitudes can be estimated by methods based on the laws of probability.

Accidental errors are the irregular discrepancies between successive measurements that always occur when a measurement is repeated *with great care*. Suppose, for example, that an ordinary meter stick is used to measure the length of a table. If one were content to get the length to the nearest centimeter, he could easily get the same result every time the measurement is repeated; but if he uses great care to get the result to the nearest tenth of a millimeter, a slightly different result will be obtained every time the measurement is repeated. Experience shows that in a long series of independent, equally trustworthy observations most of the accidental errors will be of small magnitude, and only a very few will be large. Moreover, positive and negative errors are equally probable. It follows from this last that if one takes the *arithmetic mean* of a large number of readings, the error in the final result should tend to average out.

The amount by which an individual observation in a series differs from the arithmetic mean is called the *residual* of the observation. For example, suppose that eleven measurements of the length of a certain object, made with a meter stick and expressed in centimeters, are

25.06 25.15 25.08 25.10 25.09 25.05 25.12 25.12 25.11 25.09 25.04

The arithmetic mean of the series is 25.09 cm, and the residuals, arranged in order of magnitude without regard to algebraic sign, are

0.0 0.0 0.1 0.1 0.2 **0.3** 0.3 0.3 0.4 0.5 0.6

The middle residual, **0.3**, may be taken as a criterion of the accuracy of the measurements, for, as far as this series may be taken as representative, there is as much likelihood that the error of any individual measurement will be less than 0.3 as there is that it will be greater. This criterion is called the *probable error of a single observation*. It is the quantity which when added to and subtracted from the mean gives limiting values such that, if a single measurement of the same kind is made, it is as likely to lie outside of the limits, on either the positive or the negative side, as it is to lie between them. In the theory of errors it is shown that the probable error e of a single observation in a series of n observations is given by

$$e = \pm 0.6745 \sqrt{\frac{r_1^2 + r_2^2 + \cdots + r_n^2}{n-1}},$$

in which r_1 etc. are the successive residuals. In practice the probable error is always computed from this formula rather than by the illustrative method which we have used.

If a series of n measurements is repeated, it is an even chance that the mean of this second series will differ from the mean of the first series by more or less than $e/\sqrt{n}$, and this quantity is therefore called the *probable error of the mean*.

It is to be emphasized that the probable error indicates nothing whatever with regard to constant errors, which may be very large. It only indicates the agreement of observations *among themselves* as regards the *accidental* errors.

Suggested references on the theory of errors: G. C. Comstock, *Method of Least Squares* (Ginn, 1890), pp. 1–4; M. Merriman, *Method of Least Squares* (Wiley, 1901), pp. 1–18; H. M. Goodwin, *Precision of Measurements and Graphical Methods* (McGraw-Hill, 1913).

o

3. Common Approximations

In order to avoid laborious computations, approximation formulas should be used wherever possible. Whenever an approximation suggests itself, however, the error introduced by using it should be investigated and the approximation not made unless this error is small enough to leave unaffected any figure that otherwise could be trusted in the result.

True value	Approximate value	When applicable	Approximate error introduced
$1 + a + a^2$ . . .	$1 + a$	a small	$- a^2$ *
$(1 + a)(1 + b)$ . .	$1 + a + b$	a and b small	$- ab$
$(1 \pm a)^m$	$1 \pm ma$ †	a small	$- \frac{1}{2} m(m - 1)a^2$
$\sqrt{ab}$	$\frac{1}{2}(a + b)$	b nearly equal to a	$+ (b - a)^2/8\,a$
$\sin a$	a radians	a small	$+ a^3/6$
$\cos a$	1	a small	$+ a^2/2$
$\tan a$	a radians	a small	$- a^3/3$
$\tan a$	$\sin a$	a small	$- a^3/2$

* For example, when $a = 0.1$, the error is 1 percent; when $a = 0.01$, the error is 0.01 percent.

† m may be either a positive or a negative integer or a fraction.

o

4. The Micrometer Caliper

To make a measurement with the micrometer caliper (Fig. 257), place the object between the jaws and turn up the milled head until, with light pressure between thumb and finger, the head slips through the fingers instead of rotating farther. Never crowd the screw. Without removing the object read upon the scales the separation of the jaws to 0.001 mm. In the usual type of metric instrument, graduations on the fixed scale are in millimeters and on the movable scale in hundredths of a millimeter. The pitch of the screw is ordinarily 0.5 mm; care must therefore be taken to note whether the reading is, for example, 5.068 or 4.568, since the reading on the movable scale is the same in both cases.

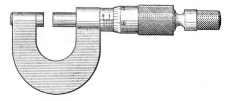

Remove the object and close the jaws, being care-

Fig. 257. A micrometer caliper

ful to use the same pressure as before, and take the *zero reading*. Correct the first reading for this zero error of the instrument.

Take great care always to exercise the same pressure on the milled head. In some types of instrument the head is equipped with a ratchet designed to slip as soon as a certain pressure is exceeded; in this case the ratchet should be caused to slip always the same number of notches.

o

5. The Vernier Scale

The *vernier*, named after PIERRE VERNIER, who gave it its present form in 1631, is an auxiliary sliding scale which enables the observer to increase the accuracy of his estimation of a fractional portion of the smallest division of the main scale. Usually verniers are so ruled that n divisions on the vernier correspond

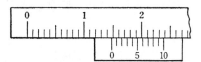

Fig. 258. The vernier. The reading is 1.48

to $n-1$ of the main-scale divisions, and the smallest indication thereby given is $1/n$ of a division. For example, in Fig. 258 nine divisions of the main scale correspond to ten on the vernier, and the smallest indication is one tenth of a division of the main scale. To make a reading, observe the left-hand division on the main scale that is nearest to the zero mark on the ver-

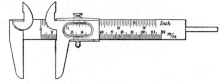

Fig. 259. Caliper equipped with vernier

nier. (In Fig. 258 this is 1.4 cm.) Then observe the mark on the vernier that lies in the same straight line as some mark on the main scale. (In Fig. 258 this happens to be the eighth line.) This gives the fractional portion of the smallest scale division which must be added to the main-scale reading. (Thus, in Fig. 258 the complete reading is 1.48 cm.)

o

6. The Mercurial Barometer

The mercurial barometer is the most reliable instrument for the measurement of atmospheric pressure. In reading it, take the following steps in order :

a. Read the temperature.

b. Adjust the level of the mercury in the cistern of the barometer by turning the screw at the bottom until the mercury surface just touches the tip of the ivory pointer in the cistern; this is best done by observing when the image of the pointer formed by the mercury surface appears just to touch the pointer itself.

c. Gently tap the upper part of the barometer tube to free the mercury surface from the wall of the tube and then adjust the vernier by means of the thumbscrew at the side until, on looking through the slit in the barometer case, the upper part of the mercury meniscus is seen to be just tangent to the line joining the sharp edges at the front and back of the vernier.

d. Read the main scale and vernier.

e. Employ Tables G, H, and I, Appendix 13, to correct the observed barometric height for a capillary depression and to reduce the observed height to 0° C and g_s, when such corrections and reductions are necessary.

o

7. Centers of Mass by Integration

In most cases the location of the center of mass must be accomplished by imagining the body divided into small elements and performing an infinite summation or integration. The element of mass Δm is $\rho \Delta V$, where ρ is the density and ΔV is the element of volume ; hence Eqs. [70], Chap. 6, take the form

$$x_c = \frac{\int \rho x\, dV}{\int \rho\, dV}, \quad y_c = \frac{\int \rho y\, dV}{\int \rho\, dV}, \quad z_c = \frac{\int \rho z\, dV}{\int \rho\, dV},$$

where $\int \rho\, dV$ is the mass M of the body and the integrals are taken over the whole body.

Case 1. Straight rod of length l, the density of which varies as the nth power of the distance from one end. Let x be the distance of any mass element Δm from the end of zero density; then $\rho = kx^n$, and hence

$$x_c = \frac{\int_0^l kx^{n+1}dx}{\int_0^l kx^n dx} = \left[\frac{\dfrac{x^{n+2}}{n+2}}{\dfrac{x^{n+1}}{n+1}}\right]_0^l = \frac{n+1}{n+2}\, l.$$

If $n = 0$, $x_c = l/2$. If $n = 1$, $x_c = 2\,l/3$; this result could also be applied to the area of a triangle or to the surface of a cone or pyramid (why?). If $n = 2$, $x_c = 3\,l/4$; this result could be applied to the case of a solid cone or pyramid (why?).

Case 2. Circular arc (Fig. 260).

$$x_c = \frac{\int_{-\alpha}^{\alpha} r\cos\phi\,(\rho r\,d\phi)}{\int_{-\alpha}^{\alpha}\rho r\,d\phi} = \frac{r\int_{-\alpha}^{\alpha}\cos\phi\,d\phi}{\int_{-\alpha}^{\alpha}d\phi} = \frac{r\sin\alpha}{\alpha} = \text{radius}\cdot\frac{\text{chord}}{\text{arc}}.$$

For a semicircle, $\alpha = \pi/2$; hence $x_c = 2\,r/\pi = 0.6366\,r$.

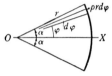

Fig. 260. Circular arc   Fig. 261. Segment of a circle   Fig. 262. Sector of a circle

Case 3. Segment of a circle (Fig. 261).

$$x_c = \frac{2}{3}\, r\, \frac{\sin^3\alpha}{\alpha - \cos\alpha\sin\alpha}.$$

Case 4. Sector of a circle (Fig. 262).

$$x_c = \frac{2}{3}\, r\, \frac{\sin\alpha}{\alpha}.$$

For a semicircular plate, $x_c = 4\,r/3\,\pi$.

○

8. Moments of Inertia

Case 1. Thin ring, or a cylinder with thin walls, about an axis through the center and normal to the circular section.

$$I = \int_0^M r^2\,dm = MR^2.$$

Case 2. Thin circular ring about any diameter. a. $I_x = I_y = \frac{1}{2}\,(I_x + I_y) = \frac{1}{2}\,\Sigma m\,(x^2 + y^2) = \frac{1}{2}\,\Sigma mr^2 = \frac{1}{2}\,MR^2$; thus, in general,

$$I_x + I_y = I_z.$$

b. This result can also be obtained by integration as follows (Fig. 263). Let σ be the mass per unit length; then

$$I = \int_0^M r^2 \, dm = 2 \int_0^\pi R^2 \sin^2 \theta \sigma R \, d\theta = 2 \, \sigma R^3 \int_0^\pi \sin^2 \theta \, d\theta$$
$$= 2 \, \sigma R^3 [-\tfrac{1}{2} \cos \theta \sin \theta + \tfrac{1}{2} \, \theta]_0^\pi$$
$$= \sigma R^3 \cdot \pi = \tfrac{1}{2} \, MR^2,$$

since $2 \, \pi R \sigma = M$.

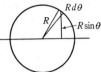

Fig. 263. Thin circular ring

Case 3. Solid disk or cylinder about an axis through the center and normal to the circular section. *a.* The cylinder may be regarded as made of a series of concentric cylinders of radii r, thickness dr, and length L; then

$$I = \int_0^M r^2 \, dm = \int_0^R r^2 \cdot 2 \, \pi r \, dr \, L\rho = 2 \, \pi L\rho \int_0^R r^3 \, dr$$
$$= 2 \, \pi L\rho \, \frac{R^4}{4} = \frac{1}{2} \, MR^2,$$

since $M = \pi R^2 L\rho$.

b. By a more general method (Fig. 264),

$$I = \int_0^M r^2 \, dm = \int_0^{2\pi} \int_0^R r^2 \cdot r \, d\theta \, dr \, L\rho$$
$$= L\rho \int_0^{2\pi} \int_0^R r^3 \, d\theta \, dr = L\rho \, \frac{R^4}{4} \int_0^{2\pi} d\theta$$
$$= \frac{\pi L\rho R^4}{2} = \frac{1}{2} \, MR^2.$$

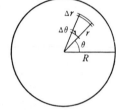

Fig. 264. Solid circular cylinder rotating about its longitudinal axis

Case 4. Solid disk about any diameter.

a. $I_x = I_y = \tfrac{1}{2}(I_x + I_y) = \tfrac{1}{2} \Sigma m(x^2 + y^2) = \tfrac{1}{2} \Sigma m r^2 = \tfrac{1}{4} \, MR^2 = \tfrac{1}{2} I_z.$

b. By integration,

$$I = \int_0^{2\pi} \int_0^R r^2 \sin^2 \theta \, r \, d\theta \, dr\rho = \rho \, \frac{R^4}{4} \cdot \pi = \frac{1}{4} \, MR^2.$$

Case 5. Thin rod or wire about an axis normal to the length, at one end. If dl be an element of length, distant l from the axis, and σ be the mass per unit length,

$$I = \int_0^L l^2 \sigma \, dl = \sigma \, \frac{L^3}{3} = \frac{1}{3} \, ML^2.$$

Case 6. Thin rod or wire about an axis normal to the length, through the center.

$$I = \int_{-\frac{L}{2}}^{\frac{L}{2}} l^2 \sigma \, dl = \sigma \left[\frac{l^3}{3} \right]_{-\frac{L}{2}}^{\frac{L}{2}} = \frac{\sigma}{3} \, \frac{L^3}{4} = \frac{1}{12} \, ML^2.$$

The student should check this result by changing it to that of the preceding case by means of Lagrange's transfer theorem.

Case 7. Rectangular bar of length L, width a, and thickness b, about an axis parallel to b through the center of mass (Fig. 265).

$$I = \int_{-\frac{L}{2}}^{\frac{L}{2}} \int_{-\frac{a}{2}}^{\frac{a}{2}} (x^2 + y^2) dx\, dy\, b\rho = \rho b \int_{-\frac{L}{2}}^{\frac{L}{2}} \left(x^2 a + \frac{a^3}{12} \right) dx$$

$$= \frac{\rho b a L}{12} (L^2 + a^2) = \frac{1}{12} M(L^2 + a^2).$$

Case 8. Solid sphere about any diameter. a. Obtain the expression for a thin spherical shell about any diameter and then regard the solid sphere as made up of a series of concentric shells.

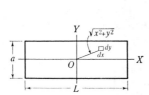

FIG. 265. Rectangular bar

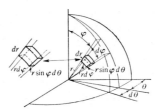

FIG. 266. Solid sphere

b. Employing polar coordinates (Fig. 266),

$$dm = r \sin \phi\, d\theta \cdot r\, d\phi \cdot dr \cdot \rho,$$

and therefore

$$I = 8 \int_0^{\frac{\pi}{2}} \int_0^{\frac{\pi}{2}} \int_0^R r^2 \sin^2 \phi \cdot r^2 \sin \phi\, d\theta\, d\phi\, dr \cdot \rho$$

$$= 8\, \rho\, \frac{R^5}{5} \int_0^{\frac{\pi}{2}} \int_0^{\frac{\pi}{2}} \sin^3 \phi\, d\phi\, d\theta$$

$$= 8\, \rho\, \frac{R^5}{5} \frac{\pi}{2} \int_0^{\frac{\pi}{2}} \sin^3 \phi\, d\phi = \frac{4\, \pi \rho R^5}{5} \frac{2}{3} = \frac{2}{5}\, MR^2.$$

c. Employing Cartesian coordinates,

$$I = \int_0^R \int_0^{\sqrt{R^2 - x^2}} \int_0^{\sqrt{R^2 - x^2 - y^2}} (x^2 + y^2) dx\, dy\, dz \cdot \rho = \tfrac{2}{5}\, MR^2.$$

Case 9. General rule. The moment of inertia about any axis of symmetry is given by

$$I = M \frac{\text{sum of squares of perpendicular semi-axes}}{3,\ 4,\ \text{or}\ 5},$$

where the denominator is to be 3, 4, or 5 according as the body is rectangular, elliptical (including circular), or ellipsoidal (including spherical).

Case 10. Solid circular cylinder about an axis perpendicular to the axis of symmetry, through the center. If the radius of the cylinder is R and its length is L,

$$I = M \left(\frac{R^2}{4} + \frac{L^2}{12} \right).$$

○

9. The Micrometer Microscope

The saw-toothed index seen in the field of view of the micrometer microscope (Fig. 267) is intended to facilitate the counting of whole turns of the micrometer screw without removing the eye from the eyepiece. The graduated disk on the micrometer screw is usually held in position by a friction washer to permit setting it to a proper position. When the moving cross hair coincides with one of the notches, this disk should be at the zero position. In using the microscope, proceed as follows:

FIG. 267. Field of view of micrometer microscope

a. By sliding the eyepiece only, bring the cross hairs into good focus. Then bring the object into focus by moving the whole microscope tube. The test of a good focus is the absence of parallax, that is, no relative motion of the cross hairs and the image when the eye is moved sideways.

b. To measure a length, make a number of settings of the movable cross hair first on one end and then on the other, and read the position for each setting. Thus determine the required length in terms of the number of turns and fractions of a turn of the micrometer screw. To avoid errors due to lost motion, the final motion of the screw in making settings should always be in the same direction.

c. Observe on a standard scale the number of turns and fractions of a turn that correspond to 1 mm.

○

10. Standardization of a Thermometer

Thermometer readings must in general be corrected (*a*) for the errors of the instrument itself and (*b*) for the length of the exposed thread of mercury. If the first correction is to be made accurately, the thermometer should be compared with a standard thermometer; a method for making this comparison is suggested in the optional laboratory problem accompanying Exp. IXA. If it is not practicable to compare the thermometer with a standard thermometer, the corrections may be obtained with a moderate degree of accuracy by observing the corrections at the steam point and the ice point, and interpolating between these two points for the corrections at other temperatures (Fig. 268).

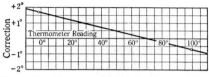

FIG. 268. This correction curve for a certain centigrade thermometer was made from the following data: when in ice the thermometer read − 1.7°; when immersed to the top of the thread in a steam bath of temperature 99.1°, it read 99.8°. The corrections for readings of − 1.7° and 99.8° are therefore + 1.7° and − 0.7° respectively

The correction for the exposed thread (see Prob. 15, Chap. 9) is obtained by adding to the observed temperature the quantity $0.00016(t − t')l$, in which 0.00016 is the apparent expansivity of mercury in glass [0.000181 (mercury) − 0.000025 (glass)], t is the observed temperature corrected according to (a), t' is the mean temperature of the exposed stem as obtained from a second thermometer whose bulb hangs near the middle of this stem, and l is the length in degrees of the exposed thread of mercury.

For further details concerning the standardization of thermometers see E. Griffiths, *Methods of Measuring Temperature* (Griffin, 1918), Chap. II, or C. W. Waidner and H. C. Dickinson, *Bulletin of the Bureau of Standards* **3**, 663 (1907).

○

11. Greek Letters Used as Symbols

α	Alpha	μ	Mu
β	Beta	ν	Nu
γ	Gamma	π	Pi
δ	Delta	ρ	Rho
Δ	Delta (capital)	σ	Sigma
ε	Epsilon	Σ	Sigma (capital)
η	Eta	τ	Tau
θ	Theta	φ	Phi
κ	Kappa	ψ	Psi
λ	Lambda	ω	Omega

12. Symbols Used in This Textbook

(See also Appendix 13, Tables K and N)

a = linear acceleration

a = linear acceleration, magnitude of

a, b = van der Waals constants

A_s = standard atmosphere

B = barometric pressure

c = specific heat

c_p = specific heat at constant pressure

c_v = specific heat at constant volume

C = a constant

e = coefficient of restitution; base of the system of natural logarithms

E = kinetic energy

f, F = forces

f, F = force, magnitudes of

g = acceleration due to gravity

g_0 = acceleration due to gravity, at sea level

g_s = acceleration due to gravity, standard

G = gravitational constant

H = angular momentum

h, H = vertical distances

I = moment of inertia

J = Joule's equivalent

k, K = constants

k = volume modulus of elasticity; radius of gyration

k_0 = Boltzmann constant

°K = degrees Kelvin (absolute centigrade scale)

l = length; fractional loss of kinetic energy in impact

L = torque

L = torque, magnitude of

L_o = constant of torsion, or torque per unit twist

$[L]$ = dimension of length

m, M = mass

$[M]$ = dimension of mass

n = unit vector normal to the path

n = shear modulus of elasticity

n_0 = Loschmidt number

N_0 = Avogadro number

N = normal force; number of moles

p = momentum

p = momentum, magnitude of

P = power; pressure

Q = quantity of heat

q = specific humidity

r = position vector

r = relative humidity; radius

R = linear impulse

R = linear impulse, magnitude of; gas constant per mole; radius

R' = gas constant per gram

$\Re$ = angular impulse

s = linear displacement

s = distance;
  length of arc
S = capillary constant
t = time elapsed;
  temperature
T = absolute temperature;
  period of an oscillation
T_0 = ice point, absolute scale
$[T]$ = dimension of time
u, U = linear speeds
v = linear velocity
v, V = linear speeds
V = total volume;
  volume per mole;
  potential energy
W = weight;
  work
x, y, z = rectangular coordinates
Y = Young's modulus of
  elasticity
α, β, γ = direction angles (angles
  between a line and
  the X-, Y-, Z-axes)
α = angular acceleration
α = angular acceleration,
  magnitude of;
  volume expansivity;
  angle of contact
β = pressure coefficient of
  expansion
γ = ratio c_p/c_v of specific
  heats
δ, Δ = small increase in $\cdots$

η = coefficient of viscosity
θ = angular displacement
θ = angle;
  angular distance
$[\theta]$ = dimension of temperature
λ = linear expansivity;
  mean free path of a gas
  molecule;
  wavelength
μ = coefficient of friction;
  coefficient of diffusion
ν = frequency
ρ = density;
  absolute humidity
σ = surface density;
  linear density;
  coefficient of surface
  tension;
  Poisson ratio
Σ = the sum of all terms
  such as $\cdots$
τ = unit vector along the
  tangent to the path
τ = time interval;
  dew point
ϕ, ψ = angles
ω = angular velocity
ω = angular speed
$\partial/\partial x$ = partial derivative with
  respect to x, all other
  variables being kept
  constant

13. Tables

TABLE A · *Approximate Densities of Various Substances* ($g \cdot cm^{-3}$)

Solids			
Aluminum	2.7	Iron, wrought	*c.* 7.8
Brass	*c.* 8.6	Lead	11.3
Brick	*c.* 2.0	Nickel	8.8
Copper	8.9	Oak	*c.* 0.8
Cork	*c.* 0.24	Pine	*c.* 0.5
Glass, common crown	*c.* 2.6	Platinum	21.4
Glass, flint	2.9–5.9	Silver	10.5
Gold	19.3	Steel	7.8
Ice at 0° C	0.917	Tin	7.3
Iron, cast	*c.* 7.6	Zinc	7.1
Liquids at 0° C			
Alcohol, ethyl	0.807	Gasoline	0.66–0.69
Benzene	0.899	Glycerin	1.26
Ether	0.736	Mercury	13.595
Gases at 0° C, 76 cm of Mercury Pressure			
Air	0.001293	Hydrogen	0.0000899
Argon	0.001784	Neon	0.000900
Carbon dioxide	0.001977	Nitrogen	0.001251
Helium	0.000178	Oxygen	0.001429

TABLE B · *Relative Density of Water*

t° C	Density	t° C	Density	t° C	Density
0°	0.99987	15°	0.99913	30°	0.99568
1	0.99993	16	0.99897	35	0.99406
2	0.99997	17	0.99880	40	0.99225
3	0.99999	18	0.99862	45	0.99025
4	1.00000	19	0.99843	50	0.98807
5	0.99999	20	0.99823	55	0.98572
6	0.99997	21	0.99802	60	0.98324
7	0.99993	22	0.99780	65	0.98059
8	0.99988	23	0.99757	70	0.97781
9	0.99981	24	0.99733	75	0.97489
10	0.99973	25	0.99708	80	0.97183
11	0.99963	26	0.99682	85	0.96865
12	0.99952	27	0.99655	90	0.96534
13	0.99940	28	0.99627	95	0.96192
14	0.99927	29	0.99598	100	0.95838

TABLE C · *Density of Mercury* $(g \cdot cm^{-3})$

$t°$ C	Density	$t°$ C	Density	$t°$ C	Density
0°	13.595	22°	13.541	40°	13.497
5	583	24	536	50	473
10	570	26	531	60	448
12	566	28	526	70	424
14	561	30	521	80	400
16	556	32	516	90	372
18	551	34	512	100	352
20	546	36	507	120	304

TABLE D · *Density of Dry Air* $(g \cdot cm^{-3})$ *at Temperature t and Pressure B Millimeters of Mercury*

$t°C$	$B = 720$	730	740	750	760	770
10°	0.001182	0.001198	0.001215	0.001231	0.001247	0.001264
11	178	193	210	227	243	259
12	173	190	206	222	239	255
13	169	186	202	218	234	251
14	165	181	198	214	230	246
15	0.001161	0.001177	0.001193	0.001210	0.001226	0.001242
16	157	173	189	205	221	238
17	153	169	185	201	217	233
18	149	165	181	197	213	229
19	145	161	177	193	209	225
20	0.001141	0.001157	0.001173	0.001189	0.001205	0.001221
21	137	153	169	185	201	216
22	134	149	165	181	197	212
23	130	145	161	177	193	208
24	126	142	157	173	189	204
25	0.001122	0.001138	0.001153	0.001169	0.001185	0.001200
26	118	134	149	165	181	196
27	115	130	146	161	177	192
28	111	126	142	157	173	188
29	107	123	138	153	169	184
30	104	119	134	150	165	180

Correction for Moisture in Above Table

Dew point	Subtract	Dew point	Subtract	Dew point	Subtract	Dew point	Subtract
− 10°	0.000001	0°	0.000003	10°	0.000006	20°	0.000010
− 8	02	2	03	12	06	22	12
− 6	02	4	04	14	07	24	13
− 4	02	6	04	16	08	26	15
− 2	03	8	05	18	09	28	16

TABLE E · *Density and Pressure of Saturated Water Vapor*

Showing the pressure P (mm of mercury) and the density ρ (g · cm^{-3}) of aqueous vapor saturated at temperature $t°$ C.; or showing the boiling point t of water and density ρ of steam corresponding to an outside pressure P.

$t°$	P	ρ	$t°$	P	ρ	$t°$	P	ρ
− 10	2.1	2.2×10^{-6}	21	18.7	18.3×10^{-6}	98.4	717.6	...
− 9	2.3	2.3	22	19.8	19.4	98.6	722.8	...
− 8	2.5	2.6	23	21.1	20.6	98.8	728.0	...
− 7	2.7	2.8	24	22.4	21.8	99	733.2	579×10^{-6}
− 6	2.9	3.0	25	23.8	23.0	99.2	738.5	...
− 5	3.2	3.3	26	25.2	24.4	99.4	743.9	...
− 4	3.4	3.5	27	26.8	25.8	99.6	749.2	...
− 3	3.7	3.8	28	28.4	27.2	99.8	754.6	...
− 2	4.0	4.1	29	30.1	28.8	100	760.0	598
− 1	4.3	4.5	30	31.9	30.4	100.2	765.5	...
0	4.6	4.8	35	42.2	39.6	100.4	770.9	...
1	4.9	5.2	40	55.4	51.1	100.6	776.4	...
2	5.3	5.6	45	72.0	65.6	100.8	782.0	...
3	5.7	5.9	50	92.6	83.2	101	787.5	618
4	6.1	6.4	55	118.2	104.6	102	815.9	639
5	6.5	6.8	60	149.6	130.5	103	845.0	661
6	7.0	7.3	65	187.8	161.5	104	875.1	683
7	7.5	7.8	70	233.9	198.4	105	906.0	705
8	8.0	8.3	75	289.3	242.1	110	1074	827
9	8.6	8.8	80	355.4	293.8	120	1489	1122
10	9.2	9.4	85	433.7	354.1	130	2025	1498
11	9.9	10.0	90	526.0	424.1	140	2709	1968
12	10.5	10.7	91	546.3	439.5	150	3568	2550
13	11.2	11.3	92	567.2	455.2	170	5936	4127
14	12.0	12.1	93	588.8	471.3	190	9404	6390
15	12.8	12.8	94	611.1	487.8	195	10480	7090
16	13.6	13.6	95	634.1	505	196	10700	7230
17	14.5	14.5	96	657.8	523	197	10940	7380
18	15.5	15.4	97	682.2	541	198	11170	7530
19	16.5	16.3	98	707.3	560	199	11410	7690
20	17.6	17.3	98.2	712.4	...	200	11650	7840

TABLE F · *Approximate Values of Moduluses of Elasticity* ($dyne \cdot cm^{-2}$)

Substance	Volume modulus, k	Shear modulus, n	Young's modulus, Y
Aluminum.	7.5×10^{11}	3.2×10^{11}	7.0×10^{11}
Brass	11.	3.5	9.0
Copper	13.	4.2	12.
Glass	4.5	2.4	6.0
Iron (drawn)	16.	8.0	20.
Iron (cast)	9.6	5.3	9.0
Mercury	2.6	. . .	. . .
Phosphor bronze . . .	. . .	4.3	. . .
Quartz fiber	. . .	2.8	. . .
Silver	10.	2.6	7.5
Steel	18.	8.2	20.
Water	0.22	. . .	. . .

TABLE G · *Reduction of Barometer Reading to 0° C*

The corrections represent the number of millimeters to be subtracted from readings observed on a mercurial barometer equipped with a brass scale.

$t°$ C	Observed reading (mm)							
	700	710	720	730	740	750	760	770
10	1.14	1.16	1.17	1.19	1.21	1.22	1.24	1.26
11	1.26	1.27	1.29	1.31	1.33	1.35	1.36	1.38
12	1.37	1.39	1.41	1.43	1.45	1.47	1.49	1.51
13	1.48	1.50	1.53	1.55	1.57	1.59	1.61	1.63
14	1.60	1.62	1.64	1.67	1.69	1.71	1.73	1.76
15	1.71	1.74	1.76	1.78	1.81	1.83	1.86	1.88
16	1.82	1.85	1.88	1.90	1.93	1.96	1.98	2.01
17	1.94	1.97	1.99	2.02	2.05	2.08	2.10	2.13
18	2.05	2.08	2.11	2.14	2.17	2.20	2.23	2.26
19	2.17	2.20	2.23	2.26	2.29	2.32	2.35	2.38
20	2.28	2.31	2.34	2.38	2.41	2.44	2.47	2.51
21	2.39	2.43	2.46	2.50	2.53	2.56	2.60	2.63
22	2.51	2.54	2.58	2.61	2.65	2.69	2.72	2.76
23	2.62	2.66	2.69	2.73	2.77	2.81	2.84	2.88
24	2.73	2.77	2.81	2.85	2.89	2.93	2.97	3.01
25	2.85	2.89	2.93	2.97	3.01	3.05	3.09	3.13

TABLE H · *Capillary Depression of Mercury in Glass (mm)*

Diameter of tube (mm)	Height of the meniscus (mm)							
	0.4	**0.6**	**0.8**	**1.0**	**1.2**	**1.4**	**1.6**	**1.8**
4	0.83	1.22	1.54	1.98	2.37	...	...	...
5	0.47	0.65	0.86	1.19	1.45	1.80	...	...
6	0.27	0.41	0.56	0.78	0.98	1.21	1.43	...
7	0.18	0.28	0.40	0.53	0.67	0.82	0.97	1.13
8		0.20	0.29	0.38	0.46	0.56	0.65	0.77
9		0.15	0.21	0.28	0.33	0.40	0.46	0.52
10			0.15	0.20	0.25	0.29	0.33	0.37
11			0.10	0.14	0.18	0.21	0.24	0.27
12			0.07	0.10	0.13	0.15	0.18	0.19
13			0.04	0.07	0.10	0.12	0.13	0.14

TABLE IA · *Reduction of Barometer to g at Sea Level*

Correction to be subtracted (mm)

Altitude, H (meters)	Observed reading (mm)						
	500	**550**	**600**	**650**	**700**	**750**	**800**
500	...	...	...	...	0.11	0.12	0.13
1000	...	0.18	0.19	0.20	22	24	
1500	0.24	26	28	30	33		
2000	31	34	38	41			
2500	39	43	47				

TABLE IB · *Reduction of Barometer to g at Latitude 45°*

The correction is in millimeters; subtract for latitudes less than 45°; add for latitudes greater than 45°.

Latitude		Observed reading (mm)					
		680	**700**	**720**	**740**	**760**	**780**
25°	65°	1.16	1.20	1.23	1.27	1.30	1.33
30°	60°	0.91	0.94	0.96	0.98	1.01	1.04
35°	55°	0.62	0.64	0.66	0.67	0.69	0.71
40°	50°	0.31	0.32	0.33	0.34	0.35	0.36

TABLE J · *Volume and Linear Expansivities (per deg C)*

Volume Expansivity, α		
Alcohol . . 1.012×10^{-3}	Glass . . . 2.5×10^{-5}	Mercury . 18.18×10^{-5}

Linear Expansivity, λ		
Aluminum . 2.3×10^{-5}	Iron, cast . 1.05×10^{-5}	Quartz . . 0.057×10^{-5}
Brass . . . 1.8	Iron, soft . 1.2	Silver . . . 1.9
Copper . . 1.7	Lead . . 2.9	Steel . . . 1.1
Gold 1.4	Platinum . 0.9	Zinc . . . 2.6

TABLE K · *Probable Values of Some Important Physical Constants*

Selected from R. T. Birge, "Probable Values of the General Physical Constants," *Reviews of Modern Physics* 1, 1 (1929).

Gravitational constant	$G = (6.664 \pm 0.002) \times 10^{-8}$ dyne $\cdot$ cm$^2 \cdot$ g^{-2}
Liter	$l = 1000.027 \pm 0.001$ cm^3
Volume of ideal gas (0° C, A$_s$)	$V_s = (22.4141 \pm 0.0008) \times 10^3$ cm$^3 \cdot$ mole^{-1}
Atomic weights . . .	H = 1.00777 ± 0.00002 N = 14.0083 ± 0.0008
	He = 4.0022 ± 0.0004 O = 16.0000
	C = 12.003 ± 0.001 Ag = 107.880 ± 0.001
Standard atmosphere.	$A_s = (1.013249 \pm 0.000003) \times 10^6$ dyne $\cdot$ cm^{-2}
Ice point (absolute scale)	$T_0 = 273.18 \pm 0.03°$ K
Joule's equivalent . .	$J_{15} = 4.1852 \pm 0.0006$ j $\cdot$ cal$_{15}^{-1}$
Planck constant . . .	$h = (6.547 \pm 0.008) \times 10^{-27}$ erg $\cdot$ sec
Acceleration due to gravity, standard .	$g_s = 980.665$ cm $\cdot$ sec^{-2}
Maximum density of water	0.999973 ± 0.000001 g $\cdot$ cm^{-3}
Density of oxygen gas (0° C, A$_{45}$)	1.428965 ± 0.000030 g $\cdot$ l^{-1}
Factor converting oxygen (0° C, A$_{45}$) to ideal gas	1.000927 ± 0.000030
Density of nitrogen (0° C, A$_{45}$)	1.25046 ± 0.000045 g $\cdot$ l^{-1}
Factor converting nitrogen (0° C, A$_{45}$) to ideal gas	1.00043 ± 0.00002
Density of mercury (0° C, A$_s$)	13.59509 ± 0.00003 g $\cdot$ cm^{-3}
Avogadro number . .	$N_0 = (6.064_{36} \pm 0.006) \times 10^{23}$ mole^{-1}
Gas constant per mole	$R = V_sA_s/T_0 = (8.3136_0 \pm 0.0010) \times 10^7$ erg $\cdot$ deg$^{-1} \cdot$ mole^{-1}
	$= 1.9864_3 \pm 0.0004$ cal$_{15} \cdot$ deg$^{-1} \cdot$ mole^{-1}
Boltzmann constant .	$k_0 = R/N_0 = (1.3708_9 \pm 0.0014) \times 10^{-16}$ erg $\cdot$ deg^{-1}
Mass of hydrogen atom	$M_H = H/N_0 = (1.6617_9 \pm 0.0017) \times 10^{-24}$ g
Number of atoms per gram of hydrogen .	$1/M_H = (6.017_{61} \pm 0.006) \times 10^{23}$ g^{-1}
Loschmidt number . .	$n_0 = N_0/V_s = (2.705_{60} \pm 0.003) \times 10^{19}$ cm^{-3} (0° C, 1 A$_s$)

TABLE L · *Specific Heats* $(cal \cdot g^{-1} \cdot deg^{-1})$

Alcohol, ethyl	0° C	0.548
	40°	0.648
Aluminum	0°	0.208
	100°	0.225
Brass	0°	0.089
Copper	0°–300°	$0.0915 + 2.4 \times 10^{-5} t$
Glass, thermometer . . .	19°–100°	0.199
Glass, crown	10°–50°	0.161
Gold	0°	0.0302
	18°	0.0312
	100°	0.0314
Ice	0°	0.487
Iron.	0°–400°	$0.1060 + 9.6 \times 10^{-5} t$
Lead	0°–300°	$0.0295 + 2 \times 10^{-5} t$
Mercury	0°	0.0335
	15°	0.0333
	30°	0.0332
	100°	0.0327
Nickel	0°–300°	$0.1020 + 1.18 \times 10^{-4} t - 6 \times 10^{-8} t^2$
Platinum	0°–1625°	$0.03162 + 6.17 \times 10^{-6} t + 2.33 \times 10^{-10} t^2$
Quartz	12°–100°	0.188
Silver	0°–400°	$0.0556 + 8 \times 10^{-6} t$
Tin	0°–200°	$0.0525 + 5.2 \times 10^{-5} t$
Wood	20°	0.327
Zinc.	0°–300°	$0.0913 + 4.4 \times 10^{-5} t$

TABLE M · *Coefficients of Surface Tension of Liquids in Contact with Air*
$(dyne \cdot cm^{-1}, \text{ or } erg \cdot cm^{-2})$

Alcohol, ethyl, 15° C	22.7	Mercury, 20° C	520.
20° C	22.3	Water, 15° C	73.4
25° C	21.8	20° C	72.7
Ether, 20° C	17.0	25° C	71.9

TABLE N · *Important Numbers and Conversion Factors*

$\pi = 3.1416$	$\pi^2 = 9.8696$	$1/\pi = 0.31831$	$\log \pi = 0.49715$

Base of the natural system of logarithms, $e = 2.7183$

1 in. = 2.54 cm	1 m = 39.37 in.	1 mi = 1.609 km
1 kg = 2.2 lb	1 oz = 28.35 g	1 grain = 64.8 mg]
1μ (micron) $= 10^{-6}$ m	$1 m\mu = 10^{-9}$ m	$1 \mu\mu = 10^{-12}$ m

1 Å (Ångstrom unit) $= 10^{-10}$ m

Weight of 1 ft³ of water = 62.4 lbwt

TABLE O · *Four-Place Logarithms*

	0	1	2	3	4	5	6	7	8	9	1 2 3	4 5 6	7 8 9
10	0000	0043	0086	0128	0170	0212	0253	0294	0334	0374	4 8 12	17 21 25	29 33 37
11	0414	0453	0492	0531	0569	0607	0645	0682	0719	0755	4 8 11	15 19 23	26 30 34
12	0792	0828	0864	0899	0934	0969	1004	1038	1072	1106	3 7 10	14 17 21	24 28 31
13	1139	1173	1206	1239	1271	1303	1335	1367	1399	1430	3 6 10	13 16 19	23 26 29
14	1461	1492	1523	1553	1584	1614	1644	1673	1703	1732	3 6 9	12 15 18	21 24 27
15	1761	1790	1818	1847	1875	1903	1931	1959	1987	2014	3 6 8	11 14 17	20 22 25
16	2041	2068	2095	2122	2148	2175	2201	2227	2253	2279	3 5 8	11 13 16	18 21 24
17	2304	2330	2355	2380	2405	2430	2455	2480	2504	2529	2 5 7	10 12 15	17 20 22
18	2553	2577	2601	2625	2648	2672	2695	2718	2742	2765	2 5 7	9 12 14	16 19 21
19	2788	2810	2833	2856	2878	2900	2923	2945	2967	2989	2 4 7	9 11 13	16 18 20
20	3010	3032	3054	3075	3096	3118	3139	3160	3181	3201	2 4 6	8 11 13	15 17 19
21	3222	3243	3263	3284	3304	3324	3345	3365	3385	3404	2 4 6	8 10 12	14 16 18
22	3424	3444	3464	3483	3502	3522	3541	3560	3579	3598	2 4 6	8 10 12	14 15 17
23	3617	3636	3655	3674	3692	3711	3729	3747	3766	3784	2 4 6	7 9 11	13 15 17
24	3802	3820	3838	3856	3874	3892	3909	3927	3945	3962	2 4 5	7 9 11	12 14 16
25	3979	3997	4014	4031	4048	4065	4082	4099	4116	4133	2 3 5	7 9 10	12 14 15
26	4150	4166	4183	4200	4216	4232	4249	4265	4281	4298	2 3 5	7 8 10	11 13 15
27	4314	4330	4346	4362	4378	4393	4409	4425	4440	4456	2 3 5	6 8 9	11 13 14
28	4472	4487	4502	4518	4533	4548	4564	4579	4594	4609	2 3 5	6 8 9	11 12 14
29	4624	4639	4654	4669	4683	4698	4713	4728	4742	4757	1 3 4	6 7 9	10 12 13
30	4771	4786	4800	4814	4829	4843	4857	4871	4886	4900	1 3 4	6 7 9	10 11 13
31	4914	4928	4942	4955	4969	4983	4997	5011	5024	5038	1 3 4	6 7 8	10 11 12
32	5051	5065	5079	5092	5105	5119	5132	5145	5159	5172	1 3 4	5 7 8	9 11 12
33	5185	5198	5211	5224	5237	5250	5263	5276	5289	5302	1 3 4	5 6 8	9 10 12
34	5315	5328	5340	5353	5366	5378	5391	5403	5416	5428	1 3 4	5 6 8	9 10 11
35	5441	5453	5465	5478	5490	5502	5514	5527	5539	5551	1 2 4	5 6 7	9 10 11
36	5563	5575	5587	5599	5611	5623	5635	5647	5658	5670	1 2 4	5 6 7	8 10 11
37	5682	5694	5705	5717	5729	5740	5752	5763	5775	5786	1 2 3	5 6 7	8 9 10
38	5798	5809	5821	5832	5843	5855	5866	5877	5888	5899	1 2 3	5 6 7	8 9 10
39	5911	5922	5933	5944	5955	5966	5977	5988	5999	6010	1 2 3	4 5 7	8 9 10
40	6021	6031	6042	6053	6064	6075	6085	6096	6107	6117	1 2 3	4 5 6	8 9 10
41	6128	6138	6149	6160	6170	6180	6191	6201	6212	6222	1 2 3	4 5 6	7 8 9
42	6232	6243	6253	6263	6274	6284	6294	6304	6314	6325	1 2 3	4 5 6	7 8 9
43	6335	6345	6355	6365	6375	6385	6395	6405	6415	6425	1 2 3	4 5 6	7 8 9
44	6435	6444	6454	6464	6474	6484	6493	6503	6513	6522	1 2 3	4 5 6	7 8 9
45	6532	6542	6551	6561	6571	6580	6590	6599	6609	6618	1 2 3	4 5 6	7 8 9
46	6628	6637	6646	6656	6665	6675	6684	6693	6702	6712	1 2 3	4 5 6	7 7 8
47	6721	6730	6739	6749	6758	6767	6776	6785	6794	6803	1 2 3	4 5 5	6 7 8
48	6812	6821	6830	6839	6848	6857	6866	6875	6884	6893	1 2 3	4 4 5	6 7 8
49	6902	6911	6920	6928	6937	6946	6955	6964	6972	6981	1 2 3	4 4 5	6 7 8
50	6990	6998	7007	7016	7024	7033	7042	7050	7059	7067	1 2 3	3 4 5	6 7 8
51	7076	7084	7093	7101	7110	7118	7126	7135	7143	7152	1 2 3	3 4 5	6 7 8
52	7160	7168	7177	7185	7193	7202	7210	7218	7226	7235	1 2 2	3 4 5	6 7 7
53	7243	7251	7259	7267	7275	7284	7292	7300	7308	7316	1 2 2	3 4 5	6 6 7
54	7324	7332	7340	7348	7356	7364	7372	7380	7388	7396	1 2 2	3 4 5	6 6 7

TABLE O · *Four-Place Logarithms* (Continued)

	0	1	2	3	4	5	6	7	8	9	1 2 3	4 5 6	7 8 9
55	7404	7412	7419	7427	7435	7443	7451	7459	7466	7474	1 2 2	3 4 5	5 6 7
56	7482	7490	7497	7505	7513	7520	7528	7536	7543	7551	1 2 2	3 4 5	5 6 7
57	7559	7566	7574	7582	7589	7597	7604	7612	7619	7627	1 2 2	3 4 5	5 6 7
58	7634	7642	7649	7657	7664	7672	7679	7686	7694	7701	1 1 2	3 4 4	5 6 7
59	7709	7716	7723	7731	7738	7745	7752	7760	7767	7774	1 1 2	3 4 4	5 6 7
60	7782	7789	7796	7803	7810	7818	7825	7832	7839	7846	1 1 2	3 4 4	5 6 6
61	7853	7860	7868	7875	7882	7889	7896	7903	7910	7917	1 1 2	3 4 4	5 6 6
62	7924	7931	7938	7945	7952	7959	7966	7973	7980	7987	1 1 2	3 3 4	5 6 6
63	7993	8000	8007	8014	8021	8028	8035	8041	8048	8055	1 1 2	3 3 4	5 5 6
64	8062	8069	8075	8082	8089	8096	8102	8109	8116	8122	1 1 2	3 3 4	5 5 6
65	8129	8136	8142	8149	8156	8162	8169	8176	8182	8189	1 1 2	3 3 4	5 5 6
66	8195	8202	8209	8215	8222	8228	8235	8241	8248	8254	1 1 2	3 3 4	5 5 6
67	8261	8267	8274	8280	8287	8293	8299	8306	8312	8319	1 1 2	3 3 4	5 5 6
68	8325	8331	8338	8344	8351	8357	8363	8370	8376	8382	1 1 2	3 3 4	4 5 6
69	8388	8395	8401	8407	8414	8420	8426	8432	8439	8445	1 1 2	2 3 4	4 5 6
70	8451	8457	8463	8470	8476	8482	8488	8494	8500	8506	1 1 2	2 3 4	4 5 6
71	8513	8519	8525	8531	8537	8543	8549	8555	8561	8567	1 1 2	2 3 4	4 5 5
72	8573	8579	8585	8591	8597	8603	8609	8615	8621	8627	1 1 2	2 3 4	4 5 5
73	8633	8639	8645	8651	8657	8663	8669	8675	8681	8686	1 1 2	2 3 4	4 5 5
74	8692	8698	8704	8710	8716	8722	8727	8733	8739	8745	1 1 2	2 3 4	4 5 5
75	8751	8756	8762	8768	8774	8779	8785	8791	8797	8802	1 1 2	2 3 3	4 5 5
76	8808	8814	8820	8825	8831	8837	8842	8848	8854	8859	1 1 2	2 3 3	4 5 5
77	8865	8871	8876	8882	8887	8893	8899	8904	8910	8915	1 1 2	2 3 3	4 4 5
78	8921	8927	8932	8938	8943	8949	8954	8960	8965	8971	1 1 2	2 3 3	4 4 5
79	8976	8982	8987	8993	8998	9004	9009	9015	9020	9025	1 1 2	2 3 3	4 4 5
80	9031	9036	9042	9047	9053	9058	9063	9069	9074	9079	1 1 2	2 3 3	4 4 5
81	9085	9090	9096	9101	9106	9112	9117	9122	9128	9133	1 1 2	2 3 3	4 4 5
82	9138	9143	9149	9154	9159	9165	9170	9175	9180	9186	1 1 2	2 3 3	4 4 5
83	9191	9196	9201	9206	9212	9217	9222	9227	9232	9238	1 1 2	2 3 3	4 4 5
84	9243	9248	9253	9258	9263	9269	9274	9279	9284	9289	1 1 2	2 3 3	4 4 5
85	9294	9299	9304	9309	9315	9320	9325	9330	9335	9340	1 1 2	2 3 3	4 4 5
86	9345	9350	9355	9360	9365	9370	9375	9380	9385	9390	1 1 2	2 3 3	4 4 5
87	9395	9400	9405	9410	9415	9420	9425	9430	9435	9440	0 1 1	2 2 3	3 4 4
88	9445	9450	9455	9460	9465	9469	9474	9479	9484	9489	0 1 1	2 2 3	3 4 4
89	9494	9499	9504	9509	9513	9518	9523	9528	9533	9538	0 1 1	2 2 3	3 4 4
90	9542	9547	9552	9557	9562	9566	9571	9576	9581	9586	0 1 1	2 2 3	3 4 4
91	9590	9595	9600	9605	9609	9614	9619	9624	9628	9633	0 1 1	2 2 3	3 4 4
92	9638	9643	9647	9652	9657	9661	9666	9671	9675	9680	0 1 1	2 2 3	3 4 4
93	9685	9689	9694	9699	9703	9708	9713	9717	9722	9727	0 1 1	2 2 3	3 4 4
94	9731	9736	9741	9745	9750	9754	9759	9763	9768	9773	0 1 1	2 2 3	3 4 4
95	9777	9782	9786	9791	9795	9800	9805	9809	9814	9818	0 1 1	2 2 3	3 4 4
96	9823	9827	9832	9836	9841	9845	9850	9854	9859	9863	0 1 1	2 2 3	3 4 4
97	9868	9872	9877	9881	9886	9890	9894	9899	9903	9908	0 1 1	2 2 3	3 4 4
98	9912	9917	9921	9926	9930	9934	9939	9943	9948	9952	0 1 1	2 2 3	3 4 4
99	9956	9961	9965	9969	9974	9978	9983	9987	9991	9996	0 1 1	2 2 3	3 3 4

TABLE P · *Trigonometric Functions*

Angle	Sine		Cosine		Tangent		Cotangent		Angle
	Nat.	Log.	Nat.	Log.	Nat.	Log.	Nat.	Log.	
0° 00′	0.0000	∞	1.0000	0.0000	0.0000	∞	∞	∞	90° 00′
10	.0029	7.4637	1.0000	0000	.0029	7.4637	343.77	2.5363	50
20	.0058	7648	1.0000	0000	.0058	7648	171.89	2352	40
30	.0087	9408	1.0000	0000	.0087	7.9409	114.59	2.0591	30
40	.0116	8.0658	0.9999	0000	.0116	8.0658	85.940	1.9342	20
50	.0145	1627	.9999	0.0000	.0145	1627	68.750	8373	10
1° 00′	0.0175	8.2419	0.9998	9.9999	0.0175	8.2419	57.290	1.7581	89° 00′
10	.0204	3088	.9998	9999	.0204	3089	49.104	6911	50
20	.0233	3668	.9997	9999	.0233	3669	42.964	6331	40
30	.0262	4179	.9997	9999	.0262	4181	38.188	5819	30
40	.0291	4637	.9996	9998	.0291	4638	34.368	5362	20
50	.0320	5050	.9995	9998	.0320	5053	31.242	4947	10
2° 00′	0.0349	8.5428	0.9994	9.9997	0.0349	8.5431	28.636	1.4569	88° 00′
10	.0378	5776	.9993	9997	.0378	5779	26.432	4221	50
20	.0407	6097	.9992	9996	.0407	6101	24.542	3899	40
30	.0436	6397	.9990	9996	.0437	6401	22.904	3599	30
40	.0465	6677	.9989	9995	.0466	6682	21.470	3318	20
50	.0494	6940	.9988	9995	.0495	6945	20.206	3055	10
3° 00′	0.0523	8.7188	0.9986	9.9994	0.0524	8.7194	19.081	1.2806	87° 00′
10	.0552	7423	.9985	9993	.0553	7429	18.075	2571	50
20	.0581	7645	.9983	9993	.0582	7652	17.169	2348	40
30	.0610	7857	.9981	9992	.0612	7865	16.350	2135	30
40	.0640	8059	.9980	9991	.0641	8067	15.605	1933	20
50	.0669	8251	.9978	9990	.0670	8261	14.924	1739	10
4° 00′	0.0698	8.8436	0.9976	9.9989	0.0699	8.8446	14.301	1.1554	86° 00′
10	.0727	8613	.9974	9989	.0729	8624	13.727	1376	50
20	.0756	8783	.9971	9988	.0758	8795	13.197	1205	40
30	.0785	8946	.9969	9987	.0787	8960	12.706	1040	30
40	.0814	9104	.9967	9986	.0816	9118	12.251	0882	20
50	.0843	9256	.9964	9985	.0846	9272	11.826	0728	10
5° 00′	0.0872	8.9403	0.9962	9.9983	0.0875	8.9420	11.430	1.0580	85° 00′
10	.0901	9545	.9959	9982	.0904	9563	11.059	0437	50
20	.0929	9682	.9957	9981	.0934	9701	10.712	0299	40
30	.0958	9816	.9954	9980	.0963	9836	10.385	0164	30
40	.0987	8.9945	.9951	9979	.0992	8.9966	10.078	1.0034	20
50	.1016	9.0070	.9948	9977	.1022	9.0093	9.7882	0.9907	10
6° 00′	0.1045	9.0192	0.9945	9.9976	0.1051	9.0216	9.5144	0.9784	84° 00′
10	.1074	0311	.9942	9975	.1080	0336	9.2553	9664	50
20	.1103	0426	.9939	9973	.1110	0453	9.0098	9547	40
30	.1132	0539	.9936	9972	.1139	0567	8.7769	9433	30
40	.1161	0648	.9932	9971	.1169	0678	8.5555	9322	20
50	.1190	0755	.9929	9969	.1198	0786	8.3450	9214	10
7° 00′	0.1219	9.0859	0.9925	9.9968	0.1228	9.0891	8.1443	0.9109	83° 00′
10	.1248	0961	.9922	9966	.1257	0995	7.9530	9005	50
20	.1276	1060	.9918	9964	.1287	1096	7.7704	8904	40
30	.1305	1157	.9914	9963	.1317	1194	7.5958	8806	30
40	.1334	1252	.9911	9961	.1346	1291	7.4287	8709	20
50	.1363	1345	.9907	9959	.1376	1385	7.2687	8615	10
8° 00′	0.1392	9.1436	0.9903	9.9958	0.1405	9.1478	7.1154	0.8522	82° 00′
10	.1421	1525	.9899	9956	.1435	1569	6.9682	8431	50
20	.1449	1612	.9894	9954	.1465	1658	6.8269	8342	40
30	.1478	1697	.9890	9952	.1495	1745	6.6912	8255	30
40	.1507	1781	.9886	9950	.1524	1831	6.5606	8169	20
50	.1536	1863	.9881	9948	.1554	1915	6.4348	8085	10
9° 00′	0.1564	9.1943	0.9877	9.9946	0.1584	9.1997	6.3138	0.8003	81° 00′
10	.1593	2022	.9872	9944	.1614	2078	6.1970	7922	50
20	.1622	2100	.9868	9942	.1644	2158	6.0844	7842	40
30	.1650	2176	.9863	9940	.1673	2236	5.9758	7764	30
40	.1679	2251	.9858	9938	.1703	2313	5.8708	7687	20
50	.1708	2324	.9853	9936	.1733	2389	5.7694	7611	10
10° 00′	0.1736	9.2397	0.9848	9.9934	0.1763	9.2463	5.6713	0.7537	80° 00′
10	.1765	2468	.9843	9931	.1793	2536	5.5764	7464	50
20	.1794	2538	.9838	9929	.1823	2609	5.4845	7391	40
30	.1822	2606	.9833	9927	.1853	2680	5.3955	7320	30
40	.1851	2674	.9827	9924	.1883	2750	5.3093	7250	20
50	.1880	2740	.9822	9922	.1914	2819	5.2257	7181	10
11° 00′	0.1908	9.2806	0.9816	9.9919	0.1944	9.2887	5.1446	0.7113	79° 00′
11° 10′	0.1937	9.2870	0.9811	9.9917	0.1974	9.2953	5.0658	0.7047	78° 50′
	Nat.	Log.	Nat.	Log.	Nat.	Log.	Nat.	Log.	
Angle	Cosine		Sine		Cotangent		Tangent		Angle

TABLE P · *Trigonometric Functions* (*Continued*)

Angle	Sine		Cosine		Tangent		Cotangent		Angle
	Nat.	Log.	Nat.	Log.	Nat.	Log.	Nat.	Log.	
11° 20′	0.1965	9.2934	0.9805	9.9914	0.2004	9.3020	4.9894	0.6980	78° 40′
30	.1994	2997	.9799	9912	.2035	.3085	4.9152	6915	30
40	.2022	3058	.9793	9909	.2065	3149	4.8430	6851	20
50	.2051	3119	.9787	9907	.2095	3212	4.7729	6788	10
12° 00′	0.2079	9.3179	.9781	9.9904	0.2126	9.3275	4.7046	0.6725	78° 00′
10	.2108	3238	.9775	9901	.2156	3336	4.6382	6664	50
20	.2136	3296	.9769	9899	.2186	3397	4.5736	6603	40
30	.2164	3353	.9763	9896	.2217	3458	4.5107	6542	30
40	.2193	3410	.9757	9893	.2247	3517	4.4494	6483	20
50	.2221	3466	.9750	9890	.2278	3576	4.3897	6424	10
13° 00′	0.2250	9.3521	.9744	9.9887	0.2309	9.3634	4.3315	0.6366	77° 00′
10	.2278	3575	.9737	9884	.2339	3691	4.2747	6309	50
20	.2306	3629	.9730	9881	.2370	3748	4.2193	6252	40
30	.2334	3682	.9724	9878	.2401	3804	4.1653	6196	30
40	2363	3734	9717	9875	.2432	3859	4.1126	6141	20
50	.2391	3786	.9710	9872	.2462	3914	4.0611	6086	10
14° 00′	0.2419	9.3837	0.9703	9.9869	0.2493	9.3968	4.0108	0.6032	76° 00′
10	.2447	3887	.9696	9866	.2524	4021	3.9617	5979	50
20	.2476	3937	.9689	9863	.2555	4074	3.9136	5926	40
30	.2504	3986	.9681	9859	.2586	4127	3.8667	5873	30
40	.2532	4035	.9674	9856	.2617	4178	3.8208	5822	20
50	.2560	4083	.9667	9853	.2648	4230	3.7760	5770	10
15° 00′	0.2588	9.4130	0.9659	9.9849	0.2679	9.4281	3.7321	0.5719	75° 00′
10	.2616	4177	.9652	9846	.2711	4331	3.6891	5669	50
20	.2644	4223	.9644	9843	.2742	4381	3.6470	5619	40
30	.2672	4269	.9636	9839	.2773	4430	3.6059	5570	30
40	.2700	4314	.9628	9836	.2805	4479	3.5656	5521	20
50	.2728	4359	.9621	9832	.2836	4527	3.5261	5473	10
16° 00′	0.2756	9.4403	0.9613	9.9828	0.2867	9.4575	3.4874	0.5425	74° 00′
10	.2784	4447	.9605	9825	.2899	4622	3.4495	5378	50
20	.2812	4491	.9596	9821	.2931	4669	3.4124	5331	40
30	.2840	4533	.9588	9817	.2962	4716	3.3759	5284	30
40	.2868	4576	.9580	9814	.2994	4762	3.3402	5238	20
50	.2896	4618	.9572	9810	.3026	4808	3.3052	5192	10
17° 00′	0.2924	9.4659	.9563	9.9806	0.3057	9.4853	3.2709	0.5147	73° 00′
10	.2952	4700	.9555	9802	.3089	4898	3.2371	5102	50
20	.2979	4741	.9546	9798	.3121	4943	3.2041	5057	40
30	.3007	4781	.9537	9794	.3153	4987	3.1716	5013	30
40	.3035	4821	.9528	9790	.3185	5031	3.1397	4969	20
50	.3062	4861	.9520	9786	.3217	5075	3.1084	4925	10
18° 00′	0.3090	9.4900	.9511	9.9782	0.3249	9.5118	3.0777	0.4882	72° 00′
10	.3118	4939	.9502	9778	.3281	5161	3.0475	4839	50
20	.3145	4977	.9492	9774	.3314	5203	3.0178	4797	40
30	.3173	5015	.9483	9770	.3346	5245	2.9887	4755	30
40	.3201	5052	.9474	9765	.3378	5287	2.9600	4713	20
50	.3228	5090	.9465	9761	.3411	5329	2.9319	4671	10
19° 00′	0.3256	9.5126	.9455	9.9757	0.3443	9.5370	2.9042	0.4630	71° 00′
10	.3283	5163	.9446	9752	.3476	5411	2.8770	4589	50
20	.3311	5199	.9436	9748	.3508	5451	2.8502	4549	40
30	.3338	5235	.9426	9743	.3541	5491	2.8239	4509	30
40	.3365	5270	.9417	9739	.3574	5531	2.7980	4469	20
50	.3393	5306	.9407	9734	.3607	5571	2.7725	4429	10
20° 00′	0.3420	9.5341	.9397	9.9730	0.3640	9.5611	2.7475	0.4389	70° 00′
10	.3448	5375	.9387	9725	.3673	5650	2.7228	4350	50
20	.3475	5409	.9377	9721	.3706	5689	2.6985	4311	40
30	.3502	5443	.9367	9716	.3739	5727	2.6746	4273	30
40	.3529	5477	.9356	9711	.3772	5766	2.6511	4234	20
50	.3557	5510	.9346	9706	.3805	5804	2.6279	4196	10
21° 00′	0.3584	9.5543	.9336	9.9702	0.3839	9.5842	2.6051	0.4158	69° 00′
10	.3611	5576	.9325	9697	.3872	5879	2.5826	4121	50
20	.3638	5609	.9315	9692	.3906	5917	2.5605	4083	40
30	.3665	5641	.9304	9687	.3939	5954	2.5386	4046	30
40	.3692	5673	.9293	9682	.3973	5991	2.5172	4009	20
50	.3719	5704	.9283	9677	.4006	6028	2.4960	3972	10
22° 00′	0.3746	9.5736	.9272	9.9672	0.4040	9.6064	2.4751	0.3936	68° 00′
10	.3773	5767	.9261	9667	.4074	6100	2.4545	3900	50
20	.3800	5798	.9250	9661	.4108	6136	2.4342	3864	40
22° 30′	0.3827	9.5828	0.9239	9.9656	0.4142	9.6172	2.4142	0.3828	67° 30′
	Nat.	Log.	Nat.	Log.	Nat.	Log.	Nat.	Log.	
Angle	Cosine		Sine		Cotangent		Tangent		Angle

TABLE P · *Trigonometric Functions (Continued)*

Angle	Sine		Cosine		Tangent		Cotangent		Angle
	Nat.	Log.	Nat.	Log.	Nat.	Log.	Nat.	Log.	
22° 40′	0.3854	9.5859	0.9228	9.9651	0.4176	9.6208	2.3945	0.3792	67° 20′
50	.3881	5889	.9216	9646	.4210	6243	2.3750	3757	10
23° 00′	0.3907	9.5919	0.9205	9.9640	0.4245	9.6279	2.3559	0.3721	67° 00′
10	.3934	5948	.9194	9635	.4279	6314	2.3369	3686	50
20	.3961	5978	.9182	9629	.4314	6348	2.3183	3652	40
30	.3987	6007	.9171	9624	.4348	6383	2.2998	3617	30
40	.4014	6036	.9159	9618	.4383	6417	2.2817	3583	20
50	.4041	6065	.9147	9613	.4417	6452	2.2637	3548	10
24° 00′	0.4067	9.6093	0.9135	9.9607	0.4452	9.6486	2.2460	0.3514	66° 00′
10	.4094	6121	.9124	9602	.4487	6520	2.2286	3480	50
20	.4120	6149	.9112	9596	.4522	6553	2.2113	3447	40
30	.4147	6177	.9100	9590	.4557	6587	2.1943	3413	30
40	.4173	6205	.9088	9584	.4592	6620	2.1775	3380	20
50	.4200	6232	.9075	9579	.4628	6654	2.1609	3346	10
25° 00′	0.4226	9.6259	0.9063	9.9573	0.4663	9.6687	2.1445	0.3313	65° 00′
10	.4253	6286	.9051	9567	.4699	6720	2.1283	3280	50
20	.4279	6313	.9038	9561	.4734	6752	2.1123	3248	40
30	.4305	6340	.9026	9555	.4770	6785	2.0965	3215	30
40	.4331	6366	.9013	9549	.4806	6817	2.0809	3183	20
50	.4358	6392	.9001	9543	.4841	6850	2.0655	3150	10
26° 00′	0.4384	9.6418	0.8988	9.9537	0.4877	9.6882	2.0503	0.3118	64° 00′
10	.4410	6444	.8975	9530	.4913	6914	2.0353	3086	50
20	.4436	6470	.8962	9524	.4950	6946	2.0204	3054	40
30	.4462	6495	.8949	9518	.4986	6977	2.0057	3023	30
40	.4488	6521	.8936	9512	.5022	7009	1.9912	2991	20
50	.4514	6546	.8923	9505	.5059	7040	1.9768	2960	10
27° 00′	0.4540	9.6570	0.8910	9.9499	0.5095	9.7072	1.9626	0.2928	63° 00′
10	.4566	6595	.8897	9492	.5132	7103	1.9486	2897	50
20	.4592	6620	.8884	9486	.5169	7134	1.9347	2866	40
30	.4617	6644	.8870	9479	.5206	7165	1.9210	2835	30
40	.4643	6668	.8857	9473	.5243	7196	1.9074	2804	20
50	.4669	6692	.8843	9466	.5280	7226	1.8940	2774	10
28° 00′	0.4695	9.6716	0.8829	9.9459	0.5317	9.7257	1.8807	0.2743	62° 00′
10	.4720	6740	.8816	9453	.5354	7287	1.8676	2713	50
20	.4746	6763	.8802	9446	.5392	7317	1.8546	2683	40
30	.4772	6787	.8788	9439	.5430	7348	1.8418	2652	30
40	.4797	6810	.8774	9432	.5467	7378	1.8291	2622	20
50	.4823	6833	.8760	9425	.5505	7408	1.8165	2592	10
29° 00′	0.4848	9.6856	0.8746	9.9418	0.5543	9.7438	1.8040	0.2562	61° 00′
10	.4874	6878	.8732	9411	.5581	7467	1.7917	2533	50
20	.4899	6901	.8718	9404	.5619	7497	1.7796	2503	40
30	.4924	6923	.8704	9397	.5658	7526	1.7675	2474	30
40	.4950	6946	.8689	9390	.5696	7556	1.7556	2444	20
50	.4975	6968	.8675	9383	.5735	7585	1.7437	2415	10
30° 00′	0.5000	9.6990	0.8660	9.9375	0.5774	9.7614	1.7321	0.2386	60° 00′
10	.5025	7012	.8646	9368	.5812	7644	1.7205	2356	50
20	.5050	7033	.8631	9361	.5851	7673	1.7090	2327	40
30	.5075	7055	.8616	9353	.5890	7701	1.6977	2299	30
40	.5100	7076	.8601	9346	.5930	7730	1.6864	2270	20
50	.5125	7097	.8587	9338	.5969	7759	1.6753	2241	10
31° 00′	0.5150	9.7118	0.8572	9.9331	0.6009	9.7788	1.6643	0.2212	59° 00′
10	.5175	7139	.8557	9323	.6048	7816	1.6534	2184	50
20	.5200	7160	.8542	9315	.6088	7845	1.6426	2155	40
30	.5225	7181	.8526	9308	.6128	7873	1.6319	2127	30
40	.5250	7201	.8511	9300	.6168	7902	1.6212	2098	20
50	.5275	7222	.8496	9292	.6208	7930	1.6107	2070	10
32° 00′	0.5299	9.7242	0.8480	9.9284	0.6249	9.7958	1.6003	0.2042	58° 00′
10	.5324	7262	.8465	9276	.6289	7986	1.5900	2014	50
20	.5348	7282	.8450	9268	.6330	8014	1.5798	1986	40
30	.5373	7302	.8434	9260	.6371	8042	1.5697	1958	30
40	.5398	7322	.8418	9252	.6412	8070	1.5597	1930	20
50	.5422	7342	.8403	9244	.6453	8097	1.5497	1903	10
33° 00′	0.5446	9.7361	0.8387	9.9236	0.6494	9.8125	1.5399	0.1875	57° 00′
10	.5471	7380	.8371	9228	.6536	8153	1.5301	1847	50
20	.5495	7400	.8355	9219	.6577	8180	1.5204	1820	40
30	.5519	7419	.8339	9211	.6619	8208	1.5108	1792	30
40	.5544	7438	.8323	9203	.6661	8235	1.5013	1765	20
33° 50′	0.5568	9.7457	0.8307	9.9194	0.6703	9.8263	1.4919	0.1737	56° 10′
	Nat.	Log.	Nat.	Log.	Nat.	Log.	Nat.	Log.	
Angle	Cosine		Sine		Cotangent		Tangent		Angle

TABLE P · *Trigonometric Functions* (*Continued*)

Angle	Sine		Cosine		Tangent		Cotangent		Angle
	Nat.	Log.	Nat.	Log.	Nat.	Log.	Nat.	Log.	
34° 00'	0.5592	9.7476	0.8290	9.9186	0.6745	9.8290	1.4826	0.1710	56° 00'
10	.5616	7494	.8274	9177	.6787	8317	1.4733	1683	50
20	.5640	7513	.8258	9169	.6830	8344	1.4641	1656	40
30	.5664	7531	.8241	9160	.6873	8371	1.4550	1629	30
40	.5688	7550	.8225	9151	.6916	8398	1.4460	1602	20
50	.5712	7568	.8208	9142	.6959	8425	1.4370	1575	10
35° 00'	0.5736	9.7586	0.8192	9.9134	0.7002	9.8452	1.4281	0.1548	55° 00'
10	.5760	7604	.8175	9125	.7046	8479	1.4193	1521	50
20	.5783	7622	.8158	9116	.7089	8506	1.4106	1494	40
30	.5807	7640	.8141	9107	.7133	8533	1.4019	1467	30
40	.5831	7657	.8124	9098	.7177	8559	1.3934	1441	20
50	.5854	7675	.8107	9089	.7221	8586	1.3848	1414	10
36° 00'	0.5878	9.7692	0.8090	9.9080	0.7265	9.8613	1.3764	0.1387	54° 00'
10	.5901	7710	.8073	9070	.7310	8639	1.3680	1361	50
20	.5925	7727	.8056	9061	.7355	8666	1.3597	1334	40
30	.5948	7744	.8039	9052	.7400	8692	1.3514	1308	30
40	.5972	7761	.8021	9042	.7445	8718	1.3432	1282	20
50	.5995	7778	.8004	9033	.7490	8745	1.3351	1255	10
37° 00'	0.6018	9.7795	0.7986	9.9023	0.7536	9.8771	1.3270	0.1229	53° 00'
10	.6041	7811	.7969	9014	.7581	8797	1.3190	1203	50
20	.6065	7828	.7951	9004	.7627	8824	1.3111	1176	40
30	.6088	7844	.7934	8995	.7673	8850	1.3032	1150	30
40	.6111	7861	.7916	8985	.7720	8876	1.2954	1124	20
50	.6134	7877	.7898	8975	.7766	8902	1.2876	1098	10
38° 00'	0.6157	9.7893	0.7880	9.8965	0.7813	9.8928	1.2799	0.1072	52° 00'
10	.6180	7910	.7862	8955	.7860	8954	1.2723	1046	50
20	.6202	7926	.7844	8945	.7907	8980	1.2647	1020	40
30	.6225	7941	.7826	8935	.7954	9006	1.2572	0994	30
40	.6248	7957	.7808	8925	.8002	9032	1.2497	0968	20
50	.6271	7973	.7790	8915	.8050	9058	1.2423	0942	10
39° 00'	0.6293	9.7989	0.7771	9.8905	0.8098	9.9084	1.2349	0.0916	51° 00'
10	.6316	8004	.7753	8895	.8146	9110	1.2276	0890	50
20	.6338	8020	.7735	8884	.8195	9135	1.2203	0865	40
30	.6361	8035	.7716	8874	.8243	9161	1.2131	0839	30
40	.6383	8050	.7698	8864	.8292	9187	1.2059	0813	20
50	.6406	8066	.7679	8853	.8342	9212	1.1988	0788	10
40° 00'	0.6428	9.8081	0.7660	9.8843	0.8391	9.9238	1.1918	0.0762	50° 00'
10	.6450	8096	.7642	8832	.8441	9264	1.1847	0736	50
20	.6472	8111	.7623	8821	.8491	9289	1.1778	0711	40
30	.6494	8125	.7604	8810	.8541	9315	1.1708	0685	30
40	.6517	8140	.7585	8800	.8591	9341	1.1640	0659	20
50	.6539	8155	.7566	8789	.8642	9366	1.1571	0634	10
41° 00'	0.6561	9.8169	0.7547	9.8778	0.8693	9.9392	1.1504	0.0608	49° 00'
10	.6583	8184	.7528	8767	.8744	9417	1.1436	0583	50
20	.6604	8198	.7509	8756	.8796	9443	1.1369	0557	40
30	.6626	8213	.7490	8745	.8847	9468	1.1303	0532	30
40	.6648	8227	.7470	8733	.8899	9494	1.1237	0506	20
50	.6670	8241	.7451	8722	.8952	9519	1.1171	0481	10
42° 00'	0.6691	9.8255	0.7431	9.8711	0.9004	9.9544	1.1106	0.0456	48° 00'
10	.6713	8269	.7412	8699	.9057	9570	1.1041	0430	50
20	.6734	8283	.7392	8688	.9110	9595	1.0977	0405	40
30	.6756	8297	.7373	8676	.9163	9621	1.0913	0379	30
40	.6777	8311	.7353	8665	.9217	9646	1.0850	0354	20
50	.6799	8324	.7333	8653	.9271	9671	1.0786	0329	10
43° 00'	0.6820	9.8338	0.7314	9.8641	0.9325	9.9697	1.0724	0.0303	47° 00'
10	.6841	8351	.7294	8629	.9380	9722	1.0661	0278	50
20	.6862	8365	.7274	8618	.9435	9747	1.0599	0253	40
30	.6884	8378	.7254	8606	.9490	9772	1.0538	0228	30
40	.6905	8391	.7234	8594	.9545	9798	1.0477	0202	20
50	.6926	8405	.7214	8582	.9601	9823	1.0416	0177	10
44° 00'	0.6947	9.8418	0.7193	9.8569	0.9657	9.9848	1.0355	0.0152	46° 00'
10	.6967	8431	.7173	8557	.9713	9874	1.0295	0126	50
20	.6988	8444	.7153	8545	.9770	9899	1.0235	0101	40
30	.7009	8457	.7133	8532	.9827	9924	1.0176	0076	30
40	.7030	8469	.7112	8520	.9884	9949	1.0117	0051	20
50	.7050	8482	.7092	8507	0.9942	9.9975	1.0058	0025	10
45° 00'	0.7071	9.8495	0.7071	9.8495	1.0000	0.0000	1.0000	0.0000	45° 00'
	Nat.	Log.	Nat.	Log.	Nat.	Log.	Nat.	Log.	
Angle	Cosine		Sine		Cotangent		Tangent		Angle

NAME INDEX

[The numbers refer to pages, except that those enclosed in brackets refer to plates.]

SUBJECT INDEX

[The numbers refer to pages, except that those enclosed in brackets refer to plates. The page numbers in black type refer to experiments for the students.]

THE M.I.T. PAPERBACK SERIES